Biotic Response to Global Change
The Last 145 Million Years

Concern about the effects of global change on our planet's future has driven much current research into the last few thousand years of Earth history. In contrast, this volume takes a much longer viewpoint to provide a historical perspective to recent and future global change.

Over 40 international specialists investigate the reaction of life to global environmental changes that have taken place from Cretaceous times to the present day. During this time the earth's climate has changed from a very warm, 'greenhouse' phase with no significant ice sheets to the 'icehouse' world of today. A wide spectrum of animal, plant and protistan life is discussed, encompassing terrestrial, shallow-marine and deep-marine realms. Each chapter considers a particular taxonomic group, looking first at the general picture and then focusing on more specialized aspects such as extinctions, diversity and biogeography.

This volume will form an invaluable reference for researchers and graduate students in palaeontology, geology, biology, oceanography and climatology.

Biotic Response to Global Change
The Last 145 Million Years

Edited by

Stephen J. Culver
East Carolina University (formerly of The Natural History Museum, London)

and

Peter F. Rawson
University College London

CAMBRIDGE UNIVERSITY PRESS
Cambridge, New York, Melbourne, Madrid, Cape Town, Singapore, São Paulo

Cambridge University Press
The Edinburgh Building, Cambridge CB2 2RU, UK

Published in the United States of America by Cambridge University Press, New York

www.cambridge.org
Information on this title: www.cambridge.org/9780521663045

First published 2000
This digitally printed first paperback version 2006

A catalogue record for this publication is available from the British Library

Library of Congress Cataloguing in Publication data
Biotic Response to Global Change / edited by Stephen J. Culver and
Peter F. Rawson
      p.   cm.
Includes bibliographical references and index.
ISBN 0 521 66304 0
1. Palaeoclimatology. 2. Climatic changes – environmental aspects.
I. Culver, Stephen J. II. Rawson, Peter Franklin.
QC884.B52 2000
577.2´2–dc21   99-16232 CIP

ISBN-13 978-0-521-66304-5 hardback
ISBN-10 0-521-66304-0 hardback

ISBN-13 978-0-521-03419-7 paperback
ISBN-10 0-521-03419-1 paperback

Contents

Contributors

Stephen J. Culver

Department of Geology
East Carolina University
Greenville
NC 27858, USA

Peter Andrews
Stephen J. Brooks
Andrew Currant
Peter Forey
Andrew S. Gale
Jeremy J. Hooker
Charlotte H. Jeffery
Norman MacLeod
Angela Milner
Noel Morris
Brian Rosen
Andrew J. Ross
Andrew B. Smith
Chris Stringer
John Taylor
Paul Taylor
Peter J. Whybrow
Jeremy R. Young

The Natural History Museum
Cromwell Road
London SW7 5BD, UK

Peter F. Rawson

Department of Geological Sciences
University College London
Gower Street
London WC1E 6BT, UK

Paul R. Bown
Jackie A. Burnett
Susan E. Evans
Kevin T. Pickering

University College London
Gower Street, London WC1E 6BT, UK

Martin A. Buzas

Department of Paleobiology
National Museum of Natural History
Smithsonian Institution
Washington DC 20560, USA

Mark R. Chapman

School of Environmental Sciences
University of East Anglia
Norwich NR4 7TJ, UK

William Clyde
Nina Fefferman

Jena MacLean
Nievez Ortiz
Christine Schulter

Department of Geological Sciences
Princeton University
Princeton
NJ 08544, USA

Margaret E. Collinson

Department of Geology
Royal Holloway
University of London
Egham
Surrey TW20 0EX, UK

G. Russell Coope

Centre for Quaternary Research
Department of Geography
Royal Holloway
University of London
Egham
Surrey TW20 0EX, UK

J. Alastair Crame

British Antarctic Survey
High Cross
Madingley Road
Cambridge CB3 0ET, UK

Peter R. Crane

Royal Botanic Gardens, Kew
Richmond
Surrey TW9 3AB, UK

Scott Lidgard
Richard Lupia

Committee on Evolutionary Biology
The University of Chicago, and
Department of Geology
The Field Museum
Roosevelt Road at Lake Shore Drive
Chicago
IL 60605, USA

Ed. A. Jarzembowski

Booth Museum of Natural History
Dyke Road
Brighton
BN1 5AA, UK

Andrew R. Milner

Department of Biology
Birkbeck College
Malet Street
London WC1E 7HX, UK

Adrian G. Parker

Department of Geography
School of Social Sciences and Law
Oxford Brookes University
Gypsy Lane Campus
Headington
Oxford OX3 0BP, UK

Robert A. Spicer

Department of Earth Sciences
The Open University
Walton Hall,
Milton Keynes
Buckinghamshire MK7 6AA, UK

Preface

This book has developed from a joint research programme on *Global Change and the Biosphere* developed by the Palaeontology Department at The Natural History Museum, London, and the Department of Geological Sciences at University College London in 1994. Research on global change has attracted a huge number of scientists around the world over the last 20 years, primarily because of many governments' concerns about human impact on our environment. Most of this effort has concentrated on the last few thousand years of Earth history, when there have been major and rapid changes in climate and sea level that are continuing today and will undoubtedly affect our immediate future. But the world has been changing continually throughout its history, and our *Global Change* project was designed therefore to bring this much longer-term, geological perspective to the issues that concern us all today.

This volume provides a comprehensive review of the response of many different animal and plant groups to global change over the last 145 million years, since the beginning of the Cretaceous Period. It brings together contributions from members of the NHM/UCL team and other invited authors, to form a volume aimed at the advanced student and the many researchers who may need a long-term view of global change and its influence on Earth's biota.

The editors are very grateful to all the contributors for meeting our deadlines so closely. Many other scientists have helped individual contributors, including J. Böcher, L. Grande, E.M. Harper, D. Jablonski, R.M. Jacobi, T.P. Jones, G. Keller, A.E. Longbottom, S. Manchester, V. Mosbrugger, K. Muller, R.P. Nicholls, B.E. van Niel, C. Patterson, S. Revets, D. Sealy, D.C. Shreve, D. Siebert, A.J. Sutcliffe and R. Wood. Access to Indian Ocean DSDP/ODP cores was granted by ODP.

To all these, and to the publishers and authors acknowledged in the appropriate figure captions, we give our grateful thanks. Lastly, we are very grateful to the editorial team at Cambridge University Press for all their help in seeing this book through the press.

STEPHEN J. CULVER
East Carolina University
(formerly of *The Natural History Museum*)

PETER F. RAWSON
University College London
September 1999

1

Introduction

STEPHEN J. CULVER AND PETER F. RAWSON

The simple desire to understand the world we live in has driven scientists for more than two centuries to study the history of life through time. It soon became evident to the early workers that this history is most certainly not a continuous one. Indeed, the recognition of sudden and fundamental changes in faunas and floras through time led to the designation of the segments of Earth history that we know as geological periods. The two greatest discontinuities in life's history, since organisms evolved fossilizable hard parts, are the mass extinctions that define the boundaries of three great segments of Earth history, the Palaeozoic, the Mesozoic and the Cenozoic eras.

What caused these and many other changes in the Earth's biota? For a long time palaeontologists have realized that faunal and floral changes are the result of either evolutionary innovations or of environmental changes at many different scales. This volume addresses biotic changes through time for a wide variety of organisms from tiny single-celled calcareous nannoplankton to the largest whale. We concentrate on the past 145 million years of Earth history because this is a large enough segment of time to allow us to evaluate both geologically instantaneous global environmental changes and environmental trends that take millions of years to unfold. It takes us from an interval when Earth was in a 'greenhouse' state with some of the highest sea levels in the Phanerozoic, through a long period of overall climatic decline, to the late Cenozoic/present-day 'icehouse' world in which sea level is very low and continents particularly emergent. Finally, it is also the period of time that is characterized by organisms not too far removed from those that inhabit the Earth today, as well as by some major groups that are now long extinct. The fossils of these organisms provide us with the only direct evidence of how life in the past has responded to global environmental change.

This knowledge, in turn, is extremely valuable when trying to predict how organisms will respond to global environmental changes in the future. Thus, we have turned around one of the basic tenets of geology, 'The present is the key to the past' to a new perspective, 'The past is the key to the future'.

The term global change does not imply globally uniform or globally synchronous environmental change. Scenarios for future global warming predict that some areas will warm more than others, some will cool, and regional variations in precipitation patterns will also occur. This variety of response has also been demonstrated for the relatively recent past. The CLIMAP project showed how, at the height of the last glaciation, 18 000 years ago, when so much water was locked up as ice, and sea level was 120 m lower than it is today, some areas of the world's oceans had average annual temperatures several degrees warmer than today's.

Thus global change is a major driving force behind regional environmental change. These changes can have various causes, but it is difficult to separate cause and effect. Plate tectonic movements, volcanism, eustatic sea level, global temperature, ocean circulation patterns are all interrelated – one cannot undergo a significant change of state without affecting the other variables. Add to this such widespread biotic phenomena as the evolution of widespread grasslands and the local, regional and/or global affects of bolides colliding with the Earth, and we have heady mix of episodic, severe, environmental shifts overprinting a background of more sedate but always changing environments.

This volume begins with accounts of the main parameters of global change during the Cretaceous (Gale) and Cenozoic (Pickering). This sets the scene for other authors to examine the fossil data for their particular group of organisms and to determine the nature, scale and rate of biotic response, if any, to these global changes. It is, of course, very difficult to identify the proximal forcing mechanism and in most cases that has not been attempted but each author has attempted to document patterns in the fossil data and, when justified and in the context of the incompleteness of the fossil record, to relate those patterns to environmental changes ranging from local to regional and global scale.

Each author was asked to review briefly the geological record of her/his specialist taxonomic group and then to address the topic of biotic response to global change as seen fit. Thus, we deliberately have only limited uniformity of treatment in the chapters that comprise this volume. Some authors provide reviews of newly published material (e.g. Rawson on Cretaceous cephalopods, Ross on Mesozoic and Tertiary insects, Currant on Quaternary mammals), some provide the results of new research (e.g. Taylor on bryozoans, Culver and Buzas on benthic foraminifera), whilst some chapters describe the methods by which the nature of past climates is elucidated by using biotic information (e.g. Coope on Quaternary insects, Spicer on Mesozoic and Cenozoic angiosperms). Some authors deal with the entire 145 million years that are the subject of this book (e.g. Crame on bivalves, Morris and Taylor on gastropods), whilst others concentrate on much shorter episodes (e.g. MacLeod on planktic foraminifera across three discrete intervals of environmental change, Chapman on planktic foraminifera of the past 3 million years). Some take a global view of their organisms (e.g. Milner and others on reptiles, amphibians and birds, Smith and Jeffery on echinoids), while others examine biotic response through a much more localized scale (e.g. Currant on Quaternary mammals of the British Isles, Parker on post-glacial pollen in southern England). We hope that this multi-faceted approach to the subject matter provides the reader with hard data about past biotic response and the methods that may be used to interpret the past, with the knowledge that, despite intensive research, we know very little about the subject of this book, and with the stimulus to address some of the questions that this book poses.

The two words 'global change', have become a rallying cry for scientists, politicians and the general public, given the warning about anthropogenically enhanced global warming. But global change is the norm. The Earth has never been static and neither have its inhabitants. Global change will continue in the future as it has in the past and organisms will adapt to it or go extinct. But what is of particular concern right now is that, before the current interglacial comes to an icy end as the next glacial period is established, our own species may well change our global environment at a rate that has no precedent in the past, with the sole exception of bolide impact. Our production of greenhouse gases may well cause global warming with concomitant rapid sea-level rise, disruption of coastal ecosystems, changed boundaries of climatic belts, etc., etc. Organisms have reacted to rapid climatic oscillations in the past (e.g. the Younger Dryas episode about 11 thousand years ago) but now human intervention has disrupted vegetation zones across most continents. Plants and animals that followed migrating climatic belts in the past may not be able to do so in the future. To predict the ultimate effects of our environmental disturbance during an episode of rapid global change is extremely difficult to say the least. The studies of past biotic response described in this book do, however, provide us with the beginning of an understanding of how life responds to changes in its environment. To predict the future of life on this planet with these data as a starting point is fraught with difficulty, but to predict the future without the evidence from past life at all would be doomed to failure.

2

The Cretaceous world

ANDREW S. GALE

2.1 INTRODUCTION

The Cretaceous Period extended from 145 Ma to 65 Ma ago, and is believed to have differed from our present world in several major respects. First, the period has been characterized as a time of globally warm 'greenhouse' conditions, in which there were no polar ice caps. Although this is generally correct, there is evidence for considerable climatic change in both temperature and patterns of humidity. Secondly, Cretaceous deep ocean water was warm and saline and derived from low latitude areas of high evaporation, in contrast to the cold polar water which occupies the deep oceans at present. Thirdly, eustatic sea levels were very high, particularly during the Late Cretaceous, and the interiors of the major continents were covered with shallow seas.

Scientists have only recently provided good explanations for these differences. Although detailed interpretations vary, there is general agreement that high levels of atmospheric carbon dioxide (a well-known greenhouse gas), perhaps eight to ten times those at present, were responsible for elevated temperatures in the Cretaceous. The resulting climates were equable, with moderately high polar temperatures and a low pole to equator temperature gradient. As a consequence, high latitude sea water was neither sufficiently cold nor saline to sink and form bottom water. The source of high carbon dioxide levels has been attributed to outgassing during periods of exceptionally fast ocean floor spreading. The increase in ocean ridge volume, in addition to the lack of polar ice-sheets, was responsible for the very high sea levels.

Although these long-term controls on environmental change in the Cretaceous world have become much better understood during the past decade, we still know very little about shorter term processes. Milankovitch cycles (18–400 ka) have been identified widely in marine Cretaceous sediments, but we do not know the feedback mechanisms which amplified the signals such that they moderated climate change.

This chapter summarizes the principal changes that took place during the Cretaceous in plate tectonics, palaeogeography, volcanism, climate and palaeoceanography, and examines the complex interplay of processes which brought about environmental change (Fig. 2.1).

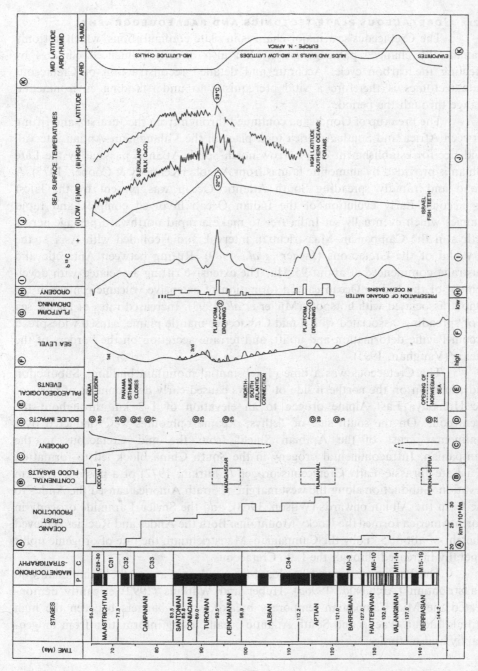

Fig. 2.1. Summary of changing environmental parameters through the Cretaceous. References to particular sources are in the relevant text. Note the importance of the latest Cenomanian–earliest Turonian in terms of maximum sea level, major carbon burial event and isotope excursion, and temperature maximum for the Cretaceous. Note also the dramatic changes during the Maastrichtian, including falling sea levels, falling temperatures, decreased carbon cycling, increased orogeny and increased volcanic activity.

2.2   CRETACEOUS PLATE TECTONICS AND PALAEOGEOGRAPHY

The Cretaceous saw major changes in plate configurations, which not only dramatically changed palaeogeography but also directly influenced climates by affecting the carbon cycle. Accurate and detailed reconstruction of Cretaceous plate tectonics is therefore a vital prerequisite to understanding environmental change through the period.

The breakup of Gondwana continued through from the Jurassic, and rifting between Africa and South America took place in the Valanginian–Aptian interval. Evidence for establishment of a narrow north–south Atlantic passage by the Late Albian is provided by ammonite faunas from Angola (Kennedy & Cooper, 1975). A broad and rapidly spreading South Atlantic Ocean was present by the latest Cretaceous. Early evolution of the Indian Ocean involved complex and rapid changes which eventually set India free to make a rapid northward passage across Tethys in the Campanian–Maastrichtian interval. India collided with Asia at the very end of the Cretaceous (Jaeger *et al.*, 1989). Rifting between Antarctic and Australia commenced at about 95 Ma. The extensive rifting associated with development of the Indian Ocean caused formation of extensive volcanic centres in the region, associated with hotspots (Müller *et al.*, 1993). Increased rates of Pacific sea floor spreading, associated with a mid Cretaceous mantle plume, caused widespread circum-Pacific deformation and uplift, and terrane accretion on the borders of the Pacific (Vaughan, 1995).

The Cretaceous was a time of substantial mountain building. Subduction and accretion on the northern side of Tethys caused early development of parts of the Himalaya–East Alpine orogen to an elevation of 1–2 km throughout the Cretaceous. On the south side of Tethys, oceanic ophiolites were obducted over continental crust of the Arabian Shield from the mid Cretaceous to the Campanian. Intracontinental orogeny in the South China block led to formation of a Late Jurassic–Early Cretaceous orogen (Churkin, 1972) of a possible 1–2 km elevation. Subduction along the west margin of South America caused the Andes to rise from the Albian onwards (Wilson, 1963), and the Sevier–Laramide orogeny in North America formed the Rocky Mountains. Both the Andes and Rockies achieved a height of about 5 km by the Campanian–Maastrichtian; the rate of orogenic uplift generally increased through the Late Cretaceous.

The Panama Isthmus was closed by a volcanic arc in the Campanian and Maastrichtian (Archibald, 1996*a*). Huber and Watkins (1992) elegantly demonstrated the Mid Campanian temporary breakdown of barriers between the high latitude Indian, Pacific and South Atlantic oceans, using information from biogeography and tectonics.

2.3   IGNEOUS ACTIVITY

2.3.1   Continental flood basalts

Major, rapid continental basaltic eruptions with volumes in excess of 1^6 km^3 have been classified as continental flood basalts (CFBs – Courtillot *et al.*, 1996). Their formation has been attributed to mantle plumes and major rifting, and characteristically lasted for less than one million years. The Cretaceous saw four such outpourings, the first associated with rifting of the South Atlantic – the Parana–

Serra traps in Brazil, and the originally contiguous Etendeka traps in Namibia, of earliest Cretaceous age. Extensive basalts of Aptian age are found at Rajmahal in eastern India, and Cenomanian CFBs occur in Madagascar (Courtillot *et al.*, 1996). However, the Deccan Traps in western India have by far the largest volume of basalt, erupted in the brief span of about 1 million years, probably across the K–T boundary. All of these volcanic events were related to hotspot development (Müller *et al.*, 1993).

2.3.2   The mid Cretaceous superplume

For the interval from 120–80 Ma (Aptian–Early Campanian) there is considerable evidence of a doubling of the rate of ocean crust formation, both by increased mid-ocean ridge spreading rates and by plateau formation in the Pacific Ocean. This coincides with the long normal Cretaceous magnetozone (34° N), and was interpreted by Larson (1991) as evidence for a mantle 'superplume' which rose through the mantle and provided thermal energy for volcanism on the Pacific sea floor. This superplume episode probably provided the CO_2 responsible for warming the Cretaceous greenhouse, which reached a temperature maximum in the Late Cenomanian (Larson *et al.*, 1993). Larson *et al.* (1993) also suggested that there may be a relationship between individual episodes of marine organic deposition (anoxic events) and volcanism. For example, formation of the Ontong–Java Plateau in the equatorial mid Pacific (Tarduno *et al.*, 1991) is exactly coincident with the Aptian oceanic anoxic event (OAE) 1A.

2.4   BOLIDE IMPACTS

Considerable effort has been put into the search for impact structures which coincide with and therefore account for major extinctions in the geological record. This was given impetus when Alvarez *et al.* (1980) discovered anomalous levels of the platinum group metal iridium (common in meteorites) in the boundary clay at the Cretaceous–Tertiary boundary at Gubbio in Italy. The subsequent search for a 'smoking gun' impact structure of latest Cretaceous age has fascinated scientists and public alike.

In a review of all known terrestrial impact structures, Grieve (1991) listed a total of 36 as being of Cretaceous age, of which 11 were datable with sufficient accuracy to be included in data used to search for periodicity. The structure at Kara, in Siberian Russia, is the largest at 65 km in diameter and is dated as Campanian. It must be remembered that impacts in areas that are now sea are mostly unknown, and that the preservation potential of structures decreases sharply with age. It is thus likely that over 100 sizeable impacts took place during the Cretaceous.

The Manson Impact Structure in central Iowa is 35 km in diameter. It is the largest impact structure of any age known in the US, and was initially dated as terminal Cretaceous in age. Biostratigraphical data subsequently showed the structure to be of early Late Campanian age (73.8 Ma; Izett *et al.*, 1993). A giant 180–200 km structure was subsequently discovered at Chicxulub on the Yucatan Peninsula in Mexico (Hildebrand *et al.*, 1991), which if it is an impact crater, is by far the largest known on earth. This structure has been dated as terminal Cretaceous. However, some workers think that the Chicxulub structure is, in fact, a volcanic complex (Meyerhoff *et al.*, 1994); it falls within a well-known Cretaceous igneous province, and boreholes show the underlying rocks to be well bedded, the opposite of the thick melt-sheet which impact would cause (Ward *et al.*, 1995). The case for Chicxulub as an impact site remains unproven (but see also Pickering, this volume).

A neglected aspect of the argument for impact-generated mass extinction is the negative relationship between extinctions and impacts for events and intervals which have been examined in detail. For example, the two largest known undisputed Cretaceous impacts (Kara at 65 km, 73 ± 3 Ma; Manson at 35 km, 73.8 ± 0.3 Ma) both fall within the Campanian Stage, for which interval no one has ever claimed high extinction levels.

2.5   PALAEOCEANOGRAPHY

2.5.1   Eustasy and the great transgressions

Global sea levels were low in the Early Cretaceous, but the Late Cretaceous saw the highest sea levels of Mesozoic time, possibly of the entire Phanerozoic, with a probable maximum stand in the Early Turonian (Haq *et al.*, 1987; Hallam, 1992). Vast shelf areas were flooded at the height of transgression and it is often difficult to position shorelines accurately because marine sediments spread far beyond the main outcrops (Hancock & Kauffman, 1979). Many of the major transgressive events are recognizable globally, although details of the third-order sea-level curve remain to be worked out and their correlation demonstrated widely. Sequence stratigraphy has vastly improved our understanding of Cretaceous sea levels, but much remains to be done. It is not possible to use the Haq *et al.* (1987) curve as an entirely reliable guide to Cretaceous sea levels partly because of errors in biostratigraphical calibration.

Data for Early Cretaceous sea levels are taken mostly from Europe, because this area is more intensively studied than elsewhere. The earliest Cretaceous was widely marked by a lowstand (Late Volgian–Early Berriasian) which is often represented by a gap or highly condensed beds over structures (Rawson & Riley, 1982). A rapid sea-level rise followed in the Late Berriasian, which continued with minor interruptions through the Valanginian and Hauterivian. A further regression in the Barremian saw the widespread development of a shallow-water anoxic facies in north-west Europe.

The Lower Aptian is strongly transgressive, and the basal part is commonly condensed. It marks the commencement of almost continuous overall sea-level rise which continued into the Late Cenomanian. In western Europe, the Early Aptian transgression commonly rests on non- or quasimarine Wealden facies. A distinctive mid Aptian break is related to a brief sea-level fall, and was succeeded by the Late Aptian *Parahoplites nutfieldiensis* Zone which is widely transgressive. The Albian saw at least three major flooding events, at or near the beginning of each of the Early, Mid and Late intervals.

The Cenomanian continued the period of overall sea-level rise, and five progressively onlapping third-order sequences can be recognized across Europe. The fifth of these sequences, in the *Metoicoceras geslinianum* Zone of the Late Cenomanian, was a rapid and major sea-level rise of 50–100 m. In the Early Turonian, sea levels remained high and may even have exceeded those of the Late

Cenomanian, but fell during the mid and later Turonian. This period of time is commonly unrepresented by sediment or highly condensed in marginal marine areas. The latest Turonian saw a further important transgression, which resulted in the spread of chalk facies into the Gulf Coast and Western Interior of the USA, for example. Above the Turonian, third order sequences are not well studied. The base of the Santonian is commonly transgressive, as across the Russian Platform, and the Late Santonian *Uintacrinus socialis* and *Marsupites testudinarius* Zones are globally developed in a deep-water chalk facies.

The Campanian is the most poorly understood interval of Late Cretaceous time, and the sea-level curves from the two regions in which it is well known differ in important ways. In Europe, maximum sea-level highs were reached in the Late Campanian, which according to Hancock and Kauffman (1979) and Hancock (1989) registered the highest levels of the Cretaceous, exceeding even those of the Early Turonian. In the United States, however, Campanian sea levels were relatively low, perhaps on account of the Laramide–Sevier orogenic uplift. Haq *et al.* (1988) showed the long-term Campanian high as significantly lower than that of the Late Cenomanian–Early Turonian maximum. The early Maastrichtian was a time of important transgression particularly on the African–Arabian Shield (Flexer & Reyment, 1989). For those parts of the Cretaceous in which a detailed third-order sea-level curve has been worked out from sequence analysis, notably the Cenomanian and the Turonian, sequence duration lies in the range of 0.8–1.0 million years, which is confirmed by an orbital timescale (Gale, 1995).

The first-order sea-level rise to an Early Turonian maximum (Haq *et al.*, 1987) can be related to increased mid-ocean ridge volume and length associated with development of the mantle superplume (Hays & Pitman, 1973; Larson, 1991) which displaced water from ocean basins. The third-order sea-level changes, often the most conspicuous because they controlled sequences in platform successions, are difficult to explain in the absence of glacioeustatic control; they were too rapid to have been caused by ocean ridge changes (Schlanger, 1986).

2.5.2   Oceanic circulation

The warm, equable Cretaceous had a very different thermohaline circulation to that which exists at present. Today, oceanic deep water forms in the high latitudes where sea-ice formation increases the salinity and therefore the density of surface water; the cold water sinks and flows along the ocean basins. In the Cretaceous, deep ocean water was probably warm and saline, and derived from low-latitude sites of high evaporation (warm saline bottom water, WSBW: Brass *et al.*, 1982), although some authors have suggested that a 'mixed' circulation of both cold and warm deep water existed in the Cretaceous (Salzman & Barron, 1982). The same authors provided oxygen isotope data from oceanic Cretaceous sites indicating bottom temperatures of 5–16 °C. At present, the most saline ocean water is formed equatorially by high evaporation, but does not sink because it is warm enough to maintain a low density (Thurman, 1985). Because the solubility of oxygen decreases with rising temperature, it is likely that WSBW was more prone to anoxia than modern cold, deep water, and this may have contributed to formation of mid Cretaceous organic-rich shales. Additionally, WSBW would have had a longer residence time than cold bottom water (Arthur *et al.*, 1990) which increased its tendency to anoxia. Using computer modelling, Barron and Peterson (1990) identified the most probable site of bottom water generation as falling in the east Tethys, although Woo *et al.*(1992) found evidence of WSBW formation in the Gulf of Mexico.

Our best information about Cretaceous ocean surface currents comes from general circulation modelling (GCM). Barron and Peterson's (1990) model differs from earlier reconstructions (e.g. Haq, 1984) in showing that Tethys was dominated by two clockwise gyres, and current flow along the northern side of the ocean was dominantly easterly. It also suggests that changes in sea level have the largest potential to produce different patterns; for example, low sea levels resulted in reversal of Tethyan current patterns. Geological evidence is needed to test this model.

2.5.3   Oceanic hiatuses and deep water circulation

The deep-sea record of Cretaceous sediments is punctuated by numerous hiatuses, many of which extend between different ocean basins. For example, the Late Albian–Cenomanian interval is represented by a major hiatus across the South Atlantic (Zimmerman *et al.*, 1987), the North Atlantic (de Graciansky *et al.*, 1987), and the Central Pacific Basin (Sliter, 1995; for review see Schlanger, 1986). Hiatuses in the Turonian and Early Santonian are ubiquitously developed in high southern latitudes (Sliter, 1995). Sliter and Brown (1993) related these breaks to episodes of warm saline bottom water production, associated with sea-level rises. During highstand, flooding of low-latitude shallow seas resulted in increased production of WSBW, which intensified oceanic circulation at intermediate depths. Huber and Watkins (1992) interpreted the Mid Campanian hiatus which is present in the southern Indian, Pacific and South Atlantic Oceans to result from establishment of marine connections between basins.

2.5.4   Global persistence of chalk facies

The widespread deposition of chalk in relatively shallow water on the Cretaceous shelves is a remarkable phenomenon in its own right because, at the present day, nannofossil oozes are characteristic of deep pelagic environments (Håkansson *et al.*, 1974). Yet more remarkable is the global persistence at midlatitudes through the Late Cretaceous of a distinctive facies suite comprising successively, glauconitic marl, marly chalk and white chalk. This succession reflects initial transgression and subsequent increased deepening of the chalk sea, with progressive fining (sand–silt–clay) and decrease of clastic flux upwards. The precise age of the basal chalk transgression was dependent upon the interplay of local subsidence history and major transgressions.

The general decrease in clastic material through time in chalk successions is largely a consequence of distance from and burial of source areas, but was assisted by post-Cenomanian aridity at mid-latitudes which decreased runoff and thus sediment input (see Section 2.5.4). In addition, shelf edge fronts, which at the present day separate coastal from oceanic water, probably disappeared during the very high sea-level stands of Turonian–Maastrichtian time (Hay, 1995). This let oceanic conditions onto the shelves and into the flooded continental interiors during the Late Cretaceous. The Cretaceous dominance of coccoliths as primary producers differs from the situation which obtains at the present day, where diatoms compete success-fully with nannoflora. This success may be related to Cretaceous evolutionary diver-sification within the calcareous nannoplankton.

2.5.5 Cretaceous upwelling and productivity events

Upwelling is the process by which deep nutrient-rich water is brought to the surface, usually by the wind divergence of equatorial currents or coastal currents that pull water away from the coast (Thurman, 1985). Upwelling zones are characterized by high productivity, and organic-rich sediments may form in these areas. Parrish and Curtis (1982) constructed qualitative atmospheric circulation maps for the Mesozoic and Cenozoic, and used the wind patterns predicted by these to identify the location of upwelling zones. They found a significant statistical correlation between predicted upwelling zones and known sites of mid Cretaceous organic rich sediments, but emphasized the role of anoxia and trangression in formation of these facies. Arthur *et al.* (1988, 1990) suggested that the rates of upwelling associated with the Cenomanian–Turonian boundary event were increased by greater rates of production of deep saline water at, or close to, the Cretaceous thermal maximum.

In shallow Cretaceous epicontinental seas, limited accommodation space led to the local formation of highly condensed sequences containing abundant nodular phosphorite. Formation of phosphate requires considerable nutrient recharge, prob-ably caused by upwelling, but fully oxygenated conditions in wave-agitated shallow water lead to complete oxidation of organic matter. Phosphate-rich horizons on platforms may correspond to organic horizons in adjacent basins, as in the Albian of the Helvetic Shelf and Vocontian Basin in south-east France (Breheret & Delamette, 1989).

2.5.6 Carbonate platform drowning

A major feature of mid Cretaceous carbonate platforms is the periodic occurrence of platform drowning events, where shallow warm water carbonate fac-tories were temporarily shut down and buried under outer shelf carbonates; nor-mally, platform carbonate production keeps pace with rising sea levels (Schlager & Philip, 1990). These events were associated with extinction in various groups like the rudists (Ross & Skelton, 1993). The most important drowning events (which occurred synchronously in different regions) were the mid Aptian and Late Cenomanian – both major transgressive periods. Hallock and Schlager (1986) noted that episodes of platform drowning coincide closely with oxygen deficiency in the ocean basins (OAEs), and proposed that mid-ocean overturn caused nutrient floods which killed off the ecologically sensitive low-nutrient benthos of platform areas. Scott (1995) suggested that low-oxygen, nutrient-rich waters periodically flooded platforms and stressed biotas, leading to elimination of some taxa and changes in reef communities. However, in the shallow water settings of carbonate platforms, wave agitation should have ensured adequate oxygenation to a depth of greater than 50 m.

2.6   CRETACEOUS CLIMATES AND CLIMATIC CHANGE

A wide range of methods have been used in attempts to reconstruct Cretaceous climates and climatic history, including faunal and floral distribution, the distribution of sedimentary facies and mineral phases controlled particularly by arid and humid climates, and temperature curves from different latitudes based on oxygen isotope data. More recently, climate modelling using either conceptual models or computerized general circulation models have been widely used. These models are constantly improving as feedback from geological data is incorporated into their construction (e.g. Barron *et al.*, 1993, 1995).

2.6.1   Cretaceous polar ice?

One of the main controversies concerning the Cretaceous world is whether or not significant ice caps existed at the poles for at least short intervals of time. For Fischer (1981) the Cretaceous fell within a greenhouse climatic episode; other workers have suggested that in the Early Cretaceous polar areas experienced episodic seasonal glaciation (Frakes & Francis, 1988, 1990; Francis, 1991).

Perhaps the most important evidence for the presence or absence of polar sea-level ice is the record of high-latitude terrestrial vegetation. This is best known from the North Slope of Alaska which was situated at about 85° N through the Cretaceous (Spicer, 1987). Here, plant macrofossils are abundant and allow both community structure and climate to be interpreted from the Albian through to the Palaeocene (Spicer & Corfield, 1992). The flora was dominated by conifers, ginkgophytes, ferns and cycads, with local angiosperm abundance. Although there is no comparable present-day flora, general palaeoenvironmental conclusions can be drawn from this material. Most taxa were deciduous, and fossil wood shows evidence of growth cessation at the start of winter, probably caused by rapid shortening of day length. Overall, the flora indicates a cool temperate regime, with a mean annual temperature varying between about 5 and 13 °C through the Cretaceous. There is no evidence of permafrost or periglacial features in palaeosols. Palaeotemperatures from southern high-latitude Cretaceous marine carbonates on James Ross Island (Pirrie & Marshall, 1990; Ditchfield *et al.*, 1994) indicate temperatures in the range of 10–20 °C through the entire period. Pole to equator temperature gradients were low in the Cretaceous, and the palaeobotanical work of Wolfe and Upchurch (1987) in North America suggests latitudinal temperature changes of about 0.3 °C per degree of latitude.

The presence of extensive high-latitude coals, for example in the Hauterivian of Arctic Canada and Spitsbergen (Kemper, 1987) and the Albian–Cenomanian of the North Slope of Alaska (Parrish & Spicer, 1988*b*), indicates that very humid conditions at high northern latitudes existed for parts of the Cretaceous. This humidity is evidence that, for part of the year at least, the Arctic region was dominated by low-pressure regions, in contrast to the permanent high pressure circulation maintained by the present glacial Arctic (Spicer & Parrish, 1990).

Direct sedimentological evidence for sea-level ice in the Cretaceous is contentious. Large exotic clasts in high-latitude Mesozoic marine shales were interpreted as dropstones from ice-rafts by Frakes and Francis (1988, 1990). These authors concluded that seasonal ice-rafting took place through at least the first half of the

Cretaceous at latitudes higher than 60° N. The problem here lies in the correct identification of dropstones as unequivocal indicators of ice-rafting; dropstones are also commonly carried by tree-roots, kelp holdfasts and as stomach-stones of reptiles, as discussed thoroughly by Bennett and Doyle (1996). In fact, there are no definitive criteria with which to identify the transporting agency of individual drop-stone occurrences; dropstones also occur commonly in mid and low latitude Cretaceous sediments. There are no known examples of tillites or striated surfaces from the Cretaceous (Frakes & Francis, 1990), so dropstones remain the solitary and rather unconvincing evidence of Cretaceous ice.

The presence of calcite pseudomorphs after ikaite ($CaCO_3.6H_2O$), an early authigenic mineral which forms in superficial sediment at temperatures of less than 6 °C (Shearman & Smith, 1985), has been taken as evidence of very cold water at high latitudes in the Cretaceous. Stellate masses of these pseudomorphs, commonly called glendonites, are common in Lower Cretaceous shales in northern Siberia, Spitsbergen and Arctic Canada and have been recorded from the Aptian–Albian Eromanga Basin in South Australia (Francis, 1991). Most of these occurrences are in shales that have also yielded dropstones. Ikaite in these situations must certainly indicate cold sea floor temperatures, but cannot be used as evidence of sea-level ice.

If significant ice had been present at the Cretaceous poles, it is likely that it would have initiated in Antarctica long before developing in the Arctic, mirroring the pattern of ice cap development through the Cenozoic which is caused by the low thermal conductivity of land as compared to the sea. There are no Cretaceous tillites known from Antarctica, and extensive cool temperate podocarp and auricarian forests grew on the Antarctic Peninsula at 70° S during the Campanian to Maastrichtian (Francis, 1986).

Spicer and Parrish (1990) considered that permanent ice may have existed only above 1000 m in Cretaceous polar regions, although the possibility that mon-tane glaciers occasionally calved into the sea cannot be totally discounted. Climate modellers using four times present atmospheric CO_2 levels (Barron et al., 1993) have suggested that permanent ice existed in the Antarctic interior even at peak Cretaceous temperatures. There is no undisputed geological evidence to support these assertions; the isotopic evidence supporting the existence of ice-sheets in the Berriasian to Valanginian provided by Stoll and Schrag (1996) can be explained solely in terms of temperature change.

The absence of firm evidence for Cretaceous polar glaciation has a number of important implications for palaeoceanography because (i) it demonstrates that sea-level changes cannot have been under glacioeustatic control, and (ii) if cold deep oceanic water did form, it would sink due to density increase from low temperature, and not because of salinity-induced high density caused by ice formation. Production of cold deep water was presumably much less vigorous when sea-ice was absent.

2.6.2 Boreal vs. Tethyan faunas

A large amount of work on Cretaceous climatic change is based on the premise that southerly migration of Boreal taxa represents cooling events, and north-wards migration of Tethyan forms demonstrates warming. This approach was cham-pioned by Kemper (1987), who created a temperature curve for the entire Cretaceous based on faunal migration and on the belief that dark sediments formed in cold water and light ones in warmer water. His curve shows a progressive overall increase in temperature through the Cretaceous, punctuated by two orders of cyclicity, and bears no similarity to any oxygen isotope curve.

North–south migration of taxa can be related to many factors other than temperature, such as the formation or breakdown of physical barriers to migration, changing paths of currents, both of which can be caused by sea-level change, for example (see Rawson, 1994). Temperature can only be verified as a control on Boreal/Tethyan faunal shifts when accompanied by oxygen isotope data indicating warming or cooling. A well-documented example is the Late Cenomanian North Boreal fauna of Jefferies (1962) which spread south and west across Europe in the *Metoicoceras geslinianum* Zone, reaching south-east France (Gale & Christensen, 1996). Oxygen isotope data show a heavy (cold) excursion exactly coincident with the southward faunal migration.

2.6.3   Oxygen isotopes from marine carbonates and fossils

The ratio of ^{16}O and ^{18}O deposited in biogenic calcium carbonate is partially determined by sea-water temperature, although some organisms deposit oxygen isotopes out of equilibrium with sea-water temperature (vital effect). The isotopic ratio may be altered by diagenetic recrystallization which is not always easy to identify (for review, see Marshall, 1994). The calcium carbonate compensation depth (CCD) was relatively shallow in the Cretaceous, and the ocean record contains widespread major hiatuses; it is difficult to find Ocean Drilling Program (ODP) sites in which long periods of Cretaceous time are continuously recorded by uncemented carbonate sediment. Douglas and Savin (1975) and Huber *et al.* (1995) generated $\delta^{18}O$ curves for the Barremian–Maastrichtian of the North Pacific region and the Aptian–Maastrichtian of high-latitude Indian and South Atlantic Oceans, respectively, based on analyses of planktic and benthic foraminifera. In spite of hiatuses and poor core recovery, the data show an overall temperature rise from the Barremian into the Albian–Cenomanian, a mid Cretaceous (Cenomanian–Santonian) high temperature plateau, and a fall through the Campanian and Maastrichtian.

Jenkyns *et al.* (1994) used chalk successions in western Europe to generate a $\delta^{18}O$ curve for the Cenomanian–Maastrichtian interval, based on analysis of bulk sediment. This shows a progressive but stepped rise in temperature through the Cenomanian to a maximum in the latest Cenomanian and earliest Turonian. From there on temperatures fell, slowly at first, then faster into the Campanian and Maastrichtian. The Cenomanian–Turonian peak is thus an important climatic turning point in the history of the Mesozoic. Kolodny and Raab (1988) produced a low-latitude oxygen isotope curve using fish teeth from Israel, which also showed a Cretaceous temperature acme in the latest Cenomanian.

There is one very important difference in the curves obtained from high latitudes and those from low and mid latitudes. In the high latitude data, the thermal maximum persisted from the Turonian to the earliest Campanian, whereas in mid- and low latitudes, temperatures fell following the Cenomanian–Turonian thermal maximum.

At present, we have only very poor resolution oxygen isotope data for the Early Cretaceous, insufficient to identify detailed events other than an overall warming from the Barremian onwards.

2.6.4   Cretaceous arid and humid zones

Numerous sedimentary, mineralogical and palaeobotanical criteria are potentially usable in the recognition of arid and humid zones in the geological record (Sellwood & Price, 1993). The best known and most widely used are coals as indicators of humidity and evaporites for aridity; Hallam (1984) has discussed their distribution in the Mesozoic. Additionally, clay mineral suites have been widely used to identify arid and humid environments, specifically, kaolinite as an indicator of humid weathering, and palygorskite and sepiolite as indicators of arid regions. Care must be taken here, since many other factors also influence the distribution of clay minerals (Curtis, 1990).

We now have a good general picture of changing relative humidity in the mid-latitude Early Cretaceous of Europe. The Early Cretaceous in southern England saw a widespread major facies change from limestones with evaporites of the Purbeck facies to non-marine clastic deposits of the Wealden facies, explained by Allen (1981) as a change from an arid to a humid climate. This change has been widely recognized within non-marine basins on the Atlantic borderlands of Europe (Ruffell & Rawson, 1994); it commenced in the Valanginian and continued into the Early Hauterivian. In the Late Hauterivian to Barremian, a semiarid phase commenced, which lasted into the Early Aptian, according to Ruffell and Batten (1990); they related the spread of shallow-water carbonates of the Urgonian facies in Tethys to this aridity. In contrast, Rat (1989) interpreted the Urgonian facies as humid. The Aptian–Cenomanian interval is thought to have been relatively humid in north-west Europe, and is represented widely by clays and marls passing up into marly limestones (Weissert & Lini, 1991). Reasonably humid Cenomanian conditions across Europe are reflected by marly chalks which locally yield terrestrial floras dominated by ferns and conifers.

The base of the white chalk facies in Europe commonly falls at the top of the Cenomanian and reflects virtually complete cut off of clastic sediment. This was partly due to high Late Cenomanian–Early Turonian sea levels (see above), but is also explained by increased aridity, as evidenced by the widespread occurrence of dolomitic chalks formed from Mg-rich brines in northern France (Hancock, 1976; Quine & Bosence, 1991) and the virtual disappearance of kaolinite above the Cenomanian. A low and mid-latitude belt of low precipitation was identified in North America by Wolfe and Upchurch (1987). To the north of the chalk facies in Europe, thick mud-dominated clastic successions accumulated in the Viking Graben and Norwegian Sea (Hancock, 1989), and thin clastic successions are present in the north of the Russian Platform (Naidin, 1960). These represent a humid northern belt.

Climate modelling (Lloyd, 1982; Parrish et al., 1982; Barron et al., 1989) has predicted that a strongly developed monsoon was present in the Cretaceous, as a means of transfer of latent heat and moisture poleward. Barron et al. (1989, fig. 13) argued that globally warm climates meant high evaporation and precipitation rates and predicted very high seasonal rainfall focused on the northern and southern borders of Tethys. However, Cretaceous low-latitude vegetation was poorly developed and xeromorphic (Spicer *et al.*, 1993), and tropical everwet vegetation (rainforest) is completely unknown. Cretaceous plant productivity was instead concentrated at mid- and high latitudes, and it seems likely that high evaporation in the low and mid-latitudes resulted in poleward transport of moisture and increased rainfall at high latitudes, allowing coals to form there (Parrish *et al.*, 1982).

Parrish *et al.* (1982) noted that, by the Cenomanian, the opening of the South Atlantic and widening of the North Atlantic would have led to increased rainfall on the eastern coasts of North and South America. With further widening of the Atlantic, greater marine influence should have resulted in more rainfall in western Europe.

2.6.5   Supertethys and hot tropical oceans

It has been claimed (Johnson & Kauffman, 1990; Kauffman & Fagerstrom, 1993; Johnson *et al.*, 1996) that the abundance and high diversity of rudist bivalves in low-latitude Tethyan seas was a consequence of very high water temperatures which restricted coral growth but favoured that of rudists; this hypothetical zone was called 'Supertethys'. The suggestion that rudists somehow competed with and displaced scleractinian corals has been dismissed by Gili *et al.* (1995*a*) who showed that rudists were gregarious soft-sediment dwellers which never constructed framework reefs. In any case, Gili *et al.* (1995*b*) and Scott (1995) showed that abundant Cretaceous hermatypic corals consistently lived in slightly deeper water than did rudists. There is no evidence from oxygen isotopes that the central Tethyan region was hotter than the present equatorial region, and the Supertethys hypothesis was based entirely on climate modelling (Barron *et al.*, 1995). (See Skelton & Donovan, 1997, for a useful discussion.)

2.6.6   The Cretaceous carbon cycle and its relation to climate

Investigation of the ratios of ^{12}C and ^{13}C ($\delta^{13}C$) in Cretaceous carbonate sediments has revealed the presence of periodic major positive excursions, which coincide with periods of organic rich black shale deposition on the outer shelves and in ocean basins. These excursions commenced in the Late Valanginian, and continued through to the Early Campanian. They were called Oceanic Anoxic Events (OAE) by Schlanger and Jenkyns (1976), and represent times of enhanced preservation of organic matter associated with ocean- and outer shelf-floor anoxia. Organic matter preferentially sequesters light carbon and OAEs are reflected by positive excursions in $\delta^{13}C$, usually registered both in the organic and carbonate carbon reservoirs. Because these events are short-lived and globally synchronous on a geological timescale (the turnover time of the carbon cycle is about 10 ka) they are excellent correlation tools (Gale *et al.*, 1993).

Anoxic events have been held responsible for Cretaceous marine extinctions, in particular the faunal turnover which occured around the Cenomanian–Turonian boundary (Kauffman & Hart, 1995). This event has been attributed to expansion of an oxygen minimum zone (OMZ) which impinged on the sea floor in relatively shallow depths causing dysoxia and disappearances of some benthic foraminiferans and ostracodes (e.g. Jarvis *et al.*, 1988). However, good sedimentary and ichnofabric evidence for anoxia or strong dysaerobia is restricted to deeper waters of the outer shelves and ocean basins.

The Early Cretaceous registers steady, rather low values of $\delta^{13}C$, continuing through from the Late Jurassic (Weissert & Lini, 1991). The first positive excursion is found in the Late Valanginan–Early Hauterivian (Lini *et al.*, 1994), followed by a major double excursion in the Aptian (OAE 1; Weissert & Lini, 1991; Erbacher, 1994). The Albian $\delta^{13}C$ curve is not well known, but minor excursions are found in the latest Albian–Early Cenomanian (Gale *et al.*, 1996; Paul *et al.*, 1994). Values rose through the Cenomanian (Jenkyns *et al.*, 1994) to a short-lived high peak in the latest Cenomanian–Early Turonian (OAE 2). A Turonian trough in $\delta^{13}C$ is followed by a broad Coniacian–Santonian peak (OAE 3). Values remained steady through the Early Campanian, but fell steadily through the Late Campanian and plunged through the Maastrichtian.

2.7   CONTROLS ON CRETACEOUS CLIMATE CHANGE

2.7.1   Long-term controls: plate tectonics

It is possible that the rise of orogens in the Americas, eastern Europe and Asia during the Campanian–Maastrichtian interval directly affected the carbon cycle, because increased chemical weathering led to decrease in atmospheric CO_2 (cf. Raymo & Ruddiman, 1992), and could thus have caused the cooling seen in the latest Cretaceous. The rise in $^{87}Sr/^{86}Sr$ ratios through the Santonian to Maastrichtian interval (McArthur *et al.*, 1994) could, therefore, be related to increased input of radiogenic strontium derived from continental uplift, in parallel with the post-Eocene rise in $^{87}Sr/^{86}Sr$ (Raymo & Ruddiman, 1992). This cooling could also have resulted in the establishment of cold, deep-water circulation in the ocean basins, causing increased oceanic oxygenation and leading to falling $\delta^{13}C$ values through the Maastrichtian.

The development of a north–south Atlantic connection (Late Albian), opening of the Norwegian Sea (Early Cretaceous) and closure of the Panama Isthmus (Campanian–Maastrichtian) probably had major consequences not only for oceanic circulation but also for climatic evolution. Using physical modelling, Luyendyk *et al.* (1972) demonstrated that closure of the Panama Isthmus resulted in cessation of cross-polar flow because warm Atlantic water no longer flowed into the Pacific. However, their conclusion that this triggered glaciation in the Arctic has been generally disputed.

2.7.2   Cretaceous atmospheric carbon dioxide

The huge pulse of submarine volcanism which began in the Aptian and continued through the mid Cretaceous in the east Pacific Ocean (mantle superplume) is thought to have increased the rate of CO_2 release and thus to have led directly to greenhouse conditions in the Aptian–Santonian interval. CO_2 levels were perhaps ten times higher than those at the present day (Berner, 1990). The increased rate of ocean crust production was also responsible for the very high sea-level stands of the mid Cretaceous. It is thus no coincidence that both temperature and sea level reached their highest levels in the Late Cenomanian to Early Turonian interval.

Periods of increased organic burial during the mid Cretaceous (seen as both organic-rich mud shales and positive excursions in $\delta^{13}C$) also coincided with increased rates of oceanic crust production. The organic burial events were perhaps initiated by sea-level rise leading to upwelling and massively increased productivity which was the direct cause of anoxia (Pedersen & Calvert, 1990). Volcanic events are also directly implicated in the initiation of anoxic events; as noted in Section 2.3.2, formation of the large Ontong–Java Plateau coincides with the first Aptian OAE. Basalt extrusion on the sea floor may have led to destabilization of the water column and advection of deep nutrient-rich waters to the surface (Bralower et al., 1994). The occurrence of organic-burial events can also be related to high temperature intervals, which reduced the pCO_2 of seawater and made the system more prone to axoxia. The presence of warm saline bottom water in the Cretaceous oceans was related to reduced pCO_2 by Herbert and Sarmiento (1991), who suggested its cause lay in increased efficiency of plankton to extract nutrients at low latitudes.

Organic burial events in the mid Cretaceous seas were themselves the cause of reduced atmospheric CO_2, which led to cooling. Burial of organic matter in the ocean basins and on the outer shelves depleted the oceanic carbon reservoir by causing drawdown of atmospheric CO_2, thus instigating global cooling. Arthur et al. (1988) calculated that the major carbon burial event at the end of the Cenomanian (OAE 2) would have completely stripped Cretaceous atmospheric CO_2 (assuming eight times present-day levels) in 100 ka. Jenkyns et al. (1994) have shown that temperature did indeed start to fall immediately after the end-Cenomanian event, and suggested that this was maintained by an unknown positive feedback mechanism. The Campanian–Maastrichtian fall in $\delta^{13}C$ can be related to decreased CO_2 caused by both falling rates of oceanic crust production, and increased erosional drawdown in the newly emerged mountain chains.

2.7.3   Short-term controls: Milankovitch band

In addition to the long-term climatic changes described above, relatively short-term climatic cycles falling in Milankovitch Band frequencies (19–100 ka) are widely described from the Cretaceous. These cycles are particularly well recorded in sediments comprising a mixture of fine carbonate and clay, where they appear as decimetre-scale alternations of marls and chalks, the ratio of components sensitively reflecting climatic changes which controlled carbonate production and detrital flux. Although such cycles were originally described as a product of clay dilution, there is now strong evidence that they are dominantly formed by variation in carbonate productivity. In deeper water facies, particularly during OAEs, redox cycles record high-frequency carbon burial events as alternations of laminated mudstone and bioturbated marl. Ditchfield and Marshall (1989) interpreted changes in $\delta^{18}O$ values across a group of chalk–marl couplets as representing changes of a few degrees centigrade. However, it is likely that isotopic ratios from rhythmic alternations of marly chalk and marl also have a diagenetic component.

Changes in net insolation from eccentricity and precession are very small, and feedbacks are required to amplify these, as the short eccentricity cycle does in the Pleistocene ice albedo feedback (Imbrie et al., 1992). It is likely that the western extent of the monsoon was moderated by precession as it has been since the Pleistocene (Herbert & Fischer, 1986).

2.8   SUMMARY

The global paleoenvironmental history of the Cretaceous can be divided into three successive episodes.

Berriasian–Barremian transitional period

This interval really represents a climatic transition from the Late Jurassic, which was exceptionally humid and driven by strong monsoonal seasons (Hallam, 1984). Rates of sea floor spreading were low, as reflected in generally low eustatic sea levels. Apart from a brief event in the Valanginian to Hauterivian, carbon cycling was slow as reflected by generally low $\delta^{13}C$ values (Weissert & Lini, 1991).

Aptian–Mid Campanian greenhouse

This interval represents the Cretaceous greenhouse world at its full development. Exceptionally high rates of oceanic crust production resulted in high sea-level stands and high atmospheric CO_2 levels, perhaps eight to ten times those at present, which led in turn to high global temperatures. Maximum sea-level stands and temperatures coincided in the latest Cenomanian. There is evidence of strong latitudinal climatic zonation, including mid latitude arid belts. The latitudinal temperature gradient was low, and there is no evidence of sea-level ice at the poles. Oceanic circulation was driven almost entirely by warm, deep, saline water (WDSW). High productivity and anoxia led to the widespread formation of organic-rich shales in the ocean basins and on the outer shelves during relatively brief episodes. These are reflected in the $\delta^{13}C$ signal by many short-lived positive excursions. The opening of major seaways, such as the North–South Atlantic connection, led to increased equability.

Late Campanian–Maastrichtian cooling

The latest part of the Cretaceous was characterized by falling temperatures and generally falling sea levels. Deep ocean circulation may have been driven partly by cold, more oxygen-rich water derived from high latitudes. This latter feature is reflected in the falling levels of oceanic organic burial, seen as significantly reduced $\delta^{13}C$ values. The causes of this climatic amelioration probably lie in the rising orogens of the Americas, and in decreased rates of oceanic crust production; both led to significantly lower atmospheric carbon dioxide levels. Additionally, palaeogeographical changes caused the cut-off of major marine connections (e.g. Panama Isthmus) and therefore reduced equability by restricting oceanic heat transfer.

3

The Cenozoic world

KEVIN T. PICKERING

3.1   INTRODUCTION

The Cenozoic represents the last 65 million years of the Earth's history, during which time there were major changes in the distribution of continents, land-surface area and oceanic basins as the Earth's plates evolved towards their present configuration. These plate tectonic processes account for the overall deterioration in global climate throughout the Cenozoic from the relatively warm and sea-ice-free Cretaceous and Early Eocene, to the Quaternary icehouse world. This change in global climate is manifest in biotic changes and as a series of stepwise changes in, for example, the oxygen, carbon, strontium and osmium stable isotopes, together with other proxy chemical and geological data (Fig. 3.1).

Once icehouse conditions were established, the high-frequency climate record of glacials and interglacials, and of stadials and interstadials, is best explained by Milankovitch cyclicity. However, higher frequency global climate change on a sub-Milankovitch scale is manifest as millennial, century, decadal, annual and even intra-annual changes. Such very high frequency events are best appreciated in the Quaternary record.

This chapter summarizes the principal changes that took place during the Cenozoic, and considers some of the processes that drove the changes.

3.2   CENOZOIC PLATE TECTONICS AND PALAEOGEOGRAPHY

The Cenozoic witnessed the continuing break-up of the supercontinent of Gondwana, the continuing widening of the Atlantic Ocean, the closure of Tethys and the opening of many marginal basins in the western Pacific Ocean. The break-up of Gondwana caused major changes. India began moving northwards during the Cretaceous, and collision with the Asian continental lithosphere ultimately resulted in the uplift of the Himalayas and the Tibetan Plateau. Throughout the Cenozoic the Alpine mountain chain was forming, stretching from Central Asia to the western-most Alps. The oblique collision of the Iberian and European plates created the Pyrenees, and farther south the collision between the Iberian and African plates created the Betics.

The continued northward movement of the African plate against the European plate (and intervening microplates) resulted in the Mediterranean 'Messinian salinity crisis', which was brought about by the isolation of the western Tethys (Ryan, 1973; Hsü et al., 1977; Aharon et al., 1993). Over about a 1 million-year interval the regional climate became drier, and without an open marine seaway connecting the Mediterranean to other oceans it virtually dried up to become a desert, with brine pools associated with the widespread deposition of thick evaporites. About 5 million years ago the dam which separated the Mediterranean from the Atlantic Ocean was finally breached. The hot desert climate at the bottom of the Mediterranean was reclaimed by the sea.

During the Messinian time interval, eustatic change in sea level periodically recharged the Mediterranean. These eustatic changes are seen in cores drilled in the carbonate cap of Niue, a mid-oceanic carbonate platform in the South Pacific, now 35–140 m below sea level. They are marked by intensely leached and chalky intervals associated with a marked ^{18}O and ^{13}C depletion (Aharon et al., 1993). At this Pacific locality Sr-isotope chronostratigraphy suggests that the Messinian salinity crisis was associated with eustatic sea-level changes of about 10 m amplitude, commencing at 6.14 Ma and culminating in a sea-level change of at least 30 m at 5.26 Ma. Such sea-level variation is ascribed to critical glacio-eustatic modulation of the connection between the Atlantic and Mediterranean water masses (Aharon et al., 1993).

The Atlantic Ocean continued to widen through the Cenozoic and is still expanding at about 2 cm per year. The movement of the Caribbean microplates and the relative motions of the North and South American plates resulted in the Late Pliocene creation of the Isthmus of Panama. This formed a land bridge that permitted terrestrial vertebrates to interchange between the Americas (Milner et al., this volume). The accompanying closure of equatorial marine connections between the Atlantic and Pacific oceans affected marine biota (Morris & Taylor, this volume).

There was a Late Oligocene Pacific-wide tectonic event associated with the creation of the modern Australia–Pacific plate boundary at ca. 28–24 Ma, the ca. 25 Ma change in the trends of the Hawaiian and Louisville hotspot chains, and the 27–25 Ma reorganization of spreading along the Antarctic–Pacific spreading ridge (Kemp, 1991). Also, at 28.5 Ma the subduction of the East Pacific Rise was initiated south of the Mendocino Fracture Zone, leading to the development of the San Andreas and related transform plate boundary. Formation of marginal basins along the western Pacific (see Taylor & Natland, 1995 and papers therein) occurred mainly through the Late Oligocene and Early to Mid Miocene (Pickering et al., 1993; Underwood et al., 1995).

Rifting of Antarctica from Australia – between Tasmania and Antarctica – began during the Late Cretaceous, but in terms of oceanographical effects reached a critical point at about 37 Ma (Section 3.7.1). At about the same time, South America began to separate from Antarctica to create the Drake Passage (see palaeogeographical reconstructions of southern Gondwana at about 40 Ma and 30 Ma – Late Eocene and Early Oligocene – by Lawyer et al., 1992).

Throughout the Cenozoic, many other regionally important plate-tectonic events occurred but which had little or no discernible effect on biota, e.g. rifting in the Gulf of Mexico; formation of the Rhine Graben.

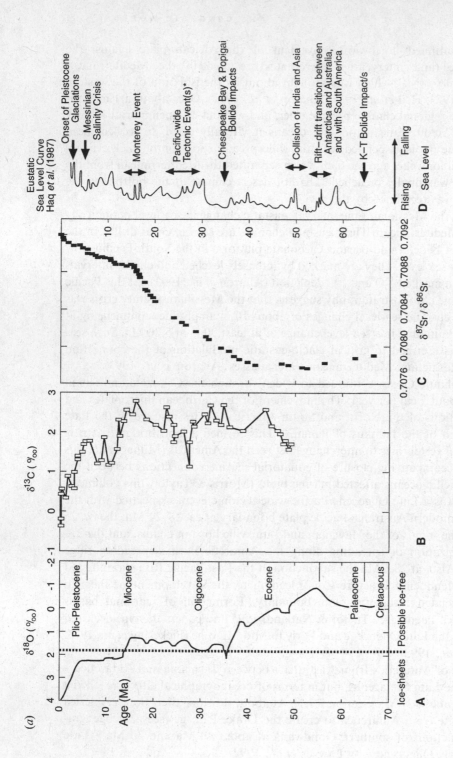

Fig. 3.1. Proxies for Tertiary global climate change.
(a) A $\delta^{18}O$ measurements from deep-sea cores in the Atlantic Ocean; B $\delta^{13}C$ for marine carbonates; C $^{87}Sr/^{86}Sr$ ratios for marine foraminifera compared with $\delta^{18}O$ PDB from bulk carbonate. Data from Shackleton (1986), reflecting surface-water temperatures; B $\delta^{13}C$ PDB from bulk carbonate. Data from Shackleton (1986), with numerical ages from Berggren et al. (1985) and Aubry et al. (1986). Fig. 3.1(b) redrawn after Thomas (1992).

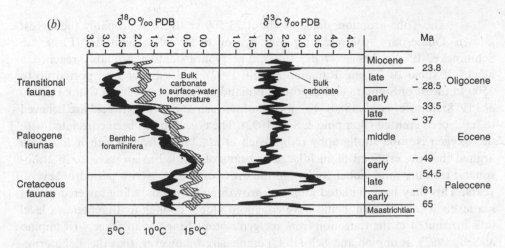

Fig. 3.1. (*cont.*)

3.3   IGNEOUS ACTIVITY

Several major igneous centres developed during the Cenozoic. The long-term eruption of flood basalts and the briefer eruptions of single volcanoes influenced global climate to varying degrees.

One of the most extensive igneous centres was the Deccan Plateau, India, where basalt extrusion took place at about the Cretaceous–Tertiary boundary some 65 million years ago (see also Gale, this volume). Here, an estimated 2–8 km^3 of igneous rock was erupted annually (Coffin & Eldholm, 1993). The Tertiary North Atlantic Igneous Province, represented above sea level by flood basalts over parts of Iceland and north-west Scotland, was erupted over a very short geological time interval at approximately 55 Ma. Other examples include the approximately 16 Ma Columbia River igneous province in the western USA.

Volcanic activity on this scale can emit enormous volumes of greenhouse gases, which may lead to substantial global warming. It has been estimated, for example, that a single flood basalt event in the Columbia River Igneous Province generated 1000 km^3 of lava, associated with the emission of 16×10^{12} kg CO_2, 3×10^{12} kg of sulphur, and 30×10^9 kg of halogens (e.g. F, Cl, Br) (Coffin & Eldholm, 1993). In addition, the estimated global network of mid-ocean ridges has produced 16–26 km^3 of new oceanic crust per annum over the past 150 million years. Since large volumes of gases such as CO_2 and SO_2 are emitted from the Earth's mantle, any large increase in the rate of generation of oceanic crust and associated mantle degassing (and/or accelerated global igneous activity) over short time intervals will have a profound forcing effect on global climate.

Individual volcanic eruptions have ejected large volumes of gases and ash which have relatively short-term effects on climate and biota, measured in time intervals from one to several years (e.g. Genin *et al.*, 1995; McCormick *et al.*, 1995). Such volcanic eruptions provide quantifiable data on their effects, as evidenced by the Late Quaternary Toba eruption in northern Sumatra.

The Toba eruption, dated at about 73 500 yr BP, was probably the largest known Quaternary volcanic eruption by an order of magnitude (Fisher & Schmincke, 1984; Chesner et al., 1991). The tephra column probably reached a height of some 20–27 km (Rampino & Self, 1992), and ash was transported up to 2500 km west of Sumatra and deposited on land as far away as India (Ninkovich et al., 1978; Stauffer et al., 1980). 10^{15} g each of fine ash and sulphuric acid are believed to have been emitted (Rampino & Self, 1992). The eruption has been correlated with the oxygen isotope stratigraphy (Ninkovich et al., 1978), from which it has been argued that the eruption of such large amounts of ash led to an increase in atmospheric turbidity and global cooling in the order of 3–5 °C over a period of several years. This may have initiated rapid ice growth and corresponding lowered global sea levels which, in turn, could have enhanced global cooling and greater sea level falls attributed to the transition from oxygen isotope Stage 5a to Stage 4 (Rampino & Self, 1992). Rampino and Self (1992) emphasized, however, that the Toba eruption occurred after the start of global sea level fall in the transition of Stage 5a to 4, suggesting that other factors were important in initiating the global climatic shift to cooler conditions. The Toba eruption, however, appears to have at least provided a contributory causal factor which probably helped drive global cooling. It may also have influenced human population numbers (Stringer, this volume).

3.4   BOLIDE IMPACTS

The suggestion that major extinctions across the Cretaceous–Tertiary (K–T) boundary were caused by a major bolide impact has led to the suggestion that such impacts could cause global climate change. It has been estimated that the Earth's global climate would be significantly affected only by the impact of meteorites greater than 1 km diameter – the size required to inject sufficient dust into the atmosphere to perturb global climate (Bland et al., 1996). In order to penetrate the lower atmosphere, a meteorite must be more than 50 m in diameter upon entry: Meteor Crater, Arizona was formed some 50 000 years ago by the impact of an object of about 60 m diameter. Meteorites with an impact crater diameter greater than 50 km are listed in Table 3.1. Shoemaker et al. (1990) estimated that there are about 1 000 Earth crossing asteroids with a diameter greater than 1 km, suggesting that a K–T boundary sized impact (i.e. producing an impact crater > 150 km in diameter) may occur once every 100 million years.

Table 3.1. *Major Cenozoic meteorite impact craters and age*

Crater	Diameter (km)	Age (Ma)
Chicxulub, Mexico	170	64.98
Chesapeake Bay, USA	85	35.45
Popigai, Russia	100	35
Meteor Crater, Arizona, USA	1.2	0.05

THE CENOZOIC WORLD   25

3.4.1   The K-T boundary event

About 65 million years ago (Table 3.2) a bolide about 10 km in diameter hit Earth (Alverez *et al.*, 1980). The estimated explosive energy of the impact is equivalent to 100 million Mt of TNT, or roughly 10 000 times the world's total nuclear arsenal (Ravven, 1991).

The impact would have created a huge crater, for which several sites have been proposed. The one currently favoured is in the Yucatan Peninsula in the southern Gulf of Mexico (Ravven, 1991), where the 170 km diameter Chicxulub impact structure has been identified. The structure is associated with andesitic igneous rocks.

The evidence in favour of a catastrophic bolide impact at the K–T boundary cames from both marine and terrestrial rocks. In the K–T boundary beds there are widespread, anomalous concentrations of materials that are often rare on earth but occur in some meteorites. These include iridium and siderophile elements (Alvarez *et al.*, 1980), and some amino acids (Zahnie & Grinspoon, 1990). Micro-diamonds occur, similar to those found in chondritic meteorites (Carlisle & Braman, 1991). In addition, 'spheroids' (droplet-shaped amorphous minerals) at the same level were created by high-temperature melting caused by the impact. Geochemical evidence suggests that the oceans suffered an unprecedented depletion in calcium, possibly due to the position of the CCD rising to within the photic zone, i.e. to less than 200 m depth.

From terrestrial areas, concentrated fluffy carbon aggregates (graphitic carbon similar to the charcoal that is produced by forest fires today) suggest global wildfires and a global flux of carbon about 10 000 times greater than the present day and 1000 times greater than in the underlying Cretaceous and overlying Tertiary sediments (Woolbach *et al.*, 1985). A study of aquatic leaves in the K–T boundary section near Teapot Dome, Wyoming, has even suggested the time of the year at which the impact occurred. There, the reproductive stages reached by the fossil aquatic plants were arrested in early June, apparently by freezing in an 'impact winter' (Wolfe, 1991).

Not all researchers support an impact theory. The biotic evidence of decline in many fossil groups before the supposed impact (see MacLeod *et al.*, 1997) has encouraged an opposing view – that the K–T boundary events resulted from the combined effects of volcanic eruptions and global sea level changes (e.g. Officer, 1993). The problem with this explanation is that it ignores the cumulative evidence for an impact 65 million years ago.

Table 3.2. *Radiometric dates for the K–T boundary impact*

Ma	Author
64.5 ± 0.1	Shoemaker, 1988
65.5 ± 3	Ravven, 1991
65.2 ± 0.4	Swisher, *et al.*, 1992
64.98 ± 0.05	Krogh *et al.*, 1993

3.4.2   Later impacts

There is evidence that several other bolide impacts occurred through the Cenozoic, though their effects, if any, on the biota are uncertain. For example, beneath the Chesapeake Bay and the adjacent Middle Atlantic Coastal Plain (US east coast) there is a 60 m thick boulder bed containing a mixture of sediments of different ages, distributed over an area of more than 15 000 km^2. It is correlated with a layer of impact material recovered from a deep-sea drilling site on the New Jersey continental slope (DSDP Site 612). The horizon is interpreted to result from a meteorite impact in the Late Eocene (Poag *et al.*, 1992, 1994). Tektite glass (part of the impact ejecta, including shocked quartz) from DSDP Site 612 has been dated by ^{40}Ar/^{39}Ar methods to be 35 ± 0.3 (Obradovich *et al.*, 1989). The candidate impact site for this Eocene event is an 85 km wide impact crater, carved through Lower Cretaceous to Upper Eocene sediments and the underlying crystalline basement, centred on the Chesapeake Bay in Virginia (Poag *et al.*, 1994).

3.5   PALAEOCEANOGRAPHY

Throughout the Cenozoic, plate tectonic processes both destroyed and created new pathways for oceanic circulation. The circumpolar Southern Ocean circulation became established after about 40 Ma as both South America and Australia finally separated from Antarctica. During the Cenozoic, the Mid Miocene (about 14 Ma) marks the critical time when major polar ice-sheets became established on Antarctica, and the Miocene may be viewed as the transition between the relatively poorly understood Oligocene oceans and the much better understood Late Cenozoic oceans (see Kennett, 1985, and papers therein). Also, the late Neogene witnessed the closure of the equatorial oceanic circulation system with closure of oceanic seaways between northern Australia and south-east Asia, then between Africa and Eurasia, and finally the creation of the Isthmus of Panama, all of which altered global oceanic circulation patterns.

The Mediterranean 'Messinian' salinity crisis (*ca.* 5 Ma) was essentially brought about as the African–Arabian and European–Middle Eastern continental lithosphere continued to collide to effectively isolate the deep Mediterranean circulation from the world oceans (see Section 3.2).

Three major episodes of palaeoeceanographical change are recorded in the deep-sea benthic foraminiferal record, which shows three major periods of turnover during the Cenozoic (Thomas, 1992): (i) latest Paleocene rapid ($< 10^4$ y.) extinctions followed by migration and diversification; (ii) late Mid Eocene through Early Oligocene gradual turnover, associated with decreased diversity, decreased relative abundance of *Nuttallides truempyi* followed by its extinction, and a decreased relative abundance or disappearance of *Bulimina* species in lower bathyal to upper abyssal depths; (iii) late Early though Mid Miocene gradual turnover, associated with decreased abundance and/or the disappearance of uniserial species from the lower bathyal to abyssal depths, the migration of miliolid species into these niches, and the evolution of *Cibicidoides wuellerstorfi*. The latest Paleocene witnessed the extinction of some 35–50% of benthic foraminiferal species also expressed in the transient and coeval 1–2% decrease in oxygen and carbon stable isotope ratios in benthic and planktonic foraminiferal tests (Thomas, 1992). This turnover has been attributed to a shift from the main deep-water formation from high to low latitudes (Thomas, 1992). It is reasonable to predict a corresponding global shift in the sites of upwelling, surface-water productivity and nutrient flux, together with temperature and oxygenation of intermediate and deep waters. The two intervals of more gradual turnover in benthic foraminifera contain two relatively rapid (*ca.* 100 000 y) shifts toward heavier $\delta^{18}O$ values in the benthic foraminiferal tests – in the earliest Oligocene and Mid Miocene – possibly associated with global processes that were too fast to be manifest in benthic foraminiferal faunas, such as the rapid build-up of polar ice-sheets. Both gradual changes in benthic foram assemblages have been attributed to gradual changes in surface-water conditions and changes in oceanic productivity (Thomas, 1992).

Through the Cenozoic there has been an overall increase in $^{87}Sr/^{86}Sr$ in marine carbonates and foraminifera (Palmer & Elderfield, 1986), and a broadly synchronous overall increase in $^{187}Os/^{186}Os$ in metalliferous sediments from the Pacific Ocean (Peucker-Ehrenbrink *et al.*, 1995). These isotopic shifts were caused by the gradual global deterioration in global climate from Mesozoic greenhouse to Neogene icehouse conditions. At the K–T boundary there was an abrupt decrease followed by recovery in the $^{87}Sr/^{86}Sr$ and $^{187}Os/^{186}Os$ values in oceanic faunas and sediments (Peucker-Ehrenbrink *et al.*, 1995), interpreted as a response to meteorite impact.

Commencing at about 37 Ma there was a transition from an Early Eocene globally warm climate to an Oligocene climate with continental ice-sheets and/or glaciers reaching sea level (e.g. Miller *et al.*, 1991; Kennet & Stott, 1990; Zachos *et al.*, 1992). In the Southern Ocean, at about 35.9 Ma, there were abrupt increases in the $\delta^{18}O$ values for both benthic and planktic foraminifera, which correlate with a sharp pulse in the deposition of ice-rafted material on the Kerguelen Plataeu (e.g. ODP Sites 689, 690, 738, 744, 748, 749; Mackensen & Ehrmann, 1992). There was also an abrupt increase (to about 30 Ma) in $^{187}Os/^{186}Os$ in metalliferous sediments from the Pacific Ocean (Peucker-Ehrenbrink *et al.*, 1995). These isotopic shifts are interpreted to reflect abrupt global cooling, changed Southern Ocean circulation (Mackensen & Ehrmann, 1992) and the establishment of continental ice.

At about 20 Ma (essentially between 25 and 15 Ma), there was an abrupt increase in $^{87}Sr/^{86}Sr$ in marine deposits from the Atlantic Ocean (Palmer & Elderfield, 1986) and an abrupt increase in $^{143}Nd/^{144}Nd$ in marine deposits from the Atlantic Ocean (Stille, 1992). These isotopic shifts are interpreted as a response to a carbon crisis in the oceans – the so-called Monterey event. During the past 16 Ma there has been a rapid increase in $^{187}Os/^{186}Os$ in metalliferous sediments from the Pacific Ocean (Ravizza, 1993; Peucker-Ehrenbrink *et al.*, 1995), reflecting the global deterioration in climate which culminated in the onset of the Quaternary glaciations.

3.6   CENOZOIC CLIMATES

3.6.1   Long-term climate change

An overall deterioration in global climate started at least 37 million years ago (Oligocene; Boulton, 1993), possibly as early as about 45 Ma (see above), and this eventually led to the Earth's global climate changing from the warm Eocene to cooler climates and eventually icehouse conditions in the Quaternary. Warming events occurred within this overall cooling trend, particularly in the Early Miocene. By the Mid Miocene, about 14 Ma, major polar ice-sheets became established.

A marked deterioration in global climate started in the Pliocene and, with fluctuations, has continued to the present day. The identification of ice-rafted debris in cores from the Antarctic deep sea, with other evidence, has placed the onset of glaciation there as far back as 3.5 Ma (Opdyke *et al.*, 1966). However, there was then an extensive deglaciation of Antarctica during Mid Pliocene times, about 3 Ma (Barrett *et al.*, 1992).

In the North Atlantic region, there is evidence for the onset of glaciation associated with progressively deteriorating climatic cycles, and ice-sheet initiation, at about 2.5 Ma (Shackleton *et al.*, 1984). Other evidence suggests that there was an abrupt change from warm to cold climatic conditions at some time between 2.6 Ma and 1.64 Ma. Hence there is some debate about the beginning of the Quaternary Period. Some would argue that it should extend back to 2.6 Ma (discussed in detail by Shackleton *et al.*, 1990), while others date it to as recently as 1.64 Ma.

Global aridity increased, with severe fluctuations, during the past 200 000 years (up until the Last Glacial Maximum) leading to the extensive spread of deserts and sand dunes in low latitudes (Sarnthein, 1978). Regions such as the Western Sahara and Sahel were, therefore, once much more extensive than at the present day. The loess record from Central Asia has provided additional support for phases of aridity associated with cooler climatic conditions (Kukla *et al.*, 1990; Xiuming Liu *et al.*, 1992). The Holocene is marked by rapid shifts in global and regional climate, with the global changes reflecting sunspot maxima and minima (Thompson, 1992; Glenn & Kelts, 1991), ENSO events (Schaaf & Thurow, 1996) and other poorly understood decadal- to millennium-scale changes (Lamb *et al.*, 1995).

3.6.2   Quaternary climate change

There are now many high-resolution records of global climate change using chemical proxy data from both high- and low-latitude land and ocean sites. It is clear that once the Quaternary ice-sheets developed, their cyclical growth and decay can be accounted for by Milankovitch cyclicity (Pickering *et al.*, 1999). However, other records also suggest sub-Milankovitch, short-term climatic shifts.

Examples of the latter include the growing evidence of rapid advances of the Laurentide ice-sheet in North America, the 5–10 ky intervals between the events being inconsistent with Milankovitch orbital frequencies. This evidence comes from the Heinrich layers (layers of ice-rafted sediment) in the North Atlantic Ocean (Heinrich, 1988). These formed when the North Atlantic was at its coldest, and the lower $\delta^{18}O$ values in the foraminifera in these layers suggest that there was a low-salinity water mass above the site where Heinrich layers accumulated. This evidence points to time intervals when the North Atlantic was covered by extensive sea ice as in the present Arctic Ocean. Heinrich events are demonstrably coincident with rapid and major changes in the thermal conditions in the North Atlantic region. They probably resulted from the periodic release and melting of massive icebergs into the North Atlantic from the Canadian margin, which input large volumes of fresh water into the oceanic conveyor belt and disrupted the formation of deep water masses (Broecker, 1994). The release of these massive icebergs apparently caused a catastrophic disruption of deep-water formation in the North Atlantic to force a switch between colder and warmer patterns of thermohaline circulation.

The most recent six of these Heinrich layers accumulated between about 70 000 and 14 000 years ago, at a frequency of 7000–10 000 years. They indicate marked decreases in sea-surface temperature and salinity, reduced fluxes of foraminifera to the sea floor, and enormous discharge of icebergs from eastern Canada, which was produced as glaciers entered the sea and began to break up (calve) over short time intervals (Bond et al., 1992). Melting of very large volumes of icebergs drifting across the North Atlantic must have been a major factor in reducing the salinity in the surface waters (implied by the $\delta^{18}O$ values): Bond et al. (1992) also noted that the salinity drop would have been sufficient to shut down the thermohaline circulation of the North Atlantic. The ice-rafted sediments on the sea floor (including detrital carbonate derived from eastern Canada) delineate the path of the icebergs and show that they must have travelled more than 3 000 km, a distance that, in itself, suggests extreme cooling of the surface waters, and substantial volumes of drifting ice (Bond et al., 1992). These Heinrich layers all show a dominance of the left-coiled planktic foraminiferal species Neogloboquadrina pachyderma, which indicates a deep southward penetration of polar water.

Each Heinrich event was followed by a pronounced global warming and then a package of higher frequency (2000–3000 years) Dansgaard–Oeschger (D–O) cooling cycles (Dansgaard et al., 1993) in a progressive cooling trend (Bond et al., 1993). Bond and Lotti (1995) showed that the D–O cycles coincided with a sudden increase in the amount of glacial ice discharged into the North Atlantic.

In East Africa, lake water levels can be correlated with Late Quaternary global climatic fluctuations. Street-Perrott and Perrott (1990) showed that periods of low lake levels generally occurred between 20 and 13 ka, 11 and 10 ka and 8 and 7 ka. They attribute the last two lowstands to prolonged periods of aridity produced during times of anomalously low sea-surface temperatures in the North Atlantic. These low temperatures may have been caused by large volumes of glacial melt water entering the North Atlantic during deglaciation and increasing the ocean salinity stratification. Such changes could then suppress the formation of North Atlantic Deep Water (NADW) and further lower the sea-surface temperature, leading to decreased rainfall and, therefore, lower lake levels in East Africa.

The rates at which global climate change can occur, together with their abruptness, is now well established. A well-documented example is from the GRIP ice-core, Greenland. Here, between about 135 and 115 ka, during the last interglacial (Eemian interglacial in Europe, correlated with the Sangamon in North America) there were intervals of severe cold conditions which began extremely rapidly, and lasted from decades to centuries (Dansgaard et al., 1992, 1993). Since 10 ka there has been a relatively stable interglacial climate, but prior to this during the last ice age which lasted about 100 ky, and in the transitional period, global climate change was abrupt and erratic. The GRIP team have shown that changes of up to 10 °C occurred within a couple of decades, possibly even less than a decade.

There is increasing evidence to show the global correlation and synchronous nature of high-frequency climate change, for example, from the Arctic (Greenland) to the Antarctic, between the major ocean basins, and between the oceanic sedimentary record and nearby ice-sheets (Bender *et al.*, 1994). A fundamental inference from this study is that there are times when the oceanic and atmospheric record are not coupled in a simple way, and that there are times when the Northern and Southern Hemispheres, and probably different continents, experience the effects of stadials and interstadials to varying degrees. Bender *et al.* (1994) proposed that such climatic differences between the Arctic and Antarctic may be a consequence of the periodic suppression in the production of NADW, etc.

During the transition from the Last Glacial Maximum to the warmer and wetter conditions that characterize the Holocene, there was a global brief cold interval, about 11 500–10 000 years ago, referred to as the Younger Dryas. This must have ended at a minimum of 10 970 years BP (Becker *et al.*, 1991). The Younger Dryas was probably the result of the sudden increased rate of melting of the Laurentian ice-sheet with large volumes of cool melt water entering the oceans and affecting atmospheric temperatures. Ice core data from Camp Century in Greenland suggest that the Younger Dryas terminated very abruptly, possibly within a few decades (Johnsen *et al.*, 1992). In effect, the Younger Dryas was a brief cool interval in a warmer period – a *stadial*. Evidence is now emerging for abrupt and rapid changes in global climate in the Quaternary, during periods of climatic instability. These climatic changes are related to glacial–interglacial cycles which, in turn, are related to changes in the global carbon cycle. This supports the view that there are strong links between climate, biogeochemical cycles and metabolic processes in organisms.

3.7   CONTROLS ON CENOZOIC CLIMATE CHANGE

Global climate change measured in hundreds to tens of millions of years is best accounted for either by cosmic processes such as overall trends in solar luminosity and the solar system's motion within our galaxy, or by processes that are controlled from within the Earth's core and mantle. Decadal- to century-scale global climate change is commonly explained as being due to one of the following: random atmospheric variability; solar variability; inherent or forced fluctuations in the production rate of the NADW; natural variations in the atmospheric concentrations of trace gases; or natural variations in volcanic aerosols.

This section looks at some of these parameters of global change in relation to the Cenozoic record.

3.7.1   Plate tectonics

It has been argued that much of the global climate change throughout the Cenozoic was strongly influenced by the uplift of the Tibetan Plateau. In particular, Ruddiman and Kutzbach (1991), and more recently Raymo and Ruddiman (1992), have proposed that the uplift of Tibet, the Himalayas, and the American south-west, caused large areas of land in low latitudes to reach a height that altered global atmospheric circulation patterns by interfering with the jet stream, which helped induce global atmospheric cooling. The uplift of the Tibetan Plateau exposed more rock, which then underwent accelerated rates of chemical and physical weathering. During many chemical weathering reactions, there is a net drawdown of atmospheric CO_2 and the formation of bicarbonate ions, particularly in the weathering of silicates. In contrast to silicate weathering, the chemical breakdown of carbonates generally has little or no net effect on atmospheric CO_2 levels as, for every molecule of CO_2 sequestered, another is returned. Also, uplift increases river gradients, physical erosion rates and fluviatile sediment loads. Uplift may also increase storminess along a mountain front leading to more rainfall, and faster flowing rivers. Any long-term net removal of atmospheric CO_2 during chemical weathering may reduce any potential greenhouse warming, and hence through negative feedback lead to a global cooling.

The problem with linking the uplift of the Tibetan Plateau with the gradual deterioration in global climate throughout the Tertiary is that the latter began not later than about 37 Ma (Oligocene), whereas indirect evidence suggests that the rapid uplift of the Himalayas occurred some 12–17 million years later, during the Miocene (Johnson, 1994; Garzanti et al., 1996). The uplift probably did not produce significantly high relief of the Tibetan Plateau until at least Mid Miocene time (15 Ma and younger), signified by the onset of the monsoons at about 12–8 Ma. It is the uplift of the Tibetan Plateau which has had the most profound influence on the establishment of the monsoons.

A more likely cause of the gradual deterioration of global climate from about 37 Ma to the Quaternary glaciations is to be found in the 'critical separation' (sufficient to permit deep thermohaline circumpolar circulation) of Antarctica from Australia and the creation of the Drake Passage between South America and Antarctica at about this time. These rifting events permitted the establishment of a circum-Antarctic oceanic circulation pattern in the newly formed Southern Ocean as well as being associated with a large continent centred over the South Pole. Major circumpolar circulation in the Southern Ocean was delayed until after 40 Ma when Tasmania and the South Tasman Rise finally cleared north Victoria Land, East Antarctica (e.g. Lawyer et al., 1992). The creation of the Southern Ocean circumpolar gyre suppressed a significant amount of the previous oceanic heat transfer to Antarctica, thereby permitting greater isolation of Antarctica from warming effects with the result that precipitation could accumulate as snow and ice. The timing of these events, rather than the uplift of Tibet and the Himalayas, are more consistent with the geochemical proxy data for overall global cooling.

3.7.2   Ocean–atmosphere interaction

The oceans play a fundamental part in controlling and changing global climate. The mixing time of the ocean waters is about 1500 years, which means that any climate change on a millennial or longer timescale has the potential to have the atmosphere and oceans in some degree of thermal equilibrium. Global climate change measured on a century to decadal scale is very unlikely to be a consequence of global oceanic circulation, but rather of high frequency fluctuations in global mean air temperature.

Many research workers now advocate a mutual interaction between global climate and ocean–current circulation. Broecker and Denton (1990) suggested that warming in the Northern Hemisphere prompts biological activity, and the consequent production or release of CO_2 from the oceans to the atmosphere. At the same time, and in turn, this changes the ocean circulation, together with the way in which heat energy is transferred through the oceans. Such changes in the thermal structure of the oceans induces the formation of the NADW, which did not flow as strongly during glacial times. The formation of the NADW involves the upwelling of north–flowing waters of high salinity from depths of about 500 m, and as these cold waters rise to the surface they replace the warmer surface waters that flow southwards, aided by the strong winter winds. The 'Atlantic Conveyer' releases vast amounts of heat energy during this process, approximately equivalent to about a quarter to one-third of the direct input of solar energy to the surface of the North Atlantic. The volume of flowing water is immense, roughly equivalent to 20 times the combined flow of the all the world's rivers. Towards the end of a glacial period, when the NADW begins to form, it fashions a different pattern of global oceanic circulation, and redistributes the heat energy in a manner different to that of the present day. Such changes in ocean circulation and heat exchange between the oceans and atmosphere may have had a profound effect on global climate and help drive the rapid climatic changes (see also the summary in Street-Perrott & Perrott, 1990).

Geochemical data has challenged the perception that rapid climate changes in the North Atlantic at the end of the last glacial were due to the switching 'on-and-off' of the thermohaline circulation (Lehman & Kelgwin, 1992; Veum et al., 1992). Rather, the oceanic circulation oscillated between a warm, deep mode and a cold, shallow mode (Rahmstorf, 1994).

Deep oceanic circulation is governed by the interaction of oceanic currents, physiochemical parameters of the surface waters, particularly in the source area for the production of bottom waters, and the nutrient flux from surface-water primary biomass production. The present-day Arctic Ocean is important in global climate change as it is the principal site of cold, deep-water production which drives thermohaline circulation, and also through its role in surface heat balance (e.g. Aagaard & Carmack, 1994, and references therein).

CO_2 and CH_4 concentrations in the atmosphere also have changed considerably during past glacials and interglacials, even down to a century-scale (e.g. Jouzel et al., 1987; Stauffer et al., 1988). During interglacials, there is approximately 25% more CO_2 and 100% more CH_4. Organic productivity appears to have been greatest during glacial periods, thereby providing an oceanic sink for carbon. The precise causes of the changes in atmospheric gas concentrations, and the threshold conditions that precipitate a switch from glacial to interglacial period, remain poorly understood.

Data for the past 160 ka from the Vostock ice core (Petite-Marie et al., 1991) suggest that tropical wetlands are a leading influence on variations in atmospheric CH_4 levels. During glacial maxima, CH_4 levels have naturally fluctuated around 350 ppbv, compared to 650 ppbv during the warm interglacial periods. The CH_4 record from the Vostock ice core shows temporal periodicity consistent with Milankovitch cyclicity. This correlation led Petite-Maire *et al.* (1991) to propose that orbitally driven changes in monsoon rainfall exert a crucial role in controlling CH_4 emissions from low-latitude, tropical wetlands. The precession and eccentricity of the Earth are the principal controls on long-term variations in insolation in the tropics, whereas obliquity or tilt becomes increasingly important with higher latitudes.

3.7.3   Milankovitch cyclicity

High-frequency global climate change on a scale that generates glacial–interglacial cycles is best explained as resulting from Milankovitch precession (19–23 ky), obliquity (41 ky) and eccentricity (100 and 400 ky) cycles. Amongst the best documented shifts in the Earth's orbital frequency in exerting a forcing effect on global change is the Mid Pleistocene climate shift dated at 920–900 ky, also known as the Mid Pleistocene Revolution (MPR) (Berger & Jansen, 1994). The MPR involved a substantial change in the mean state, in amplitude of ice-mass build-up and decay, and in the frequency of variation of ice-mass. It is marked by a pre-shift to post-shift change to more important precession-driven variations associated with the 100–ky cycle becoming dominant. The cause of the MPR remains controversial, with explanations including a gradual (Ruddiman *et al.*, 1989) vs. abrupt change (Berger & Wefer, 1992), and the origin of the 100 ky cycles as due to non-linear positive feedback on the direct but minor insolation effects related to the eccentricity cycle, non-linear ice dynamics (Oerlemans, 1982), or a beat phenomena (Wigley, 1976).

3.8   SEA-LEVEL CHANGE: CAUSES AND RATES

A study of the Cenozoic world shows how the stratigraphical record is fashioned by many interactive processes which can be grouped into internal and external Earth forcing mechanisms. In essence, these causal or controlling parameters can be perceived as principally tectonic and climatic processes. Since the most complete stratigraphical record is marine, and because the transgressive–regressive signature underpins many inferences about global change, it is important to consider the main causes and rates of sea-level change at geologically important scales. Global and/or regional sea-level change results from the following:

(i)   Changes in the volume of mid-oceanic ridges and oceanic spreading centres, possibly associated with superplume activity, typically occur on a timescale of tens of millions of years;

(ii)   Tectonic (orogenic) activity typically occurs on a timescale of tens of millions of years, but locally may be instantaneous, as when a major earthquake results in coastal uplift/subsidence. A facet of tectonic activity is isostacy associated with erosion and uplift, etc.;

(iii)   The transmission of intraplate and interplate stress, possibly at a high frequency, may occur on a timescale to millions of years (cf. Cathles & Hallam, 1991), but this hypothesized process remains poorly understood;

(iv)   Glacio-eustacy. Major continental ice caps develop at timescales of 10^5 years and melt in 10^4 years. If the Greenland ice cap were to melt completely, then sea level would rise by about 6 m, whereas if the entire Antarctic ice-sheet melted, the rise would be approximately 60 m. Isostatic rebound, for example from the melting of extensive ice-sheets and glaciers, will cause relatively local changes in land elevation, on a time scale of 10^3–10^4 years;

(v)   Thermal expansion and contraction of the ocean waters associated with global climate change will exert a relatively small but significant control on sea level. Estimated sea-level changes would be of the order of tens of cm per 10^4 years.

3.9   SUMMARY

The global palaeoenvironmental history of the Cenozoic can be divided into two successive episodes.

Paleocene–Eocene greenhouse period

After global perturbations at the Cretaceous–Tertiary boundary, including bolide impact and extensive volcanism, the Paleocene–Eocene interval represents the continuation of the Cretaceous greenhouse world and exhibits the warmest global climate of the Cenozoic. Plate tectonic processes led to land–sea configurations that resulted in a major change in oceanic circulation and global climate at approximately 37 Ma when the Earth entered an icehouse phase.

Oligocene–Recent icehouse period

This interval is characterized by general global climatic cooling with some short-lived warmer episodes. The formation of the circum-Antarctic current at about 37 Ma led to thermal isolation of Antarctica and the formation of south polar ice-sheets. The closure of Tethys led to the cessation of an equatorial current system and the rise of the Isthmus of Panama deflected warm currents to the North Atlantic which resulted in increased high latitude precipitation and the onset of Northern Hemisphere glaciation at about 2.5 Ma. Climatic oscillations since that time are accounted for by Milankovitch cyclicity but sub-Milankovitch climatic shifts during the Quaternary have been globally and regionally significant and extremely rapid.

4

Calcareous nannoplankton and global climate change

JACKIE A. BURNETT, JEREMY R. YOUNG AND PAUL R. BOWN

4.1   INTRODUCTION — THE NATURE OF NANNOPLANKTON

Calcareous nannofossils are the smallest (typically 5–10 µm in length), routinely studied fossils. In a pure nannofossil chalk, they are present in concentrations of several billions to a gram. They include coccoliths, the definite remains of coccolithophorid algae, and nannoliths, such as discoasters and nannoconids, which are of similar size and composition but which have less-certain biological affinities. Coccoliths play a major role in global climate-change research. This is partly because they are one of the most widely distributed, biostratigraphically useful groups in pelagic (both shelf and oceanic) environments, and have thus acted as primary chronometers for palaeoceanographical studies from the Late Triassic to the Recent. However, in addition to this, they constitute a valuable tool with which to monitor global change, through the study of accumulation rates, diversity and distribution changes, stable isotopes, and organic biomarkers. Furthermore, and most intriguingly, coccolithophores may be major agents of global change through their intimate relationship with the carbon cycle. Although minute individually, nannoplankton are highly visible *en masse*, producing blooms that are detectable from space (Holligan *et al.*, 1983), and vast accumulations of pelagic sediment (most notably the Late Cretaceous chalks) which crop out in most parts of the world. These properties have combined to draw considerable attention from global change studies (e.g. Charlson *et al.*, 1987; Lovelock, 1991).

In this chapter, we briefly review nannoplankton as agents of global change, and focus on an investigation of how detailed knowledge of the geological history of the group can be interpreted in terms of climatic evolution. We approach this through biogeography, describing the key controls on biogeography and providing a geological example of how climate change has been interpreted from nannofossil palaeobiogeographical patterns in the Indian Ocean. This draws on our research efforts on both Mesozoic and living nannoplankton, most notably through our participation in the 'Global *Emiliania* Modelling Initiative' (Westbroek *et al.*, 1993, 1994) and our involvement with the Ocean Drilling Program.

4.2   NANNOPLANKTON AS AGENTS OF GLOBAL CLIMATE CHANGE

4.2.1   Primary production

Coccolithophores are important primary producers in the modern world ocean and were almost certainly more important in the Mesozoic oceans, prior to the evolution and rise to dominance of diatoms in the phytoplankton. Brummer and van Eijden (1992) have shown that there is a significant correlation between open-ocean, pelagic-carbonate accumulation rates which are dominated by coccoliths and surface-water primary productivity, and argue that, in many circumstances, pelagic-carbonate accumulation rates may provide the best available primary productivity indicator. Siesser (1995) applied this method to investigate changes in productivity through the Tertiary of the Indian Ocean, highlighting marked fluctuations in productivity directly correlatable with climatic change. On a shorter timescale, carbonate accumulation rates often seem to be correlated with other indicators of increasing productivity in sediments interpreted as showing Milankovitch cyclicity (e.g. Erba, 1992; Paul, 1992).

However, coccolithophores are only one of several phytoplankton groups. Consequently, fluctuations in coccolithophore production may occur independently of changes in total primary production due to shifts in the relative importance of coccolithophores, diatoms, and other phytoplankton groups. In addition, coccolith accumulation rates are a function of preservation processes as well as production rates. Thus, assumptions of a simple correlation between coccolith accumulation rates and primary production rates should be regarded with caution. Furthermore, total phytoplankton productivity is essentially limited by nutrients, and so coccolithophore productivity should be viewed as an alternative source of productivity rather than as an additional one. Even if all coccolithophores (species and individuals) became extinct, as almost occurred at the Cretaceous/Tertiary boundary, the direct effect on global primary productivity would probably be minimal. Extinction of coccolithophores would, however, have a significant impact on climate due to their role as calcifying organisms.

4.2.2   Calcification

Sedimentary carbonate forms a massive reservoir of carbon but the predominant effect of calcification is to increase the concentration of atmospheric carbon dioxide (CO_2) (Fig. 4.1). The sea-water carbonate system is chemically complex (e.g. Varney, 1996) but calcification can be usefully simplified to the single equation $Ca^{2+} + 2HCO_3^- \rightleftharpoons CaCO_3 + CO_2 + H_2O$ (Berger, 1982). Thus, precipitation of calcium carbonate ($CaCO_3$) results in an increase in the concentration of dissolved CO_2. This relationship holds over a wide range of temporal and spatial scales. At the cellular level, calcification may allow photosynthesis to utilise bicarbonate (HCO_3^-) instead of dissolved CO_2 (e.g. Paasche, 1962; Anning et al., 1996). On an ecological scale, blooms of coccolithophores have been shown to have far smaller CO_2 drawdown effects than blooms of non-calcifying phytoplankton (Robertson et al., 1994). On the global biogeochemical scale, limestone formation will tend to result in an increase in atmospheric CO_2 thereby triggering or exacerbating global warming. As noted by Holligan (1992), this process was explicitly formulated by Chamberlin (1898) and has been well tested in models such as those of Berger and Kier (1984). None the less,

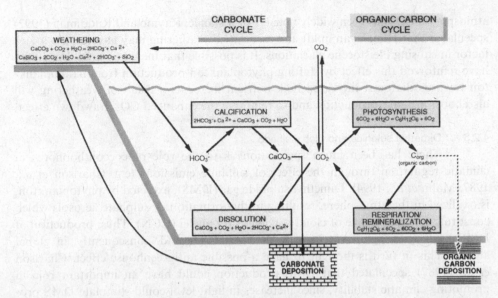

Fig. 4.1. The marine organic carbon and carbonate cycles. Fluxes are lower in the carbonate cycle than in the organic carbon cycle but the former is a less equilibriated cycle, and with a larger depositional flux. Coccolithophores play a major role in the carbonate cycle and a significant, but probably non-critical, role in the organic carbon cycle.

since the effect is counterintuitive, it is common to see erroneous casual speculation suggesting exactly the reverse process: that calcification, through burial of carbon, will result in a decrease in atmospheric CO_2.

Coccolithophores are only one of many calcifying biotic groups, but dominate pelagic calcification (Brummer & van Eijden, 1992). Accumulation rates for pelagic carbonates are much lower than for coastal carbonates but they occur over much larger areas and, on a gross scale, the two reservoirs are probably of comparable magnitude (Wollast, 1994). Prior to the evolution of nannoplankton in the Late Triassic (220 Ma), pelagic carbonates were of negligible importance. Volk (1989) noted that pelagic carbonates are liable to be subducted, unlike shelf and platform carbonates. He therefore argued that a significant effect of the shift toward pelagic carbonate deposition must have been to increase the rate of metamorphic recycling of carbonates (simplifiable to $CaCO_3 + SiO_2 \rightarrow CaSiO_3 + CO_2$). In this way, he argued, coccolithophore production has resulted, in the very long term, in an increase in global CO_2 levels.

A major control on the bulk distribution of Cenozoic to Present Day nannoplankton is that diatoms typically outcompete them when silicon is present (e.g. Smetacek et al., 1991; Egge & Aksnes, 1992). Hence, peak nannoplankton abundances occur in mesotrophic, not eutrophic, waters. In global change terms, this is an interesting phenomenon since it suggests that fluctuations in the ratio of silica (SiO_2) to bulk nutrients could cause significant differences in the open-ocean carbonate flux and so to atmospheric CO_2. Weathering of crystalline rocks causes both CO_2 drawdown and increases SiO_2 fluxes to the oceans. The effect of increased weathering on atmospheric CO_2 has been widely noted. For instance, Raymo and Ruddiman (1992) speculated that Himalayan uplift and associated weathering may have been a major factor in causing Pleistocene glaciations. It is possible that increased SiO_2 fluxes may have reinforced this effect by shifting phytoplankton production from nannoplankton towards diatoms. Increased bulk nutrient fluxes as a result of weathering will also boost primary production and so reinforce the enhanced CO_2 drawdown effect.

4.2.3   Dimethyl sulphide and global albedo

There has been much speculation as to the role of coccolithophores in climatic regulation through the effect of sulphide emissions (e.g. Charlson *et al.*, 1987; Malin *et al.*, 1994). Dimethyl sulphide gas (DMS), excreted by phytoplankton, is oxidized in the atmosphere, resulting in the formation of sulphate aerosols which constitute a major source of cloud condensation nuclei (CCN). Thus, production of DMS leads to a significant increase in cloud cover and, consequently, in global albedo. This in turn is thought to have a possible antigreenhouse effect. Charlson *et al.* (1987) speculated that DMS production could have an important role in promoting climatic stability, since increases in light level could stimulate DMS production and so produce a compensating increase in albedo. Among the phytoplankton, coccolithophores and other haptophytes are known to be major DMS producers and so have attracted special interest in this context. In addition, coccolithophore blooms, especially of *Emiliania huxleyi,* produce high-reflectance waters as a result of light dispersion by loose coccoliths in the water. This both increases net albedo and reduces light penetration, causing heat to become concentrated in surface-waters (Holligan *et al.*, 1993). This might also have a significant climatic impact (Holligan, 1992). Intriguing though these mechanisms are, their net effects on global climate are unquantified, even in the Recent, so it is probably wise to disregard them for geological purposes.

4.3   NANNOPLANKTON AS RECORDERS OF GLOBAL CLIMATE CHANGE: CONTROLS ON BIOGEOGRAPHY

4.3.1   Evolutionary dispersal/vicariance

The biogeographical distribution of any taxon is primarily a product of two types of control, its history of evolutionary dispersal and the geographical development of the ecological habitat to which it is adapted. For plankton in general, evolutionary dispersal is easy, so that wide biogeographical ranges are normal rather than exceptional. In nannoplankton, this effect appears to be developed to an extreme degree. In the modern nannoplankton, there are no obvious examples of species whose biogeographical range is restricted as a result of limited dispersal. For instance, all species known from the Pacific Ocean also occur in the Atlantic. This even applies to high-latitude-restricted taxa. Species such as *Wigwamma arctica* and *Coccolithus pelagicus* were described from high northern latitudes but have also been found in the Antarctic region. This absence of vicariance implies rapid and effective dispersal. The majority of fossil evidence also supports this, with abundant direct evidence of very rapid dispersal. For instance, there is no resolvable diachroneity in the first occurrence of *Emiliania huxleyi* (260 ka) between the Pacific, Indian and Atlantic Oceans, at the resolution of stable-isotope chemostratigraphy (Thierstein *et al.*, 1977). Thus, it appears that it had dispersed across the world ocean within, at most, a few thousand years of first evolving. This is a particularly well-documented Recent case but available evidence (e.g. Dowsett, 1989) suggests that it is a wide-spread pattern among nannoplankton, as argued by Berggren *et al.* (1995).

So, the biogeography of nannoplankton at any given time can be interpreted as a function of ecological controls, without a vicariance effect. Changes in biogeographical range through time must primarily reflect changes in the global distribution of ecological conditions, although with the complicating possibility of evolution in the adaptation of taxa or of their competitive success under particular conditions.

4.3.2   Environment

Environmental controls operate on a broad scale within the plankton and perhaps especially so in the nannoplankton. Fig. 4.2 shows the nannoplankton distribution zones mapped out by McIntyre and Bé (1967) in the Atlantic Ocean. Three decades of subsequent research on surface-sediments and plankton samples have not significantly changed or refined this distribution pattern (Winter *et al.*, 1994), nor is this likely to change since, although the zones are quite clear, there is significant overlap, with most species occurring in several zones. The zones show an obvious latitudinal distribution and so have been predominantly interpreted in terms of temperature control. Most nannoplankton species show more or less well-defined temperature preferences, both in the ocean and in culture (e.g. Watabe & Wilbur, 1966; McIntyre *et al.*, 1970; Fisher & Honjo, 1991; Brand, 1994).

The strongest single control on the bulk distribution of phytoplankton is, however, the distribution of nutrients, particularly nitrates and phosphates. Present Day latitudinal distributions can, in large part, be read in these terms: an equatorial divergence belt of moderate productivity being flanked by the oligotrophic subtropical gyres which ultimately pass into the higher-productivity, temperate and high-latitude water masses.

In addition to controlling the bulk distribution of Present Day phytoplankton, nutrient content/trophic level is a dominant control on the content of phytoplankton assemblages, and the distribution of individual species (Kilham & Kilham, 1980). Nutrient-rich, eutrophic waters are characterized by high-abundance, low-diversity assemblages of rapidly reproducing, r-selected species. Eutrophy may occur due to upwelling, riverine nutrient input, or storm-mixing of surface-waters but assemblages show strong similarities. Conversely, nutrient-depleted, oligotrophic waters show low-abundance assemblages with distinct, and often diverse, assemblages of K-selected species. The application of K- and r-selection to nannoplankton is discussed by Aubry (1992), Brand (1994), and Young (1994).

The relative roles of temperature and nutrients in Present Day oceans is somewhat blurred, as there is a strong tendency for oligotrophic waters to be warmer than eutrophic waters, which typically form from upwelling of cool subsurface waters. Thus, the observed distributions closely parallel temperature, at least within one ocean. None the less, amongst r-selected species, a reasonably clear temperature sequence can be determined including, from cool to warm, *Coccolithus pelagicus, Gephyrocapsa muellerae, Emiliania huxleyi, Gephyrocapsa oceanica*. Thus, providing

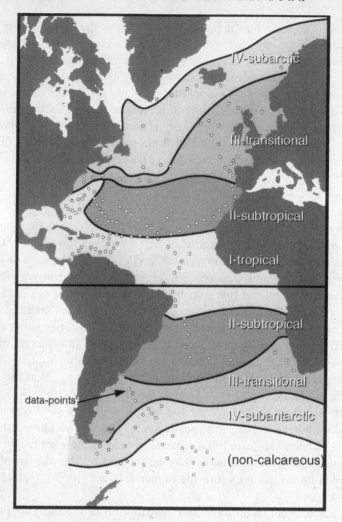

Fig. 4.2. Nannoplankton biogeographical zones, Present Day Atlantic Ocean. Redrawn from McIntyre and Bé (1967).

care is taken to avoid strongly oligotrophic species, a meaningful temperature signal can be identified.

4.3.3   Depth structure

Mixed-layer nannoplankton assemblages usually only show weak depth stratification. Often, however, a very distinct deep-photic assemblage is developed within the thermocline (Okada & Honjo, 1973; Young, 1994; Winter *et al.*, 1994). This assemblage is characterized by taxa not found in surface-water assemblages, and in particular by the rather unusual species, *Florisphaera profunda* (Fig. 4.3). This assemblage is adapted to the low-light, high-nutrient conditions of the thermocline. Populations in the deep photic zone can be very high, and *Florisphaera profunda* liths are highly abundant in the sediment. These assemblages are best developed in deep

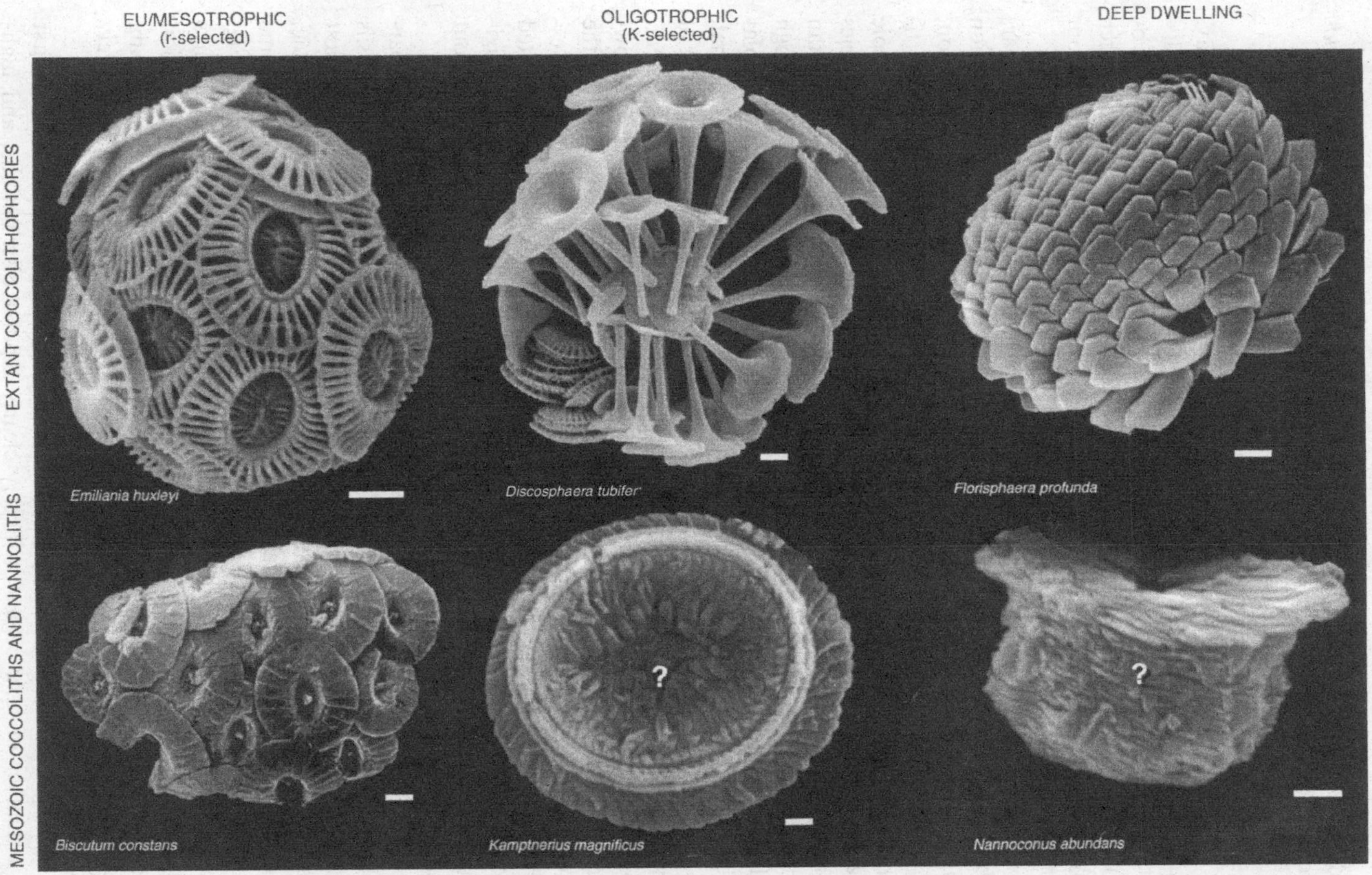

DEEP DWELLING

OLIGOTROPHIC
(K-selected)

EU/MESOTROPHIC
(r-selected)

Florisphaera profunda

Nannoconus abundans

Discosphaera tubifer

Kamptnerius magnificus

Emiliania huxleyi

Biscutum constans

EXTANT COCCOLITHOPHORES

MESOZOIC COCCOLITHS AND NANNOLITHS

Fig. 4.3. Typical examples of extant nannoplankton and extinct Mesozoic nannofossils representing different ecological adaptations. Ecological interpretations for the nannofossils are of varying reliability. Scale bars = 1 μm.

thermoclines below clear oligotrophic surface-waters, thus *Florisphaera profunda* abundance appears to be an excellent indicator of surface-water oligotrophy.

4.4   MESOZOIC AND CENOZOIC NANNOFOSSIL PALAEOBIOGEOGRAPHY

Evidence from the fossil record suggests that coccolithophores have always displayed broadly similar ecological tolerances, and that palaeobiogeographical distributions are generally comparable with those of the Present Day, characterized by dominantly cosmopolitan distributions and low endemicity. However, there is much less systematic, quantitative data available for the pre-Quaternary record and our understanding of palaeobiogeography is still rather general.

One of the most significant advances in nannopalaeontology in recent years has been the recognition of considerable palaeobiogeographical differentiation throughout the Mesozoic (e.g. Thierstein, 1976; Wise, 1988). This initial recognition, and its subsequent refinement, is largely a result of DSDP and ODP research, particularly the drilling of sedimentary sequences in the Southern Hemisphere. Mesozoic nannoplankton provincialism is generally recognized by limited endemism at species and, rarely, generic level, together with variations in proportions of cosmopolitan assemblage components. The number of endemic taxa varies considerably through the Mesozoic but a consistent feature is the presence of species which display bipolar distributions, and palaeobiogeographical zones which are broadly palaeolatitudinal.

A small number of Triassic sections have provided limited understanding of the initial appearance and diversification of calcareous nannoplankton. At present, the only undisputed reports of Triassic nannofossils come from sites with low-palaeolatitude positions, from the northern margin of western Tethys (Alps), the southern margin of eastern Tethys (NW Australian shelf and Timor), and the western Pacific margin (Queen Charlotte Islands) (Bown, 1992). From this albeit limited dataset, it appears that earliest Mesozoic nannofossils had their evolutionary originations in low latitudes, a pattern which appears to have been repeated throughout the Jurassic.

Following significant extinctions at the Triassic/Jurassic boundary, the earliest Jurassic saw the rapid geographical and evolutionary expansion of nannofossils (Bown, 1987). Limited Boreal/Tethyan distinctions are recognized from the earliest Jurassic, but distributions are virtually cosmopolitan through the Mid Jurassic. Provincialism was gradually re-established in the Late Jurassic, and became extreme by the Tithonian following a series of evolutionary inceptions, e.g. *Nannoconus* (Fig. 4.3), with distributions initially restricted to low latitudes (specifically western Tethys and the early North Atlantic Ocean). *Nannoconus* remained restricted in distribution throughout its range (Tithonian–Campanian), often occurring in rock-forming abundances in a belt from western Tethys to the Caribbean, but occurring rarely or sporadically elsewhere.

In the earliest Cretaceous, provincialism was less marked but, for the first time, a number of taxa began to display clearly bipolar distributions, and high-latitude-affinity taxa were common for the first time. Bipolar distributions are best displayed by *Crucibiscutum salebrosum* (Berriasian–?Hauterivian), *Seribiscutum primitivum* (Albian–Campanian) and *Repagulum parvidentatum* (Mutterlose, 1992a). This degree of provincialism generally persisted into the Late Cretaceous.

However, biogeographical differentiation intensified in the Late Campanian–Maastrichtian, an event characterized by moderate numbers of northern high-latitude (boreal) taxa (e.g. Burnett, 1991) and high numbers of southern high-latitude (austral) taxa (e.g. Wise & Wind, 1977; Wise, 1988).

Cenozoic biogeographical patterns are similarly understood in terms of broad general patterns. Haq and Lohmann (1976; see also Haq et al., 1977; Haq, 1980; and summarized in Haq & Malmgren, 1982) established the presence of latitudinal biogeographical variations in the Palaeogene Atlantic Ocean, identifying characteristic 'assemblages' using factor analysis, and interpreting migrations of these 'assemblages' in terms of palaeotemperature changes. Relatively low-latitude assemblages are characterized by taxa such as *Fasciculithus, Discoaster* and *Sphenolithus*, mid-latitude assemblages by *Reticulofenestra* and *Cyclicargolithus floridanus*, and high-latitude assemblages by *Prinsius martinii, Prinsius bisulcus, Chiasmolithus* and *Coccolithus pelagicus* (post-Early Eocene). More recently, the simple link between temperature and distribution has been questioned for a number of these taxa, most notably discoasters which are often low in abundance at equatorial sites (Wei & Wise, 1990). The distribution of discoasters is now thought to reflect adaptation to oligotrophic environments (Aubry, 1992; Chepstow-Lusty, 1996), with both cooler temperatures and upwelling conditions suppressing relative abundances.

4.4.1 Environment

By analogy with modern nannoplankton, Mesozoic palaeobiogeographical distributions most likely reflect temperature and nutrient controls. However, most studies have primarily cited the former as dominant, as this most obviously explains palaeolatitude-parallel patterns, but also because the circulation of Mesozoic oceans is poorly understood, as are the nutrient sensitivities of most Mesozoic taxa. Accordingly, bipolar or high-latitude species are interpreted as cold-water taxa and low-latitude taxa as warm-water forms. Whether these broadly equate to eu/mesotrophic and oligotrophic taxa, as in the Present Day, is untested.

A link between palaeobiogeography and nutrients has, however, been proposed for a number of Cretaceous coccolith species. Roth and Bowdler (1981) recognized enhanced abundances of *Biscutum constans* and *Zeugrhabdotus erectus* in predicted upwelling zones (along the western side of continents at low latitudes). This interpretation has been strengthened by stratigraphical studies which observed abundance variations of the same two species through Milankovitch-frequency cycles, interpreted as reflecting surface-water fertility fluctuations, although the magnitude of these may be low (e.g. Watkins, 1989; Erba et al., 1992; Windley, 1995). In addition, sharp increases in these proposed fertility indices were observed in Pacific Ocean ODP Sites 800 and 801, in sediments deposited as the sites drifted across the palaeo-Equator, again supporting the link between abundance and upwelling, in this case across an ocean divergence zone (Erba, 1992).

Therefore, as expected, temperature and nutrients have been demonstrated to have been the major controls on Mesozoic nannoplankton distributions, but the relative roles of the two parameters are as yet far from clearly understood.

4.5   NANNOFOSSILS AND CLIMATE CHANGE IN THE CRETACEOUS INDIAN OCEAN

4.5.1 Background

The Indian and Atlantic Oceans came into being as the supercontinent, Pangaea, broke up and dispersed. The Panthalassa and Tethys oceans shrank to accommodate these new oceans, and global circulation was able to adopt a longitudinal régime which helped in the transfer of heat between the palaeo-poles and the palaeo-Equator. Climate changed naturally in response to these events. The Cretaceous was also characterised by raised sea levels and shallow epicontinental seas (Haq et al., 1987), a lack of permanent polar ice (particularly from the Albian onwards: Frakes et al., 1992), and four to six times the Present Day concentrations of atmospheric CO_2 (e.g. Barron, 1983). Limited palaeobiogeographical evidence (floral, faunal, coral and sedimentary), and $\delta^{18}O$-isotope palaeotemperatures show the Cretaceous world to have been warm and humid with palaeotemperatures of 27–32 °C (minimum–maximum) at the palaeo-Equator and 0–15 °C at the palaeo-poles, constituting a palaeotemperature gradient of between 17 °C and 27 °C (summarized in Barron, 1983; Hallam, 1985; Frakes et al., 1992; Barron et al., 1993). On a finer timescale though, evidence suggests that the Cretaceous experienced a wide range of climatic regimes (Frakes et al., 1992; Francis & Frakes, 1993), with periods of cooling (to a palaeotemperature low) in the Valanginian–Hauterivian (Frakes et al., 1992) and warming (to a palaeotemperature peak) in the Albian–Cenomanian (Barron, 1983; Frakes et al., 1992; Francis & Frakes, 1993), and with a peak in nannoplankton endemism in the Campanian–Maastrichtian (e.g. Bown et al., 1991).

4.5.2 Case study

In the last decade, numerous works have been undertaken, most concentrating on the Southern Ocean, which have identified palaeolatitudinally restricted Cretaceous taxa (e.g. Burnett, 1991 (northern high latitudes); Wind, 1979; Wise, 1988; Shafik, 1990; Watkins et al., 1996; Burnett, unpublished data). Palaeotemperature of the water masses containing such taxa has been invoked as the major controlling factor in the palaeobiogeographical distributions, since it is virtually impossible to assess the nutrient control and since, as described above, nutrients and temperature may be intrinsically linked. Burnett (unpublished data) has taken these interpretations one step further, and reconstituted nannofossil palaeobiogeographical zones for 5 my increments through the Late Cretaceous of the Indian Ocean, thus illustrating the evolution of 'palaeobiogeographical fronts'. Fig. 4.4 illustrates the positions of these 'fronts' for one of these increments (at 75 Ma, Late Campanian). These palaeobiogeographical zones are similar to those determined by McIntyre and Bé (1967) for the Present Day Atlantic Ocean and by Okada and Honjo (1973) for the Pacific Ocean in that the assemblages contain a large proportion of cosmopolitan taxa but also include spatially restricted taxa, or taxa with spatially restricted high abundances. From the Upper Cretaceous alone, out of a total of almost 400 species preserved in the Indian Ocean, over 80 display endemic characteristics. Herein, the palaeobiogeographical patterns of nine such Cretaceous taxa exhibiting high abundances and palaeo-

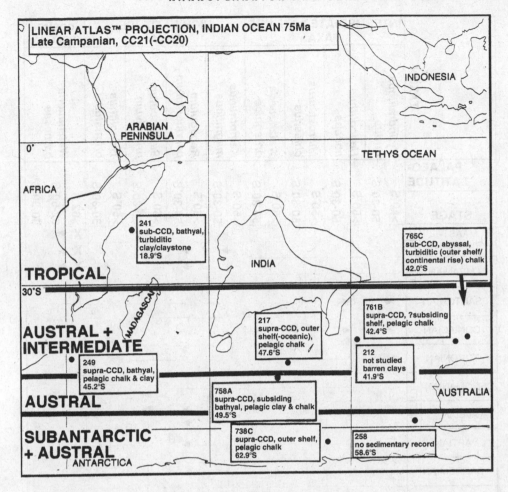

Fig. 4.4. Palaeobiogeographical zones in the Late Campanian Indian Ocean, Nannofossil Zone CC21 (after Burnett, unpublished data). Horizontal lines represent 'palaeobiogeographical fronts', or boundaries which separate sites with distinct palaeobiogeographical assemblages at 75 Ma. These are drawn at palaeolatitudes intermediate between such sites.

latitudinal dependences are examined through time (stage-scale). These patterns are compared with high- and low-latitude palaeotemperature curves and other climatic indications, and the implications of this comparison in terms of the evolution of the climate in the Cretaceous Indian Ocean is discussed.

Fig. 4.5 is a plot of the palaeobiogeographical distributions (forming a composite of the Cretaceous Indian Ocean from DSDP Sites 212, 217, 241, 249, 256, 257, 258, 259, 260, 261, 263 and ODP Sites 738, 758, 761, 765 and 766) of nine climatically influenced, Cretaceous nannofossil taxa against time (stages are those as determined by nannofossil correlations: see Burnett, 1996). These taxa have been identified previously as representative of either high- or intermediate-latitudes (Burnett, unpublished data), and were 'abundant' or 'common' in the nannofloras, i.e. these taxa formed the most significant proportion of the nanno-

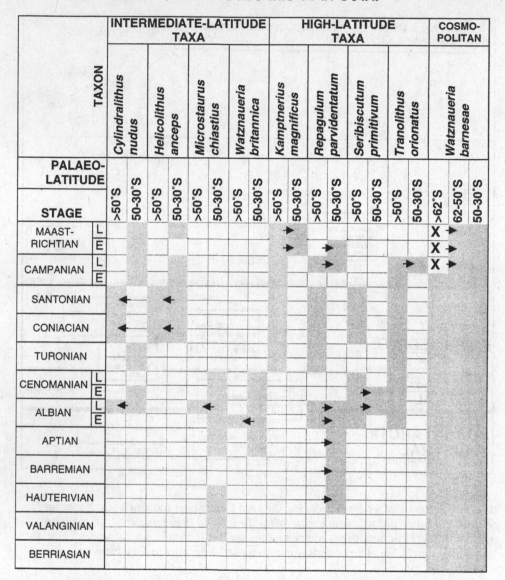

Fig. 4.5. Nannofossil climate-change indicators, Cretaceous Indian Ocean. Taxa were interpreted as palaeotemperature-sensitive, based on their palaeobiogeographical distributions and abundance patterns through the Cretaceous. Shading indicates presence of abundant/common nannofossils at palaeolatitudes determined as ultra-high (>62° S), high (>50° S, or 62–50° S in the Late Campanian–Maastrichtian) and intermediate (50–30° S). X indicates absence of *Watznaueria barnesae*. Arrows indicate periods where a taxon has moved beyond its usual palaeolatitudinal range, hence arrows pointing right indicate expansion of the Austral PBZ into lower palaeolatitudes, i.e. cooling, and arrows pointing left indicate expansion of the Intermediate PBZ into higher palaeolatitudes, i.e. warming.

floras at the stages illustrated. The palaeolatitudes shown have been determined by Burnett (unpublished data) as broadly defining southern high-latitude (Austral Palaeobiogeographical Zone (PBZ) > 50° S) and intermediate-latitude (Intermediate PBZ, 50–30° S) zones through most of the Cretaceous. Towards the end of the Cretaceous, a further (Subantarctic PBZ) zone was identified at > 62° S. Shifts in the abundant occurrence of a taxon from one palaeolatitude into another, or its encroachment into a wider spatial zone, are interpreted as either a cooling or a warming episode. For example, the intermediate-latitude *Cylindralithus nudus* made an excursion from intermediate latitudes into high latitudes in the Late Albian, and again in the Coniacian and Santonian, indicating that higher latitudes were warmer and more hospitable to the species during these periods. Conversely, an expansion of the Austral PBZ (cooler surface-waters, by analogy) into intermediate latitudes is indicated by the encroachment of *Repagulum parvidentatum* and *Tranolithus orionatus* in the Late Campanian. At the timescale used, the interpretations are broad but this was the easiest way to make a comparison with the δ^{18}O-isotope palaeotemperature curves. These episodes, constituting the evolution of climate/oceanography through the Cretaceous Indian Ocean, are summarized in Fig. 4.6, where a comparison is made with high- and low-latitude palaeotemperature curves (as presented by Kolodny & Raab, 1988, fig.3), and other evidence, as summarized in Frakes *et al.* (1992). The low-latitude palaeotemperature curve was determined from fish-teeth from Israel (thus providing a degree of relevance for the Indian Ocean) by Kolodny and Raab (1988). The high-latitude curve is a composite for northern high latitudes combining the works of Lowenstam and Epstein (1954), Lowenstam (1964) and Savin (1977). Together, these define a palaeotemperature gradient which fluctuated between a maximum of 16 °C (Cenomanian) and a minimum of 8 °C (Maastrichtian). These gradients provide a line of evidence for Barron's (1983) 'coolest' Cretaceous model.

Although few nannofossil data are available for the Berriasian–Valanginian interval, it appears, from the distribution of *Microstaurus chiastius,* that the period was one of relative palaeobiogeographical stability, with weak Austral and Intermediate PBZs indicating distinct cooler high-latitude waters and warmer intermediate-latitude waters. Low-latitude palaeotemperatures appear to have been gently rising, although there are no data points to support this. Through the Hauterivian–Early Cenomanian, the nannofossil palaeobiogeographical patterns indicate that intermediate palaeolatitudes were occupied by cooler-water nannofossils (e.g. *Repagulum parvidentatum, Seribiscutum primitivum*). This northward expansion of the Austral PBZ is coincident with a gradual, slight isotopic warming trend at low latitudes and a cooling trend at northern high latitudes (data was only available for the Albian to Early Cenomanian). This points to an expansion of the Austral water mass, but perhaps with an insignificant palaeotemperature change at lower latitudes. Intervals of southern high-latitude warming during the Albian were expressed by southward excursions of *Watznaueria britannica, Microstaurus chiastius* and *Cylindralithus nudus.* Evidence to support this extensive overall 'cool mode' includes the discovery of possible ice-rafted deposits in a number of Middle Jurassic to Lower Cretaceous sediments (Frakes *et al.*, 1992), although the absence of glacial tillites from the Mesozoic record (Frakes & Francis, 1990) has been taken

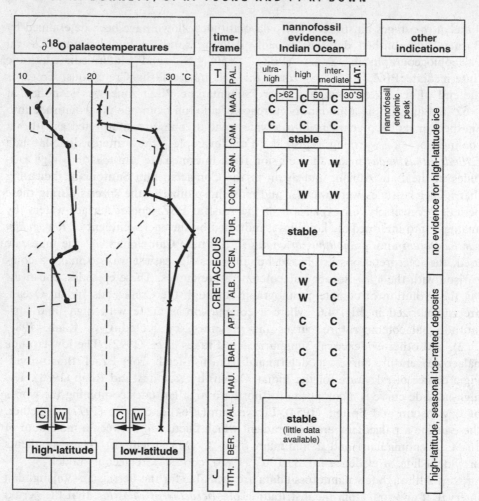

Fig. 4.6. Cretaceous marine palaeoclimate indications with particular reference to the Indian Ocean. Palaeotemperature curves redrawn from Kolodny and Raab (1988, Fig. 3). Dashed nannofossil climate–proxy curves generated from evidence presented in Fig. 4.5 and summarized here. Ultrahigh, high and intermediate palaeolatitudes correspond to those in Fig. 4.5. C denotes presumed cool surface-waters, W denotes presumed warm surface-waters, as determined from nannofossil indications in Fig. 4.5. Stable intervals represent periods where, for example, high-palaeolatitude nannofossils are restricted to high palaeolatitudes. Nannofossil endemic peak as described by Bown et al. (1991). Occurrence of ice-rafted deposits as summarized in Gregory et al. (1989) and Frakes et al. (1992).

as an indication that glaciation was a seasonal, rather than a permanent, phenomenon. In the Indian Ocean region in particular, Frakes and Francis (1988) identified Lower Cretaceous glacial deposits in association with glendonites (in central Australia, $>65°$ S palaeolatitude), whilst δ^{18}O-isotope palaeotemperature estimates from the Valanginian to Albian of SE Australia (Gregory et al., 1989) suggest polar freezing. As Frakes et al. (1992) postulated, cold polar temperatures may either have been very localized, thus not affecting lower latitudes, or the cold phases were cyclical on a scale which cannot be resolved from available palaeotemperature data (as suggested by Epshteyn, 1978; Kemper, 1987). Kemper (1987) inferred that cool/warm climates oscillated on a 10^3 to 10^6 year scale during the Cretaceous.

As both low- and high-latitude palaeotemperature curves rose to a peak in the Late Cenomanian–Early Turonian, the nannofossil palaeobiogeography seems to have restabilized, after an Early Cenomanian cooling at intermediate latitudes (manifested by a northward excursion of *Seribiscutum primitivum*). Both high- and low-palaeolatitude temperatures began to decrease in the Middle to Late Turonian, apparently gradually at high latitudes, which may also indicate that the global palaeotemperature gradient was returning to a 'normal' condition. The Coniacian–Santonian nannofossil indications are of warming at high-latitudes (expansion of the Intermediate PBZ, indicated by *Cylindralithus nudus* and *Helicolithus anceps*). The high-latitude palaeotemperature curve has no data points for this interval, and therefore the palaeotemperature evolution through this time is obscure, although the curve shows a very slight decreasing trend. Meanwhile, at low latitudes, palaeotemperatures were falling through the Early Coniacian, and then started to increase again through the Late Coniacian and Santonian. The Austral and Intermediate PBZs were again stable in the Early Campanian, during which time low- and high-latitude palaeotemperatures were starting to fall. This was followed, in the Late Campanian, by a sharp drop in global palaeotemperature. This is echoed quite elegantly by the increase in endemism exhibited by nannoplankton at all palaeolatitudes, by the excursion of Austral taxa into Intermediate palaeolatitudes (*Repagulum parvidentatum, Tranolithus orionatus*), and by the exclusion of the previously cosmopolitan taxon, *Watznaueria barnesae*, from ultrahigh palaeolatitudes. This trend continued into the Maastrichtian, with *Watznaueria barnesae* continuing to be excluded from the new Subantarctic PBZ, *Repagulum parvidentatum* and *Kamptnerius magnificus* thriving at Intermediate palaeolatitudes, and the global isotope-palaeotemperatures continuing to fall, albeit at a slow rate.

In summary, by defining nannofossil palaeobiogeographical zones it has been possible to plot these zones as they oscillate between palaeolatitudes through time. In this way, we have shown that nannofossil palaeobiogeographical patterns provide a record of marine climate change which can be correlated across the globe. In this particular example, we have described a Hauterivian to Cenomanian cool mode at intermediate to high latitudes. At this time, cooler-water/higher-latitude nannofossils were common to abundant at intermediate palaeolatitudes, (northern) high-latitude isotope-palaeotemperatures were apparently falling, although low-latitude palaeotemperatures were gradually increasing, and possible high-latitude glaciation was seasonal. This cool period was interrupted by a warmer Albian, wherein intermediate-latitude nannofossils sporadically migrated into high latitudes. 'Stable' modes, where the nannofossil palaeobiogeographical patterns indicate that high latitudes were distinctly cool, and low latitudes warm, punctuate the record between the cool/warm modes. These periods appear also to be times of isotope-palaeotemperature peaks, particularly at low latitudes. Thus, a Late Cenomanian–Early Turonian stable mode (paralleled by a low-latitude palaeotemperature peak in the Early Turonian) heralded the onset of the Late Turonian–Santonian warm mode. Another stable mode, in the Early Campanian (with palaeotemperatures decreasing from relative peaks at the end of the Santonian at both palaeolatitudes) preceded the

Late Campanian–Maastrichtian cool mode, which appears to have been a truly global event when combined with the isotope-palaeotemperature evidence.

The interpretations described above are general but can be further resolved, depending on the timeframe applied: Burnett (unpublished data) has used a 5 my increment to determine gross climate changes, but the nannofossil biostratigraphical timeframe can provide much greater resolution. Herein, we use stages which vary in length, but which are the primary basis for interdisciplinary correlations, and because this is the timescale against which the palaeotemperature curves were plotted, and which is used in summary texts such as Frakes *et al.* (1992). However, the situation highlights a major drawback in palaeoclimate research: in this instance, a reliable nannofossil timeframe was used to correlate the palaeobio-geographical interpretations with the stages, whilst Kolodny and Raab (1988) did not identify the source of their timeframe, and have presented their data in terms of stages and absolute ages, the latter constituting a questionable correlation exercise. This introduces a margin of error between correlating different datasets, which can represent millions of years. In addition, whilst the nature of Kolodny and Raab's study required them to use spot samples, the nannofossil study was done on sequences of samples which, whilst not entirely complete (the geological record being one of omission), had the advantage of providing a more even, higher-resolution data suite. The way forward is through collaboration between these disciplines. Primarily, pre-Quaternary isotope-derived palaeotemperature curves need to be generated in the context of higher-resolution timeframes, such as those which nannofossil biostratigraphy can provide. It is also apparent that a link might be established between particular nannofossil taxa and isotope palaeotemperature ranges, so that nannofossils themselves can be used as palaeotemperature proxies in the absence of isotope data or facilities, or where sample material has undergone diagenetic processes which might be considered as producing spurious isotopic values.

4.6   CONCLUSIONS

The fossil record holds the key to determining process analogues for global climate change. Nannoplankton are intimately associated with climate on a variety of levels, and appear to both influence and record climate change. It can be assumed that their fossil counterparts acted in a similar way. Nannofossils are abundantly distributed, both spatially and temporally, and have already lent themselves to a number of global change studies. Current research tends to concentrate on nannofossils as monitors of post-Tertiary change but it also provides the ground data to test predictions from models of nannoplankton as agents of global change. However, there are still numerous opportunities for multidisciplinary studies to realize the potential for nannofossils in this area. In particular, studies of the correlation between nannofossil palaeobiogeography and isotopic palaeotemperatures are needed, as are global scale studies of Mesozoic nannofossil biogeographies, which would provide an independent source of palaeoclimate data for comparison with current global palaeoclimate models.

5

Phenotypic response of foraminifera to episodes of global environmental change

NORMAN MACLEOD, NIEVEZ ORTIZ, NINA FEFFERMAN, WILLIAM CLYDE, CHRISTINE SCHULTER AND JENA MACLEAN

5.1 INTRODUCTION

No organismal group has contributed more to the origin and development of global change research than foraminifera. Foraminiferal biostratigraphical data provide many of the finest subdivisions on the clockface of geologic time. As such, foraminiferal faunas are routinely used to provide temporal control for global change research in the last 250 Ma of Earth history. Foraminiferal biogeography has also been important in the documentation of tectonic plate geometries and in the reconstruction of palaeoceanographic circulation patterns, both of which play decisive roles in controlling the distribution of climatic belts. In addition, stable isotopic analyses of foraminiferal tests (= shells) have been extensively used as palaeothermometers and palaeoproductivity indicators.

While these types of foraminiferal data have been employed (along with other independent lines of evidence) to identify, characterize, and date global change events, patterns of foraminiferal diversification have also clearly been affected by these events. Cretaceous through Recent genus-richness data for this group (Tappan & Loeblich, 1988) show maxima in the Albian–Cenomanian, Campanian–Maastrichtian, Mid Eocene, and Miocene, with minima occurring in the Early Paleocene, Oligocene, and Pliocene for both benthic and planktonic forms (Fig. 5.1). These patterns correspond to periods of generally increasing diversity that were successively interrupted by well-established global change events in the latest Maastrichtian–earliest Danian (= the so-called K–T boundary mass extinction), the Late Eocene–Oligocene (= the Paleogene–Neogene greenhouse-to-icehouse transition), and in the Late Miocene–Pliocene (= the establishment of a permanent Antarctic ice-sheet, strong global cooling, and the Mediterranean salinity crisis). Since most of these major global change events elicited biotic responses across the taxonomic spectrum, the fact that foraminiferal diversity patterns reflect these events is not unexpected. The Cretaceous–Recent foraminiferal fossil record for this interval also reveals the presence of smaller-scale events that appear to have affected only parts of the global foraminiferal fauna. One of the best examples of these is the decrease in benthic foraminiferal diversity that occurred across the Late Paleocene–Early Eocene interval (Miller et al., 1987b; Thomas, 1990a,b; Thomas & Shackleton, 1996).

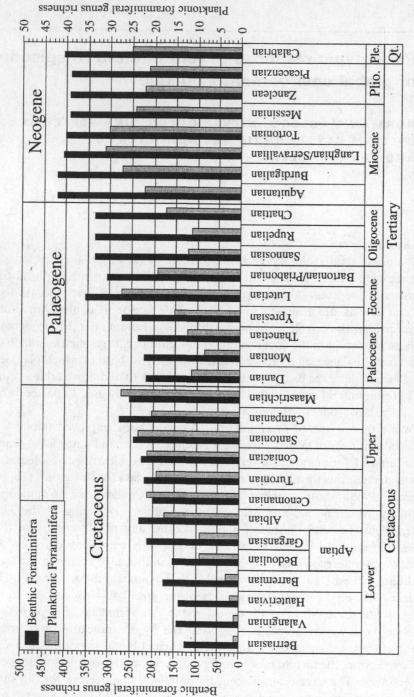

Fig. 5.1. Genus richness of planktonic (grey) and benthic (black) foraminifera from Cretaceous to Recent. Redrawn from Tappan and Loeblich (1988).

Given the prominent role foraminifera have played in Mesozoic and Cenozoic global change research, it is somewhat surprising to realize that we still know precious little about the specific effects of major environmental change episodes on foraminiferal populations. One currently popular approach to the study of biotic response to environmental change seeks to subdivide a fauna (or flora) into survivor species, disaster species, recovery species, etc. based on the geometry of these species' first or last appearance datums relative to an event horizon (see Kauffman & Harries, 1996; Harries *et al.*, 1996). This narrative approach seeks to reclassify organisms into a series of taxonomic–phylogenetic–palaeoecological categories and then attempts to describe stratigraphically the patterns of variation that characterize these new systematic classes. However, in practice, very few new insights have been gained in this way. Such an approach cannot provide an explanation for the effect of the global change events on biotic populations unless one is willing to assume any or all of the following: (i) that all species were primarily responding to 'event' (as opposed to 'background') factors throughout the turnover interval, (ii) that species exhibiting similar speciation/extinction times were necessarily responding to similar physical-ecological perturbations, (iii) that all species respond to various physical/ecologic perturbations in similar ways, and (iv) that the quality of individual species' biostratigraphical records is comparable across the taxonomic spectrum and throughout the entire stratigraphical interval covered by the event. The attraction of the narrative approach is that it can be used to 'explain' an entire group's response to any global change event without reference to any underlying biological–ecological theory and without needing to consult any formal hypothesis testing procedure. Nevertheless, such explanations are usually little more than redescriptions of raw data and, as such, relatively content free with respect to distinguishing between alternative causal models.

This situation becomes even worse when one attempts synoptic analyses that seek to address questions of similarity and differences among the biotic responses of ecologically diverse organismal groups (like foraminiferal) to different global change events. Since the narrative approach provides no way to identify modes of selection that might be responsible for originations, extinctions, and changes in relative abundance among species that lived at different times and endured different types of global change events, there is no way to decide whether similar (or different) patterns imply different (or similar) causes. What is needed is a more hypothetico-deductive, rather than purely narrative, approach to the analysis of biotic response to global change.

As an alternative to the narrative approach, one can develop specific hypotheses that make specific predictions about patterns that might (or might not) be preserved in the fossil record, and then evaluate the fossil record for its degree of conformance to those predictions. For instance, owing to the widely commented upon occurrence of iterative evolution in foraminiferal test morphotypes (see Cifelli, 1969; Lipps, 1979; Norris, 1990, 1991), it has been widely assumed that foraminiferal test shape is a prominent target of natural selection. Several alternative hypotheses have been advanced to explain what functions foraminiferal tests might serve (see Hallock *et al.*, 1991 and references therein). However, in all but a very few specialized cases (e.g. larger benthic foraminifera, see Hallock *et al.*, 1991) or for a few specialized structures (e.g. elongate tests, elevated apertural openings, see DeLaca *et al.*, 1980) these studies either fail to address the general model of shell morphotype as a specific target of differential selection pressure, or fail to explicitly test the functional explanation(s) they propose.

Do other hypotheses for the response of foraminiferal phenotypes to environmental selection pressures exist? They do. The importance of developmental and life-history characters as targets for selection has received much attention in the recent evolutionary literature (see Raff & Kauffman, 1983; Raff, 1992, 1996; Stearns, 1992). Palaeontologists can use morphology and geochemistry (especially stable isotopic compositions) to test developmental and life-history hypotheses for organisms that preserve juvenile growth stages in their shells (such as foraminifera). These studies are usually discussed under the rubric of heterochrony (see DeBeer, 1958; Gould, 1977; Alberch *et al.*, 1979; Raff & Kauffman, 1983, McKinney, 1988; McKinney & McNamara, 1991; Raff, 1996). Palaeontologists have been unusually active contributors to this literature.

Whereas Gould (1977) and Alberch *et al.* (1979) believed it possible to use heterochronic patterns to infer the operation of particular developmental processes, it has recently become appreciated that heterochrony actually represents a series of patterns existing between ancestral and descendant species and/or populations that can be the result of a variety of alternative developmental processes (see McKinney & McNamara, 1991; Raff, 1996). Although this fact imposes a severe limitation on the ability of palaeontologists to identify specific developmental process from ontogenetic patterns, it in no way compromises the utility of a developmental or life-history approach as an alternative interpretation for the origin and maintenance of morphological novelty.

In an effort to take the first steps toward formulating a more explicitly hypothetico-deductive approach to the investigation of the effects of global change events on biotic populations, this study compares patterns of phenotypic response in selected benthic and planktonic foraminiferal populations as they made their way through the Cretaceous–Tertiary, Paleocene–Eocene and Late Eocene intervals. The specific question under consideration is whether individuals within foraminiferal populations were selected on the basis of gross morphotype or on the basis of developmental or life-history factors during times of severe or sustained environmental perturbation. Morphotype-directed selection will be recognized as a controlling factor if it can be shown that foraminiferal populations exhibit a common and consistent pattern of morphotypic response to any pair of independently identified environmental events. Similarly, developmental or life-history-directed selection will be recognized as a controlling factor if it can be shown that these populations exhibit a common and consistent pattern of developmental or life-history response to any pair of global environmental change events. The null hypothesis is that foraminiferal populations exhibit random morphotypic and developmental or life-history responses to all environmental perturbations. If developmental or life-history patterns are found to be a consistent target of environmental selection for all (or even a majority of) foraminiferal taxa studied within the three target events, this study will serve to clarify the relation of morphology to selection in foraminiferal evolution, help explain large-scale interclade patterns of foraminiferal extinction and survivorship, and facilitate the development of new and more detailed hypotheses designed to provide insight into the effects of global change on other organismal groups.

5.2   MATERIALS AND METHODS

This study summarizes, extends, and synthesizes results obtained in several previous investigations of phenotypic response in foraminiferal populations across major environmental event horizons, including published papers and abstracts (MacLeod et al., 1990; MacLeod, 1990; MacLeod & Kitchell, 1990; MacLeod & Ortiz, 1995) and student theses (Clyde, 1990; Schulter, 1992; MacLean, 1992). The K–T, Late Eocene, and P–E events were chosen for this analysis because they represent a sequential series of well-documented global change events in the Paleogene, but vary in their causes, timing, and taxonomic scope (see below). Samples used in this study come from both deep-sea cores and land sections. Thus, a variety of marine depositional settings and environments are represented. A wide variety of benthic and planktonic foraminiferal taxa were also used in order to determine whether observed patterns of variation represented a consistent group-wide response to environmental variation, or simply a species-specific phenomenon.

5.2.1   Morphotype analyses

The study of foraminiferal morphotype diversification patterns was under-taken in order to assess hypotheses relating to the direct selection of particular test shapes within the three target global change intervals. It is in this context that the results of studies such as those of Corliss (1985) and Corliss and Chen (1988) – who have argued that benthic foraminifera exhibit strong and consistent patterns of morphotype differentiation among epifaunal and infaunal species in modern deep-sea faunas – must be viewed. In order to quantitatively assess general patterns of morphotype diversification within the benthic and planktonic species a morphotype classification system based on the categories recognized by Corliss and Chen (1988) was devised (Fig. 5.2). This classification was then used to classify the globally distributed planktonic species included in the following monographic treatments: Caron (1985), Toumarkine and Luterbacher (1985), and Bolli and Saunders (1985) with stratigraphical ranges for K–T boundary taxa following the biostratigraphical summary of MacLeod (1996). Like the similar summary used by Norris (1990), these systematic treatments do not include all taxa present within the interval covered. However, they should provide an adequate sample of major morphotypic trends present within the Cretaceous–Paleogene fauna. All common benthic foraminiferal taxa occurring within the Cenomanian to Early Eocene interval of the Trinidad standard section (Bolli et al., 1994) were used as the basis for the benthic forami-niferal morphotype analysis.

5.2.2   Heterochronic analyses

In order to test the alternative hypothesis of selection on developmental or life-history characters, a series of planktonic and benthic foraminiferal species occur-

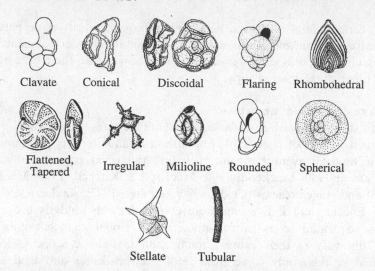

Fig. 5.2. Generalized foraminiferal morphotypic classes used in this study. This classification is based on that used by Corliss (1985) and similar to that used by Norris (1990, 1991). No distinction is made between planispiral and trochospiral modes because this distinction has dubious functional significance relative to overall test shape. Similarly, no distinction is made between biserial, triserial, and multiserial modes for flaring morphotypes. These shape categories can be used to characterize both planktonic and benthic foraminiferal test shapes.

ring throughout the K–T, P–E, or Late Eocene intervals were selected for detailed morphometric analyses. For this phase of the analysis the preprocessed sediment was checked for evidence of reworking (e.g. graded bedding, cross-laminations). In addition, all faunal samples were obtained using an Otto microsplitter and all specimens belonging to the target species encountered during examination of the sample splits, regardless of size, were picked, identified, and measured. All measured specimens were mounted on paper slides for reference and currently reside within the foraminiferal collections of Princeton University.

Patterns of intraspecific size and shape variation across the three target event horizons were documented by measuring the location of landmarks (Late Eocene and K–T studies) and boundary outlines (P–E study) for individual tests in plan, apertural, and/or umbilical views. Size variations were quantified used the Bookstein (1986) centroid size coefficient (landmark studies) or the outline area (outline study) of the tests.

Once these morphometric data had been assembled and the size/shape comparisons carried out the heterochronic modes of Alberch et al. (1979, see also McKinney, 1988) were identified by comparing intraspecific size/shape distributions among and between pre-event and post-event populations. For trochospiral test geometries (where large parts of previous chamber and test outlines are hidden) these comparisons were necessarily restricted to adult chamber morphologies. However, in the case of the K–T study (in which biserial species were included), it was possible to obtain morphometric data from the entire ontogeny for each biserial specimen.

5.3   TARGET GLOBAL CHANGE EVENTS

5.3.1   The Cretaceous-Tertiary (K-T) event

The latest Cretaceous–earliest Tertiary interval has often been characterized as being a time of warm, equable climates, much like those of the Early and mid Cretaceous (e.g. Scholle and Arthur, 1980; Barron and Washington, 1984). Nevertheless, recent high-resolution biostratigraphical and palaeoceanographic investigations have shown that 4 to 6 my prior to the K–T boundary Cretaceous climates entered a period of profound cooling (Barrera & Huber, 1990; Barrera, 1994; Barrera & Keller, 1994). Stable isotopic analyses of benthic foraminiferal tests from high-latitude deep-sea cores record the lowest marine water temperatures for the entire Late Cretaceous, coupled with pronounced fluctuations in the $\delta^{13}C$ curve (Barrera, 1994; Barrera & Keller, 1994). Placed in its appropriate context, this Maastrichtian cooling event should be seen as the culmination of a trend toward cooler marine water temperatures (and thus progressively cooler climates) that began in the Early Turonian. The proliferation of the highly species-rich Late Cretaceous planktonic foraminiferal fauna took place entirely within this progressive climatic cooling event.

Climatic fluctuations intensified in the latest Maastrichtian run up to the K–T boundary itself. Oxygen stable isotopic data from the Weddell Sea (Stott & Kennett, 1990) record a sharp southern high-latitude warming event near the lower part of Chron 29R, about 400 000 years before the K–T boundary (G. Keller, personal communication, 1996). The Late Maastrichtian interval was also a time of marked eustatic sea-level fluctuation (Haq et al., 1987), increasing $^{87}Sr/^{86}Sr$ ratios in marine sediments (indicating increased riverine input and high rates of continental erosion; Barrera, 1994), widespread and intense volcanic activity (Courtillot, 1990, Courtillot et al., 1996; Kauffman, 1984; Johnson & Kauffman, 1996), a progressive change in $\delta^{13}C$ values suggesting an increase in marine productivity (Barrera & Keller, 1990, 1994; Schmitz et al., 1992), several large bolide impacts (Grieve et al., 1996) and progressive extinctions in many marine and terrestrial groups (MacLeod et al., 1997).

5.3.2   The Paleocene-Eocene (P-E) thermal maximum event

After the Late Maastrichtian climatic temperature low, marine water temperatures grew warmer throughout the interval from the Mid Paleocene to the Early Eocene (Miller et al., 1987b; Kennett & Stott, 1990; Zachos et al., 1992, 1993). Superimposed on this progressive Paleogene warming trend, however, the Earth's climate underwent a more severe, though short term, intense warm event, called the Late Paleocene Thermal Maximum (LPTM, Zachos et al., 1993). Oxygen isotopic data suggest that deep marine waters warmed by as much as 4–6 °C during the LPTM, though tropical and subtropical surface water temperatures appear to have remained constant even as bottom water temperatures warmed by as much as 5 °C (Stott, 1992; Bralower et al., 1995; Lu & Keller, 1995a). Lu et al. (1995) have argued that a reduction in upwelling intensity during the P–E transition is indicative of weakened atmospheric circulation and a reduced meridional thermal gradient. Carbon isotopic data also record a dramatic negative shift over the Late Paleocene–Early Eocene interval in both surface and deep waters (see Thomas & Shackleton,

1996 and references therein), as well as in terrestrial settings (Koch *et al.*, 1992, 1995). A fully sufficient explanation for the LPTM and the coeval negative excursion in carbon isotopic values has yet to be proposed, but the mechanism undoubtedly involves temperature changes in the deep oceans.

From a biotic point of view, the P–E transition was a time of widespread diversification in terrestrial and marine biotas. Many established clades also greatly expanded their geographical ranges throughout this interval. During the LPTM event, however, benthic foraminifera experienced the most profound extinction event in their evolutionary history. In many sections and cores as much as 50% of Mid Paleocene species richness values were lost in the Late Paleocene–Early Eocene transition (Thomas, 1990*a,b*; Thomas & Shackleton, 1996; Ortiz, 1995). Planktonic foraminifera also underwent a series of species-level turnover events at this time, but did not experience a net decrease in global species richness among the common and/ or abundant taxa (Lu & Keller, 1993, 1995*b*, 1996)

5.3.3   The Eocene–Oligocene (E–O) event

In many ways climatic trends from the Mid Eocene through the Oligocene were the antithesis of those prevailing during the P–E event. Rather than global warming, global cooling was the order of the day. During this interval, marine deep waters experienced a progressive cooling of as much as 10 °C (Miller, 1992). This resulted in severe climatic, atmospheric, and palaeoceanographic disruptions, including: a steepened latitudinal temperature gradient (Miller, 1992), intensified low latitude precipitation patterns (Sloan & Barron, 1992), a change from thermospheric to thermohaline (psychrospheric) circulation (Benson, 1975), intensified upwelling (Parrish & Curtis, 1982), intensified atmospheric circulation (Bartek *et al.*, 1992), and increased marine productivity (McGowran, 1989). The first ice-sheets in Antarctica are now believed to have formed in the Mid Eocene, as early as 42 Ma by some reports of ice rafting, but certainly in the Late Eocene by 38.2 Ma (Barrera & Huber, 1991).

Biotically, the E–O event affected virtually all marine and terrestrial groups, including plants (see Prothero & Berggren, 1992 and references therein). Groups that show particularly high turnover patterns include: calcareous nannoplankton, foraminifera (planktonic and benthic), diatoms, dinoflagellates, radiolaria, ostracodes, molluscs, echinoderms, fish, reptiles, mammals, and birds. The overall rate of environmental change is widely thought to have been lower during the E–O event than during the K–T event, though the overall magnitude of the latter extinction event was much more profound.

5.4   PATTERNS OF MORPHOTYPIC VARIATION
5.4.1   Planktonic foraminifera

Species and morphotype richness diagrams for the composite planktonic foraminiferal dataset are shown in Fig. 5.3(*a*). The most striking aspect of morphotypic variation in these data is the complex mixture of patterns both between and within the Late Cretaceous and Paleogene radiations. Whereas a few dominant morphotypes are present more-or-less continuously throughout the entire Cenomanian–Oligocene interval (e.g. rounded; flattened, tapered), others are confined to particular radiation events (e.g. spherical). Still others are dominant in one radiation, but play a minor role in the other (e.g. discoidal, conical, spinose) and a few are present at low levels within (but not between) both radiations (e.g. spinose, clavate).

While it might be tempting to 'explain' the morphotypic aspects of these radiations in a narrative sense, this would amount to little more than a redescription of the patterns just outlined. In a more hypothetical–deductive vein, these data might provide direct evidence of functional adaptation on generalized test shape if a particular morphotype discontinuously occurred within each radiation event and consistently persisted into the subsequent global change (= extinction) event for a longer period of time than other morphotypes. While an interpretation of morphotype-centred adaptive advantage would be consistent with such a biostratigraphical pattern, it would by no means be the only possible interpretation. However, if such a pattern recurred over several global change events and a hierarchy among test shapes could be established, this would provide strong circumstantial evidence for foraminiferal morphotypes as being the targets of environmental selection.

None of the morphotype distribution patterns within these data exhibit the predicted geometries. Granted, the sample is not large (two radiations) and there are problems with taxonomic completeness and temporal resolution. In order to confirm the interpretation of no differential morphotype advantage, subsequent studies should be conducted on single sections or cores. Nevertheless, the taxonomic–biostratigraphical bias in these data favours the discontinuous morphotype distribution patterns that would support an interpretation of adaptive advantage.

It is also instructive to examine the distribution of morphotypes across the K–T, P–E, and E–O events themselves. Within these intervals, differences in the morphotypic composition of the pre-event and post-event faunas are striking. Both the K–T and E–O events are characterized by progressive patterns of species richness decline that extend over the course of several planktonic foraminiferal biozones – though chronostratigraphically the K–T event is more rapid. Morphotypically, the K–T event is also the more complex with several characteristic Late Cretaceous morphotypes disappearing from the stratigraphical record over a broad interval centered on the K–T boundary, but with most morphotype last appearances occurring within the Late Cretaceous.

By contrast, the Paleocene–Eocene planktonic foraminiferal radiation appears to have a much simpler morphotypic structure. The Mid Eocene species richness maximum contains a broad and fairly uniform distribution of test types that persist with almost constant relative abundance between zones P9 and P14. Although species richness declines in a fairly uniform manner throughout the Late Eocene and Oligocene biozones (P15–P22, see also MacLeod, 1990), morphotype richness declines in a stepped manner with planoconvex and discoidal morphotypes disappearing in Zone P14 (with the extinction of several *Truncorotaloides* species) and spinose planispiral, rounded planispiral, flattened, tapered trochospiral, and conical trochospiral morphotypes suffering severe reductions at the E–O boundary (base of Zone P18). It seems clear that the perception of a planktonic foraminiferal

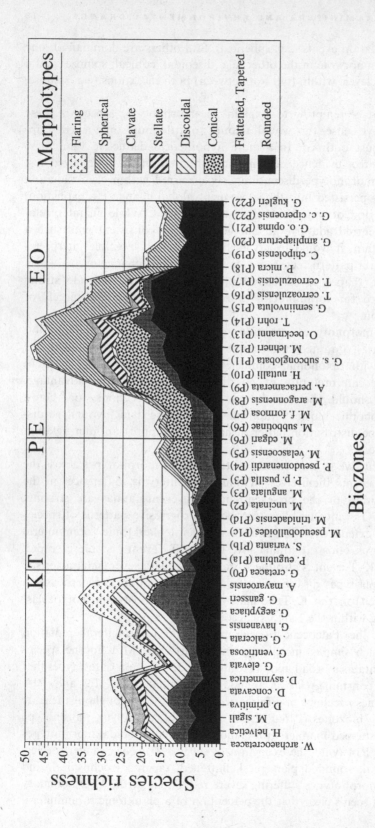

Fig. 5.3. Patterns of low- and middle-latitude planktonic (a) and benthic (b) foraminiferal morphotypic richness across the Cretaceous–Tertiary (K–T), Paleocene–Eocene (P–E), and Late Eocene–Oligocene (E–O, planktonic only) global change events. Note the complexity of morphotypic origination, diversification, and extinction patterns within the Cretaceous and Paleogene planktonic radiations. This contrasts strongly with benthic morphotype richness patterns over the same interval. Lack of a consistent morphotypic response by planktonic and benthic foraminiferal faunas to global environmental change events suggests that gross test shape plays very little role in mediating environmental selection pressures. See text for additional discussion.

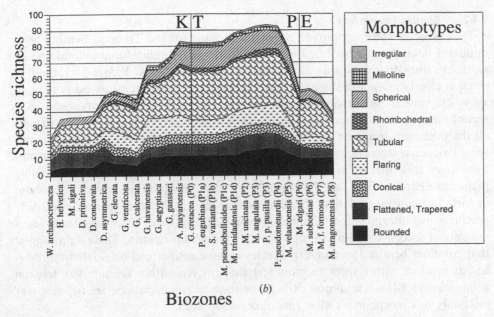

Fig. 5.3. (*cont.*)

mass extinction event at the E–O boundary (also known as the Terminal Eocene Event, see Wolfe, 1978) has as much to do with the reduction in morphotype richness as species richness.

Although planktonic foraminifera are not regarded as having been affected by the P–E event (Thomas, 1990*a,b*) these planktonic foraminiferal data suggest that, despite a modestly rising species richness through the Paleocene and into the Early Eocene, rounded (planispiral and trochospiral) morphotypes exhibited atypically low richness values through much of this interval. Rounded trochospiral planktonic foraminiferal taxa have long been regarded as being differentially resistant to extinction (see Cifelli, 1969; Norris, 1990, 1991, 1992). However, these data suggest that not only were members of the rounded morphotype group atypically rare during the Mid Paleocene–Early Eocene interval, but their relative abundance fell over an interval during which other morphotypes (e.g. conical, flattened–tapered, spherical) were diversifying. The extended interval of relatively low planktonic foraminiferal species richness following the K–T event appears to have been primarily due to a Mid Paleocene depression in the species richness of rounded planispiral and trochospiral taxa. Since this interval of depressed rounded morphotype species richness is centred on the P–E boundary, it is tempting to speculate that the LPTM event may have been responsible for this lag in the pattern of Paleogene planktonic foraminiferal diversification, though a cause–effect relationship remains to be demonstrated. The clearly depauperate Paleocene planktonic foraminiferal species richness shown in this summary stands in stark contrast to the statements of Smit (1996) who characterizes the post-K–T recovery of diversity within this group as being extraordinarily rapid.

5.4.2   Benthic foraminifera

Species and morphotype richness diagrams for the Trinidad benthic foraminiferal dataset are shown in Fig. 5.3(*b*). Although benthic foraminiferal species
richness values do not show a pronounced decline across the K–T boundary, the P–E
event is clearly evident in these data. While there is a small literature purporting to
show differential success of particular benthic foraminiferal morphotypes in local
sections across the K–T boundary, these are local ecological events in which virtually
all the taxa that disappear at or near the boundary subsequently reappear in overlying Paleocene strata (see Thomas, 1990*a*). Benthic foraminiferal morphotypic patterns over this interval are also more simply structured than the corresponding
planktonic patterns. The fact that virtually all major morphotype groups are present
in virtually constant relative abundances throughout this time period effectively
precludes any interpretation of morphotype-based differences being significant determinants of success or failure to cope with environmental stresses. These data suggest
that common benthic foraminiferal species became extinct and originated more-or-
less at random with respect to gross test shape. Even within the interval between
zones P6 and P4, where almost 50% of benthic foraminiferal species richness was
lost, only one morphotypic class (milioline) disappears.

5.5   PATTERNS OF DEVELOPMENTAL AND LIFE-HISTORY VARIATION

By morphologically or morphometrically comparing relations between an
organism's size and shape at various stages in its ontogeny, it is possible to characterize shifts in the predominant patterns of ontogenetic variation through time that
may be indicative of selection for alternative developmental or life-history pathways
and/or their ecological correlates (see McNamara, 1988; MacLeod *et al.*, 1990).
Although heterochronic comparisons are typically made between ancestral and des-
cendent species, the comparison of intraspecific variations in single-species populations can be conducted in a similar manner (McKinney & McNamara, 1991). The
intraspecific analysis of heterochrony has a series of distinct advantages, including
the avoidance of uncertainties in the determination of true ancestor–descendant
species couplets (see Fink, 1988).

5.5.1   Heterochronic response across the K–T boundary

Despite the fact that the K–T event is widely believed to have differentially
affected planktonic and benthonic foraminifera (see Thomas, 1990*a*; MacLeod &
Keller, 1994), the analysis of selection on developmental and life-history patterns
(via heterochrony) for these two groups reveals several striking similarities. For
example, studies of intraspecific variation in three planktic and one benthic species
from the Brazos Core (Fig. 5.4) shows that all four species experienced statistically
significant test size decreases in the latest Maastrichtian, coupled with static variation about a relatively smaller mean test size throughout the earliest Danian interval.
It is also interesting to note that, in all four instances, test size peaks within the
highest 50 cm of the Maastrichtian core and then falls throughout the last few tens of
centimetres of Maastrichtian strata. This repeated pattern suggests that the environmental perturbations responsible for the shift toward smaller-sized planktonic and

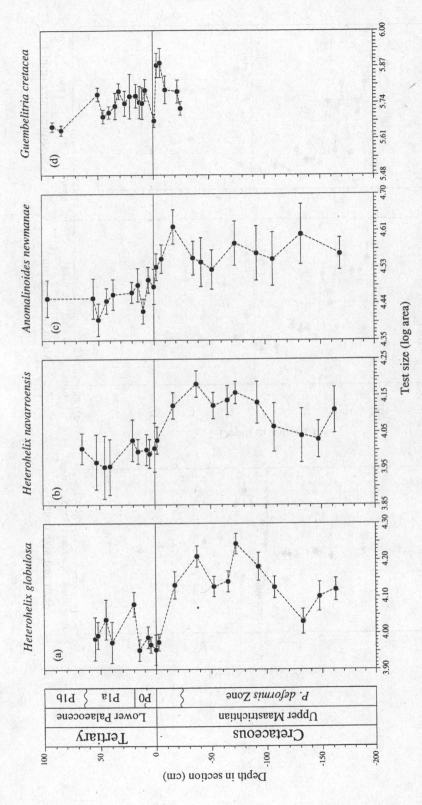

Fig. 5.4. Patterns of foraminiferal test size variation across the Cretaceous–Tertiary boundary in the Brazos Core, Brazos River, Texas (see Keller, 1989 for locality details). Data points represent sample means, error bars represent 95% confidence intervals on the mean. Solid horizontal line represents the K–T boundary horizon as defined by an Ir anomaly and biostratigraphy. Note common morphological response of all populations to environmental changes beginning in the late Maastrichtian and continuing into the lowermost Tertiary. These four species represent three planktonic ((a), (b), (d)) and one benthic (c) species.

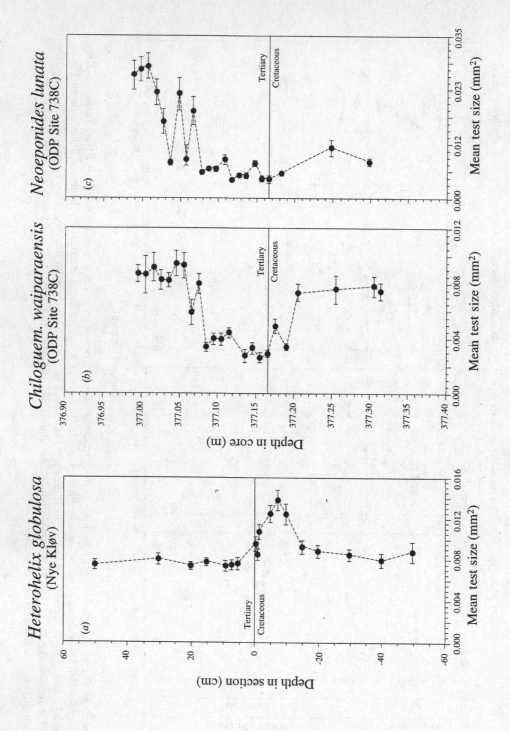

benthic foraminiferal populations began within the latest Maastrichtian rather than at the K–T boundary itself.

Both *Guembelitria cretacea* (planktonic) and *Anomalinoides newmanae* (benthic) are universally acknowledged to have survived the K–T mass extinction event owing to their presence in highest lower Danian sediments and beyond. *Heterohelix globulosa* and *H. navarroensis* had formerly been regarded as exclusively Cretaceous species, though both are now known to occur consistently in lowest Danian sediments, often ranging several metres above the local K–T boundary horizon (see Keller, 1988a, 1989, 1993; Liu & Olsson, 1992; Canudo, 1997; Masters, 1997; Olsson, 1997; Orue-extebarria, 1997). In addition, stable isotopic data for *H. globulosa* from the Brazos Core unambiguously establishes the K–T survivorship of this species (see Barrera & Keller, 1990; MacLeod & Keller, 1994). Survivorship status for *H. navarroensis* is strongly implicated by similarities in the patterns of occurrence, relative abundance variation, and morphometric variation evident between it, *H. globulosa*, and other non-controversial early Danian taxa.

Are these patterns of phenotypic variation unique to the Brazos foraminiferal populations? Analyses of the coeval intervals from high northern and southern latitudes shows the same patterns for these and other species. For example, *H. globulosa* populations from Nye Kløv (Denmark) and *Chiloguembelina waiparaensis* populations from ODP Site 738C (Kerguelen Plateau, southern Indian Ocean) exhibit patterns of test size change that are similar to those found at Brazos River (Fig. 5.5). In addition, the benthic species *Neoeponides lunata* from ODP Site 738C also shows patterns of test change consistent with those for its geographical cogener *C. waiparaensis*, and reminiscent of the patterns recorded by coeval populations of *A. newmanae* in the Brazos Core. It seems highly unlikely that such detailed similarities in the pattern of phenotypic response in both planktonic and benthic foraminiferal populations could be the result of random events or global marine reworking. The most parsimonious explanation for these data is that Late Maastrichtian environmental events were eliciting a similar response from these (and perhaps other) then extant foraminiferal populations.

Do patterns of shape variation between pre-response and post-response populations shed any light on the developmental nature of this transformation? Are the smaller-sized post-response populations simply scaled-down versions of their pre-response ancestors or has test shape, as well as test size, been affected? The three candidate heterochronic modes that involve a reduction in size between

Fig. 5.5. Patterns of foraminiferal test size variation across the Cretaceous–Tertiary boundary in the Nye Kløv (Denmark) section (*a*) and the ODP Site 738C (Kerguelen Plateau) core ((*b*) and (*c*)). Data points represent sample means, error bars represent 95% confidence intervals on the mean. Solid horizontal line represents K–T boundary horizon. Note common morphological response of these populations to environmental changes beginning in the late Maastrichtian and the similarity of these patterns to those documented for foraminiferal populations in the Brazos Core (Fig. 5.4). This similarity suggests that the environmental changes responsible for the size decrease affected planktonic and benthic foraminiferal populations on a global scale. These three species represent two planktonic ((*a*) and (*b*}) and one benthic (*c*) species.

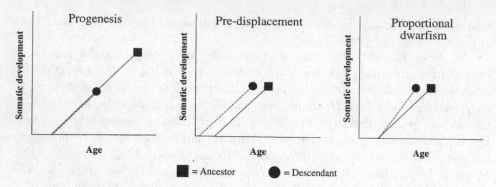

Fig. 5.6. Patterns of developmental heterochrony that require a decrease in the development position of the offset signal. This position is represented by age in studies of modern organisms and usually inferred based on body size in palaeontological investigations (but see Jones, 1988). (Note: In some planktonic species this inference can be independently checked via observations of test wall histology; see MacLeod *et al.*, 1990.) Since a variety of developmental mechanisms can result in these patterns, the existence of the patterns alone is insufficient evidence on which to base an interpretation of process (see Raff, 1996). However, the documentation of these patterns in fossil material is sufficient evidence on which to conclude that natural selection was operating through the phenotypic medium provided by the developmental programme.

ancestor and descendant populations are shown in Fig. 5.6. Since foraminifera create a morphological record of their ontogenetic development as they add chambers to the test, a series of simple measurements can quantify general aspects of morphological development for individuals, populations, and/or species. These measurements are most easily taken on those species whose chamber morphologies are externally exposed (e.g. heterohelicids), however, a variety of (albeit time-consuming) techniques are also available for the measurement of ontogenetic trajectories in involute planispiral and/or trochospiral species (see Huang, 1981; Huber, 1994). Fig. 5.7 shows two alternative test measurement strategies for heterohelicid species. Use of alternative strategies is necessary since it is not known whether the addition of individual chambers or the attainment of overall test size thresholds are better estimators of planktonic foraminiferal growth stages, though there is at least some empirical support for preferring the latter (see Hemleben *et al.*, 1988). Using these measurements it is possible to construct population-specific growth trajectories for pre-response and post-response faunas (Figs 5.8(*a*), 5.8(*b*)). Fig. 5.8(*c*)–(*f*) shows typical results for comparisons between pre- and post-response samples. In all cases the dominant heterochronic signal appears to involve a progenetic juvenilization of test shape.

These data show that, in at least three widely separated localities, at least four planktonic and two benthic species responded to K–T environmental perturbations by favouring individuals who achieved sexual maturity at an early stage in their developmental programme relative to patterns typical of ancestral populations. Moreover, since several of these populations within the study interval (and all populations ultimately) returned to their former, pre-response size and shape values by

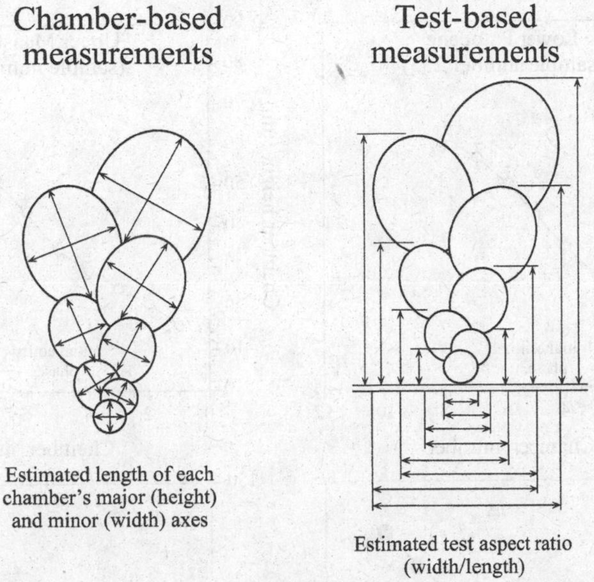

Chamber-based measurements

Test-based measurements

Estimated length of each
chamber's major (height)
and minor (width) axes

Estimated test aspect ratio
(width/length)

Fig. 5.7. Alternative test measurement strategies for identifying heterochronic patterns in flaring, multichambered foraminiferal tests. These alternative strategies were employed because it is presently uncertain whether the addition of individual chambers or the attainment of overall test size thresholds represents the better index of ontogenetic change in foraminiferal Baüplane (see Hemleben *et al.*, 1988).

the mid to late Danian, this selection event was evidently a short-term strategy that enabled these populations to cope successfully with a global and ecologically prolonged environmental shift. Owing to the fact that the Late Maastrichtian–Early Paleocene interval is also the locus of a major extinction event for planktonic foraminifera (see Figs. 5.1, 5.3(*a*)), it might be concluded that populations unable to make this phenotypic shift for either developmental and/or ecological reasons may have been at a selective disadvantage relative to these and other coeval planktonic populations.

5.5.2   Heterochronic response across the P–E boundary

Present interpretations suggest that the P–E event was very different from the K–T event in terms of the type of environmental perturbation (cold [K–T] vs. warm [P–E]) and affected faunas (planktonic [K–T] vs. benthic [P–E]). Unpublished data suggests that the K–T event may also have been triggered by a warm event of comparable magnitude to the P–E event (Keller, personal communication, 1996). Nevertheless, these events are similar in that both events involved a geologically rapid (yet ecologically prolonged) temporary intensification of environmental trends. There was no common pattern of response evident in the morphotype data for this event (see above). However, just as in the case of the K–T event at least some benthic foraminiferal populations appear to have responded to the P–E event by a paedomorphic juvenilization of the test shape.

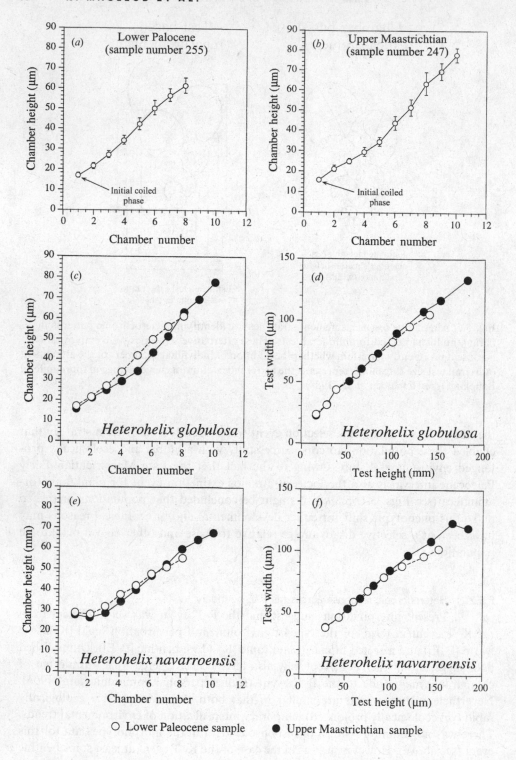

○ Lower Paleocene sample   ● Upper Maastrichtian sample

Typical of these is the well-known Paleogene benthic foraminiferal species *Nuttalides trumpeyi*. In ODP site 401 a series of *N. trumpeyi* faunas were measured across the P–E boundary. Test size results (Fig. 5.9(*a*)) show highly variable patterns in the Late Paleocene (= the time of maximal environmental instability), culminating in a significant increase in test size within the latest Paleocene followed by a significant drop in test size across the P–E boundary. At no other time in the study interval does this species' test size exhibit such extremes in rates of variation. In addition, this interval of rapid morphological change corresponds to the maximum excursion in the $\delta^{13}C$ and $\delta^{18}O$ curves (= the LPTM).

Test shape also underwent a coordinated response across the P–E boundary. In this species a pronounced excursion in the sample mean shape index is centred on the P–E boundary (Fig. 5.9). Interestingly, in this example test shape variation appears to mirror the pattern of test size variation. Although the nature of test geometry in this species precludes the routine construction of growth trajectories for individuals, it is possible to subdivide a sample into size classes and calculate a mean test shape for each class. Assuming, as is typically the case, that smaller-sized individuals represent juvenile morphologies, this size/shape distribution should serve as an accurate proxy for the sample-specific growth trajectories used in the K–T analysis (see above).

Fig. 5.10 compares the size/shape distribution for a pooled sample of pre-LPTM populations (Fig. 5.10(*a*)) with the mean shape of a larger-sized population that existed at the time of maximum environmental change (Fig. 5.10(*b*)). Despite the fact that the LPTM population is markedly larger than its pre-LPTM ancestors, the mean shape of this population is more reminiscent of the smaller-sized (and presumably juvenile) end of the pre-LPTM shape distribution. This combination of juvenilization of somatic development via a decrease in the developmental rate over time is characteristic of heterochronic mode termed neoteny (Fig. 5.10(*c*)).

5.5.3 Heterochronic response in the Late Eocene

The case for progenesis in Late Eocene populations of the planktonic foraminiferal species *Subbotina linaperta* has been discussed in detail in MacLeod *et al.* (1990) and in MacLeod & Kitchell (1990) and will only be outlined here. Once again, recall that the Late Eocene interval showed no morphotypic response to environmental perturbation in common with the K–T or P–E events (see Sections 5.5.1 and 5.5.2 above). Using a series of 19 landmark points (Fig. 5.11) and the centroid size index (see Bookstein, 1986, fig. 13), size analyses for samples from DSDP sites 612 (western North Atlantic), 94 (Gulf of Mexico), and 363 (eastern South Atlantic)

Fig. 5.8. Ontogenetic trajectories for *Heterohelix globulosa* populations in Lower Paleocene (*a*) and upper Maastrichtian (*b*) samples from the Brazos Core (see Keller, 1989 for precise sample locations). Data points represent sample means and error bars represent 95% confidence intervals on the sample mean. When these ontogenetic paths are compared (*c*), the progenetic nature of the test size change is evident. The progenetic nature of the heterochronic signal in *H. globulosa* ((*c*) and (*d*)) and *H. navarroensis* ((*e*) and (*f*)) is consistent regardless of whether the species' ontogenetic trajectory is inferred on the basis of chamber number (*c*) and (*e*) or overall test size (*d*) and (*e*).

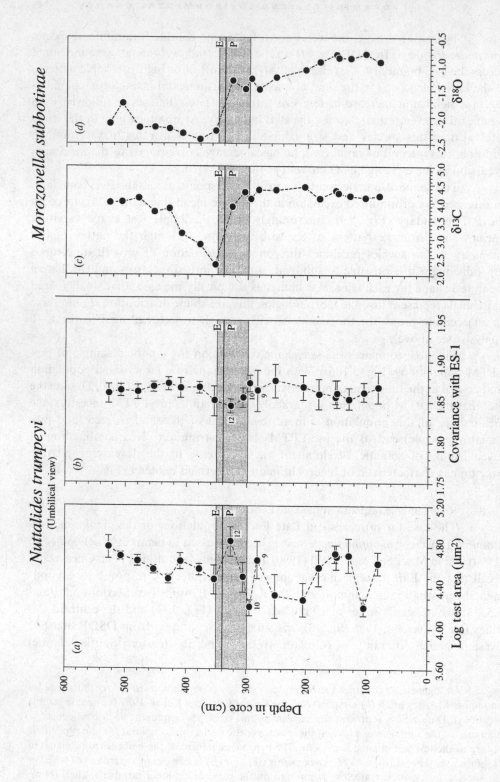

Nuttalides trumpeyi
(Umbilical view)

Morozovella subbotinae

(a) Log test area (μm²)

(b) Covariance with ES-1

(c) δ¹³C

(d) δ¹⁸O

Depth in core (cm)

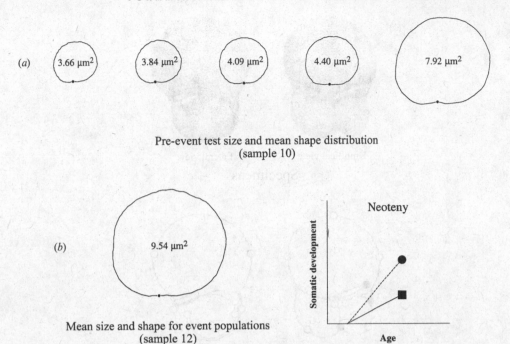

(a)   3.66 μm²   3.84 μm²   4.09 μm²   4.40 μm²   7.92 μm²

Pre-event test size and mean shape distribution
(sample 10)

(b)   9.54 μm²

Neoteny

Mean size and shape for event populations
(sample 12)

Fig. 5.10. Patterns of test shape variation (umbilical view) in late Paleocene populations of *Nuttalides trumpeyi* for series of pre-LPTM size classes ((a), sample 10) and mean shape for and LPTM sample ((b), sample 12). Pre-LPTM size classes defined by subdividing the size-ranked sample into five groups such that each group represents an equal number of specimens (not an equal size range). This manner of subdivision provides a more accurate picture of the intrasample size–shape distribution when that distribution approximates an exponential function (as is the case here). Although the mean size of the LPTM sample is larger than those of the pre-LPTM samples (see Fig. 5.9) the overall shape of this sample is much more reminiscent of smaller (presumably juvenile) size classes in pre-LPTM strata. This change in offset signal (as inferred by test size) and stage of somatic development (as inferred by test shape) can be explained as a relative decrease in the developmental rate; a paedomorphic mode termed neoteny (c). Dots along the outlines represent starting points for the digitization procedure.

Fig. 5.9. Patterns of test size (a), shape (b) for the benthic foraminiferal species *Nuttalides trumpeyi* and coeval patterns of stable isotopic variation ((c) and (d)) across the Paleocene-Eocene boundary in the DSDP Site 401 core. In (a) and (b) data points represent sample means, error bars represent 95% confidence intervals on the mean. Stable isotopic data represent $\delta^{13}C$ (c) and $\delta^{18}O$ (d) values relative to PDB. Solid horizontal line represents P–E boundary horizon. The shaded area represents the interval of stable isotopic excursion (inferred to be the Late Paleocene Thermal Maximum or LPTM) in this core. Note response of these benthic foraminiferal populations to the environmental event. Size variation is highly unstable prior to the LPTM, reaches a maximum in terms of the rate and amplitude of change within the event itself, and then stabilizes in post-LPTM strata. Moreover, note that the general character of variation in the size and shape curves are mirror images of one another. Faunal samples 9, 10, and 12 identified by number (see caption to Fig. 5.10).

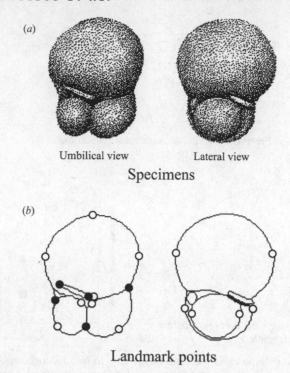

Fig. 5.11. Landmarks used to determine estimate test size and shape parameters in Late Eocene population of *Subbotina linaperta*. (*a*) Typical *S. linaperta* morphology. (*b*) Location of landmarks. Solid dots represent homologous points, open dots represent landmarks defined by extremal points.

show a statistically significant decrease in test size for sites 612 and 94 (Fig. 5.12). While the actual timing of the test size change cannot be determined for the Site 612 populations because of a hiatus in that core, based on the data from Site 94 it seems reasonable to suspect that this size decrease occurred between the Late Eocene *G. semiinvoluta* and *T. cerroazulensis* biozones. This places the test size decrease in the middle of the long Late Eocene–Oligocene global cooling event. It is also interesting to note that in both sites 612 and 94, test size reduction begins in the samples containing a well-dated microtektite horizon that, on the basis of tektite size distributions, must have occurred proximal to the location of Site 612.

Shape variation in *S. linaperta* populations from Site 612 was assessed using a series of 13 landmarks visible in umbilical view (Fig. 5.11). Comparison of mean shapes for the smaller-sized (but developmentally adult, see MacLeod *et al.*, 1990) post-event populations with the adult size/shape distribution obtained from pre-event populations once again reveals a juvenilization of test shape (Fig. 5.13). The combination of a reduction in somatic development attained without any significant change in the rate of development over time (see MacLeod *et al.*, 1990) identifies progenesis as the heterochronic mode most typical of these results.

Paedomorphic changes in developmental timing (e.g. progenesis and neoteny) have often been observed to be typical of organisms whose ecology undergoes a

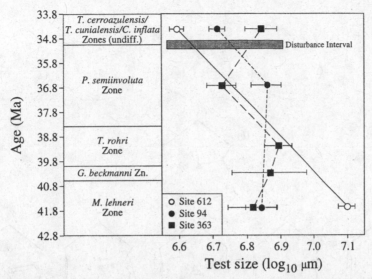

Fig. 5.12. Pattern of test size variation in *Subbotina linaperta* populations in DSDP sites 612 (western Atlantic), 94 (Gulf of Mexico), and 363 (Walvis Ridge) throughout the Late Eocene. Symbols represent sample means, error bars represent 95% confidence intervals on the sample mean. Note the similarity in the character of the response to this (apparently localized) environmental perturbation in sites 612 and 94. Because of a hiatus in Site 612 the precise timing of the size change is uncertain. However, based on the geographic ordering and timing of size changes in Site 94 it is likely that the change came about between 35.0 Ma and 34.8 Ma.

pronounced shift during development (see Gould, 1977; Alberch *et al.*, 1979). Since ecology can be regarded as an integral part of a species holomorphology and can therefore be a target of selection in its own right, the implication of finding paedo-morphic modes of change predominant in foraminiferal responses to global change events is that ecology, rather than morphology, is the proximal target of selection. This interpretation is supported in the instance of *Subbotina linaperta's* response to global environmental events in the Late Eocene via an isotopic analysis of preferred depth habitats (Fig. 5.14). These data show that the smaller-sized, progenetic, post-response morphs of this species changed their preferred depth habitat from inter-mediate to surface waters at a time coeval with that of the phenotypic response.

5.6   DISCUSSION

In all three intervals examined, there appears to have been a much stronger association between patterns of heterochronic change within populations (specifi-cally the juvenilization of adult morphologies) and global change events, than between generalized patterns of morphotype dominance and the same global change events. The prevalence of consistent heterochronic (as opposed to morphotypic) patterns of response by planktonic and benthic foraminifera to global change events suggests that the target of natural selection during these environmental perturbations was not morphology *per se*, but a group of developmental or life-history characters that covary with morphology. This interpretation was confirmed in the case of Late

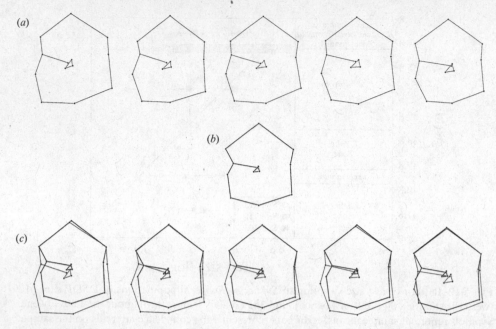

Fig. 5.13. Patterns of test shape variation (umbilical view) in Late Eocene populations of *Subbotina linaperta* for a series of pre-event sample size classes ((*a*), sample 21-6:11-13 cm, see MacLeod *et al.*, 1990; test sizes decrease from left to right) and mean shape for and post-event sample ((*b*), sample 21–5:107–108 cm). Dots represent positions of landmarks points (see Fig. 5.11(*b*); interlandmark chords are provided as a guide to relating the landmarks to the underlying shape and assessing the various shape changes. In this instance the mean size of post-event sample is much smaller than that of the pre-event sample (see Fig. 5.12), but the overall shape of this sample is much more reminiscent of smaller size classes characteristic of pre-event strata (see overlays in (*c*)). Since all of these individuals are developmental adults, this change in offset signal (as inferred by test size) and stage of somatic development (as inferred by test shape) can be explained as coordinated decrease without any significant accompanying change in developmental rate. This style of paedomorphism is termed progenesis (see Figs. 5.6, 5.8).

Eocene *Subbotina linaperta* populations because the morphologic response was accompanied by a pronounced habitat shift. Isotopic data for the K–T species used in this study are inconclusive because of the relatively shallow water depths that characterized the Texas Coastal Plain and Denmark during the latest Maastrichtian–earliest Danian (see Barrera & Keller, 1990). However, the model proposed by this study – that instances of planktonic foraminiferal heterochrony should be accompanied by some type of ecological shift – can, in principle, be tested.

In judging the potential role of test shape vs. developmental or life-history characters in mediating selection in foraminifera it is important to note that the foraminiferal test never comes in direct contact with the environment. Additionally, in most benthic and planktonic species the test does not impart its shape to the cytoplasmic mass and exists in a size range at which many of our 'intuitions' regarding functional distinctions among various shapes in aquatic envir-

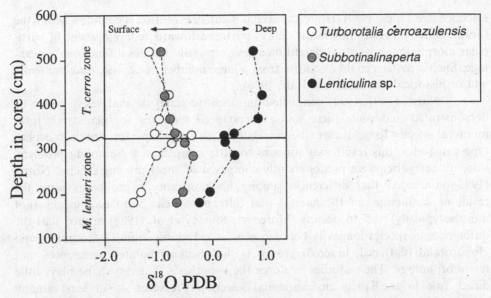

Fig. 5.14. Oxygen isotopic data for three species from ODP Site 612. These data show that the test size and shape changes that characterized *Subbotina linaperta* populations in this core were also accompanied by changes in the preferred depth habitat for this species (= isotopic values that are similar to those of the relatively shallow-dwelling species *Turborotalia cerrozaulensis*). This implies that the actual target of environmental selection for these populations was an ecological factor (e.g. preferred depth habitat) that was linked to the phenotype via developmental and life-history factors rather than morphology *per se*. Under this interpretation the morphological changes that were observed would have played no role in mediating environmental selection pressures. Rather, this morphological response may be most properly seen as a simple byproduct of the developmental mechanism used to alter the species' ecology. These data suggest that foraminiferal morphology – in the sense of gross morphotype – may not have been the target of selection in this example of foraminiferal population-level response to global environmental change.

onments are invalid (see Lipps, 1979). A number of hypotheses regarding the purpose of foraminiferal tests (e.g. sources of ballast used to control the organism's position in the water or on [or within] the sediment, anchoring points for pseudopods) are also compatible with a wide variety of test shapes, thus calling into question the functionality of test shape at least for these purposes.

The morphotype data presented above indicate that, while examples of iterative evolution within the foraminifera do exist, this phenomenon may have more to do with the generality of morphotype classifications than with the geometric structures actually present among foraminiferal species. Cifelli's (1969) morphotype classification appears to be based on particular taxa (hence his use of terms like 'pulleniatine' and 'hastigerine') while Norris' (1990, 1992) classification draws functionally dubious distinctions between 'dorsal conical' and 'ventral conical' morphotypes. These problems (to which the classification used in this study is also susceptible) are inherent in the fact that we are making very crude judgements as to what constitutes a significant deviation in morphology based on little empirical evidence (see Lipps, 1979). However, even if iterative evolution is common within the foraminifera, this does not mean that every trend toward redevelopment of particular morphologies within different lineages necessarily implies a functional advantage. Such a pattern could also arise from a large number of genetic, developmental and/or historical processes (see Raff, 1996).

Norris (1991, 1992) conducted an extensive series of analyses in which he demonstrated that while there was a pronounced tendency for planktonic foraminiferal species longevities to be correlated with test shape for certain morphotype categories, this result may amount to little more than a 'sampling problem' since all categories were predominantly composed of short-ranging species. Norris (1992) concluded that differences among his morphotype groupings were the result of distinctions in the habitat that different species were ecologically (not morphotypically) able to occupy. Moreover, Stanley et al. (1988) argued that the differences in species longevity between generalized globigerinid (= rounded) and globorotalid (flattened, tapered) groups is the result of habitat preferences, not test morphology. These studies reinforce the hypothesis that test shape plays little direct role in mediating environmental selection pressures — at least among planktonic species.

The data presented herein are consistent with the general results of Stanley et al. (1988) and Norris (1991, 1992), but provide a mechanistic means for explaining the role of development in this process. Under this hypothesis, foraminiferal populations respond to major changes in their environment by attempting to alter their ecological tolerances at the population level within limits set by the species-specific developmental patterns, augmented by variation in life-history parameters. Morphology is altered during this process, but only as a passive covariant of the developmental/life-history program. Accordingly, species longevities should be correlated with flexibility of the species' developmental programme, which, in turn, should be a heritable within individual foraminiferal lineages, but not strictly correlated with test shape.

In addition to allowing paedomorphic populations to capitalize on the benefits of altering body size and shape by altering developmental timing, paedomorphosis also provides control over many aspects of a species' biology and environment, including: generation times (and thus more rapid genotypic response to fluctuating or variable environmental conditions), dispersal capabilities, morphological variability, and routes of escape from morphological specializations. In the case of planktonic foraminifera, whose ontogenetic cycle also corresponds to a depth cycle (see Hemleben et al., 1988), paedomorphosis may well be the most effective way to adjust a species' or population's ecology so that its environment remains within defined tolerance limits. Unfortunately, the biology of modern benthic foraminifera is quite variable and does not provide a simple analogue to the planktonic foraminiferal relationships between development, habitat, and ecology. However, there are a variety of anecdotal reports of benthic foraminiferal species responding to periods of environmental stress by dwarfing (see Boltovskoy & Wright, 1976 and references therein). It remains to be seen whether these (and other) instances of size reduction contain progenetic patterns of concomitant shape variation and whether benthic foraminiferal ontogenies contain the strong and consistent ecological contrasts that apparently facilitate a developmentally mediated response to ecological selection in their planktonic kin.

5.7 CONCLUSIONS

Attempts to characterize biotic response to global change events by simply describing or developing ad hoc classifications for biostratigraphical patterns of species turnover cannot, by themselves, be used to make inferences about underlying causal processes or distinguish between alternative causal models. Instead, insights into the causes of mass extinction and modes of biotic response to global change events are best gained from an understanding of the biology of the taxa involved and based within a hypothetico-deductive framework. In this study morphotype and morphometric data from planktonic and benthic foraminiferal populations were used to assess whether test shape or developmental and life-history characteristics were the more likely targets of natural selection during three well-known Paleogene intervals for dramatic environmental change. Results suggest the following:

(i)   There was no consistent pattern of morphotype preference in either the planktonic or benthic datasets across the K–T, P–E, or Late Eocene global environmental change events. Planktonic morphotype data show that while some morphotypes persisted through the entire Turonian–Paleogene interval, others were confined entirely to the Late Cretaceous or Paleocene–Eocene radiations. A small number of morphotypes occur independently in both radiations, but these often exhibit strikingly different relative abundances, as well as first and last appearance patterns. On the whole the Cretaceous radiation appears more complexly structured than the Paleocene–Eocene radiation in terms of morphotype composition. Benthic morphotypic patterns within the Turonian–Early Eocene interval exhibit a striking pattern of almost constant morphotype composition and abundance up to and including the Late Paleocene benthic foraminiferal mass extinction event.

(ii) There is a consistent pattern of heterochronic test size-shape variation within the fossil records of several planktonic and benthic species as they passed through each of the three global change events. Pronounced paedomorphic patterns of variation characterize both planktonic and benthic species within the K–T event and a planktonic species within the Late Eocene event. Benthic foraminiferal variation within the P–E event appears to be more neotenic in character. These patterns occur over a variety spatial and temporal scales, but seem to be deterministically related to intervals of extraordinarily rapid environmental change. Intervals of prolonged environmental instability may also be associated with intraspecific patterns of size and shape instability within local populations.

(iii) The developmental mechanisms by which progenesis occurs in foraminifera are unknown. Much basic biological and genetic research on modern foraminifera will need to be undertaken before these mechanisms can be clarified and we may never be able to be certain which mechanisms operated in extinct species. However, the commonality of this pattern of response to global environmental change events exhibited by both benthic and plank-tonic foraminiferal species strongly implies that ecological aspects of the foraminiferal holophenotype are typically the targets of environmental selection. Foraminiferal populations respond to these selection pressures by altering the developmental and life-history patterns to accomplish changes in habitat and/or ecology. Although micropaleontologists typically study foraminifera through detailed morphological analysis, morphological change during rapid environmental perturbations most likely occurs as a byproduct of this selection for non-morphological aspects of the phenotype.

6

The response of planktonic foraminifera to the Late Pliocene intensification of Northern Hemisphere glaciation

MARK R. CHAPMAN

6.1   INTRODUCTION

Planktonic foraminifera are microscopic, free-floating protists that are widely distributed throughout the surface waters of the world's oceans. Unlike other members of the marine micro-zooplankton, planktonic foraminifera secrete a calcareous test during the course of their life cycle and, as a result, are a major contributor to the marine sediments that accumulate at the sea floor. Substantial deposits of this 'Globigerina ooze', more correctly termed foraminiferal ooze, are preserved within the geological record, often reaching thicknesses of several hundred metres. The excellent representation of planktonic foraminifera within Cenozoic marine sediments, whilst being disproportionate to their actual abundance levels within the total biomass of the marine zooplankton in the surface waters, is one of the prime reasons for the extensive usage of this particular microfossil group in both biostratigraphical and palaeoecological investigations.

Wei and Kennett (1986) documented an overall decline in the diversity of planktonic foraminiferal faunas over the last 5 million years, with species diversity levels in the modern ocean being approximately two-thirds those of the Neogene maximum. The most significant climatic event during this period of time was the rapid expansion of the Northern Hemisphere glaciations around 2.5 million years ago (Shackleton *et al.*, 1984). It has been shown that ocean circulation patterns were significantly affected by the environmental changes that accompanied the Late Pliocene build-up of the Northern Hemisphere ice-sheets (Loubere & Moss, 1986; Raymo *et al.*, 1989). However, previous studies have not provided evidence of any evolutionary change in the planktonic foraminiferal faunas coincident with this climatic deterioration (Wei & Kennett, 1986; Malmgren & Berggren, 1987). In general, the onset of Northern Hemisphere glaciation has been regarded as having little effect on planktonic foraminiferal populations.

In this chapter, I examine the evidence for a relationship between palaeoclimatic and palaeoceanographic changes at 2.5 Ma and evolutionary turnover within the planktonic foraminiferal faunas at several sites in the North Atlantic (Fig. 6.1). The effect of climate change on the survival of Late Pliocene planktonic foraminiferal populations is evaluated at the species level using presence/absence criteria. In addition, detailed species relative abundance records spanning the interval 3.2 Ma

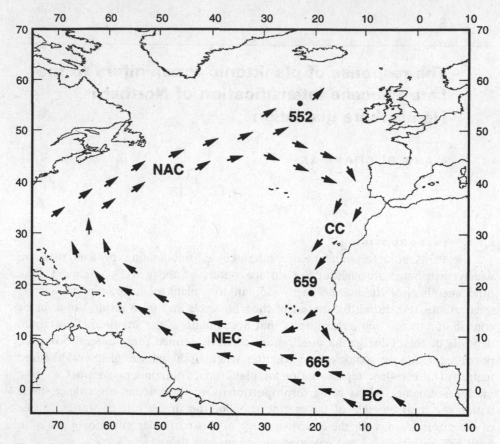

Fig. 6.1. North Atlantic DSDP/ODP sites investigated in this study. Location of sites: DSDP 552 (56°02.6′ N, 23°18.9′ W; depth 2301 m), ODP 659 (18°04.6′ N, 21°01.6′ W; depth 3070 m), ODP 665 (02°57.1′ N, 19°40.1′ W; depth 4740 m). Arrows depict the generalised pattern of surface circulation; BC = Benguela Current, CC = Canary Current, NAC = North Atlantic Current, NEC = North Equatorial Current.

to 1.9 Ma are used to provide a quantitative record of the impact of environmental changes and the extent of glacial–interglacial climatic oscillations in the 600 ky immediately before and after the 2.5 Ma event. Because Late Pliocene palaeoclimatic records are characterized by alternating glacial and interglacial periods of a few tens of thousands of years duration and remarkably rapid (in geological terms) transitions between glacial and interglacial climatic states, it is necessary to generate high resolution data if the responses of planktonic foraminifera to environmental changes are to be accurately determined. The sampling interval of the data considered in this study is in the order of 5–10 ky and hence should be sufficient to establish relationships between individual glacial or interglacial climatic events and faunal changes.

6.2   PLANKTONIC FORAMINIFERA AS PALAEOCLIMATIC AND PALAEOCEANOGRAPHIC INDICATORS

Well-defined planktonic foraminiferal biogeographical provinces have been established on the basis of data collected from surface plankton tows and ocean floor sediment samples (Bé & Tolderlund, 1971; Cifelli & Bénier, 1976; Kipp, 1976; Bé, 1977; Molfino et al., 1982). These studies show that the geographical distribution of most planktonic foraminiferal taxa are broadly circumglobal and hence can be equated with latitudinal variations in sea-surface water temperature (Fig. 6.2). Consequently, the study of fossil planktonic foraminiferal assemblages (CLIMAP, 1976; 1984; Crowley, 1981; Chapman et al., 1996b) and the stable isotope composition of foraminiferal tests (Shackleton & Opdyke, 1973, 1977) have proven to be especially valuable tools for palaeoclimatic and palaeoceanographic reconstruction.

An extraordinary amount of the information contained in the fossil record of planktonic foraminifera can be retrieved through detailed investigations of stratigraphically continuous deep sea sediment cores. In the Late Quaternary, it is possible to generate palaeoclimatic proxy data with a temporal resolution of a few hundred years (Bond & Lotti, 1995). For longer intervals of time spanning several hundred thousand years (Ruddiman et al., 1986b) or even several million years (Shackleton et al., 1990; Tiedemann et al., 1994) slightly lower resolutions (< 5 ky) are readily achievable. One consequence of this is that the numerous studies exploring the evolutionary history of the planktonic foraminifera following their appearance in the Early Cretaceous have yielded one of the most complete and continuous stratigraphical records of any major fossil group (e.g. Berggren, 1969; Cifelli, 1976; Banner, 1982; Kennett & Srinivasan, 1983; Cifelli & Scott, 1986; Wei & Kennett, 1986).

Neogene planktonic foraminiferal lineages have been widely used to investigate the mode and the tempo of evolutionary change (Jenkins & Shackleton, 1979; Malmgren & Kennett, 1981; Malmgren et al., 1983; Wei & Kennett, 1983; Banner & Lowry, 1985; Hodell & Kennett, 1986; Malmgren & Berggren, 1987; Norris, 1992). Results from these studies suggest that climatic change is the principal instigator of evolutionary changes within planktonic foraminiferal faunas and, more specifically, that there is a fundamental linkage between palaeoceanographic fluctuations in the spatial partitioning of the surface waters and periods of speciation or extinction. The structural limitations of test morphology and the primary function of the foraminiferal test to maintain the required position in the upper water column have produced a general history within the Globigerinacea that is characterized by a number of similar but distinct iterative trends superimposed on non-iterative, unidirectional evolutionary changes (Banner, 1982).

6.3   THE ONSET OF THE LATE CENOZOIC ICE AGE: LATE PLIOCENE CLIMATIC COOLING

The occurrence of localized ice rafted debris along the margins of the Norwegian–Greenland Sea and in the northern North Atlantic indicates that ice cover, at least in the form of glaciers, has existed intermittently since the Late

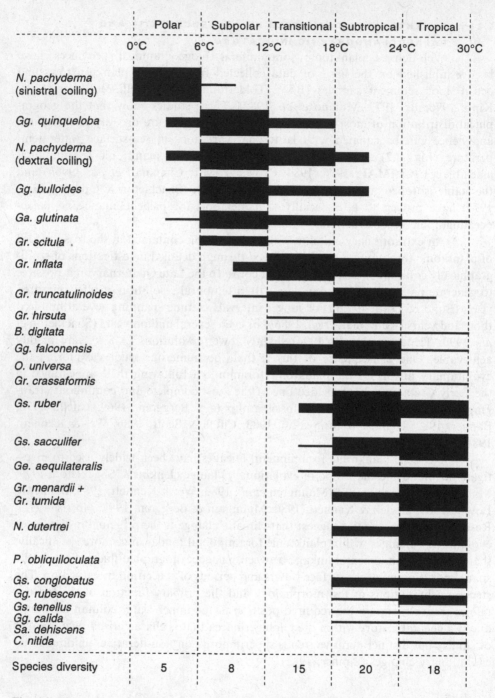

Fig. 6.2. The modern distribution of planktonic foraminiferal species within the five major North Atlantic faunal provinces. Variations in line thickness indicate differences in the relative importance of each species within the various faunal assemblages. Based on data from Bé and Tolderlund (1971) and Kipp (1976). The taxonomy and generic abbreviations used here follow Kennett and Srinivasan (1983): *N.* = *Neogloboquadrina, Gg.* = *Globigerina, Ga.* = *Globigerinita, Gr.* = *Globorotalia, B.* = *Beella, O.* = *Orbulina, P.* = *Pulleniatina, Gs.* = *Globigerinoides, Ge.* = *Globigerinella, Sa.* = *Sphaeroidinella, C.* = *Candeina.*

Miocene (Jansen & Sjoholm, 1991). However, the development of the first true pan-continental ice-sheets in the Northern Hemisphere high latitudes does not appear to have occurred until around 3 million years ago (Jansen & Sjoholm, 1991; Baumann & Meggers, 1996). The expanded biogeographical distribution of extant, warm adapted marine and terrestrial biotas in mid- and high-latitude regions provides strong evidence that climatic conditions were generally warmer in Mid Pliocene times (3–5 Ma) than at present (Dowsett et al., 1994). In the mid-latitude North Atlantic, sea surface temperature estimates suggest that conditions were warmer during the Mid Pliocene than at any time during the past 3 my (Dowsett & Loubere, 1992).

A significant but short-lived climatic cooling occurred around 3.4 Ma (3.2 Ma on the timescale used by Prell, 1984) which can be recognized in the benthic $\delta^{18}O$ records from all three major oceans. This climatic event has been attributed to a sustained but temporary cooling of bottom water temperatures following a period of ice build-up in the Northern Hemisphere (Prell, 1984). Associated changes are also evident in the North Atlantic and Mediterranean planktonic foraminiferal records. Most notably, there was a southward expansion in the biogeographical distribution of cold-tolerant species towards the low latitudes (Loubere & Moss, 1986; Zachariasse et al., 1989).

The crucial stage in the enlargement of Late Cenozoic ice-sheets commenced around 3.2 Ma (Tiedemann et al., 1994), although it was several hundred thousand years before the ice- sheets were of sufficient volume to impact forcefully on global climate. High resolution benthic $\delta^{18}O$ records document a progressive enrichment of glacial $\delta^{18}O$ values between 3.2 Ma and 2.5 Ma and primarily reflect a gradual increase in glacial intensity (Raymo et al., 1989; Tiedemann et al., 1994). The widespread increase in the amount of ice rafted debris found in the North Atlantic (Raymo et al., 1989) and the Norwegian Sea (Jansen & Sjoholm, 1991) around 2.7 Ma (isotope stage 110) is the earliest evidence for the presence of substantial continental ice-sheets. However, the onset of the Northern Hemisphere ice age is usually considered to be coincident with the first of three larger $\delta^{18}O$ excursions at 2.5 Ma (isotope stage 100). This climatic threshold is associated with a further 30% increase in continental ice volume (Raymo et al., 1989) and a dramatic expansion of the North Atlantic ice rafting belt to the northern margin of the subtropical gyre (40° N).

Long-term fluctuations in ice volume, characterized by periodic variations at Milankovitch orbital forcing frequencies, have been an important global climatic control since the Late Pliocene (Ruddiman et al., 1986a). The development of the Late Cenozoic ice age can be subdivided into two distinct phases, each delimited by a critical increase in the size and a shift in the temporal resonance of the Northern Hemisphere ice-sheets (Shackleton et al., 1988). From the Late Pliocene to the Early Pleistocene (3 to 1 Ma), the expansion and contraction of the ice-sheets were characterized by smaller amplitude $\delta^{18}O$ oscillations, and hence smaller ice-sheets, with a pronounced 41 ky cyclicity (Ruddiman et al., 1986a). By comparison, the much larger continental ice-sheets that have developed over the last million years have undergone a phase of growth and decay approximately every 100 ky (Hays et al., 1976; Imbrie et al., 1984).

The causes of the large-scale transformation of high latitude climate in the Northern Hemisphere are not fully understood. One theory is that tectonic changes led to long-term changes in the atmospheric and oceanic circulation systems. The two most important possible tectonic forcing mechanisms are (i) the gradual closing of the Central American Seaway and the emergence of the Isthmus of Panama (Berggren & Hollister, 1977; Keigwin, 1978; Berger *et al.*, 1981) and (ii) the regional uplift of mountain plateaux in Tibet and North America (Ruddiman & Raymo, 1988). An increase in the strength of the poleward flowing Gulf Stream, resulting from the shoaling of the Panama Isthmus and the cessation of equatorial surface water exchange between the Pacific and Atlantic Oceans, may have enhanced the penetration of warm saline water into the Arctic region and provided the increased moisture supply that is one prerequisite for rapid ice-sheet growth (Berger *et al.*, 1981). General circulation model results also suggest that uplift-induced changes in atmospheric circulation patterns could lead to widespread cooling for most high latitude regions in the Northern Hemisphere (Ruddiman & Kutzbach, 1989). Furthermore, the increase in chemical weathering rates which resulted from plateau uplift may provide an important mechanism for lowering atmospheric CO_2 concentrations (Raymo & Ruddiman, 1992).

Changes to the geographical configuration of ocean basins are also likely to have altered the equilibrium of internal feedback processes within the climate system. Because the Earth's climate is very sensitive to the presence of ice-sheets in the Northern Hemisphere, it is possible that internal feedback mechanisms related to contrasting surface albedo, the intensity of the thermohaline circulation and atmospheric CO_2 levels, have played an important role in amplifying the external climatic forcing patterns related to orbitally driven changes in insolation. Additional evidence is required to fully assess the relative importance of external climate forcing by factors such as tectonic and insolation variations, and internal feedback mechanisms within the Earth's climate system. However, it is apparent that the cumulative impact of these changes induced a major transformation of the Earth's climate through the establishment of a regular rhythm of alternating and strongly contrasting warm and cold conditions (glacial–interglacial cycles) which have persisted from Late Pliocene times through to the present.

6.4   FAUNAL RESPONSE TO THE ONSET OF NORTHERN HEMISPHERE GLACIATION

6.4.1   Comparison of faunal and $\delta^{18}O$ palaeoclimatic records

The composition of modern planktonic foraminiferal assemblages in the eastern subtropical Atlantic reflects the proximity of the warm waters of the subtropical gyre and the influence of cool surface waters that are transported southwards in the distal flow of the Canary Current (Cifelli & Bénier, 1976). Palaeoceanographic reconstructions indicate that the Equatorward migration of cool water species increased during glacial times because of the southward movement of the polar front and the strengthening of latitudinal thermal gradients (Crowley, 1981; Chapman *et al.*, 1996b). Analysis of planktonic foraminiferal assemblages obtained from Ocean Drilling Program (ODP) Site 659 in the eastern subtropical Atlantic should therefore reveal whether comparable changes in the surface circulation were associated with the accentuation of glacial–interglacial climatic variability in the Late Pliocene.

Quantitative faunal counts were generated for 137 samples selected from the 1.9–3.2 Ma time interval. The faunal census data are based on counts of approximately 400 whole planktonic foraminifera in the >150 μm size fraction. The age model for Site 659 is constructed by tuning the $\delta^{18}O$ stratigraphy to Milankovitch orbital variations (Tiedemann et al., 1994). Perhaps more importantly, the benthic $\delta^{18}O$ data also provide a proxy record of glacial–interglacial fluctuations in ice volume, so temporal variations in the planktonic foraminiferal populations can be directly compared to an independent palaeoclimatic record. The relative abundance records of important cold-adapted and warm-adapted planktonic foraminifera and the benthic $\delta^{18}O$ record obtained from Site 659 are shown in Fig. 6.3. The overall correlation between the two palaeoclimatic proxies is impressive, both in terms of the magnitude and the frequency of the changes depicted. The degree to which the oscillations in the $\delta^{18}O$ records are paralleled by variations in the relative abundance of the cold water species *Neogloboquadrina pachyderma* after the onset of Northern Hemisphere glaciation 2.5 million years ago is particularly striking. Both records are dominated by cyclical fluctuations at the 40 ky obliquity orbital rhythm.

The planktonic foraminiferal faunal record (Fig. 6.3) can be separated into three distinct intervals: (i) an interval characterized by relatively stable, warm conditions (3.2–2.7 Ma); (ii) a transition phase (2.7–2.5 Ma); and (iii) an interval characterised by pronounced climatic variability (2.5–1.9 Ma). The climatic deterioration during the Late Pliocene is most clearly reflected by variations in the extent and composition of the cold-water taxa. In the first interval, the cold-water assemblages were dominated by *Globorotalia puncticulata* whilst abundances of *Globigerina bulloides* and *N. pachyderma* were generally low. The palaeobiogeographical distribution of *Gr. puncticulata* suggests that the ecological tolerances of this extinct species were broadly similar to those of its modern, extant descendant *Gr. inflata*. Typically, both *Gr. puncticulata* and *Gr. inflata* have maximum abundances in the temperate region (Malmgren & Kennett, 1981), whereas the other two cold-adapted species, *Gg. bulloides* and *N. pachyderma*, are tolerant of colder conditions and form an important part of the subpolar fauna (Bé, 1977). The low abundance of these two species and the dominance of *Gr. puncticulata* in the faunal record suggests that sea surface conditions remained relatively warm even during glacial events prior to 2.7 Ma. This implies that any significant climatic cooling during the glacial events preceding oxygen isotope stage 110 was limited to higher latitudes.

The composition of the cold-water fauna began to alter around 2.7 Ma, when there was a marked decline in the abundance of *Gr. puncticulata* and an increase in the abundance of *Gg. bulloides*. This event also coincided with the first substantive peak in abundance of *N. pachyderma*, a subpolar species which was previously unimportant in this area. The development of a more extreme cold-water fauna at Site 659 occurred during a well-defined glacial episode in the benthic $\delta^{18}O$ records (Fig. 6.3). The high benthic $\delta^{18}O$ values during this glacial event (isotope stage 110) indicate a significant increase in the size of the Northern Hemisphere ice-sheets. This is supported by a coeval increase in the amount of ice- rafted debris in high-latitude North Atlantic sediments (Raymo et al., 1989; Jansen & Sjoholm,

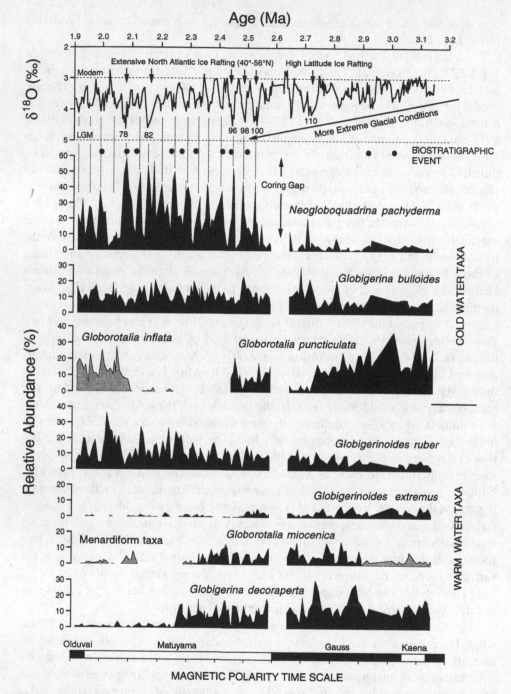

Fig. 6.3. Late Pliocene planktonic foraminiferal abundance variations at ODP Site 659 in the eastern subtropical Atlantic (18°04.6′ N, 21°01.6′ W; depth 3070 m). Benthic oxygen isotope data are from Tiedemann *et al.* (1994).

1991). The faunal changes associated with this climatic event constitute the earliest reliable evidence for any marked increase in the intensity of the Canary Current circulation along the eastern Atlantic margin. Downcore relative abundance records show that the cold-water taxon *N. pachyderma* became increasingly dominant during glacial climatic periods after the major expansion of Northern Hemisphere ice-sheets around 2.5 Ma (Fig. 6.3). The large fluctuations in the abundance of *N. pachyderma* in the interval 2.5–1.9 Ma are the overriding feature of the faunal data. The periodic influx of *N. pachyderma*, which occurred during each interval of higher ice volume, indicates that the impact of glacial cooling extended much further south after 2.5 Ma than earlier in the Late Pliocene and that the intensity of the eastern boundary current circulation fluctuated in parallel with variations in ice volume. The coherent relationship between glacial–interglacial fluctuations in global ice volume and the extent of entrainment and southward advection of cold-water taxa within the Canary Current observed from 2.5 Ma and 1.9 Ma is remarkably similar to the palaeoceanographic pattern found in late Quaternary records from the same region (Crowley, 1981).

Although glacial ice volumes varied appreciably through the Late Pliocene, it was only after the rapid expansion of Northern Hemisphere glaciation at 2.5 Ma that high-latitude climatic forcing is clearly identifiable in the subtropical faunal record from Site 659. Prior to 2.5 Ma, changes in the abundance of cold-tolerant planktonic foraminiferal species do not appear to parallel fluctuations in the high-latitude climatic signal to the same extent, even though the benthic $\delta^{18}O$ data suggest that glacial ice volumes were of similar magnitude. This disparity may in part be explained by the absence of competition from temperate species, such as *Gr. puncticulata* and *Gr. inflata*, between 2.4 Ma and 2.1 Ma and a correlative increase in the abundance of *N. pachyderma* within the temperate waters of the North Atlantic. However, this cannot be the underlying reason for the different relationship between *N. pachyderma* relative abundances and glacial $\delta^{18}O$ values before and after 2.5 Ma because the glacial abundances of *N. pachyderma* documented after the arrival of *Gr. inflata* at 2.1 Ma remained much higher than those prior to 2.5 Ma. It is therefore probable that the establishment of a coherent climate signal in the subtropical region at 2.5 Ma signifies the attainment of a fundamental palaeoclimatic threshold which altered the palaeoceanographic response to climate change.

6.4.2   Faunal turnover: tempo and mode of extinctions

The planktonic foraminifera in the North Atlantic experienced heavy extinctions during the Late Pliocene (Table 6.1). The impact of these changes is clearly evident at Site 659, where nearly 40% of all the taxa encountered in the older part of the studied time interval (2.8–3.2 Ma) had been eliminated from the fauna by the end of the Pliocene. Key members of the Late Pliocene planktonic foraminiferal faunas affected by these changes are shown in Fig. 6.4. When viewed in quantitative terms, these faunal changes represent a decrease in the extinct component of the foraminiferal population from a maximum value of 40% at 2.8 Ma to around 2% at 2.0 Ma. Most extinction events are very well defined (Figs. 6.3, 6.5). For example, the extinctions of *Globorotalia miocenica* and *Gr. puncticulata* at Site 659 are marked by an abrupt decline in abundance and a complete disappearance from the fossil record

Table 6.1. *Planktonic foraminiferal faunal change in the North Atlantic during the Late Pliocene*

Datum	Age (Ma)	Isotope stage	Bioprovince
FAD *N. pachyderma* (sinistral)	1.78	64	Polar
FAD *Gr. truncatulinoides*	2.03	74	Subtropical
Acme *Gr. inflata*	2.10	78	Temperate
FAD *Gr. menardii*	2.13	80	Tropical
FAD *Gr. inflata*	2.24	86	Temperate
Reappearance *Pulleniatina* spp.	2.27	88	Tropical
LAD *Gr. exilis*	2.28	88	Tropical
LAD *Gr. miocenica*	2.34	90	Tropical
LAD *N. atlantica*	2.42	95	Polar
LAD *Gr. puncticulata*	2.44	96	Temperate
Decline *Gs. extremus*	2.49	99	Subtropical
LAD *Gr. limbata*	2.93	120 (G 16)	Tropical
LAD *D. altispira*	3.01	124 (G 20)	Subtropical
LAD *S. seminulina*	3.17	132 (KM 4)	Tropical
FAD *Sa. dehiscens*	3.51	(MG 6)	Tropical
Disappearance *Pulleniatina* spp.	3.65	(GI 4)	Tropical

Biostratigraphic data are from Raymo *et al.* (1989), Chapman *et al.* (1996*a*) and unpublished data. Ages are calculated using the astronomical timescale of Shackleton *et al.* (1990). Isotope stages are reported using conventional notation and the new nomenclature defined in Tiedemann *et al.* (1994); glacial events are marked by even numbers in both schemes. FAD = first appearance datum, LAD = last appearance datum.

in less than 3 ky. However, *Globigerinoides extremus* and *Globigerina decoraperta* have more complex distribution patterns. Both of these species were intermittently present in Site 659 sediments at 1.7 Ma, several hundred thousand years after their initial decline in abundance.

An important feature of these Late Pliocene extinctions is the way in which warm-adapted species from the tropical–subtropical ocean, for example *Dentoglobigerina altispira, Globorotalia limbata, Gr. miocenica,* and *Globorotalia exilis,* were affected just as much as cold-adapted taxa, such as *Neogloboquadrina atlantica* and *Gr. puncticulata,* from the mid- to high latitudes. Furthermore, it is evident that these changes affected both surface-dwelling and deeper-dwelling populations and numerically abundant and rare elements of the fauna (Chapman *et al.,* 1996*a*). This pattern is particularly clear within the more diverse tropical–subtropical faunas, and demonstrates the magnitude of ecological pressures in the Late Pliocene ocean and the pervasive nature of palaeoceanographic changes. Assessing the extent of these faunal changes within the different regions of the North Atlantic must take into account the interdependency between species diversity and geography; species diversity generally decreases with increasing latitude (Fig. 6.2). When the Late Pliocene species are separated, albeit somewhat arbitrarily, into cold ($< 12\ °C$) and warm ($> 12\ °C$) assemblages the extinction rates are 25% and 37%, respectively. However, if the impact of these extinctions is evaluated in a manner which assesses the importance of the affected taxa within the population structure, then the magnitude of change is similar for both cold and warm assemblages.

Fig. 6.4. SEM photographs of important members of Late Pliocene planktonic foraminiferal faunas.

A *Gg. bulloides* (× 200) B *N. pachyderma* (× 250)
C *Gr. puncticulata* (× 150) D *Gr. inflata* (× 150)
E *D. altispira* (× 150) F *Gr. limbata* (× 100)
G *Gs. ruber* (× 150) H *Gr. miocenica* (× 150)
I *Gg. decoraperta* (× 200) J *Gr. truncatulinoides* (× 150)
K *Gs. extremus* (× 200) L *Gr. exilis* (× 100)

The last occurrence of most planktonic foraminiferal species occurred during glacial isotope stages (Table 6.1) and presumably therefore coincided with a geographical expansion of cold water masses. Although this relationship does not necessarily prove that cooling caused extinctions, it does imply that extinction events and the environmental processes ultimately responsible for extinction selectivity were enhanced during episodes of global cooling. There is no evidence for an immediate acceleration in the extinction rate of planktonic foraminifera during the first major pulse of northern hemisphere glaciation at 2.5 Ma (isotope stage 100), but there is a decline in previously numerically important taxa in the 200–300 ky following this event. It is difficult to directly link the disappearance of warm-adapted taxa from the tropical–subtropical faunas to the enhanced intensity of glacial events during the build-up and establishment of Northern Hemisphere ice-sheets, with the decline of *Gs. extremus* at the end of isotope stage 99 being a possible exception. The evidence for a direct linkage between extreme glacial

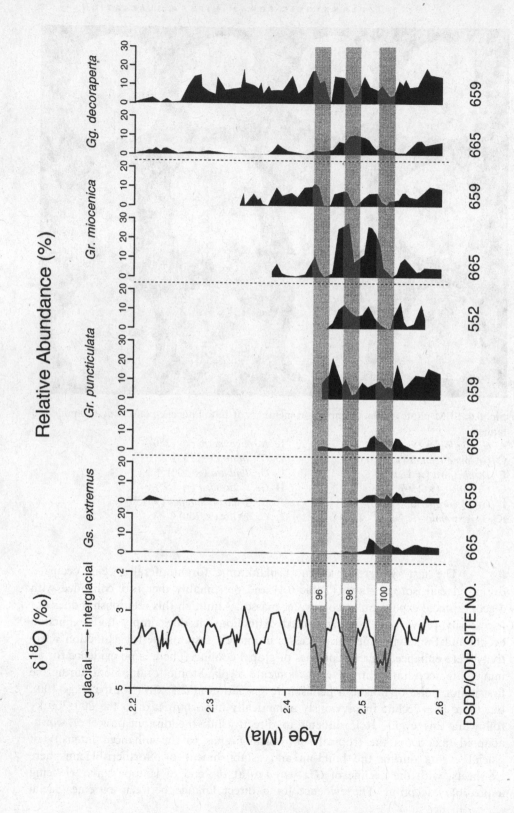

events and extinctions is stronger within the colder water assemblages. Two dominant taxa disappeared during and immediately after the third expansive Northern
Hemisphere glacial at 2.5 Ma, *Gr. puncticulata* in isotope stage 96 and *N. atlantica*
very shortly afterwards in isotope stage 95.

Competition from new planktonic foraminiferal species can be discounted
as a mechanism for bringing about large-scale changes within the planktonic foraminiferal faunas because no new taxa appear in the geological record during the
period of change. It is quite remarkable that the whole series of changes within the
North Atlantic faunas seem to have been completed prior to the first occurrence of
any 'replacement' species. The nature of the faunal turnover strongly suggests that
these modifications were a direct response to abiotic changes within the environment
and that the extinctions within the planktonic foraminiferal populations were the
preliminary phase of the establishment of a new biotic equilibrium in the ice-age
world.

As planktonic foraminiferal biogeography is primarily controlled by surface
water mass patterns without any rigidly defined barriers to movement (Cifelli &
Bénier, 1976), it should be possible for an ecologically disadvantaged species to
overcome an adverse change in the local environment by migrating to more suitable
habitats. However, detailed consideration of the patterns of faunal turnover and
their precise relationship to global climatic changes are complicated by the non-
uniform geographical response to climate change and regional aspects of species
biogeography. For instance, the CLIMAP reconstruction for the last glacial maximum (18 ka) revealed that sea surface temperatures in the mid- to high-latitude
Atlantic cooled by 8–10 °C while tropical temperatures were only lowered by around
2 °C. Furthermore, the magnitude of cooling experienced by the Late Pliocene surface ocean would almost certainly have been less pronounced, especially in parts of
the ocean at a great distance from the area of active ice-sheet growth and decay,
because Late Pliocene glacial ice volumes were much lower (50–70%) than those at
the last glacial maximum (Raymo *et al.*, 1989). Those taxa with a more northerly
distribution may have been ecologically disadvantaged because climate instability
would be greatest in their biogeographical area. This impairment would be com-
pounded because the decrease in the total area of subpolar and temperate biogeographical provinces during glacials is disproportionately large (CLIMAP, 1976) and
an adaptation to cold conditions may have made it difficult for these species to
migrate elsewhere in pursuit of more suitable environmental conditions. In these
circumstances it is conceivable that extinction selectivity may have been primarily
determined by the magnitude of climate change and sea surface temperature cooling.

As the amplitude of glacial–interglacial temperature variations would have
been appreciably smaller in the low latitude ocean, the likely extent of any ecological
reorganization on account of sea surface temperature fluctuations would presumably

Fig. 6.5. Planktonic foraminiferal extinction patterns following the expansion of Northern
Hemisphere glaciation at 2.5 Ma. The shaded lines indicate the three well-developed glacial
events (isotope stages 100, 98 and 96) that were associated with extensive ice rafting. Location
of sites: DSDP 552 (56°02.6′ N, 23°18.9′ W; depth 2301 m), ODP 659 (18°04.6′ N, 21°01.6′ W;
depth 3070 m), ODP 665 (02°57.1′ N, 19°40.1′ W; depth 4740 m).

have been much reduced. Furthermore, the wide spatial distribution of warm-adapted taxa prior to their disappearance from the tropical–subtropical ocean suggests that these species were not particularly sensitive to small changes in surface temperature. The extinct elements of the tropical–subtropical fauna had such a wide spatial distribution that it is probable that the variability in sea surface temperature encountered at any one instant in time over this large geographical area would have far exceeded the magnitude of temperature fluctuations experienced as a result of Pliocene glacial–interglacial circulation changes. It is therefore unlikely that restrictive thermal tolerances were the sole reason for the ecological crisis within low-latitude faunas during the Late Pliocene. The processes responsible for extinction within these low latitude populations must therefore reflect more complex interactions related to abiotic changes, affecting the hydrographic structure of the upper water column, and ecological factors such as nutrient availability and interspecific competition.

More pronounced depth stratification in the low latitude ocean permits greater niche separation and specialization. Schweitzer and Lohmann (1991) used oxygen and carbon isotopes to study the changes in depth habitat of *Globorotalia menardii* and *Globorotalia tumida* during the course of their life-cycle. Juvenile representatives of both these species grow in shallow water (less than 50 m) during the summer months when a seasonal thermocline is well developed, but when mature these species sink to greater depths (50 to 100 m) where they secrete a thick calcite crust. Late Pliocene menardiform taxa such as *Gr. miocenica, Gr. exilis* and *Gr. pertenuis*, which were particularly abundant in the Caribbean region (Kennett & Srinivasan, 1983) and later migrated into the open equatorial Atlantic, do not possess a thick calcite crust (Fig. 6.4). These morphological differences may indicate that these taxa were not able to inhabit cold subthermocline waters and as a consequence were restricted to tropical environments with a deep, semipermanent thermocline.

If this were the case, these taxa would have been particularly susceptible to environmental fluctuations and, in particular, the strengthening of latitudinal thermal gradients and accompanying circulation changes, which profoundly alter the stability of surface environments during glacial conditions (CLIMAP, 1976). Any permanent or seasonal reduction in the depth of the mixed layer would affect niche partitioning within the mixed layer by changing the extent of surface stratification and prevailing nutrient levels, and hence alter competition pressures within the shallow dwelling fauna. Therefore, it is possible that the Late Pliocene cooling of subsurface temperatures and the steepening of thermoclinal gradients may have exceeded the ecological thresholds of the deeper dwelling members of the fauna such as the globorotalids.

6.4.3   Recolonization of the North Atlantic

The decline and elimination of numerous taxa in the 800 ky interval prior to 2.3 Ma left the North Atlantic faunas biotically depleted. In spite of this, there is no evidence for any increase in the evolutionary rates of Late Pliocene planktonic foraminiferal lineages. The only species origination which may be related to the advent of large scale and high frequency palaeoenvironmental fluctuations is the first occurrence of *Globorotalia truncatulinoides*. Whereas the first appearance of

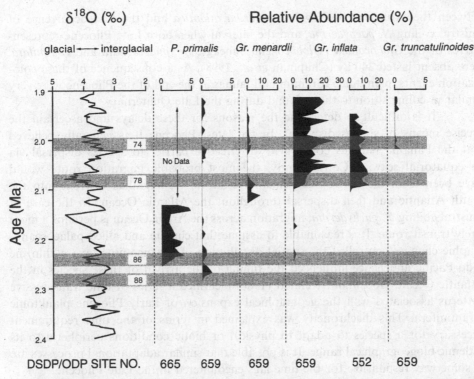

Fig. 6.6. Planktonic foraminiferal migration patterns within the Late Pliocene North Atlantic. Location of sites: DSDP 552 (56°02.6′ N, 23°18.9′ W; depth 2301 m), ODP 659 (18°04.6′ N, 21°01.6′ W; depth 3070 m), ODP 665 (02°57.1′ N, 19°40.1′ W; depth 4740 m). Note the expanded relative abundance scale used for *Pulleniatina primalis* and *Globorotalia truncatulinoides*.

this taxon in the North Atlantic at 2.0 Ma post-dates the expansion of Northern Hemisphere glaciation, the evolutionary first occurrence of *Gr. truncatulinoides* in the south-west Pacific has been documented some 500 ky earlier, stratigraphically adjacent to the Matuyama/Gauss magnetic reversal (Dowsett, 1989). Although the planktonic foraminiferal record illustrates that abiotic changes profoundly altered pre-existing ecological interactions during the last 1.5 my of the Pliocene, it appears that these changes were not sufficient to induce strong evolutionary pressures within the faunas.

Planktonic foraminiferal biotic diversity remained impoverished for a significant interval of time after the last extinction event at 2.3 Ma. The replenishment of North Atlantic faunas took place over 500 ky and was effected by the phased introduction of immigrant taxa from the Indo-Pacific oceans (Fig. 6.6). Levels of faunal diversity had recovered significantly by 1.8 Ma. The time taken for the replacement of individual elements of the fauna varied enormously. For example, the genus *Pulleniatina* was absent from the Atlantic for almost 1.4 my whilst the longest interval of time that the menardiform group was unrepresented lasted only 150 ky. In the polar and temperate regions of the North Atlantic, there was a gap of 600 ky between the extinction of *Neogloboquadrina atlantica* and the first occurrence of sinistral coiling *N. pachyderma*, and the interval when both Late Pliocene representatives of the *Globorotalia (Globoconella)* lineage, *Gr. puncticulata* and *Gr. inflata*, were absent lasted 340 ky (Chapman *et al.*, 1998). As a consequence of these colonization events, planktonic foraminiferal faunas at the end of the Pliocene were very similar in composition to those found during the Late Quaternary.

It is difficult to determine the reasons for these delays and ascertain the precise means of species dispersal. In the latest Pliocene it is generally believed that the Panama Isthmus constituted an effective barrier to species dispersal via an equatorial seaway. Consequently, the most probable migration route would have been from the subtropical Indian Ocean via the Agulhas Current into the South Atlantic and then dispersal throughout the Atlantic Ocean. In the case of sinistral coiling *N. pachyderma*, migration across the Arctic Ocean is perhaps a more likely transit route. It is reasonable to assume that climate and allied palaeoceanographic changes controlled the spatial distribution of the incoming taxa within the Indo-Pacific and hence influenced the timing of the arrival of these species in the Atlantic Ocean. Hodell and Kennett (1986) identified a series of time transgressive patterns associated with the geographical expansion of Early Pliocene planktonic foraminifera. This diachroneity was explained in terms of the time requirement necessary for a species to adapt to physical or biotic conditions peripheral to its endemic biogeographical range. It is possible that similar adaptational processes are in some way responsible for the time lags encountered in the Late Pliocene.

Evidence from the fossil record suggests that the emergence of the Panama Isthmus and cessation of species interchange between the Atlantic and Indo-Pacific biogeographical provinces influenced the spatial development of planktonic foraminiferal faunas. Coiling direction patterns in Atlantic and Indo-Pacific *Pulleniatina* populations were similar prior to the regional disappearance of *Pulleniatina* from the Atlantic at 3.65 Ma. However, after the reappearance of *Pulleniatina* at 2.3 Ma, the changes observed in the coiling direction of Atlantic populations were no longer correlated with those in the Indo-Pacific. This discordancy suggests that the two major faunal provinces have been spatially segregated since the Late Pliocene (Saito, 1976) and it may also signify that the migration of *Pulleniatina* stock into the Atlantic occurred on a single occasion. The impediment created by the closure of this biotic gateway may also have been an important factor contributing to the inability of isolated North Atlantic foraminiferal populations to accommodate the enhanced levels of climatic variability in the Late Pliocene and could, in turn, have promoted a decline in faunal diversity.

All of the new migrant taxa (*N. pachyderma* (sinistral), *Gr. inflata*, *Gr. truncatulinoides*, and *Pulleniatina primalis*) are deep dwellers capable of living at, or below, the thermocline. The ecological preferences of these species suggest that the most profound changes in Late Pliocene faunas occurred in response to changes in the intermediate and subthermocline environments. This general pattern towards colonization of deeper habitats is reinforced by the few faunal changes which have occurred since the beginning of the Pleistocene. In the low latitudes, speciation has been restricted to the *Pulleniatina* and *Gr. menardii* lineages, but deep-dwelling species such as *Gr. truncatulinoides* and *Globorotalia hirsuta* have expanded their biogeographical ranges into the mid latitudes (Weaver & Clement, 1986). Changes within the surface dwelling tropical–subtropical populations are quite suppressed in comparison; no new taxa appear following the expansion of the Northern Hemisphere glaciation. However, there was a compensatory and progressive increase in the dominance of *Globigerinoides ruber* in the interglacial periods following the decline in abundance of *Gs. extremus* and *Gg. decoraperta* (Fig. 6.3).

6.5   CONCLUSIONS

The progressive intensification of Northern Hemisphere glaciations during the Late Pliocene and the inception of large-scale ice-sheets signal a fundamental shift in the global climate system. Moreover, the development of strongly contrasting glacial–interglacial climatic cycles since 2.5 Ma mark the establishment of Milankovitch-driven ice volume variations as the dominant mechanism controlling global climate. High resolution faunal records obtained from Late Pliocene deep sea sediments show that this enhanced climatic variability coincided with a period of dramatic upheaval within North Atlantic planktonic foraminiferal populations. Surface and deep water circulation patterns appear to have been particularly sensitive to Late Pliocene environmental changes, primarily because the high latitude North Atlantic was bordered by large regions that were influenced by the dynamic processes of ice-sheet growth and decay. Even in the eastern subtropical Atlantic, periodic increases in the biogeographical distribution of cold-tolerant species parallel the pattern of fluctuating ice volumes and highlight the widespread deterioration of climate.

The scale and frequency of these palaeoceanographic changes adversely affected some members of the planktonic foraminiferal community. Around 40% of all the species encountered in the North Atlantic between 3.2 Ma and 1.8 Ma became extinct. This striking decline in species numbers occurred irregularly over an interval of 800 ky rather than in a single extinction pulse, although attrition rates in the subpolar–temperate faunas were highest immediately after the three pronounced glacial events centred around 2.5 Ma (isotope stages 100, 98, and 96). The inferred depth habitats of the extinct taxa suggest that extinctions affected both surface and deeper water species, seemingly at random. These extinctions also affected both ecologically dominant taxa and the rarer elements of the fauna. Similarly, when latitudinal diversity gradients and population structures are taken into account, there is no clear evidence of preferential extinction in any one region. Late Pliocene extinction patterns within the North Atlantic are therefore best characterized as pervasive, because planktonic foraminifera from all the major biogeographical realms and depth habitats were affected. It is also clear from the nature and sequence of events that the extinctions did not result from adverse interactions with newly evolved species or the arrival of opportunist immigrants; rather they were related to abiotic changes within the environment which presumably exceeded critical ecological thresholds.

An impoverished fauna with low levels of faunal diversity persisted for several hundred thousand years. Although the timing and pattern of the subsequent faunal diversification was different for each biogeographical province, one consistent trend does emerge: after 2.3 Ma faunal diversities were periodically augmented by the arrival of pre-existing migrant taxa from the Indo-Pacific province. The Late Pliocene faunal diversification in the North Atlantic therefore was not due to any accelerated or climatically induced speciation. It is uncertain exactly why the time delay associated with these migrations should have varied between 150 ky and 1.4 my. This delayed faunal response may be partly due to the emergence of the Panama Isthmus and the barrier that this imposed to species dispersal in the equatorial region. Complex changes in the structure and physical properties of the surface ocean may also have been necessary preconditions for the successful establishment of these species. It seems probable, however, that the timing of arrival of migrant species in the Atlantic was largely controlled by the pattern of climatic events and the impact of allied palaeoceanographic changes on the spatial distribution and abundance of these taxa.

In summary, climatically vulnerable species progressively disappeared from North Atlantic planktonic foraminiferal faunas during the Late Pliocene and were replaced by migrant species which could tolerate the climatic fluctuations that typify the Late Cenozoic ice age world. The importance and the extent of the biotic reorganization undergone during the Late Pliocene and the role that this played in the evolution of modern planktonic foraminiferal faunas is clearly emphasized by the fact that no comparable transformation of the planktonic foraminiferal populations accompanied the mid-Pleistocene climatic revolution and the further accentuation of glacial–interglacial climatic variability that has occurred over the last 1 million years. As a result, the faunal associations and biogeographical ranges of species found in the latest Pliocene are essentially the same as those observed in the modern ocean.

7

The response of Cretaceous cephalopods to global change

PETER F. RAWSON

7.1 INTRODUCTION

The Cephalopoda are an exclusively marine class of the Phylum Mollusca
with a history extending back to Cambrian times. Their fossil record is remarkable
for both its abundance and the diversity of shell form displayed. Thus the cephalopods can be divided into three informal groups, the nautiloids, ammonoids and
coleoids. These were formerly regarded as three orders or subclasses, but the nautiloids include a very diverse array of forms that nowadays are divided into several
orders. The nautiloid cephalopods reached their peak of diversity during Early
Palaeozoic times and by the beginning of the Mesozoic most groups had long
become extinct (Flower, 1988, fig. 1). Conversely, the Ammonoidea were a Late
Palaeozoic to Mesozoic group while the Coleoidea are predominantly a Mesozoic
to Recent subclass.

This chapter firstly reviews briefly the main changes in cephalopod distribution over the last 145 million years, in particular showing their varying response to
the K/T boundary event. It goes on to concentrate on the Cretaceous interval (145–
65 Ma), when two cephalopod groups, the ammonites and the belemnites, were
among the most common marine macrofossils. Both groups appear to have been
very sensitive to several parameters of global change. Thus, interpreting their ecology and determining the changing patterns in their distribution through space and
time provides fundamental clues to the nature of Cretaceous palaeogeography, climate and oceanography. The influence of sea-level change and ocean currents is
highlighted.

7.2 THE LAST 145 MA: AN OVERVIEW

The Cretaceous to Recent history of the Cephalopoda shows dramatic
changes in the relative abundance of different types. During the Cretaceous, the
ammonites (Ammonoidea) and belemnites (Coleoidea) were the dominant cephalopods. Other coleoids were apparently less common (though the early squids, cuttlefish and octopods were not easily fossilized, so their record is very patchy), while
nautiloids were represented only by the Suborder Nautilina. Then, in late Cretaceous
times, both the ammonites and belemnites began to decline in diversity, though
remaining at least locally common almost to the end of the Cretaceous.

A similar decline in the Ammonoidea had occurred at least three times in their earlier history (House, 1993), each time followed by a new radiation. But the end Cretaceous event was different. Both the Ammonoidea and the belemnites became extinct – though other coleoid groups, together with the Nautilina, survived the extinction event apparently unscathed.

Although the teleost fish probably took over many of the biotopes vacated by the ammonites and belemnites, other coleoids continued to thrive and indeed diversified considerably at suprageneric level through the Cenozoic (Doyle *et al.*, 1994, fig. 1). There is thus a very rich present-day coleoid record, in which the squids, cuttlefish and octopods have successfully colonized both shallow shelves and oceanic areas, from equatorial to polar waters. Conversely, the diversity and area of distribution of the morphologically simpler and more conservative Nautilina diminished through the Cenozoic until, at the present day, they are represented by a single genus, *Nautilus*. This is now limited to the south-west Pacific and parts of the Indian Ocean.

7.3   CRETACEOUS CEPHALOPOD PALAEOECOLOGY

Among the cephalopods, the ammonites are the group that have always attracted most attention because of their diversity, abundance, rapid evolution and, not least, the aesthetic appeal of their shells. They have provided a relative timescale for the Mesozoic that has yet to be surpassed, reflecting an almost unparalleled speed of evolution.

In general, ammonites occupied a wide variety of environments, from deep water to shallow shelf seas, and at least some could tolerate departures from waters of normal marine salinity. How far they occupied the open ocean is a matter for speculation – only a limited number of DSDP/ODP holes have yielded ammonites, and those records are of sediments deposited quite near to the shorelines of the opening Atlantic. However, other holes have yielded aptychi (ammonite jaw parts) representing two important, long-lived groups, the phylloceratids and lytoceratids (Westermann, 1996). This supports the general belief that these showed a strong preference for the open oceanic waters of Tethys – though they sometimes migrated to shallower water areas, as in the Early Cretaceous of East Africa. Another group, the Haplocerataceae, probably shared a similar habitat (Rawson, 1981; Wright, 1996, p. 9). Thus these three groups may be the most 'oceanic' of the ammonite groups. They also include some of the most widely dispersed of Cretaceous taxa.

The potential for dispersal is clearly dependent on the lifestyle of the organism. It is generally agreed that Ammonoidea had a planktonic early growth stage, lasting several weeks to a few months (Landman *et al.*, 1996). But the modes of life of many adult, normally coiled ammonites remain uncertain. Although they are often regarded as free-swimming nektonic or nekto-benthonic predators, an alternative interpretation that has become increasingly popular is that many were probably sluggishly moving, nekto-benthonic organisms – possibly scooping up food from the sea floor rather than actively predating. However, in the latest, very comprehensive review, Westermann (1996) argued that many ammonoids were pelagic, 'divided about equally among swimmers, drifters and vertical migrants'. While those taxa limited to epicontinental shelf seas may have been dispersed only locally, neritic or open ocean species could have been very widely distributed.

In latest Jurassic times a group of uncoiling (heteromorph) forms evolved, and through the Cretaceous these formed a major addition to the ammonite record. Their varied, sometimes bizarre, shell shapes indicate adaptations to a wide variety of more specialized environments, from planktonic to benthonic. Some forms achieved a remarkably wide distribution, outdistancing many of the normally coiled forms.

The belemnites were apparently nektonic, probably living mainly in schools on the continental shelf rather like many modern squids. However, some may have been adapted to a more oceanic environment (Doyle, 1992).

7.4 CRETACEOUS CEPHALOPOD PALAEOBIOGEOGRAPHY

It is over 100 years since Neumayr (1883) first recognized that globally, two Mesozoic faunal realms could be distinguished on the basis of ammonite distributions – the Boreal (northern) and Tethyan Realms. Since then there has been much debate on whether other realms could be distinguished and, in particular, whether there was an Austral (southern) Realm to 'counterbalance' the Boreal one. The ammonite evidence remains equivocal, but the belemnites (and several microfossil groups) indicate that an Austral Realm was developing during the late Early Cretaceous. But the sharpest distinction remains that between the Boreal and Tethyan realms.

In the Northern Hemisphere, the palaeobiogeographical pattern that characterized much of the Early Cretaceous was inherited from the Jurassic. Differentiation into Boreal and Tethyan Realms became marked early in Mid Jurassic times and continued to about the end of the Barremian (Rawson, 1981). Through the Early Cretaceous (Berriasian–Barremian) interval, Tethyan areas were characterized by very diverse faunas at both family and generic level. There was a marked decrease in diversity northwards, and the more northerly parts of the Boreal Realm (the Boreal or Arctic Province) were characterised by a single ammonite superfamily, the Perisphinctaceae, and by the belemnite families Cylindroteuthidae (Mid Jurassic to earliest Barremian) and their replacement, the Oxyteuthidae (Barremian).

The boundary between the two realms was quite sharply defined (Fig. 7.1), though it oscillated latitudinally and there was some overlap of faunas along marginal areas. Northward migration ('Tethyan spread') of belemnites was particularly noticeable in the Hauterivian, when *Hibolithes* penetrated as far north as Svalbard in the Boreal Province (Doyle & Kelly, 1988). Ammonite exchanges were more frequent but often short-lived; their significance is discussed in Section 7.5 below.

Both ammonite and belemnite diversity generally decreased again southwards from mid-latitudes but the taxa were all Tethyan or Tethyan-derived local endemics. Although some provinciation is apparent (Rawson, 1981; Doyle, 1992; Page, 1996) there is no evidence of a distinctive Austral Realm, though this may be because, as in the Jurassic, the most southerly preserved faunas were still a significant distance from the pole, possibly beyond the range of distribution of any Austral fauna.

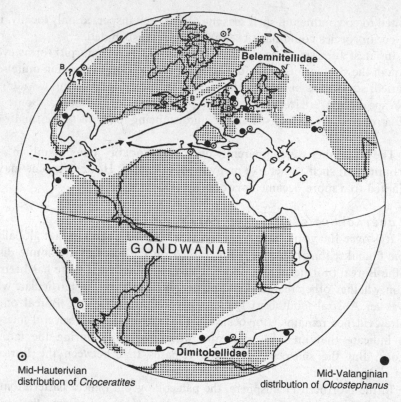

Fig. 7.1. Valanginian global palaeogeography and selected aspects of Cretaceous biogeography. Map modified from Rawson (1993, Fig. 10.1); position of continents drawn from the Cambridge Paleomap Atlas Programme (1990), perspective projection, 130 Ma. B—T indicates the approximate boundary of Boreal and Tethyan Realms in mid Valanginian times. The solid arrows indicate conjectural migration routes of phylloceratids and lytoceratids to East Greenland (Early Valanginian) and of *Heteroceras* to the North Sea Basin (Late Barremian). The dash-and-dot arrow indicates the possible origin of *Aconeceras*, which migrated to the North Sea Basin at the same time as *Heteroceras*. ? indicates that the age of *Olcostephanus* in California and *Crioceratites* in Arctic Canada is uncertain.

During the mid Cretaceous this pattern began to change. The distinctively boreal ammonite and belemnite families died out by early Aptian times, and in the few high latitude areas where mid Cretaceous cephalopods occur they are derived from Tethys. At times these gave rise, in turn, to genera that spread more widely southwards, such as the Albian hoplitids, but, although these are sometimes called 'boreal', they extended far beyond the confines of the former Boreal Realm to reach the marginal shelves of Tethys.

In Tethys ammonites continued to thrive but belemnites started to decline and by the end of the Cenomanian had vanished from the area. In contrast, to the south of Tethys a new belemnite family, the Dimitobelidae, evolved during the Aptian to form a distinctive Austral fauna distributed around Gondwana (Doyle, 1992). The associated ammonites were still of Tethyan type.

With the continuing spread of shelf seas over so much of the continental crust as sea levels reached their Mesozoic peak during the Late Cretaceous, many ammonites achieved a very wide, sometimes global, distribution. Nevertheless, although there is no clearly defined Boreal Realm, some genera still remained limited to high to mid-latitudes in the Northern Hemisphere, for example the Turonian collignoceratids (Wiedmann, 1988). Such genera showed a southerly limit comparable to that of the Albian hoplitids.

In contrast, the later Cretaceous belemnites exhibited a markedly bipolar distribution (Doyle, 1992). In the Northern Hemisphere, the Belemnitellidae were distributed over much the same area as some of the 'Boreal' ammonites, extending as far south as southern Europe and Texas, while the Dimitobelidae continued to distinguish an Austral Realm (Fig. 7.1). There were no belemnites in the intervening mid-latitudes. This pattern of distribution continued to the Maastrichtian, when both groups apparently became extinct.

Superimposed on this broad, gradually evolving biogeographical pattern were a series of short-lived, episodic migrations of faunas from one biogeographical area to another. This sometimes resulted in particular genera spreading over much of the world for a brief period of time – thus providing important markers for inter-regional correlation. Often these genera became extinct quickly outside their area of origin, but in other cases they evolved in a particular basin by allopatric speciation, to form local endemic populations.

7.5   MAJOR CONTROLS ON CRETACEOUS CEPHALOPOD DISTRIBUTIONS

Of the various parameters of global change that may have influenced the distribution and evolution of Cretaceous cephalopods, three appear to have played a particularly significant role: their varying effects are discussed here.

7.5.1   Plate tectonics and palaeogeography

The break-up of Gondwana started in the Early Cretaceous and some of the fragments began to drift towards the South Pole. This is reflected in the belemnites (and other fossil groups) by the development of a distinct Austral Realm during the Aptian. The opening of the Central Atlantic started late in the Jurassic and rifting spread northwards and southwards during the Cretaceous. A North Atlantic seaway had developed well before the breakup and provided a route of migration for ammonites from the Caribbean area to north-west Europe and East Greenland (see Section 7.5.3 below). By late in the Cretaceous there was a sizeable ocean between North America and Europe, which facilitated North–South migration of some ammonite taxa. The South Atlantic seaway began to open during the mid Cretaceous, allowing European ammonites to start migrating into the South Atlantic basins during the Albian (Wiedmann, 1988).

7.5.2   Climate

Climate has played a major role in the later Cenozoic, where temperature, seasonality and diurnal illumination all play a part. It is reflected in the latitudinal distribution patterns of many present-day organisms, with a marked bipolarity in the distribution of marine taxa. The Boreal/Tethyan latitudinal distribution of many

Cretaceous taxa, the development of a bipolar belemnite distribution from mid Cretaceous times and the general decrease in taxonomic diversity from Equator to poles all suggest some climatic control on cephalopod distribution at a global scale. This is supported by evidence from other fossil groups too. But the very broad belt of Tethyan faunas indicates a wider temperate to tropical belt than at the present day and there is no evidence of significant polar ice. Nevertheless, many authors have suggested that boreal faunas were cooler-water ones. If so, then the oscillation of the Boreal/Tethyan boundary through Early Cretaceous time may reflect shifting climatic belts as well as changing geographies.

7.5.3   Palaeoceanography

Two palaeoceanographical factors had a major influence on the evolution and distribution of Cretaceous cephalopods. One was sea-level change at varying scales, the other the development of new ocean current systems as Gondwana began to split and the Atlantic started to open.

Changes in sea level affected the distribution of cephalopods on two different scales. Long-term change (2nd order cyclicity) was a significant control on the overall evolution of the main biogeographical areas. The development of a clear Boreal/Tethyan dichotomy in the Mid Jurassic apparently reflected a slight fall in sea level coupled with uplift in the North Sea area, which caused near isolation of the Arctic seas. Sea levels rose again later in the Jurassic but fell markedly across the Jurassic–Cretaceous boundary. The fall severely restricted the movement of ammonites between Boreal and Tethyan areas at that time, and even though sea levels began to rise again early in the Cretaceous the geographical connections between the two realms were limited and their boundaries generally remained tightly defined. Conversely, the more open connections between Tethys and some of the basins fringing Gondwana allowed a more frequent interchange of faunas between those areas, so that differences between Tethyan and more southerly faunas were usually at a lower taxonomic level.

The breakdown of the sharp Boreal/Tethyan dichotomy began when the great mid Cretaceous sea-level rises and associated plate movements started. Although the Belemnitellidae and some ammonite genera remained limited to mid- to high latitudes in the Northern Hemisphere never again was there such a sharply defined Boreal Realm.

There is also a clear link between these long-term sea-level changes and ammonite diversity, high diversity coinciding closely with high sea level (Fig. 7.2: House, 1993; Rawson, 1993).

On a shorter timescale, there were periodic, rapid worldwide rises in sea level through the Cretaceous at about 2–3 million year intervals. This is at a frequency comparable to the third-order sequences of some sequence stratigraphers. They had a profound effect on the ammonite faunas at a variety of scales.

- At a global scale they led to the rapid spread of particular genera across much of the globe. Examples are discussed below. At some levels there was also a global turnover at family/subfamily level.

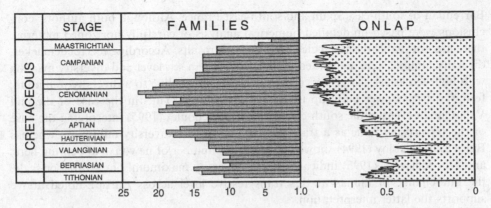

Fig. 7.2. Relationship between abundance of families and sea-level change (modified from House 1993, Fig. 2.4).

- Regionally they facilitated the interchange of taxa between adjacent biogeographical regions.
- In individual basins they have been linked to important turnovers in the local ammonite faunas.

Three important horizons of rapid sea-level rise are discussed below to exemplify their varying influences:

The mid Valanginian event

A mid Valanginian sea-level rise is evident in many parts of the world and especially well documented in Europe and Argentina. Although it had only a very limited effect on global ammonite turnover at family/subfamily level, it markedly influenced the distribution and evolution of many genera (Rawson, 1993). The event was marked initially by the spread and 'bloom' of *Olcostephanus*, which achieved an extraordinarily wide and abundant distribution at this time (Fig. 7.1). Immediately afterwards (at the base of the *Saynoceras verrucosum* Zone of the Mediterranean Region) some short-ranged ammonite genera, notably *Valanginites*, *Saynoceras*, *Karakaschiceras* and *Neohoploceras*, spread over much of the Tethyan Realm and into the West European Province of the Boreal Realm (Kemper *et al.*, 1981; Rawson, 1993). At the same time some boreal taxa extended southwards across Europe into the northern margin of Tethys.

This was one of the levels examined by Reboulet (1995) in his detailed studies on the Valanginian–Lower Hauterivian of SE France. Over this time interval he demonstrated a series of 'renewals' of the ammonite faunas, each marked by a period of evolution of new taxa, followed by a phase of 'disappearance' (local extinction). The renewal phase was associated with lowstand and the extinction with the transgressive phase of a third-order sequence. Evolutionary stasis was reached during the maximum flooding phase, when the ammonite fauna became impoverished and of low diversity. Hoedemaeker (1995) was looking at a similar scale of event when he highlighted seven turnover levels through the Berriasian to

Barremian of south-east Spain and south-east France. Although both authors' conclusions were based on detailed numerical analysis of carefully horizoned material, there are noticeable discrepancies between their results. According to Hoedemaeker, diversity minima were associated with a rapid fall in sea level and diversity maxima with high sea levels – after which there was a rapid decline to a diversity minimum, followed by a slower build-up to the next maximum. Thus in the case of the mid Valanginian event in the south-east of France, Reboulet (1995) interpreted the base of the *verrucosum* Zone as a transgressive high with a diversity minimum, whereas Bulot and Thieuloy (1994) showed a significant number of new taxa appearing here, and Hoedemaeker (1995) indicated it as a diversity maximum! On a broader scale, the interchange of faunas at this transgressive level across Europe noted above supports the latter interpretation.

The mid Hauterivian event

A mid Hauterivian rise in sea level is equally well documented, and triggered a significant turnover at family/subfamily level (Fig. 7.3: Rawson, 1993). In Europe, there was an important influx of West Mediterranean faunas into the West European Province (Bulot *et al.*, 1993, Rawson, 1995). The same event is recognized as far afield as the Neuquén Basin in Argentina, where *Spitidiscus* and *Crioceratites* closely comparable with European forms appear in black shales immediately above a non-marine sandstone (Aguirre Urreta & Rawson, 1997). The same event was apparently responsible for the spread of *Crioceratites* across much of the world at this time (Fig. 7.1); earlier records of this genus have only been proved from Europe.

The late Barremian event

Another important sea-level rise has been recognized recently in various parts of Europe and dated to early in the *giraudi* Zone in south-east France (Arnaud-Vanneau & Arnaud, 1990). Rawson (1995) has suggested that it was responsible for a flood of *Aconeceras* into north-west Europe. On a global scale the same event appears to have carried the short-lived heteromorph genera *Heteroceras* and *Colchidites* to many parts of the world. Increasing awareness of the widespread influence of this event suggests that it should be regarded as the first of the great 'mid Cretaceous' sea-level rises that accelerated during the Aptian and Albian to flood so many continental areas.

In summary, a sea-level rise may simply facilitate the interchange of taxa between adjacent areas, but it may also lead to the global spread of particular genera. However, I have already indicated (Rawson, 1973) that, while the main migrations took place at such levels, the connecting seaways often remained open during the intervening time intervals. Thus changes in sea level cannot be the sole control on the migrations. Rawson (1973), Kemper and Wiedenroth (1987) and Mutterlose (1992b) have all suggested that climate was important, boreal spreads indicating cooler water influxes to adjacent parts of Tethys, Tethyan spreads warmer water influxes to the Boreal area. Reboulet and Atrops (1995) linked climate with sea level when they related the brief appearance of boreal *Dichotomites* at three distinct levels in the Upper Valanginian of south-east France to a cooler water influx during

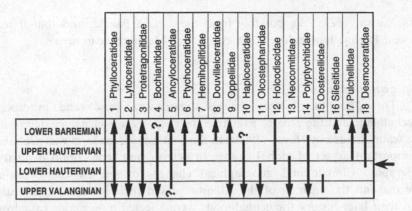

Fig. 7.3. Influence of the mid Hauterivian event (arrowed) on global turnover of ammonite families/subfamilies (modified and updated from Rawson, 1993, Fig. 10.2).

sea-level highs in third-order cycles, whereas the immigration of Tethyan forms from Spain was linked to warmer influxes during sea-level lows.

Although the rapid long-distance migration of particular taxa can so often be linked with a particular sea-level rise, longer-term migratory trends may also reflect the development of a particular seaway and associated ocean currents. The opening of the Atlantic during the Early Cretaceous provides an interesting example of how the careful analysis of ammonite distributions can provide evidence for deducing the existence of a connection and the direction of migration.

Rifting probably started in the Caribbean area in late Jurassic times, and by early in the Cretaceous there was an extensive 'western Tethys' – an area of predominantly shallow marine water extending eastwards and southwards from Mexico, where rich Lower Cretaceous ammonite faunas are preserved. Further to the north-east, rifting only started in the northernmost Atlantic in mid to later Cretaceous times. But, there is faunal evidence in East Greenland and north-west Europe for the existence of a seaway linking western Tethys with the Arctic from very early in the Cretaceous. Ager (1971) first noted the possibility when explaining the occurrence of a single pygopid (Tethyan) brachiopod in the Valanginian sequence of Traill Island, East Greenland. Rawson (1973, 1981) showed that, in the same succession, there is a distinctive Tethyan pelagic element in the ammonite faunas, with a much higher proportion of *Phylloceras* and lytoceratids than has ever been recorded in north-west Europe – the alternative migration route for these forms. A 'proto-Gulf Stream' would be the most likely route (Fig. 7.1).

More recently, a distinctive, short-lived but widely distributed hetermorph ammonite, *Heteroceras*, was discovered in the Upper Barremian of eastern England. At that time the North Sea Basin was cut off from Tethyan Europe through closure of the Polish Furrow. Accompanying this form are abundant *Aconeceras*, which is not known in other parts of Europe (the Mediterranean province) until Late Aptian times. Rawson (1995) thus deduced that *Aconeceras* too must have come through an

Atlantic seaway (Fig. 7.1), possibly from the eastern Pacific, and that it reached north-west Europe long before it penetrated the Mediterranean area.

7.6   CONCLUSIONS

Throughout their history, the cephalopods waxed and waned in response to sea-level change and the area of available shelf sea, as was elegantly demonstrated for the Ammonoidea by House (1993). In the Cretaceous, sea level was probably the most important aspect of global change in determining cephalopod diversity and biogeography. Climate and geographical changes caused by the break-up of Gondwana and the opening of the Atlantic played their roles too. And, at least once in their later history the cephalopods were affected by a major catastrophe – the apparent impact at the end of the Cretaceous (Gale, this volume; Pickering, this volume) saw the final demise of the ammonites and belemnites, while the nautiloids apparently survived unscathed.

8

Global change and the fossil fish record: the relevance of systematics

PETER FOREY

8.1   INTRODUCTION

The modern fish fauna is dominated by two very different kinds of fishes, the neoselachians (sharks, skates and rays) and teleosts (e.g. herring, salmon, cod and perches). Both of these major groups originated before the Cretaceous but the age of differentiation (Hennig, 1966) of teleosts is approximately coincident with the temporal limit of this volume – the Late Jurassic/Early Cretaceous – and that of neoselachians is Early Jurassic. The last 145 million years have seen the diversification of neoselachians, resulting in a modern tally of 815 species contained in 164 genera and 42 families. In the same timespan teleosts have shown more impressive diversification to 23 637 species within 4061 genera and 426 families (counts taken from Nelson, 1994). Other, minor groups of fishes inhabiting the modern world include the jawless hagfishes and lampreys (73 species), ratfishes or chimaeriforms (30 species), polypteriforms or bichirs (11 species), lepisosteiforms or garpikes (7 species), amiiforms or bowfins (1 species), the coelacanth and the lungfishes (6 species). Together these make up less than 0.5% of Recent species diversity.

The brief given to contributing authors of this volume includes an invitation to outline the geological history of individual groups with reference to global changes which have taken place. This is rather daunting as it is all too easy to slip into scenario building, assuming cause and effect with little substantive evidence. Such evidence includes an assessment of the nature of the fossil record. In order to associate fish evolution with global change, we must assure ourselves that this fossil record can document singular events which may be matched with geological theories of earth history. Those singular events are phylogenetic histories which must be approached through systematics. We may also think of fragmentations and collisions of landmasses and continental blocks as a geological phylogeny (Young, 1986). A recurrent theme throughout this chapter leading to a single message is that any speculation about the influence of global change and fish evolution requires phylogenetic hypotheses as the starting point.

8.2   THE NATURE OF THE FOSSIL FISH RECORD

The fossil record of both neoselachians and teleosts is relatively rich compared with that of other vertebrate groups and some idea of the diversification is given in Fig. 8.1. The teleost record shows very clearly the effect of Lagerstätten, as mentioned by Patterson (1993a) and Patterson and Smith (1989). Those authors recorded at least two peaks of teleost diversity, one in the Campanian and another in the Eocene. This shows up most obviously in Fig. 8.1(a), for instance, as increases in numbers of families against Elopomorpha and Acanthomorpha and, in the Campanian, as first family occurences for Otophysi, Esociformes and Salmoniformes. The first peak is due to a diverse fauna occurring at Sedenhorst, Westphalia. The second is due primarily to the occurrence of the famous fish-bearing deposits of Monte Bolca, Italy and the London Clay of the Anglo-Paris Basin. This is set against the relative paucity of Paleocene fish localities below and Late Eocene localities above. A little closer look at the Eocene teleost record is even more revealing as to how estimations of standing diversity (and implied extinctions/originations) are sensitive to 'fortuitous localities'. This is more clearly demonstrated by considering origins of groups rather than extinctions.

Taking the Ypresian (Early Eocene) first, then 61 families originated within this time interval. Of these, 23/61 are found in the London Clay, 29/61 are found (as otoliths) in the Argile de Gan, Gan, Pyrenees (6/61 are found in both the London Clay and the Argile de Gan). So 75% of the 'origins' are accounted for by the occurrence of two deposits. Within the Mid Eocene (Lutetian) the situation is no better. Here, representatives of a further 55 families are seen for the first time. But of these 29/55 (52%) are found at Monte Bolca, Italy and 10/55 (18%) are found (as otoliths) in the Calcaire Grossier, France.

Such number-crunching may seem adversarial to discussions of patterns and causes of diversification and we might ask the question: what patterns would we infer if we had no knowledge of the limestones of Monte Bolca or the London Clay? However, the record is there so what may we make of it?

The inferred history of neoselachians and teleosts shares similarities but also some differences. For neoselachians, representatives of all modern orders are present in earliest Cretaceous times and some assumed primitive members of modern groups occur in the Early and Mid Jurassic (members of Heterodontidae, Brachaeluridae, Hexanchidae, Squatinidae and Rhinobatidae). This implies that the age of differentiation was considerably earlier than that of teleosts (see below) and that, if the phylogeny shown in Fig. 8.1(b) is correct, this must have occurred in the Late Triassic (but see below).

For teleosts, the record as judged against our current ideas of phylogeny shows a different pattern. First, very few members of modern teleost groups are known prior to the Cretaceous. The exceptions are primitive osteoglossomorphs, primitive elopomorphs and one stem species belonging to the Clupeocephala – and all these are found in the Late Jurassic (Fig. 8.1(a)). Secondly, there are several extinct stem teleost taxa (Patterson, 1993a, 1994) which occupy the span of time between the age of origin (Carnian – 225 Ma) and the age of differentiation (Kimmeridgian – 150 Ma). These are shown as the box numbered '6' in Fig. 8.1(a). Thirdly, different terminal taxa shown in Fig. 8.1(a) appear at different

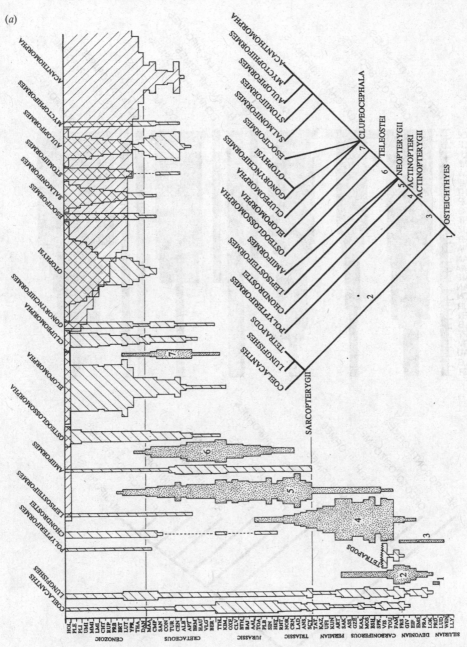

(a)

(*cont. on page 110*)

Fig. 8.1. Plots of fish families through time. (*a*) Plot of family diversity of bony fishes with phylogeny as inset (taken from Patterson, 1994, with permission). Only the clade Teleostei is relevant to this chapter where it is seen that the age of appearance of the major groups closely matches the phylogenetic sequence. The stippled boxes represent stem 'groups': that is, paraphyletic assemblages of taxa which occupy internodes on the phylogeny of Recent Osteichthyes. The number within each box corresponds with the numbers in the phylogeny and denotes the taxonomic rank. The stippled box numbered '6' shows stem-teleosts (that is, those fossil teleosts which are plesiomorphic to modern members and which occupied the time between the age of origin and the age of differentiation of the modern members).

(b)

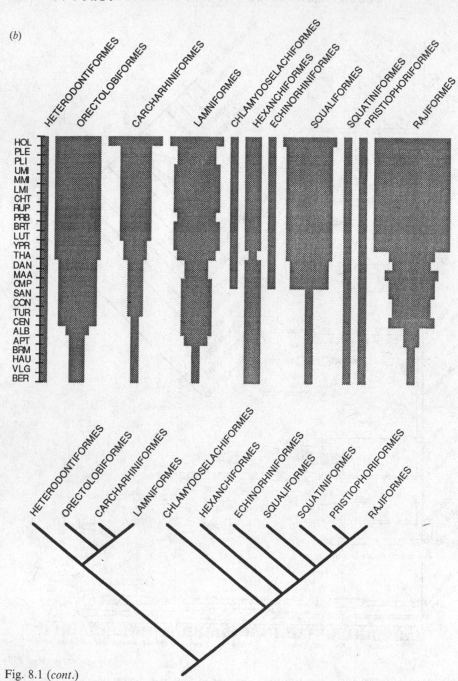

Fig. 8.1 (cont.)

(b) Plot of neoselachian families, plotted back to the beginning of the Cretaceous with phylogeny below taken from Shirai (1996) and Carvahlo (1966) with assignments of fossil taxa after Capetta et al. (1993). The width of the narrowest bar equals one family. For the neoselachian plot it should be noted that fossil members of the heterodontiforms, orectolobiforms and rajiforms are known from the Toarcian and hexanchiforms from the Sinemurian, meaning that the age of differentiation was probably Late Triassic. The abbreviations down the left-hand side are the standard stage abbreviations as used in *The Fossil Record 2* (Benton, 1993). See text for discussion.

times and, with the possible exception of Esociformes and Salmoniformes, there is compatibility between the phylogeny and the stratigraphic record of first occurrences.

These differences may melt away should our ideas concerning the higher-level phylogeny of neoselachians and/or teleosts change radically. I doubt that this will happen for teleosts. For neoselachians there is some general agreement between ideas of the ordinal relationships (Shirai, 1996; Carvahlo, 1996) – the chief differences concern alternative ideas on the relationships of chlamydoselachiforms and hexanchiforms. What is of greater concern here is the difficulty of assigning fossil sharks to modern groups. Most fossil sharks are known only from isolated teeth and taxonomic placement is notoriously difficult. Two examples serve to illustrate this point. The first concerns the identification of very large Miocene teeth ('megalodon') which are sometimes referred to the modern Great White genus *Carcharodon*, therefore implying relationship between the fossil and Recent forms, and sometimes to a different genus *Carcharocles*, implying only distant relationships to one another. Chief arguments for the first alternative stress similarity in size and the presence of serrations along the cutting edges. Chief arguments for the second alternative are centred on inferences of fossil lineages traced through the Cenozoic where the common possession of serrations is seen as accidental similarity of no phylogenetic significance. The latter view, which stresses the independent derivation of common feeding adaptations, reflected in tooth morphology, is commonly accepted by shark workers (Cappetta, 1987, p. 14), meaning that most of the visible features of shark teeth are disregarded by some but accepted as phylogenetically significant by others. The second example follows on from this. Turner and Young (1987) described a tooth of Early/Mid Devonian age and questionably referred to the modern group Hexanchiformes because of strong similarities on crown morphology. Accepting this assignment would place the age of differentiation of neoselachians in the Early Devonian, leaving an unexplained stratigraphic gap of some 200 Ma.

The correspondence between age and clade rank within the teleost record gives greater confidence that, even though the record is patchy and distorted by Lagerstätten, it is a representative sample of relative diversity through time. Comparison between the neoselachian and teleost records shows some similarities in the apparent diversification of family-level taxa in the Cenomanian and another more marked diversification in the Eocene. The Cenomanian rise is interesting to the extent that it marks the first appearance of acanthomorphs, recognized by Patterson (1993b) as those teleosts in which there are one or more spines in both dorsal and anal fins. Acanthomorphs (14 650 modern species) constitute about 60% of all modern teleosts. The Cenomanian also saw a rapid diversification of rajiform neoselachians (456 modern species) which today account for a little over half of modern neoselachians.

Patterson (1993b) reviewed the early history of acanthomorphs and pointed out that most of the 23 Cenomanian genera belong to the Polymixiformes and the Beryciformes – two relatively plesiomorphic members of the Acanthomorpha (Johnson & Patterson, 1993). Five of the remaining are members of the acanthomorph stem-lineage, leaving just three genera of doubtful affinity (see Patterson, 1993b, for possible relationships with acanthomorph subgroups). It appears that the congruence between clade rank and first stratigraphic appearance holds good here as well and that the appearance of Cenomanian acanthomorphs signifies a very brief period of acanthomorph history prior to this time. Absence of acanthomorphs from older Albian localities is probably a real phenomenon (Patterson, 1993b).

Our knowledge of the phylogeny of teleost fishes derived from study of comparative anatomy matched with stratigraphic first appearance suggests that phylogenetic hypotheses may be tests of the quality of taxonomic sampling of the fossil record. As far as the Cretaceous is concerned, the record seems to pass this test. On the other hand, the tremendous increase in the number of families in both neoselachians and teleosts in the Thanetian, Ypresian and Lutetian cannot be so neatly explained. First, there is the scale of increase (see above). Secondly, there is far less phylogenetic control for this time. Members of many of the Recent families appeared in the Late Paleocene/Early Eocene record. Many of these were members of the Percomorpha (by far the largest subgroup of acanthomorphs) and our idea of relationships between these families is still very imperfect (Patterson, 1993b: 29). Perhaps all we can say at present is that the modern appearance of so many morphologically disparate forms suggests a presently undiscovered history that must have extended back to the Late Cretaceous.

There are very few Maastrichtian and Danian fish deposits currently known and none which may be described as Lagerstätten. Parenthetically, I may note that there is no evidence within the fossil record to suggest an end Cretaceous extinction of genera or families of fishes that is any greater than normal turnover (Smith & Patterson, 1988; Patterson & Smith, 1989; Macleod et al., 1997).

8.3   IMPACT OF GLOBAL CHANGE ON FISH EVOLUTION

Probable factors affecting the pattern of fish evolution include plate tectonics and changing coastlines, transgressions/regressions, intracontinental tectonics leading to changes in drainage patterns and formation and evanescence of lakes, changing ocean circulation routes, expansion and contraction of oceanic anoxic layers and glaciations.

The most obvious manifestation of global change affecting fish evolution is geographic distribution. This is also the easiest to analyse without dipping into a large bag of assumptions. Biogeographic distribution is often said to have historical and ecological explanations. Ecological perturbation (e.g. glaciation, aridity) may have effects such as local extinction, or range extensions which become obvious over relatively short periods of time ($< 10^4$ yr). Historical explanations seek to explain disjunct distributions through vicariance and/or dispersal which have arisen over much longer periods of time ($> 10^6$ yr) and are usually linked to theories of plate movements, land bridges, island chains, etc. There is overlap between these two types of explanation and ecological change must have a history (even effects of a bolide impact have a brief history). But historical explanations are accessible through phylogenetic studies of organisms, whereas ecological explanations are not.

In the examples which follow, I illustrate through the work of others, the role that systematics plays in choosing between historical explanations. Some of these agree with traditional theories of Earth history, some do not. The use of systematics in choosing between historical explanations of distribution has flourished on the heels of panbiogeography (Croizat, 1958) and cladistic methods of systematics and vicariance biogeography (Nelson, 1985; Nelson & Platnick, 1981). Simply put, if systematic relationships of different taxa inhabiting presently disjunct areas are congruent this may provide evidence of a historical common cause of that disjunct distribution. Introduction of fossil representatives, as long as their relationships can be unambiguously assigned, can provide the time control necessary to link global events to organism history (Grande, 1985).

8.3.1   Africa–South American rift

The modern distribution of characoids and lepidosirenid lungfishes which occur in South America and Africa reflects the fragmentation of western Gondwana (Novacek & Marshall, 1976). Rifting between Africa and South America began in the earliest Cretaceous and separation was completed by the Santonian (Reyment & Dingle, 1987). Initial rifting formed a series of rift valley lakes which became inundated by the sea from the south as the southern tips of South America and Africa rotated away from each other. By the end of the Albian a seaway was established between the embryonic South Atlantic and Tethys to the north (Gale, this volume).

Similarity between the fossil fish faunas of the Neocomian of West Africa (Guinea and Gabon) and eastern Brazil is very marked and there are at least two fish groups which have matching transatlantic sister-group pairings (Fig. 8.2) suggesting that the original area of endemism was north-east Brazil and West Africa. The Brazilian and West African fishes referred to in Fig. 8.2 are found in freshwater deposits.

To some extent, the similarity in the fish fauna of Africa and north-east Brazil persisted into the Albian where some elements of the famous fish faunas of the Santana Formation of the Araripe Basin, Brazil (Maisey, 1991) are very similar to the fishes of the Kem Kem beds (?Albian–Cenomanian) of Morocco. Common fishes include the coelacanth *Mawsonia* and the amioid *Calamopleurus*. The Moroccan fish fauna remains to be described and it is uncertain if there are specific sister-group relationships as in the underlying strata.

However, there is some evidence to suggest that the overall faunal relationships had changed by the Mid Albian. In addition to the two forms mentioned above, the Santana Formation contains many other species which appear to have relationships elsewhere. It has yielded a very diverse fauna of some 24+ fish species (Maisey, 1991), some of which occur in the Sergipe–Alagoas Basin to the east and the Parnaiba Basin to the west (Maisey, 1991: figure on p.32). Perhaps more importantly there is a very strong resemblance between these fishes and a fauna of, as yet, undescribed fishes from the Middle Albian deposits of Tepexi, Mexico (Espinosa-Arrabarrena & Applegate, 1996). Within this Mexican fauna I have identified species of *Rhacolepis*, *Notelops*, *Parelops* and *Vinctifer* which, if not conspecific, are probably sister-species of the Brazilian fishes. There is doubt about the palaeoenvironment of the Santana Formation (Martill, 1993; Maisey, 1991) but the fish bearing sediments of the Tlayúa Formation of Mexico are clearly marine as they contain ammonites and echinoderms (Espinosa-Arrabarrena and Applegate, 1996). The similarity of the fish faunas between Mexico, the Parnaiba Basin, the Araripe Basin and perhaps the upper deposits in the Sergipe–Alagoas Basin suggests a

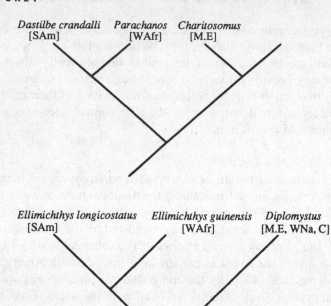

Fig. 8.2. African–South American rift. Sister-group relationships of two fish taxa suggest an immediate contiguous area of endemism which became separated in the mid-Albian. The top figure shows three gonorhynchiform taxa. The lower figure shows three clupeomorphs. By mid-Albian the relationships of taxa in South America and West Africa shows more widespread affinities. C = China, M.E = Middle East, WAfr = West Africa, SAm = South America, WNa = western North America. Most are freshwater fishes but *Charitosmus* and the Middle East occurence *Diplomystus* are found in marine deposits.

marine incursion from the northern Tethys which may well have separated Brazil and Africa by the Mid Albian.

The fishes from the Santana Formation (and presumably Mexico) show no special relationship with West Africa even though there are two taxa shared at generic level. I conclude that systematic study shows that marine incursions coincident with or immediately following initial separation of Africa and South America were sufficient to cause vicariance between the continents by Mid Albian times.

8.3.2   Eocene freshwater fishes of North America and Europe

The record of Eocene freshwater fishes of North America (Grande, 1989; Wilson & Williams, 1992) and Europe (Gaudant, 1993; Grande & Micklich, 1993) is quite extensive. The classic localities are the Green River Shales of Wyoming, Utah and Colorado in western North America (Grande, 1984) and Messel and Geisaltal in Germany (Schaal & Ziegler, 1992) but both areas are supplemented by surrounding fish-bearing deposits which suggest that their respective faunas are more widespread (Wilson & Williams, 1992; Gaudant, 1988).

The distribution of these fishes has prompted much discussion. A traditional explanation for apparent similarity between the faunas is that they originated in both North America and Europe and that there was free interchange between the two in the Eocene when land connections (the Thule Bridge [Europe–Greenland–Baffin Islands] and the de Geer route [Norway–Spitzbergen–Ellesmere Island]) were established. The assumed history of these fishes is tied to the traditional theory of mammal palaeobiogeography (Storch & Schaarschmidt, 1992) and supported by the fact that many modern North American freshwater fishes have their closest relatives in Western Europe. Gaudant (1988) has suggested that representatives of the most ancient lineages found at Messel (lepisosteids, amiids, ?notopteroids and umbrids) imply a European–North American connection sometime in the Late Cretaceous. However, from the point of view of reconstructed coastlines and land masses this would seem counterintuitive (Smith *et al.*, 1995).

Gaudant reached his conclusion by acknowledging the gross overall similarity between fishes found at Messel and Green River. However, a more rigorous and phylogenetic approach taken by Grande (1985, 1994) suggests that many of the fish species in the Green River fauna show sister-taxa in east Asia (specifically China) and/or Indonesia, a view which is supported by Wilson and Williams (1992). Of the 15+ species of teleosts occurring at Green River, five show immediate transpacific relationships (Grande, 1994); the remainder are either confined to North America or systematic relationships are too imprecisely determined to be informative. However, the fact that there are five congruent patterns of relationships (Fig. 8.3) (and our knowledge of the others does not contradict them) appears to be strong evidence that the phase of Earth history responsible for the appearance of Green River fishes involved continuity between western North America and eastern Asia, rather than between North America and Europe. Our current understanding of palaeocoastlines is more favourable to this view since the mid-continental seaway separated western North America (including land which the Green River Lakes later occupied) from eastern North America until the end of the Cretaceous. In terms of vicariance biogeography, these sister-taxa imply that the original area of endemism relevant to these particular fishes was eastern Asia plus western North America and that the causative event was the opening up of the North Pacific. Conventional palaeocontinental reconstructions (Smith *et al.*, 1994) show that no connection existed between eastern Asia and North America until latest Cretaceous or even Paleocene times. In terms of dispersal biogeography explanations, east Asia⇌North America dispersal could have taken place. However, dispersal is very much an opportunistic phenomenon and unlikely to have affected representatives of five independent groups of freshwater teleosts. The more phylogenetic relationships that we discover, and which show congruence with these five, will only serve to make the dispersal hypothesis more unlikely. Other geological explanations may be that Asia and North America were much closer in the Cretaceous than is commonly reconstructed (Owen, 1976), or that fragments of an ancestral continental block (Pacifica) became accreted to the Pacific margins of Asia, North and South America (Nur & Ben-Avraham, 1981). Such a view is favoured by Grande (1994). (See also Nelson

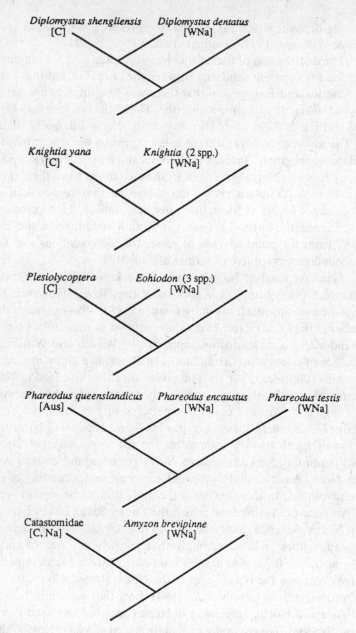

Fig. 8.3. Eocene freshwater fish relationships. Freshwater fishes of western North America show close affinities with those of east Asia rather than with Europe. Sister-group relationships of at least five different kinds of fishes suggest that western North America and east Asia were part of the same area of endemism. Data taken from Grande (1994), Chang and Chow (1986), Li (1994) and Li and Wilson (1996). Aus = Australia, C = China, Na = North America, WNa = western North America.

(1985) for a variety of organisms which show circum-Pacific distributions and Parenti (1991) for theories of Tethyan closure and the distribution of Indo-Pacific freshwater fishes.)

What of the other fishes (lepisosteids, amiids, notopteroids and umbrids) at Messel which led Gaudant (1988, 1993) to suggest a faunal continuity between North America and Europe? Representatives of all these higher taxa except the last are found at Green River. The Messel lepisosteid is *Atractosteus strausi* which, as Grande and Micklich (1993) pointed out, was regarded by Wiley (1976) as the most plesiomorphic member of the genus otherwise known from species found today in North and Middle America as well as by fossils in the Cretaceous of western North America and Africa. This very wide distribution of more derived species including representation in Africa speaks for a vicariant event which is much older than those affecting the teleosts discussed above. Wiley (1976) suggested that the break-up of Gondwana may have been responsible for the early distributional history of *Atractosteus* species. For the Messel amiid – *Cyclurus keheri* – we have insufficient systematic information (Grande & Micklich, 1993). In the latest revision of amiid fishes (Grande & Bemis, 1998) *C. keheri* is one of eight species in the genus *Cyclurus*. Grande and Bemis demonstrate the monophyly of the genus *Cyclurus* but are uncertain about the interrelationships of the species which, apart from European species, include one from the Paleocene of Siberia, one from the Upper Cretaceous and one from the Eocene of western North America. In the absence of a species phylogeny we can only speculate but in the light of the species track (western North America–Siberia–Europe) it is tempting to reject a transatlantic connection for these fishes.

For the umbrid (*Paleoesox fritzschei*) and the ?notopteroids (*Thuamaturus intermedius* and *T. spannuthi*) so little is known of the relationships that any implication for biogeographic hypotheses is premature. *Thaumaturus* is particularly enigmatic and has, at one time or another, been referred to salmonids, esocoids or notopteroids (Gaudant, 1988).

The purpose of reiterating discussions here of the relationship between the Eocene freshwater fish faunas is not to establish who is right and who is wrong. After all, both explanations might be wholly or partially correct or incorrect: and it is worth pointing out that we lack a good Early Tertiary freshwater fish record from eastern North America. Instead, I wish to draw attention to the relevance of understanding a detailed phylogeny of the species present against a time frame imposed by the fossils before we can speculate on the effect that global change may have had on distributions of freshwater fishes.

8.3.3 Intracontinental tectonism

Intracontinental tectonism has clearly had effects on the distributions and species compositions of freshwater fish faunas. For example, the formation and subsequent histories of the African Rift valley lakes, which range in age from 10–0.75 my, has been judged pivotal in explanations of the explosive speciation of cichlid fishes (Greenwood, 1974). Similarly the building of the Rocky Mountains and the Andes must have had profound effects on the evolution of the North and South American freshwater fishes. Precisely how such tectonic events contribute to explanations of modern distributions require some correlation of phylogenetic studies of organisms and geological terranes. Unfortunately very few studies have gone this far. Two authors have attempted to correlate species phylogenies with geological terrane phylogenies. Young (1986, 1990) has attempted this on a global scale for Devonian fishes and the landmasses they occupied. And Rosen (1976, 1978, 1985), in a series of papers which have now become classics in studies of vicariance biogeography, compared area phylogenies for organisms living in the Caribbean and adjacent coasts of North, Central and South America with a geological phylogeny of the Caribbean. Arguably, these studies may be said to involve oceanic plates rather than intracontinental deformation so I give a more limited but more pertinent example here taken from the works of Minckley *et al.* (1986) and Smith (1992). This example concerns the evolution of a group of freshwater fishes of the western United States.

The Catostomidae (suckers) are freshwater fishes inhabiting lakes and rivers throughout North America, northeast Siberia and China. The genus *Catostomus* consists of about 22 species, most of which occur in drainage systems west of the Rocky Mountains south to Mexico. The western United States has been subject to repeated periods of tectonism following the major Laramide orogeny involving relative motions of the many microplates and faults which make up the western US (see Minckley *et al.*, 1986, for a geological summary). A phylogeny of *Catostomus* spp. has been proposed by Smith and Koehn (1971) and Smith (1992), and for the more derived species (but not the more plesiomorphic members) the phylogenetic hypotheses agree very closely. The derived species, which are often placed taxonomically within a subgenus (*Pantosteus*) are ecologically distinct as hill-stream dwelling species as opposed to lake and larger river species. The concordance between phylogeny and tectonic history is very good. The phylogeny, current distributions and cartoons suggesting a vicariant history are given in Fig. 8.4. Since the present day distributions of *C. platyrhynchus* and *C. discobolus* overlap those of other species, then some dispersal following vicariance must have occurred. Such a study, correlating a single phylogeny with the geological history of the area needs to be tested against phylogenies of other groups of organisms inhabiting the same areas and I hope this will be done in the near future. However, the study by Minckley *et al.* (1986) does show the value of considering phylogeny in our attempt to recognize the influence of earth history on biotas. Parenthetically, it is worth noting that a study of lungless salamanders inhabiting the western United States utilized the congruence between phylogeny of organisms and geological terranes (Hendrickson, 1986). That study associated salamander vicariance to sequential northern movement of separate terranes which became accreted to the western US continental block throughout the Tertiary.

8.3.4   Ecological factors

There is no doubt that ecological changes have influenced the patterns of fish distribution but it is not so clear that they have caused speciation. Pleistocene glaciation was the endpoint of a gradual but intermittent global cooling throughout the Late Paleogene and Early Neogene and this cooling was coincident with a number of extinctions of fishes in northern latitudes (Cavender, 1986). The Pleistocene glaciation caused widespread extinction in the northern hemisphere

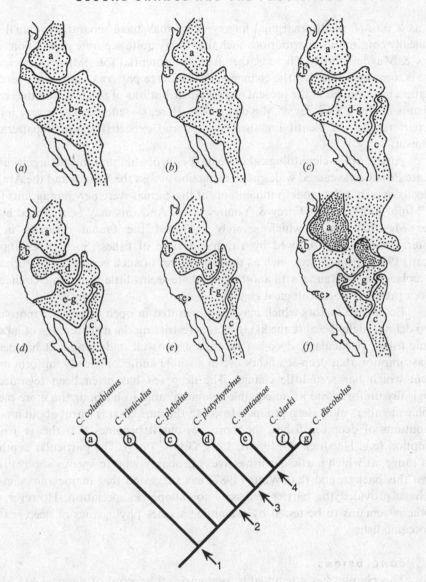

Fig. 8.4. Intracontinental tectonism. Biogeographical history of western North American hill-stream suckers of the genus *Catostomus* simplified and modified from the work of Smith (1992) and Minckley *et al.* (1986). Above: cartoons representing likely sequence of vicariant events indicated in diagrams (*a–f*) leading to the present distribution (*f*) of suckers. The phylogeny of seven species of catostomids (abbreviated as a–g) is given below. According to Minckley *et al.* (1986) some of the vicariance events (numbered 1–4) may be tied to specific geological events in the following way: 1 (Fig. 8.4(*a*)) = general uplift causing large alluvial fans (Tyee Formation) in the Late Eocene, 2 (Fig. 8.4(*c*)) = initiation of the Rio Grande Rift and Sierra Madre occidental volcanics (late Oligocene), 3 (Fig. 8.4(*d*)) = collapse of Great Basin Region (Oligo-Miocene), 4 (Fig. 8.4(*e*)) = development of southern margin of Colorado Plateau (early Miocene). To explain the present distribution it is necessary to hypothesise that *Catostomus platyrhynchus* (d) and *C. discobolus* (g) expanded their ranges by dispersal following a vicariance event. The phylogeny of *Catostomus* species on which this scenario is based is shown below (from Smith, 1992, some taxa omitted).

and, as it pushed south, marginal fingers of ice may have separated original geographically continuous distributions such that today quite separate populations exist (Wiley & Mayden, 1985). There is, therefore, the potential for allopatric speciation. There is some evidence that the change in the drainage patterns caused as a result of glaciation has resulted in the present disjunct distributions of fishes within the central highlands of the US (Wiley & Mayden, 1985). Here, systematics has a very important part to play in identifying sister groups and expected repeated patterns of endemism.

At the other climatological extreme it is probable that increasing aridity in the Late Pleistocene caused widespread extinction across the Sahara and the Arabian Peninsula and that the modern inhabitants of these areas were new immigrants from Asia Minor and Europe (Forey & Young, 1999). And this may be matched by the earlier Messinian crisis which severely restricted the faunal diversity in the Mediterranean, to be followed by a different suite of fishes (Sorbini & Tirapelle Rancan, 1979). In instances such as these the effect of such ecological perturbation is to replace the fish fauna with another and there seems little systematic connection between pre- and post-ecological change.

Ecological factors which may have operated in open oceans are more difficult to define chiefly because the likely factors restricting the distributions of modern oceanic fishes, particularly deep-sea fishes, are not well understood. It has been a tacit assumption that deep-sea fishes live in a stable and ecologically uniform environment which has seen little change. The deep sea has often been regarded as ecologically uniform and as an equable refugium, able to harbour the more plesiomorphic members of modern groups. However, the more that is learnt about modern distributions of deep sea fishes, the more obvious it becomes that this is a naive assumption (e.g. Haedrich & Merrett, 1988; Gibbs, 1985). The particular depth, or depth range, at which particular fishes live is probably vital to species survival. It is against this background that White (1987) has suggested that major anoxic events may have provided the barrier necessary for allopatric speciation. However, this hypothesis remains to be tested by examining species phylogenies of deep sea and mid-oceanic fishes.

8.4   CONCLUSIONS

This chapter has attempted to summarize the record of the past 145 Ma of the two dominant groups of modern fishes (neoselachians and teleosts) which account for well over half of all Recent vertebrate species. The record appears reasonably good at detecting the age of origin of the major groups of neoselachians and teleosts to the extent that there is good correspondence between the stratigraphic ages of clades and the sequence of phylogenetic relationships (clade ranks). The record, however, is patchy and shows patterns of extinction/origination which may be misled by occasional lagerstätten as well as by the fact that most species of fishes are found at a single locality. Such phenomena may hinder our attempts to correlate fish speciation and extinction patterns with global change. Global change must have had influence on the distribution and composition of fish faunas through time, but equations of cause and effect can usually only be offered in the most general terms.

At present the most significant contribution that the fish record can make to ideas of the effect of global change is by way of helping reconstruct phylogenies which deal with singular events affecting small parts of the world. It is possible to correlate theories of phylogenetic history with ecological or tectonic history. Three examples, taken from the work of others, are presented here involving the rifting between Africa and South America, the relationship between the Eocene freshwater fishes of Europe and North America and the distributional history of hill stream-dwelling catostomids in western North America. In each of these examples, phylogenetic history is seen as central to interpreting cause (global change) and effect (phylogenetic history). Such examples give encouragement to future attempts to determine the influence of global change on the history of fishes.

9

Response of shallow water foraminiferal palaeocommunities to global and regional environmental change

STEPHEN J. CULVER AND MARTIN A. BUZAS

9.1 INTRODUCTION

How do communities of organisms respond to environmental change? In this chapter we address this question using a group of organisms with an excellent fossil record, shallow water benthic foraminifera (Fig. 9.1). To set the scene, we first provide an overview of the Cretaceous to Recent history of these organisms in shallow water settings and then summarize the environmental context for our detailed studies of Cenozoic foraminiferal palaeocommunities in eastern North America. We then show that the temporal framework of major environmental perturbations, as well as the evolutionary framework within which these changes occurred, affect the way in which shallow water benthic foraminifera have responded to regional and global environmental change. We note that there is no simple pattern of cause and effect. There are many exceptions to any generalities and no set of conditions is ever repeated exactly through geological time.

9.2 OVERVIEW OF CRETACEOUS TO RECENT HISTORY OF SHALLOW WATER BENTHIC FORAMINIFERA

Two major episodes of global perturbations of shallow water benthic foraminifera occurred during the Cretaceous, one in the middle and one at the end. The oceanic anoxic events of the Late Barremian to Early Campanian affected water as shallow as 100 to 150 m (Hart, 1985). The Late Cenomanian event (Fig. 9.2; Gale, this volume) has been particularly well studied and has been shown to have greatly affected shallow water benthic foraminifera around the world (e.g. Western Europe, Hart & Bigg, 1981; Hart, 1996; Peryt & Lamolda, 1996; Brazil, Koutsoukos et al., 1991; south-east India, Tewari et al., 1996). In England, at this time, mixed calcareous/agglutinated assemblages in chalks were rapidly replaced by low diversity agglutinated assemblages in black bands (Hart & Bigg, 1981). Short-lived floods of a small benthic foraminifer (*Bulimina* sp.), characteristic of low-oxygen environments, also occur (Hart, 1996). In northern Spain and south-east India, similar reductions in benthic diversity are accompanied by extinctions (Peryt & Lamolda, 1996; Tewari et al., 1996); 35 to 45 % of benthic species did not survive the Late Cenomanian oceanic anoxic event (Tewari et al., 1996). Several families of larger benthic foraminifera in shallow water, tropical carbonates also became extinct at this

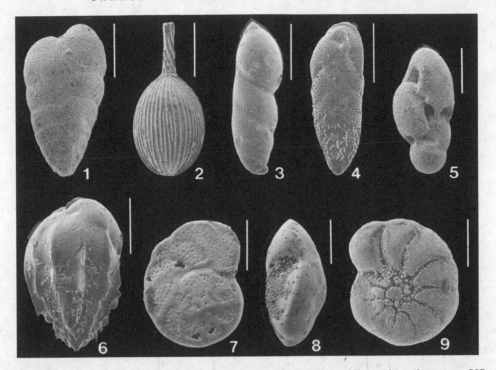

Fig. 9.1. Shallow water benthic foraminifera from the Salisbury–Albemarle embayment, US mid-Atlantic coastal plain. 1, *Textularia* sp., Eastover Formation, scale bar 270 µm. 2, *Lagena substriata* (Williamson), Eastover Formation, scale bar 136 µm. 3, *Marginulina* cf. *M. colligata* (Brückmann), Nanjemoy Formation, scale bar 231 µm. 4, *Bolivina midwayensis* (Cushman), Nanjemoy Formation, scale bar 136 µm. 5, *Virgulina gunteri curtata* (Cushman and Ponton), Eastover Formation, scale bar 86 µm. 6, *Bulimina jacksonensis cuneata* (Cushman), Nanjemoy Formation, scale bar, 136 µm. 7, *Cibicides lobatulus* (Walker and Jacob), Eastover Formation, scale bar 100 µm. 8, *Buccella inusitata* (Cushman and Todd), Eastover Formation, scale bar 176 µm. 9, *Elphidium excavatum* (Terquem), Eastover Formation, scale bar 100 µm.

time (Bilotte, 1984; Brasier, 1988). Recovery from this globally synchronous extinction event was slow; the Early Turonian was characterized by low diversity faunas represented by long-ranging taxa (Hart, 1996).

The end Cretaceous environmental perturbations and shallow water benthic foraminiferal response is more contentious. Debate centres on sudden mass extinction vs. a more progressive decline in diversity. Early in the debate, some authors suggested that there was a mass extinction of shallow water benthic foraminifera at the Cretaceous–Tertiary boundary, whilst deeper-dwelling assemblages were generally unaffected (e.g. Lipps & Hickman, 1982). In contrast, Miller (1982) concluded that neither shallow nor deep-dwelling species exhibited mass extinction. Discrepancies in interpretations continue. More recent work in the Tethys region (Keller, 1988b; Coccioni & Galeotti, 1994), the South Atlantic (Widmark & Malmgren, 1992) and the high southern latitudes (Thomas, 1990a) support the latter view whereas Adams (1989) and Speijer and Van der Zwaan (1996) inferred impact-related mass extinction in the Tethys.

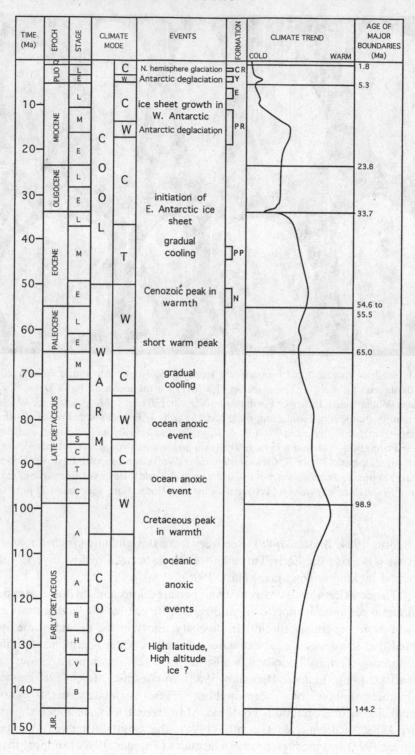

TIME (Ma)	EPOCH	STAGE	CLIMATE MODE	EVENTS	FORMATION	CLIMATE TREND	AGE OF MAJOR BOUNDARIES (Ma)
	PLIO Q	L	C	N. hemisphere glaciation	CR	COLD — WARM	1.8
		E	W	Antarctic deglaciation	Y		5.3
10	MIOCENE	L	C	ice sheet growth in W. Antarctic	E		
		M	W	Antarctic deglaciation	PR		
20		E					
	OLIGOCENE	L	C				23.8
30		E		initiation of E. Antarctic ice sheet			
		L					33.7
40	EOCENE	M	T	gradual cooling	PP		
50		E		Cenozoic peak in warmth	N		
	PALEOCENE	L	W				54.6 to
60		E		short warm peak			55.5
		M		gradual cooling			65.0
70	LATE CRETACEOUS	C	C				
80		S	W	ocean anoxic event			
90		C	C	ocean anoxic event			
		T					
		C	W				
100				Cretaceous peak in warmth			98.9
110	EARLY CRETACEOUS	A	C	oceanic			
120		A	O	anoxic			
		B	O	events			
130		H	L	High latitude, High altitude ice ?			
		V	C				
140		B					
150	JUR.						144.2

The chronostratigraphical control necessary to evaluate the inferred sudden disappearance of larger foraminiferal lineages (Adams, 1989) is lacking (MacLeod *et al.*, 1997). This is not the case for the smaller benthic foraminiferal record but, nevertheless, very different interpretations exist. Speijer and Van der Zwaan (1996), working at El Kef, Tunisia, described a sudden, impact-related ecosystem collapse at the Cretaceous–Tertiary boundary with the disappearance of more than 50% of the benthic foraminiferal taxa. They considered that sea-level variations played a role, but that changes in the benthic fauna were mainly the result of major ocean-wide perturbations in oxygen and food supply. They speculated that the oceanographic changes that diminished oxygen supply were the result of climatic perturbations amplified by various feedback mechanisms triggered by the impact of a bolide at the K–T boundary.

In strong contrast, Keller (1988*b*), using independent data from El Kef, found it difficult to explain the faunal changes that she documented prior to, at, and after, the K–T boundary, to a single impact event but rather to a series of environmental changes. Keller and Lindinger (1989), described relatively warm temperatures in the late Maastrichtian, followed by sudden cooling associated with reduced surface water productivity. Unstable environmental conditions with low productivity continued into the early Danian until some 300 000 to 400 000 years after the K–T boundary when conditions stabilized to generally pre K–T boundary conditions. N. Ortiz and G. Keller (unpublished data) suggested that, following a late Maastrichtian sea-level fall and commencing 100 000 years before the boundary, the oxygen minimum zone expanded while sea level rose during an interval of global warming. These environmental perturbations led not to a mass extinction but to a progressively decreased number of species with an immediate post-K–T boundary assemblage dominated by low oxygen tolerant species.

Patterns of benthic foraminiferal response to global environmental (palaeoceanographic) changes during the Cenozoic are well documented for deep-sea benthic foraminifera (e.g. Douglas & Woodruff, 1981; Thomas, 1990*b*; Kennett & Stott, 1991). Unfortunately, although there are vast amounts of data on Cenozoic shallow water benthic foraminifera, the data are scattered and diverse and difficult to summarize in a global context. Both gradual and sudden environmental change can be recognised in the Cenozoic (Fig. 9.2; see Pickering, this volume) but, as we shall see later in this chapter, shallow marine benthic foraminiferal species respond individually to environmental change (Buzas & Culver, 1994), with the result that changes in foraminiferal palaeocommunities took place whenever a major environ-

Fig. 9.2. Climate modes, climate trends and major environmental events during the Cretaceous and Cenozoic; C = cool, W = warm, T = transitional. Temporal framework of US mid-Atlantic coastal plain formations; N = Nanjemoy Formation, PP = Piney Point Formation; PR = Pungo River Formation; E = Eastover Formation; Y = Yorktown Formation; CR = Chowan River Formation. Cretaceous stages (from oldest to youngest), Berriasian, Valanginian, Hauterivian, Barremian, Aptian, Albian, Cenomanian, Turonian, Coniacian, Santonian, Campanian, Maastrichtian. Q = Quaternary, JUR = Jurassic, L = Late, M = Middle, E = Early. Timescale from Gradstein *et al.* (1995) and Berggren *et al.* (1995). Climate data mainly from Frakes *et al.* (1992).

mental perturbation occurred. For shallow water benthic foraminifera, eustatic sea-level rises and falls represent the proximal forcing agent for such changes, although their scale is dependent on the temporal and evolutionary framework within which that agent operates (Buzas & Culver, 1998).

Some generalities, however, are possible. In a study which incorporated the worldwide stratigraphical and geographical distribution of hundreds of modern continental margin benthic foraminiferal species from offshore North America, Buzas and Culver (1989) noted that the average species duration for both commonly and rarely occurring species is about 21 Ma. About half of the species originated in the Miocene with approximately equal numbers originating in the Pleistocene, Pliocene, Oligocene and Eocene. The lone exception was the low number of rarely occurring species originating in the Paleocene. The early Mid Miocene was a period of global warmth (Fig. 9.2) so it could be hypothesized that this triggered increased species originations. But this scenario was not reflected in the even warmer Early Eocene. Indeed, the hypothesis that increased originations in the Miocene are related to the generally decreasing temperatures during the Cenozoic could be promoted. There appears to be a trend of more high latitude originations through the Neogene, as climatic cooling progressively occurred, but the distribution of outcrops may well give this perception (Buzas & Culver, 1989).

Major global perturbations of shallow water benthic foraminifera, such as those at the Cenomanian–Turonian and Cretaceous–Tertiary boundaries, are not immediately evident during the Cenozoic although the major oceanographic events that affected deep-sea benthic foraminifera did have minor effects on shelf assemblages (e.g. McGowran et al., 1992). Even the rapid climatic oscillations with attendant sea-level rises and falls of the past two million or so years do not seem to have resulted in fundamental changes to shallow water foraminiferal communities.

Following this overview of the history of Cretaceous to Recent shallow water benthic foraminifera we now turn to a specific example from the Cenozoic of eastern North America. Changes in palaeocommunities and diversity patterns are described in the context of both regional and global environmental changes.

9.3   GLOBAL AND REGIONAL ENVIRONMENTAL CONTEXT OF CENOZOIC SHELF FORAMINIFERA OF EASTERN NORTH AMERICA

9.3.1   Depositional history

Cenozoic shelf deposits on the US Middle Atlantic continental margin (Fig. 9.3) are preserved in a series of onshore-offshore basins on the western margin of the Baltimore Canyon Trough, a deep offshore locus of Cenozoic sedimentation (Schlee, 1981; Poag, 1985; Olsson et al., 1988). This study involves six of the many Cenozoic formations that accumulated in the Salisbury and Albemarle embayments (SAE) of Delaware, Maryland, Virginia and North Carolina (see Ward, 1984); the Nanjemoy, Piney Point, Pungo River, Eastover, Yorktown and Chowan River formations. Poag and Ward (1987), Poag and Commeau (1995) and Miller et al. (1996) have shown how the deposition of Coastal Plain units, with the obvious exception of the Late Eocene (35 Ma) asteroid impact-generated Exmore boulder bed in the Salisbury embayment (Poag et al., 1992, 1994), can be related to global high sea-level stands related to polar ice volume and general climate trend during the Cenozoic (Figs. 9.2

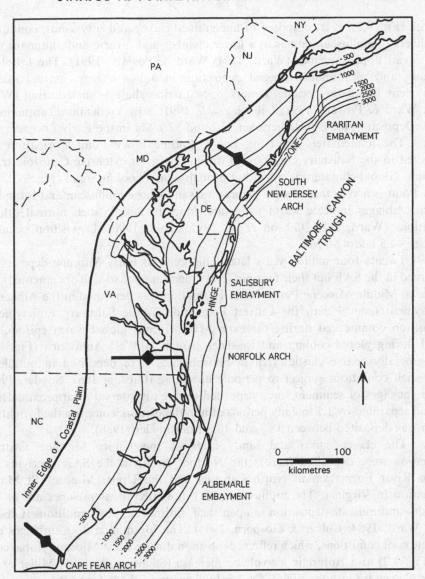

Fig. 9.3. Sedimentary basins and structural highs of the U.S. mid-Atlantic continental margin. Structure contours (m) are of pre-Mesozoic basement. PA, Pennsylvania; NY, New York; NJ, New Jersey; DE, Delware; MD, Maryland; VA, Virginia; NC, North Carolina. Modified from Benson (1984) and Olsson *et al.* (1988).

and 9.3). Sedimentary geometry was, however, affected by periodic rotational realignment of structural axes causing differential subsidence of crustal segments (Brown *et al.*, 1972). The loci of deposition varied through time and the Norfolk Arch, a structural high separating the Salisbury and Albemarle embayments, was sometimes breached (Ward, 1984).

The Nanjemoy Formation was deposited during an Early Eocene transgression into the Salisbury Embayment that coincided with the Cenozoic peak in global warmth (Fig. 9.2). It is composed of fine-grained clayey and silty sands, containing low diversity molluscan faunas in a lower member and diverse and abundant molluscs in an upper member (Ward, 1984; Ward & Powars, 1991). The lithology, molluscs and foraminifera suggest deposition in a low energy, normal salinity embayment that later became a more open marine shallow shelf setting (Ward, 1984; Ward & Powars, 1991; Gibson et al., 1991; S.E. Vredenburg, unpublished data). Deposition took place between 57.2 and 52.7 Ma in the Early Eocene.

The immediately overlying Lower Eocene Piney Point Formation was deposited in the Salisbury Embayment during the most extensive Cenozoic transgression. Global climate was warm but probably on a cooling trend (Fig. 9.2). The Piney Point is a very fossiliferous glauconitic sand whose molluscan and foraminiferal assemblages indicate warm temperate, inner to middle shelf normal salinity conditions (Ward, 1984; Gibson et al., 1991; Jones, 1990). Deposition occurred between 44.5 and 42.0 Ma.

Twenty-four million years later (Oligocene to Early Miocene deposits are preserved in the SAE but their foraminiferal faunas are not so well documented), the Lower to Middle Miocene Pungo River Formation was deposited in the Albemarle Embayment (coeval with the Calvert Formation in the Salisbury Embayment). Deposition commenced during the early Mid Miocene global warm episode but ended during global cooling and ice-sheet growth in West Antarctica (Fig. 9.2). The generally coarse clastics rich in phosphorites were deposited in middle to outer shelf conditions subject to periodic upwelling (Gibson, 1967; Snyder, 1988). Three packages of sediment, each deposited during an interval of approximately 1 my and separated by 1.5 to 2 my periods of non-deposition, comprise the formation, which was deposited between 18.3 and 10.8 Ma (Snyder, 1988).

The clayey sands and sandy clays of the Upper Miocene Eastover Formation were deposited (across the Norfolk Arch) in the SAE overlying the Pungo River Formation in North Carolina and the Upper Miocene St Mary's Formation in Virginia. The molluscan and foraminiferal assemblages are of low diversity and indicate deposition in open shelf, normal salinity conditions (Gibson, 1983; Ward, 1984; Culver & Goshorn, 1996). The foraminiferal assemblages also indicate cool conditions, which reflects deposition during latest Miocene global cooling (Fig. 9.2) and Antarctic ice-volume increase (Savin et al., 1985. Miller et al., 1987a; Culver & Goshorn, 1996). Sea-level fall around 5.4 Ma led to the cessation of Eastover deposition (Keigwin, 1987; Krantz, 1991).

Global warming and reduced Antarctic ice cover at 4.5 Ma led to sea-level rise and deposition of the Lower Pliocene Yorktown Formation in the SAE (Krantz, 1991; Webb et al., 1984). The warmer conditions are reflected in the high diversity molluscan and foraminiferal shelf faunas characterizing the four members that were deposited during three ice-volume mediated Yorktown transgressions (Ward & Blackwelder, 1980; Gibson, 1983; S.W. Snyder, unpublished data).

A global cooling event and major glacial phase at 2.4 Ma (Fig. 9.2) was followed by global warming and sea-level rise that resulted in deposition of the Upper Pliocene Chowan River Formation in the Albemarle Embayment between 1.9 and 2.4 Ma (Krantz, 1991). These silty sands contain normal marine, shallow neritic molluscan and foraminiferal assemblages. Molluscan faunas are different and of lower diversity than those of the Yorktown Formation (Stanley, 1986) but the foraminiferal faunas are very similar (S. W. Snyder, unpublished data; Buzas & Culver, 1998).

9.3.2   Foraminiferal response to environmental change
Species diversity

It is generally accepted that species diversity of benthic foraminifera is related to temperature and water depth (distance from shore). For example, a random sample of 300 foraminifera specimens from 10 m water depth off Puerto Rico will contain some 60 species (Culver, 1990) whereas a sample from the same depth off New York will contain only 20 species (Buzas, 1965). Similarly, on the Atlantic continental margin of North America Buzas and Gibson (1969) and Gibson and Buzas (1973) showed that species diversity increases offshore from the shoreline to the continental slope where it may stay constant or decline. Thus if the same physiographic location on the shelf is compared for each transgression into the SAE one would expect variations in species diversity to reflect broad temperature variations. As we have seen in the previous section, not surprisingly the sediments we have sampled from successive formations do not represent repeats of identical environments. However, all formations were deposited in open marine inner to outer shelf environments. Thus species diversity variations, if considerable, could reasonably be expected to reflect variations in marine climate (i.e. temperature).

Unfortunately, it is not that straightforward because taxonomic diversity for fossilizable marine animal species, including foraminifera, has generally increased throughout the Cenozoic, the final stage of the trend of generally increasing taxonomic diversity throughout the Phanerozoic (Sepkoski, 1993). Thus, it must be asked whether Cenozoic diversity trends in the SAE reflect regional changes of marine climate, global changes of marine climate, or the apparently unrelated general pattern of increase in taxonomic diversity throughout the Cenozoic.

Table 9.1 lists values of Fisher's α (a measure of diversity) for the benthic foraminiferal faunas in the six SAE formations. The generally increasing values through time, with the exception of the Eastover Formation, could be interpreted

Table 9.1. *Values of species diversity (Fisher's α) for six formations in the Salisbury–Albemarle Embayment*

Formation	Number of samples	Number of individuals	Number of species	Fisher's α	Confidence limits
Chowan River	9	2 033	92	19.83	± 1.97
Yorktown	33	8 525	122	20.17	± 1.45
Eastover	66	10 391	53	7.30	± 0.70
Pungo River	115	11 627	100	15.04	± 1.12
Piney Point	59	16 323	88	12.23	± 0.92
Nanjemoy	72	17 103	64	8.40	± 0.71
Totals	354	66 002	356		

as reflecting generally increasing temperatures through time. But as Fig. 9.2 indicates, the opposite temperature trend is the case. The Nanjemoy Formation was deposited during the Cenozoic peak in warmth yet it has almost the lowest value for α, a value whose confidence limits overlap the value for the Eastover Formation, which does indeed reflect the generally cool climatic conditions at the time of its deposition. The Piney Point and Pungo River formations also reflect the marine climate conditions while they were deposited – quite high species diversities during relatively warm intervals. The Yorktown Formation, similarly deposited during a period of relative global warmth, exhibits the highest values of α. But the Chowan River Formation, deposited during a relatively cool interval, which is reflected in its low diversity molluscan faunas (Stanley, 1986), exhibits species diversity as high as that of the Yorktown Formation.

So what generalities can we draw from the species diversity data? In the context of temperature-related patterns, species diversity in four formations is about what one would predict. But, in two formations it is not. Diversity is lower than expected in the Nanjemoy Formation and higher than expected in the Chowan River Formation. So, even the exceptions to the expected pattern are not consistent. Further, the general pattern of species diversity through time is opposite to the expected temperature-related trend. The data are, in fact, more consistent with the overall trend of increasing marine diversity throughout the Cenozoic (Sepkoski, 1993). Superimposed on this trend is the single regional and global climatic overprint of low diversity related to low temperatures in the Eastover Formation.

Palaeocommunities

Ecological communities have been considered as highly integrated units with strong biotic interactions (e.g. Clements, 1916) or as loosely integrated associations of species occupying the same space because of similar abiotic requirements (e.g. Gleason, 1926; Andrewarta & Birch, 1954). An intermediate position is often adopted by many more recent ecologists (e.g. MacArthur, 1972). To which of these three models do the Cenozoic foraminiferal paleocommunities of the SAE conform?

The data we utilize consist of 356 species groups identified in a total of 66 002 specimens extracted from 354 samples (Table 9.1). The question at hand can be addressed by documenting the geographical location, worldwide, of every published record of each taxon and the age (in million years) of each of those records that prove to be of conspecific material (further details of methodology are given in Buzas & Culver, 1994, 1998).

The worldwide stratigraphical ranges of all of the taxa recorded in the six SAE formations are illustrated in Fig. 9.4. Note that the values along the base of the figure (the number of species that have that particular stratigraphical range) are, almost without exception, low. In other words, the stratigraphical ranges of most species are unique in time. Is their distribution also unique in space? To answer this question, in Fig. 9.5 we expand the lower part of the left portion of Fig. 9.4 and list the locations where the 41 species with records prior to 57.2 Ma occur. No species has exactly the same distribution in space and time. For example, nine species (the fifth entry in Fig. 9.4) have a range of 64.7 to 52.7 Ma and occur and terminate in the

Table 9.2. *Similarly (as indicated by Jaccard coefficients) for foraminiferal occurrences in the six SAE formations*

Formation	Nanjemoy	Piney Point	Pungo River	Eastover	Yorktown	Chowan River
Nanjemoy	1.00					
Piney Point	0.08	1.00				
Pungo River	0.01	0.03	1.00			
Eastover	0.02	0.02	0.13	1.00		
Yorktown	0.02	0.02	0.19	0.14	1.00	
Chowan River	0.02	0.03	0.18	0.16	0.62	1.00

Nanjemoy Formation. The boxed section in Fig. 9.5 shows that before their assembly in the Nanjemoy Formation all of these species had a distinct spatial distribution. Thus, although their temporal distribution is the same (i.e. grouped together in Fig. 9.4), because their spatial distribution differs, each species does have a distinct distributional history.

Immigrants to each SAE formation are a mixture of former SAE inhabitants and species that lived elsewhere, mostly in the shelf environments of the Atlantic and Gulf of Mexico continental margins (e.g. see Fig. 9.5). Emigrants can be newly evolved or passing through and may or may not return. Thus, the neritic species of the Atlantic and Gulf margins constitute a species pool (Buzas & Culver, 1994) from which species are recruited each time a significant sea-level rise results in the transgression and flooding of the SAE. But Figs. 9.4 and 9.5 indicate that, during any transgression, only a portion of available species migrate into the SAE. These figures also indicate that different taxa in the species pool migrate and inhabit the SAE during each and every transgression. Only a small number of species habitually return to the SAE; of the 356 species only two occur in all six formations, one in five and 15 in four. The fact that only two species inhabit the SAE during all of the six episodes addressed in this study is not surprising given the 50-million-year interval involved and the 20-million-year average species duration for neritic benthic foraminifera (Buzas & Culver, 1989).

It is unlikely that species assemble into a new SAE community via a random draw from the species pool. In addition to chance, dispersal ability, barriers, distance to and nature of the newly formed environment all play a part. The regional environment, usually reflective of the global environment, is, however, one of the more important factors in controlling community composition and structure. For example, 70 species emigrated from the Pungo River Formation but only 18 of these immigrated into the SAE during deposition of the succeeding (only 3 million years later) Eastover Formation. However, 35 emigrants from the Pungo River Formation appear in the Yorktown Formation (7 million years later). The similarity between the Pungo River and Yorktown Formations is greater than between the Pungo River and Eastover Formations (Table 9.2). Thus, environmental regime is apparently more important than elapsed time in determining community composition (if the elapsed time is not too great compared with species durations).

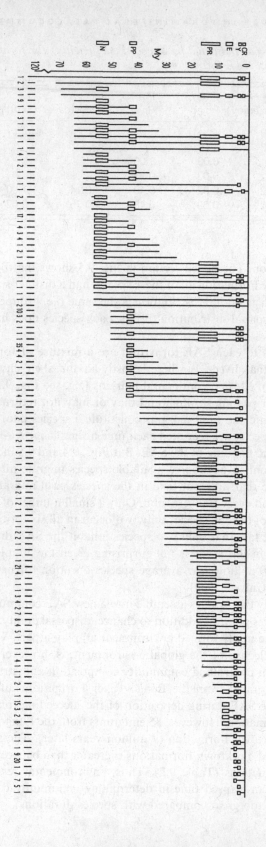

FORMATION	AGE (Ma)	LOCATION	SPECIES OCCURRENCE
Nanjemoy	57.2-52.7	VA	X X X X X X X X X X X X X X X X X X X X X X X X X X X X
Bashi	57.2-55.6	AL	X X X X
Lodo	58.5	CA	X X X X
Martinez	58.5	CA	X
Wilcox	59.9-54.3	AL	X X X X X X X X X X
Aquila	61.6-58.5	MD-VA	X X X X X X X X X X X X X X X X X X X X X
Salt Mountain	61.6-58.5	AL	X X X X X
Vincentown	61.6-58.5	NJ	X X X X X
Nanafalia	61.6-58.5	AL	X X X X X X X
Naheola	61.6-58.5	AL	X X X X X X X X X
Clayton	64.7	GA	X X
Clayton	64.7	KY	X X X X
Clayton	64.7	TN	X X X
Clayton	64.7	AL	X X
Porters Creek	64.7	AR	X
Midway	64.7-62.6	TX	X X X X X X X X X
Midway	64.7-62.6	AL	X X
Midway	64.7-62.6	AR	X X X X
Midway	64.7-62.6	MS	X
Hornerstown	64.7-62.6	NJ	X X X X X X X
Soldado	64.7-62.6	Trinidad	X
Red Bank	70-65	NJ	X X
Kemp Clay	70-65	TX	X
Dos Palos Sh	70-65	CA	X
Saratoga Chalk	72-68	AR	X
Selma Chalk	72-68	TN	X
Grayson	108-92	TX	X
Duck Creek	120	OK	X

Fig. 9.5. Spatial and temporal distribution of the 41 species with records prior to 57.2 Ma (immigrants and potential immigrants to the Nanjemoy Formation). Each column represents the temporal distribution of a species. See text for further explanation. VA, Virginia; AL, Alabama; CA, California; MD, Maryland; NJ, New Jersey; GA, Georgia; KY, Kentucky; TN, Tennessee; AR, Arkansas; TX, Texas; MS, Mississippi; OK, Oklahoma. Modified from Buzas and Culver (1998).

The Eastover Formation, deposited during an interval of relatively cold global climate (Fig. 9.2), has a low diversity community (Table 9.1), but a closer examination of the species composition indicates an abundance and dominance of cold-adapted species (Culver & Goshorn, 1996) several of which are characteristic of communities occurring today off the Atlantic and Arctic coast of Canada. Again, however, simple generalizations are not appropriate. The Chowan River Formation was also deposited during an interval of relatively cold global climate yet it contains a high diversity fauna of great similarity to the preceding Yorktown Formation that accumulated during an episode of relative global warmth.

9.4   CONCLUSIONS

Shallow water benthic foraminiferal palaeocommunities preserved in the Cenozoic deposits of the Salisbury–Albemarle Embayment of eastern USA do not

Fig. 9.4. World-wide stratigraphical ranges of all of the benthic foraminiferal species recorded in six Salisbury–Albemarle Embayment formations (N, Nanjemoy Formation; PP, Piney Point Formation; PR, Pungo River Formation; E, Eastover Formation; Y, Yorktown Formation; CR, Chowan River Formation). Temporal range of each formation is indicated on the vertical axis and correlative boxes within the diagram indicate presence of species in a formation. Number of species with each particular stratigraphical range and pattern of occurrence is given at the base of the diagram. Modified from Buzas and Culver (1998).

exhibit a simple relationship with global and regional environmental change. Observed species diversity agrees with predicted diversity (if diversity is temperature related) for only four of six communities. Further, the trend of species diversity is opposite to that predicted by the general pattern of climate cooling but consistent with the general trend of increasing taxonomic diversity for marine organisms throughout the Cenozoic. Environmental regime is, however, important in determining community composition but again, there are exceptions. Thus, although it would be possible to predict the response of shallow water benthic foraminiferal communities to future regional or global environmental change it would be unwise to attach a high level of reliability to those predictions.

10

Intrinsic and extrinsic controls on the diversification of the Bivalvia

J. ALISTAIR CRAME

10.1 INTRODUCTION

At the present day the Class Bivalvia comprises some 8000 species, subdivided into approximately 700 genera, 107 families, and 41 superfamilies (Morton, 1996, and references therein). It represents one of the most diverse invertebrate groups within the marine realm and is often portrayed as the end-product of a long-term, adaptive radiation (e.g. Stanley, 1977). Over the last 250 Ma, in particular, there has been a steady but inexorable rise in the number of bivalve taxa (Fig. 10.1), and today they have come to occupy a very broad spectrum of benthic habitats and trophic categories (e.g. Bottjer, 1985). This is what Morton (1996, p.348) has referred to as the 'expanding success of the Bivalvia'.

Although the possible effects of the 'pull of the Recent' cannot be ignored (Hallam & Miller, 1988), it would appear that there was a particularly steep rise in the number of bivalve taxa through the latest Mesozoic and Cenozoic eras (Fig. 10.1). Other benthic groups, such as the echinoids, gastropods, decapod crustaceans and fishes, show this pattern too (e.g. Vermeij, 1977), and it becomes a matter of some importance to determine why this should be so. Geologists are now almost certain that, over the greater part of the last 90 Ma, marine climates have been deteriorating (Pickering, this volume); glacial climates may now be traced back over 40 Ma in Antarctica (see below). Somehow we seem to intrinsically link phases of diversification with benign climates, and yet here we have one of the most striking examples of biotic radiation linked to one of the most striking examples of climatic decay. It is also apparent that the bivalves were only temporarily deflected from their continual rise by the global mass extinction event at the Cretaceous–Tertiary boundary (Fig. 10.1).

Hence it remains debatable how much, if at all, bivalves actually responded to global change. It may be that their diversification was driven largely by evolutionary developments within the class. Whatever the prevailing environmental conditions may have been, a succession of adaptive breakthroughs in the design and function of the animal has ensured progressive colonisation of unoccupied ecospace. In other words, continual refinement of the bivalve *bauplan* has led to continual evolutionary success. If this is not the case, and if environmental change is unimportant, an alternative would be to consider the influence of other groups of benthic

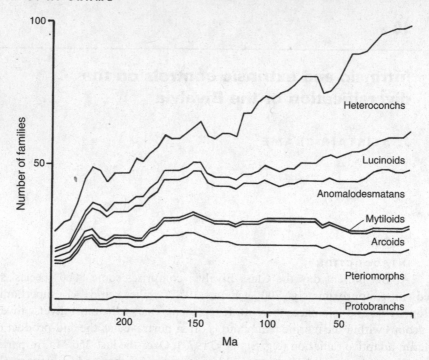

Fig. 10.1. Mesozoic–Cenozoic radiation of the seven principal bivalve clades. Redrawn from Skelton *et al.*, 1990, Fig. 5.1.

marine organisms. Has the pervasive presence of either competitors or predators (or both) driven the radiation of the bivalves? Finally, we should consider the proposition that environmental change is important, but in ways that are perhaps not immediately apparent.

10.2   TEMPORAL PATTERNS

10.2.1   Some general considerations

If the overall trend depicted in Fig. 10.1 is replotted with a logarithmic ordinate scale, it can be shown that the adaptive radiation of the Bivalvia is indeed approximately exponential (Miller & Sepkoski, 1988, fig. 1b). It is possible to fit a linear regression to these data, the slope of which (0.0025) is proportional to the per taxon rate of diversification. Nevertheless, it is apparent that there were at least three times in Earth history when the per genus rate of diversification was substantially greater than the background rate: (i) Ordovician (the initial radiation), (ii) Early–Mid Triassic (the aftermath of the Late Permian mass extinction), and (iii) Early Cenozoic (the aftermath of the Late Cretaceous mass extinction). Miller and Sepkoski (1988) have argued that a pattern of intermittent bursts superimposed upon a background of slower exponential diversification is best explained by analogy to a coupled logistic model. This conclusion is important, for such models are based upon the evolutionary diversification of two or more component phases. It suggests strongly that the adaptive radiation of the Bivalvia has involved something more than the progressive occupation of previously unoccupied ecospace. Interactions with other taxa may have played a key role throughout the evolutionary history of the group.

It should also be emphasized that, whatever its precise nature, the adaptive radiation of the Bivalvia has been extremely slow. For example, Stanley (1977) has shown that it took approximately 100 million years for the first relatively large (> 10 cm) burrowing bivalve to evolve (the Mid Silurian *Megalomoidea*); the ability to burrow rapidly and develop shell ornamentation to assist burrowing took even longer. Using a variety of techniques, it has been possible to demonstrate that the class as a whole is characterized by comparatively low rates of speciation (e.g. Stanley, 1985). A mean species duration of approximately 11 million years compares well with that of other extant benthic marine groups such as the Gastropoda, but may be as much as an order of magnitude greater than that for terrestrial insects and mammals (Stanley, 1985).

10.2.2   Taxonomic trends through time

The broad features of bivalve evolution are now reasonably well understood. From extremely modest, probably infaunal origins in the early Cambrian, the class underwent a pronounced phase of predominantly epifaunal expansion through the Palaeozoic, followed by a phase of predominantly infaunal expansion in the Mesozoic–Cenozoic. It would seem reasonable to attribute this major shift in life-habit styles largely to a progressive increase in benthic predation intensity (Stanley, 1977; Vermeij, 1977). Morton (1996) has suggested that the most primitive bivalves were pedal feeders, with gills that were entirely respiratory in function. However, with the adoption of an epifaunal, byssally attached mode of life new food-collecting methods would have had to be adopted, and this led to the rapid development of a filtering ctenidium from a respiratory gill. The enlarged, exclusively food-collecting eulamellibranch gill was perhaps the key adaptive breakthrough, which guaranteed the continued evolutionary success of the Bivalvia.

Most modern classifications subdivide the Bivalvia into approximately five Subclasses: the Palaeotaxodonta (essentially the protobranchs; taxodont dentition), Pteriomorphia (a large grouping of anisomyarian and monomyarian forms which includes most modern epifaunal bivalves), Isofilibranchia (a small group comprising the mussels and their ancestors), Heteroconchia (a very large group characterized by heterodont dentition and comprising the majority of post-Palaeozoic burrowing forms), and Anomalodesmata (a coherent group of largely infaunal taxa with edentulate hinges and thin, nacreo-prismatic shells (Skelton & Benton, 1993, and references therein). In their analysis of post-Palaeozoic bivalve radiation, Skelton *et al.* (1990) regarded each of these categories as informal 'clades'. In addition, they followed previous authors in removing the arcoids (which are generally isomyarian, equivalved, and have to some extent reverted to a shallow-burrowing lifestyle) from the pteriomorphs, and separated the lucinoids (which have developed an elongated foot to produce an inhalant mucus tube) from the heteroconchs. The relative frequencies of these seven clades through the Mesozoic and Cenozoic are shown in Fig. 10.1.

Analysis at the family level reveals that the numbers of protobranchs, arcoids and mytiloids have remained relatively constant over the last 145 million

Table 10.1. *Some important living families of heteroconch bivalves and their approximate times of origin*

Cardiliidae	Eocene (Lutetian/Bartonian)
Petricolidae	Eocene (Lutetian)
Condylocardiidae	Eocene (Lutetian)
Gaimardiidae	Eocene (Ypresian)
Scrobiculariidae	Palaeocene (Thanetian)
Kelliellidae	Palaeocene (Thanetian)
Semelidae	Palaeocene (Thanetian)
Solenidae	Palaeocene (Thanetian)
Psammobiidae	Palaeocene (Danian)
Solecurtidae	Late Cretaceous (Campanian)
Glossidae	Late Cretaceous (Cenomanian)
Mesodesmatidae	Late Cretaceous (Cenomanian)
Pharellidae	Early Cretaceous (Albian)
Trapeziidae	Early Cretaceous (?Aptian)
Donacidae	Early Cretaceous (Aptian)
Mactridae	Early Cretaceous (Aptian)
Tellinidae	Early Cretaceous (Hauterivian)
Veneridae	Early Cretaceous (Valanginian)
Pharidae	Late Jurassic (Tithonian)
Arcticidae	Late Triassic (Rhaetian)
Cardiidae	Late Triassic (Norian)
Corbiculidae	Late Triassic (Carnian)
Tancrediidae	Late Triassic (Carnian)
Carditidae	Late Permian (Tatarian)
Crassatellidae	Early Permian (Artinskian)
Astartidae	Early Devonian (Pragian)

Based on data contained within Skelton and Benton (1993).

years (Fig. 10.1). There has been a slight decline in the number of pteriomorphs but this is counterbalanced by slight to moderate rises in both the lucinoids and anomalodesmatans. However, there is no doubting that the principal expansion over this time period occurred within the heteroconchs, with a particularly sharp rise in the early Cenozoic era (Fig. 10.1; Table 10.1; see also Hallam & Miller, 1988, fig. 6.3).

In accounting for the Mesozoic–Cenozoic success of the heteroconchs, much emphasis has rightly been placed on the development of mantle fusion and posterior siphons (Stanley, 1977). Such a process undoubtedly led to the more efficient exploitation of the infaunal adaptive zone and enabled many veneroid and tellinoid clades to proliferate. Nevertheless, it has to be borne in mind that siphons are also characteristic of both the lucinoids and anomalodesmatans, and it can be argued that they represent a secondary adaptation within groups that were already efficient burrowers (Skelton *et al.*, 1990). Possibly of equal importance to the development of siphons was possession of an elongate, C–spring (or parivincular) form of ligament. Necessary functional constraints on the growth of the ligament in an adult shell are overcome in one of two basic ways: they may be of either the 'break' type (i.e. alivincular, duplivincular or multivincular), or of the 'fold' type (i.e. parivincular). Seilacher (1984) has demonstrated that, although a parivincular ligament system permits the continuous generation of both external lamellar and inner fibrous layers, it imposes severe restrictions on both shell thickness and degree of shell curvature. Those taxa employing it (i.e. the heteroconchs, lucinoids and anomalodesmatans) are necessarily comparatively small, thin and streamlined.

It would now seem that all seven of the bivalve groups under discussion were originally aragonitic in shell composition (Taylor, 1973). Two of them, the protobranchs and anomalodesmatans, are still entirely aragonitic at the present day, but the other five all show some degree of incorporation of calcite into the shell. This trend is most apparent amongst epifaunal bivalves or those with a proven epifaunal ancestry (i.e. principally the pteriomorphs and mytiloids, and the extinct rudists); arcoids, lucinoids and heteroconchs typically only show traces of calcite in the outermost layers of certain cold-water taxa (Carter, 1980). Aragonitic shell structure within the heteroconchs usually consists of either a three-layer arrangement, with outer composite prisms and middle and inner layers of crossed-lamellar or complex crossed-lamellar structure, or a two-layer arrangement of just crossed-lamellar and complex crossed-lamellar (Taylor, 1973). Some veneroids show even further reduction of the crossed-lamellar layers to simple homogenous structure only (Taylor, 1973). The marked persistence of crossed-lamellar and complex crossed-lamellar shell structures within the Heteroconchia may well be linked to its predominantly infaunal lifestyle. Microhardness tests have shown these two shell structural types to be particularly resistant to abrasion, and thus of potential value in enhancing the mechanical process of burrowing.

Important as the heteroconch radiation has been over the last 145 million years, it would not appear to have been at the wholesale expense of other taxonomic groups. For example, even though functional categories such as the epi- and endobyssate suspension feeders declined sharply over this time (e.g. Hallam & Miller, 1988, fig. 6.4), certain pteriomorphs have continued to flourish. Families such as the Inoceramidae, Entoliidae, Oxytomidae and Bakevelliidae became extinct at the Cretaceous–Tertiary (K–T) boundary, or very shortly afterwards (Skelton & Benton, 1993). Nevertheless, their demise was, to at least some extent, counterbalanced by the Cenozoic radiation of certain pectinids, limids, oysters and other groups. Such types have been particularly successful in exploiting a variety of cryptic and marginal habitats in warm temperate and tropical marine environments; in addition, two of them (the pectinids and limids) have developed a number of free-swimming forms. Morton (1996, p.347) has rightly emphasized the importance of pteriomorphs in the shallow-water, 'modern bivalve adaptive zone'.

The arcoids provide an interesting case in point too. Even though they possess filibranch gills, lack siphons, and are, at best, only sluggish burrowers, certain members of the Subfamily Anadarinae were clearly able to radiate through the Cenozoic (Stanley, 1977). Many lucinoids are now known to be restricted to oxygen-deficient environments, where they employ a chemosymbiotic mode of feeding (i.e. they obtain nutrients by means of endosymbiotic sulphide-oxidizing bacteria contained in the gills) (Reid & Brand, 1986). It is thought likely that their radiation through the latest Mesozoic and Cenozoic was triggered by the evolution of seagrass communities.

Many taxa within both the protobranchs and anomalodesmatans were displaced from comparatively shallow into deep seas over the last 145 million years (e.g. Vermeij, 1987). The precise reasons for this process, which occurs in many other marine invertebrate groups too, are unknown (Skelton *et al.*, 1990, and references therein). However, it may represent nothing more than a simple diffusion process. Many specialized clades of shallow-water origin moved into deeper waters because there was, in a sense, nowhere else for them to go. Amongst the deep sea anomalodesmatans to develop at this time was a remarkable radiation of carnivorous verticordiids, poromyids and cuspidariids. Such types feed by means of raptorial inhalent siphons and modified labial palps which can transfer prey items to the mouth. In shallower waters, anomalodesmatans have prospered by adapting to specialized habitats and employing simultaneous hermaphroditism, self-fertilization and short larval lives. Examples here would include deep burrowing in muds (Laternulidae), cemented (Myochamidae, Cleidothaeridae), and coral-boring (some Clavagellidae) (Morton, 1996).

It would be possible to conclude then, at this stage, that the evolutionary radiation of the Bivalvia over the last 145 Ma was attributable largely to internal controls. As new adaptive breakthroughs, such as mantle fusion, crossed-lamellar shell structures and the parivincular ligament in the heteroconchs, or byssal attachment and foliated shell structure in the pteriomorphs, evolved, they led to the progressive exploitation of new ecospace. The fact that some of these breakthroughs were of early Mesozoic, or even Palaeozoic, origin may not be significant. It undoubtedly took considerable passages of time for their full effects to be felt, and in one respect the Mesozoic–Cenozoic radiation of the Bivalvia may represent nothing more than the full emergence of a series of clades of considerable antiquity (Stanley, 1977).

At the family level, it would seem that the Bivalvia was little affected by the terminal Cretaceous (K–T) mass extinction event (65Ma; Fig. 10.1). The largest reduction in numbers occurs within the heteroconchs, but this can be very largely attributed to the extinction of seven families within the Order Hippuritoida (i.e. the rudists). Nevertheless, at lower taxonomic levels estimates of genus level extinctions range as high as 50%, and species level as high as 70–80%. Besides the rudists and those pteriomorph groups listed previously, other important taxa to become extinct at this time include the pectinid Subfamily Neithiinae, the palaeolophid oysters, and nearly all trigoniids (Macleod *et al.*, 1997).

However, we should not necessarily conclude at this stage that neither the environment nor other groups of organisms played an insignificant role in the radiation of the Bivalvia. To see why this is so, we need first of all to learn something more about the process of diversification in a spatial sense. As more and more taxa were added, how were they accommodated on the face of the Earth?

10.3   SPATIAL PATTERNS: THE LATITUDINAL DIVERSITY GRADIENT

At the present day, the Bivalvia provides one of the most striking examples of a gradient in taxonomic diversity (or richness) from low- to high-latitude regions. Such a pattern was first established fully by Stehli *et al.* (1967), who contoured the distribution of living taxa at 36 regional localities. Their map showed the two foci of

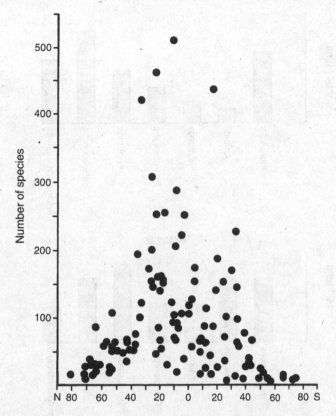

Fig. 10.2. The latitudinal gradient in bivalve species diversity at the present day. Redrawn from Flessa and Jablonski (1995, Fig.1). Plotted points represent the number of species at individual localities (see also text for further explanation).

highest diversity to be almost exactly coincident with the equator (Stehli *et al.*, 1967, fig. 3). Such a pattern has recently been substantiated by Flessa and Jablonski (1995), who plotted numbers of living species and genera at 115 regional localities (Fig. 10.2). There is some evidence to suggest that this gradient has persisted over the last 145 Ma, and may even have its origins as far back as the late Palaeozoic era (Crame, 1996*b*, and references therein).

There can be no doubt that a substantial part of the reason for the tropical high diversity of bivalves must lie in their strong association with the habitat complexity associated with coral reefs. Nevertheless, it is becoming apparent that many living reefs, and in particular those developed around oceanic islands, support only limited bivalve diversity (Flessa & Jablonski, 1995). Today, the richest tropical bivalve assemblages are associated with the lagoons of certain large Polynesian atolls, and high islands and continental margins where primary productivity is enhanced. Morton (1996, p. 347) has referred to the soft, nutrient-rich sediments on continental margins as 'the seat of modern bivalve diversity'. Assemblages from such settings typically comprise types such as arcoids, pholadomyoids and endobyssate mytiloids. However, they are dominated numerically by the heteroconchs, and

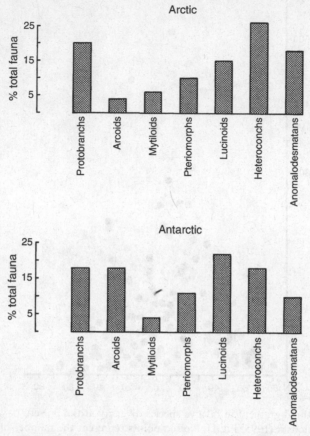

Fig. 10.3. Composition of the Arctic and Antarctic bivalve faunas at the present day. Histograms depict the number of species in each of the seven principal bivalve clades as a percentage of the total fauna. Arctic fauna comprises all taxa (both shallow and deep-water) north of approximately 60° N. Antarctic fauna comprises all taxa from south of approximately 50° S. Data sources available from the author on request.

in particular by members of the orders Veneroida and Tellinoida (Morton, 1996). It would appear that the latitudinal gradient in heteroconch bivalves must be particularly steep.

In a recent compilation of the living Antarctic bivalve fauna, a total of 171 species, from some 33 families, was obtained (Fig. 10.3; see Crame, 1996a for full details of this study). A comparable compilation of the living Arctic bivalve fauna has now been undertaken, and this yielded 199 species from 37 families (Fig. 10.3). Such studies are perhaps closer to the concept of provincial, or ξ-, diversity than that of regional, or γ-, diversity (*sensu* Magurran, 1988; see below). As yet, there is no simple, rigorous way of comparing diversity values on such a scale, but it is fairly obvious that the polar provinces are considerably less rich in numbers of taxa than all their tropical and warm-temperate counterparts (Crame, 1997, and references therein). There are at least six low-latitude, regional localities, each of which is comparable in scale to those established by Stehli *et al.* (1967), which is now

Table 10.2. *A comparison of the number of species in each of the seven bivalve clades between polar and low-latitude regions*

Province	Protobranchs	Arcoids	Mytiloids	Pteriomorphs	Lucinoids	Heteroconchs	Anomalodesmatans	Total	Reference
Arctic	39	8	12	21	17	67	35	199	Crame, 1996a
China	53	103	74	217	30	553	144	1176	Bernard et al., 1993
Panamic	66	53	39	74	35	400	126	793	Keen, 1971
Antarctic	31	31	6	18	5	63	17	171	Crame, 1996a

known to comprise 400 or more bivalve species. These are: China, Philippines, northern Queensland, the Red Sea and the Panamic province (Flessa & Jablonski, 1995) (Table 10.2).

The sharp reduction in number of species in both polar regions has affected all seven bivalve clades (Table 10.2). Nevertheless, in comparison with many other regions, the protobranchs are relatively prominent in both the Arctic and Antarctic; the Nuculanidae is the single most important family in the former region, and is joined by representatives of both the Malletiidae and Siliculidae in the latter. The present day strongholds of the largely deposit-feeding protobranchs are very much the polar and deep seas (Vermeij, 1987). The predominantly epifaunal clades, the arcoids, mytiloids and pteriomorphs, are generally poorly represented in both the Arctic and Antarctic. It should be pointed out that, although the arcoids are relatively conspicuous in the Antarctic fauna (Fig. 10.3), this is due to the development of just two groups: the limopsids, which comprise 11 species (36% of the arcoid fauna), and the philobryids (19 species; 61%). The latter represents an unusual group of tiny, mytiliform arcoids which have radiated largely, but not exclusively, within the Southern Ocean (Crame, 1996a, and references therein).

Whereas some common epifaunal bivalve groups, such as the oysters (i.e. both ostreids and gryphaeids), isognomonids, pinnids, spondylids and plicatulids, are completely missing from both polar bivalve faunas, others, such as the arcids, mytilids, limids and pectinids (*sensu lato*), are in demonstrably short supply. A substantial part of the reason for these missing taxa may simply be a lack of suitable substrates. Due to phenomena such as constant ice scour and anchor-ice formation, both the intertidal and shallow subtidal zones are severely limited in extent in the polar regions. However, whether a lack of suitable shallow-water, hard substrates can account for all the observed decline in epifaunal bivalves is far from certain. Many of the missing taxa are much thicker-shelled forms with a prominent outer shell layer of prismatic and/or foliated calcite. Given that there are at least some indications of a latitudinal cline in rates of calcification (Clarke, 1990), might it not also be that it is genuinely more difficult for many epifaunal types to secrete shells in cold polar waters? It is certainly striking how high-Arctic and Antarctic pectinids are significantly smaller and thinner-shelled than their immediate boreal and subantarctic counterparts.

Within the lucinoids, the Thyasiridae is the most prominent family in both the Arctic and Antarctic, and within the anomalodesmatans, Arctic forms are represented by a range of taxa which includes three large species of *Mya*, the prolific *Hiatella arctica* (Linnaeus), and representatives of *Thracia, Periploma, Lyonsia* and *Cuspidaria* (Fig. 10.3). There is somewhat less variety within the latter category in the Antarctic, but forms such as *Thracia meridionalis* Smith and *Laternula elliptica* (King and Broderip) are amongst the commonest bivalves in shallower waters (i.e. < 100 m).

Tiny heteroconchs belonging to the superfamilies Galeommatoidea and Cyamioidea occur in both polar regions. Montacutids are the most conspicuous group, with the genus *Mysella* being another bipolar taxon. Over 50% of the Antarctic heteroconchs (Fig. 10.3) can be assigned to just three families: the Gaimardiidae, Carditidae and Condylocardiidae; all other major groupings (Table 10.1) are poorly represented in Antarctic seas. There are, for example, only five species of venerid, two astartids and two tellinids; families such as the Mactridae and Cardiidae are completely unrepresented. Within the Arctic heteroconchs, it can be demonstrated that 53% of the total number of species belong to just two genera: *Astarte* and *Macoma*. There are, in addition, seven carditids and four venerids, and either one or two species within the following families: Mactridae, Cultellidae, Psammobiidae, Semelidae and Arcticidae (Table 10.1).

In summary, all seven clades of bivalves at the present day are significantly reduced in numbers in the polar regions (Table 10.2). However, the trend would seem to be particularly marked within both the pteriomorphs and heteroconchs, the two groups which, globally, were so prominent in the latest Mesozoic–Cenozoic radiation of the Bivalvia (Fig. 10.1). Whereas the demise of the former of these groups may be attributable, at least in part, to lack of suitable substrates, the same is not necessarily true for the latter. These shallow- to deep-burrowing clams are composed largely of comparatively thin shell layers of crossed-lamellar and complex crossed lamellar aragonite. It may well be that secretion of this type of shell material too is significantly impeded in cold, polar waters (i.e. because of the slightly greater solubility of aragonite than calcite in cold water).

We should also consider the possibility that the latitudinal gradient in bivalve taxonomic diversity at the present day reflects a cline in productivity. Even though the latter can be high in polar regions, it is strongly seasonal (e.g. Clarke, 1988). Primary production in the polar regions is restricted to just a short summer season and for most of the rest of the year phytoplankton is in short supply. In this respect, the polar regions are in sharp contrast to the 'seats of modern bivalve diversity' on tropical/subtropical high islands and continental margins.

Substantiation of steep latitudinal gradients in bivalve diversity back through the Cenozoic and Mesozoic eras will provide an important test of the productivity hypothesis. This is so because as we go back in time climates become progressively milder (Gale, this volume), and thus it is increasingly difficult to invoke some form of temperature control. Seasonality, however, has always been a prominent factor, for no matter how warm the polar regions may have been in the past, they have always been subjected to long periods of winter darkness.

10.4   DISCUSSION

If the marked rise in global bivalve diversity over the last 145 Ma reflects only a significant increase in within-habitat, or α-, diversity, then we could attribute it largely to the operation of internal controls. Perhaps it has taken tens, or even hundreds, of millions of years for the full evolutionary effects of adaptive breakthroughs in both soft- and hard-part anatomy to take hold? Stanley (1977) has emphasized consistently how such 'macroevolutionary lags' are an integral part of all bivalve radiations.

This is not to say that environmental changes could not have promoted the exploitation of new ecospace and increasingly fine partitioning of available resources. Bambach (1977), for example, has suggested that a steep early Cenozoic rise in α-diversity within marine level-bottom communities could have been triggered by a boost in trophic resources. Both the coeval radiation of the angiosperms and a major reorganization of oceanic circulation patterns brought about by generation of cold polar bottom waters may have significantly broadened the base of marine food chains. Perhaps this is the time when a difference in productivity levels between high- and low-latitude regions was enhanced significantly?

Interactions with other groups of organisms undoubtedly promoted the process of α-diversification too. In particular, the bivalves may have been involved in some form of intensive coevolutionary process with their principal predators (certain prosobranch gastropods, decapod crustaceans and teleost fish) (Vermeij, 1977, 1987). As all these types radiated extensively through the latest Cretaceous and early Cenozoic, bivalves would have been forced to continually exploit new marginal-marine, cryptic and infaunal habitats. If, as some authorities have suggested, predation pressure is more intense in the Tropics (e.g. Vermeij, 1987), then this may form an important part of the explanation for the latitudinal diversity gradient. Nevertheless, it has to be pointed out that our knowledge of benthic predation in the high-latitude and polar regions is still far from complete. At least some Antarctic bivalve communities are known to suffer intense predation from groups such as ophiuroids and asteroids.

We should also consider other evidence which brings a controlling influence of α-diversity patterns on the generation of latitudinal gradients into question. Although some studies have demonstrated a significant difference between tropical and temperate α-diversity values in the terrestrial realm (e.g. forest birds), this is less apparent in the marine one. In particular, a range of investigations has suggested that there may be no latitudinal variation in the numbers of macrobenthic infaunal taxa inhabiting shallow, muddy sediments. The recent use of k-dominance curves has shown no significant difference in α-diversity values obtained from Spitsbergen, the North Sea and Java (Kendall, 1996).

What we also need to consider here are the geographical components of diversification, namely β- and γ-diversity. The former of these is a measure of taxonomic differentiation between localities or communities; it gives an indication

Table 10.3. *Estimated number of provinces in the marine realm*

Late Cenozoic–Quaternary	31
Early Cenozoic	6
Late Cretaceous	6
Mid-Cretaceous	5
Early Cretaceous	4
Late Jurassic	5

Based on Valentine *et al.* (1978) and references therein.

of habitat breadth or specialization. Viewed in a historical context, we might regard α-diversity as the packing of species within a habitat or community, and β-diversity as the packing of communities within a higher level biogeographical entity such as a province (Valentine *et al.*, 1978). The concept of γ-diversity is a more controversial one (e.g. Sepkoski, 1988), but in this study it is simply regarded as the number of taxa occurring within a large geographical region (*sensu* Magurran, 1988). Strictly speaking, systematic changes in γ-diversity should be investigated by deriving δ-diversity patterns (δ-diversity being a higher level equivalent of β-diversity), but in practice this has rarely been attempted. Gamma diversity reflects the degree of provincialisation within the biosphere at any one time (Valentine *et al.*, 1978).

A number of lines of evidence indicate that α- and β-diversity covary widely in nature. This is, at least to some extent, only to be expected, for as an organism differentiates its niche, so it defines its habitat too. Concurrent increases in α- and β-diversity have been demonstrated on an evolutionary timescale for both Palaeozoic marine invertebrates (e.g. Sepkoski, 1988) and Cenozoic terrestrial vertebrates; clear trends of habitat specialization through time can be linked to greater geographical differentiation between communities. Unfortunately, we still know virtually nothing about systematic variations in between-habitat diversity with latitude. Thus, although β-diversity may have been an important factor in generating the global rise of marine taxa through the late Mesozoic–Cenozoic, its full role in the evolution of taxonomic diversity gradients cannot yet be assessed.

In any event, it is unlikely that the α- and β-components of diversity alone are sufficient to account for the global patterns that we see on either temporal or spatial scales. Attention must be paid to the role of regional, or γ-, diversity, and Valentine (1973, and references therein) has stressed consistently the link between Phanerozoic diversity patterns and the degree of global provinciality. In particular, the sharp rise in numbers of marine invertebrate taxa through the Cenozoic era can be matched with the progressive subdivision of the marine realm into a unique set of faunal provinces. With the gradual rise to dominance of the polar regions over oceanic circulation systems, essentially north–south trending continental shelves have been partitioned into a distinct series of latitudinal provinces (Valentine *et al.*, 1978). It is surely no coincidence that both the number of provinces in the shallow marine realm and the latitudinal temperature gradient are at Phanerozoic maxima at the present day (Table 10.3; Pickering, this volume).

As climates deteriorated from their mid Cretaceous zenith (Gale, this volume), thermally controlled, latitude-parallel marine provinces became progressively more distinct. From comparisons made between the geographical ranges of Pleistocene and Recent bivalves along the west coast of North America, we know that, during periods of climatic amelioration, distinct southern forms extended much farther north, and during periods of climatic deterioration, distinct northern forms extended much farther south (Roy *et al.*, 1995). If this was just a one-off event it would probably not have had much effect on the generation of large-scale diversity patterns. However, we now know that, during the last 2.4 Ma alone, there has been something in the order of 50 complete climatic cycles (Pickering, this volume), and the ranges of benthic taxa such as bivalves must have fluctuated continually in concertina fashion.

Although the precise mode of operation of such 'diversity pumps' is unknown, it is clear that it involves some sort of diffusion of taxa between adjacent provinces. If we look at the pattern of build-up of taxa through time (Fig. 10.1), this would appear to have been a rather gradual process over the last 100 Ma or so; however, if we look at the palaeotemperature record (Pickering, this volume), it may have been somewhat more abrupt. There is now evidence to suggest that it was as recently as just 15 Ma ago that polar–equatorial thermal gradients approaching those of today developed. Studies based on the fossil record have yet to pinpoint the precise time of change from the early Cenozoic state of low global provinciality to the late Cenozoic one of high provinciality, but it may have been as late as Mid to Late Miocene (Table 10.3).

In a series of classic papers, Valentine (1973, and references therein) suggested that the net effect of the operation of a Cenozoic diversity pump was to accumulate more marine invertebrate taxa in the Tropics than at the Poles. During a phase of climatic amelioration, many mid- and high-latitude taxa will be displaced towards the low latitudes, leaving, in some instances, vacant sites for significant evolutionary innovation during a subsequent phase of climatic amelioration. In this respect the Tropics may be regarded as a large-scale refugium for taxa that originate in mid- to high-latitude regions. Nevertheless, such a view has to be balanced against one which suggests that the majority of species are generated in the tropical regions. As these constitute by far the largest biome on the surface of the Earth, both at the present day and through the geological past, it has been argued that they represent natural sites for both enhanced speciation and reduced extinction. In essence, the larger the area, the greater the chance of generating peripheral isolates (and thus allopatric speciation), and the greater the chance of finding a refuge in an environmental crisis. However, powerful as these sorts of argument may be, they are not entirely borne out by the fossil record. Within the marine realm, there is evidence to suggest that certain cold-temperate and polar molluscan clades have radiated just as rapidly as tropical ones (Crame & Clarke, 1997). The latter are more diverse simply because they are older.

It may be that it is the age of the Tropics, as much as their size, which controls taxonomic diversity (Crame, 1997). In a way it is not surprising that the Tropics appear to be the preferred site of origin of many major marine clades, for they are of demonstrably great antiquity. Whereas there has been some form of circumequatorial ocean since the late Palaeozoic era, parts of the cold-temperate and polar regions are of only late Cenozoic origin (Pickering, this volume). At least part of the reason for the present-day exceptionally steep latitudinal diversity gradient in bivalves could simply lie in the respective ages of major biomes. Clades such as the heteroconchs and pteriomorphs became established in the formerly widespread Tropics, and are only now disseminating slowly into the geologically 'recent' polar regions.

10.5   CONCLUSIONS

In concert with other benthic marine groups such as the echinoids, gastropods, decapod crustaceans and fish, there has been a steep rise in the number of bivalve taxa over the last 145 Ma. Although some key groups did become extinct at the K–T boundary, the overall impression is very much one of an inexorable rise through time. At first sight this is something of a paradox, for it coincides with one of the most dramatic temperature declines in Phanerozoic history.

In all probability this major evolutionary diversification of the Bivalvia can be attributed largely to some form of macroevolutionary lag. Only now may the full evolutionary consequences of adaptive breakthroughs such as the eulamellibranch gill, mantle fusion and posterior siphons, the parivincular ligament, and crossed-lamellar shell structures be coming to fruition. Seven major 'clades' participated in the latest Mesozoic–Cenozoic diversification of the Bivalvia, but the trend is most apparent in the predominantly shallow-burrowing and suspension-feeding heteroconchs.

It is also the heteroconchs which show by far the steepest latitudinal gradient in taxonomic diversity at the present day; there is some evidence to suggest that they may also have done so over the greater part of the last 145 Ma. To understand fully the processes which governed the radiation of the bivalves, we need to know how taxa accumulated on a variety of spatial scales. Although variations in within-habitat (α-) diversity are important contributors, they cannot satisfactorily explain the generation and maintenance of major global diversity patterns. Geographical components (β- and γ- diversity) must be considered too.

The overall deterioration in climate through the Cenozoic in fact played an important part in the process of taxonomic diversification. As temperatures declined, a unique set of thermal provinces became progressively established on continental shelves. Serial range expansions and contractions in concert with individual climatic cycles promoted the diffusion of taxa between these provinces, and in time more of them were concentrated in the Tropics than at the Poles.

11

Global events and biotic interaction as controls on the evolution of gastropods

NOEL MORRIS AND JOHN TAYLOR

11.1 INTRODUCTION

Gastropods are one of the most diverse groups of animals, encompassing a wide variety of life habits from deposit feeding to hunting predation. They are abundant in most marine and many non-marine habitats and because of the relatively robust shells have an excellent fossil record. Because of their diversity and abundance as fossils, gastropods have been widely used in palaeobiogeographical and palaeoenvironmental studies, and some of the best examples of the influence of global events on the marine fauna involve these molluscs.

Although there are many examples to support the coincidence of global events and macroevolutionary history, there is evidence that the evolution of gastropod molluscs may, in addition, have been driven by biotic interaction.

In this chapter we discuss the development of predation in the Mesozoic before describing some aspects of the Cretaceous and Cenozoic development of the gastropods. Our examples demonstrate that both global events and biotic interaction have played a role but that we are a long way short of a definitive answer to what has been the relative importance of each as a controlling factor in the evolution of the present-day biota.

11.2 NATURE AND LIMITATIONS OF THE FOSSIL RECORD

Gastropods first appeared early in the Cambrian Period and, because of the relatively robust nature of their shells, they have remained one of the most species-rich groups that are preserved as fossils. Well over 50 000 living marine species of gastropod have been described and, on average, about 650 new species, both living and fossil, are described each year (Bouchet, 1997).

The last decade has seen a revolution in our understanding of gastropod anatomy and evolutionary relationships (Haszprunar 1988; Bieler, 1992; Ponder & Lindberg, 1996). A surge in research activity has stemmed largely from the development of cladistic methodology, new techniques for investigating fine anatomy, the discovery of new gastropod faunas in the deep sea and at hydrothermal vents, and importantly the development of molecular techniques for determining relationships. A stable classification has yet to emerge from this activity, but it is clear that some of the old traditional classificatory groupings of gastropods are poly- and paraphyletic and will have to be abandoned. Thus far, fossil gastropods have only partly been incorporated into these new phylogenetic schemes, although Bandel (1991, 1993, 1996) and his associates have reappraised the status of many Mesozoic gastropods using larval shell characters.

However, taxonomic problems remain, despite the relative abundance of gastropods in the fossil record. The gastropod shell has relatively few characters that can be used in taxonomy, particularly when compared with the complex organism living inside. Convergent shell shapes are common and often shells of living taxa cannot be classified without corroborative evidence from anatomy. There are, therefore, few clearcut characters from which the relationships of fossil gastropods can be determined. For example, the similarity of the fusiform shells with columellar folds of the neogastropod families Mitridae and Costellariidae was once taken as an indication of close relationship, but anatomical studies have shown them to be quite different in foregut anatomy and diet. The simple, cap-shaped, limpet shell form has evolved many times over in totally unrelated clades and for this reason some late Mesozoic and younger marine limpets are difficult to ascribe to a particular order or subclass, even when the shells are well-preserved. These include the Cretaceous genera *Anisomyon*, *Gigantocapulus* and Mesozoic members of the Patellogastropoda, Siphonariidae and other limpets.

A further problem with the fossil record lies in the mineral composition of the gastropod shell, made entirely of aragonite in the vast majority of species, which means that gastropods are often poorly-preserved in carbonate platform environments.

11.3   PREDATORS AND FEEDING: THE MESOZOIC MARINE REVOLUTION

An overriding phenomenon that affected marine animal evolution during the Mesozoic was recognized by Vermeij (1977, 1978, 1987, 1993a). It was marked by an evolutionary radiation of predatory animals in many different clades with a concomitant general increase in predation pressure. Vermeij termed this the 'Mesozoic marine revolution' (MMR). It is believed that predation pressure led directly to some evolutionary changes in gastropod shell morphology. By the end of the Mesozoic many gastropods had produced shells with constructional features which served as armour against predators, while some, as an alternative, exuded noxious substances from exposed tissues as a defensive mechanism. A cryptic life habit was taken up by others.

By comparison with the habits of living gastropods, we are able to consider how some of these modifications arose in the evolutionary history of the group. At the present day gastropods exhibit a plethora of shell features, such as elaborate spines, tubercles, ribs, varices, apertural plications and teeth which are considered as having a mechanically defensive role (Fig. 11.1). One example that Vermeij (1987) pointed out was that umbilicate and planospiral or near planospiral shells are seen today to be more vulnerable to crushing predation by crabs or fish. Such shells formed a much smaller proportion of the Vetigastropoda after the Mesozoic, and those which survived into the Tertiary and on to the present day are often very small and cryptic.

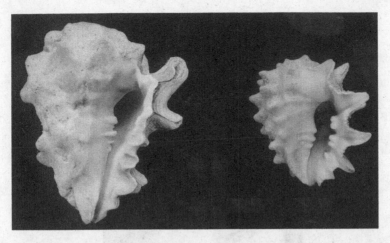

Fig. 11.1. Two living gastropods, *Drupa ricinus* and *Vasum turbinellus*, showing shell features (narrow apertures, apertural teeth, spiny and tuberculate shells) associated with resistance to predation.

A second feature of the vetigastropods is that a higher proportion of them have developed colabral (parallel to the aperture) sculpture which Vermeij judged to have been a protection against predatory crushing attacks, largely by crabs. Additionally, Vermeij *et al.* (1981) demonstrated an increased incidence during the late Mesozoic of shell damage attributed to crabs. Crabs first appeared at the beginning of the Mesozoic but diversified into their present families during the Cretaceous. A particular family, the Calappidae, dating from the Early Cretaceous (Hauterivian), developed the ability to cut marine gastropod shells following the coiling backwards from the outer lip (Fig. 11.2). The first appearance of shell damage caused in this way is from the Paleocene. Adaptations specifically against this type of attack include the development of the narrow, strengthened and armoured apertures (Fig. 11.1) found in adult cowries (Cypraeoidea) which appear at the begining of the Cretaceous (Kay, 1996).

The development of strong chelae in crabs and other Crustacea, and the parallel development of more resistant shells, has been viewed with some justification as an evolutionary arms race. Vermeij (1977, 1978, 1987, 1993b) and others noting the prevalence of attack by crushing fishes and Crustacea today in the warm shallow waters of the Tropics, looked to the essentially warm climate of the Mesozoic and the high sea levels with concomitant widespead epicontinental seas to have allowed the physical space in which the MMR took place. Mesozoic times saw ever increasing diversity of gastropod taxa. Examples of family origins during the Cretaceous from the Vetigastropoda include the Haliotidae, Umboniidae, Angariidae, Liotiidae, Stomatiidae; from the Caenogastropoda they include the Turritellidae, Campanilidae, Cerithiidae, Cypraeidae, Ranellidae, Cassidae and all the families of Neogastropoda. It was also the likely time of origin for the nudibranchs (Opisthobranchia) and probably the air breathing land snails (Stylommatophora).

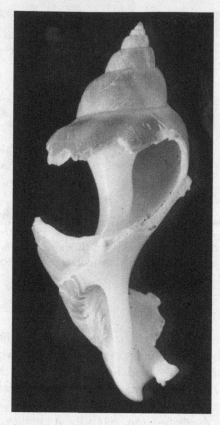

Fig. 11.2. Shell of *Babylobia lutosa* from Hong Kong with typical shell damage resulting from the peeling predation of calappid crabs.

The MMR included the evolution of carnivorous habits in the gastropods themselves (Figs. 11.3, 11.4). Gastropod carnivory includes not only grazing on the tissue of sedentary animals, such as sponges and cnidarians, but also the active hunting of often mobile prey. The largely carnivorous opisthobranchs, as at present recognized, first appeared in the Early Carboniferous and came to prominence during the mid and later Triassic (Bandel, 1996). The Naticoidea also appeared in earnest during the Early Jurassic, with the placement of some Triassic taxa in this group uncertain (Bandel, 1993). Their habit of drilling the shells of bivalves and gastropods, however, was not conclusively developed before the Early Cretaceous (Albian) (Taylor *et al.*, 1983), although naticoidean-like drill holes are reported from the Triassic (Fursich & Jablonski, 1984). This is also the time of the appearance of the Neogastropoda and Tonnoidea, with isolated Aptian species and the beginnings of their diversification apparent by the Mid Albian.

As a result of developing various styles of predation to add to earlier methods of feeding, modern marine gastropods encompass a wide variety of different life habits. They include, for example, algal grazers, suspension, mucus net and deposit feeders, grazing carnivores, hunting predators on a wide variety of invertebrate and

Fig. 11.3. The Recent vetigastropod *Ninella torquata* from Western Australia with robust shell and heavily calcified operculum. Damage to the upper part of the operculum results from drilling by the predatory dogwhelk *Discathais orbita*.

5 cm

Fig. 11.4. Predatory gastropod *Tonna zonatum* from Hong Kong feeding on a holothurian (from Morton, 1991). Species of Tonnidae seem to be specialist feeders on holothurians. Reproduced by permission of the Malacological Society of London.

even vertebrate prey, parasites and commensals. Most recently, a diversity of novel gastropods has been found associated with hydrothermal vent and cold seep communities where they exploit suphide-oxidizing bacteria (Warén & Bouchet, 1993). Some tropical gastropods have extremely specialized diets, for example, many Sacoglossa species feed on particular genera of green algae (Jensen, 1993), species of the family Mitridae seem to feed on nothing else but sipunculan worms (Taylor, 1989) and members of the family Cassidae on echinoids and Tonnidae on holothurians (Riedel, 1996). A remarkable recent discovery is the feeding of Colubrariidae species on sleeping parrot fishes (Bouchet & Perrine, 1996). Most gastropods are benthic in habit, but a few groups, namely the heteropods, pteropods, Janthinidae and some nudibranchs, such as *Glaucus*, have radiated into planktonic lifestyles and occupy the pelagic realm.

11.4   CRETACOUS GASTROPOD HISTORY
11.4.1   The invasion of non-marine habitats

At the present day, the major groups of non-marine gastropods (i.e. land and freshwater snails) are in the Neritopsina, a number of families of Caenogastropoda (particularly Cyclophoridae, Viviparidae, Cerithiodea and Hydrobiidae) and from the Heterobranchia, most Basommatophora and Stylommatophora, but also many Valvatoidea. The origin and relationships of many of these groups are very uncertain and the fossil record is very patchy.

However, there is some evidence that the invasion of non-marine habitats and the resulting evolutionary radiation of non-marine gastropods were encouraged by the formation of large tracts of non-marine sediment near to sea level. One such interval occurred in response to a global fall in sea level across the Jurassic–Cretaceous boundary. In north-west Europe this is represented by the development of a semiarid sequence, the Purbeck Group, followed by a sequence of more humid origin, the Wealden Group. Several non-marine gastropod groups originated or expanded at about that time. The Valvatoidea have marine representatives in the Triassic (Bandel, 1996) but brackish and fresh-water species date from the Late Jurassic (Bandel, 1991). In the freshwater Basommatophora, the Planorbidae date from the Early Jurassic but other groups such as the Lymnaeidae appeared close to the Jurassic–Cretaceous boundary. Among the Aracaeopulmonata, the Ellobiidae, which today mainly live in supratidal marginal marine habitats, can be traced back to the Late Jurassic (Bandel, 1991).

Among the caenogastropods, the Viviparidae are first recorded with certainty from the Mid Bathonian (Mid Jurassic), but hydrobiids and procerithiids first appeared in brackish water sediments at the end of the Jurassic (Bandel, 1993). The inshore, intertidal to supratidal Cretaceous family Cassiopidae also first appear at this time (Cleevely & Morris, 1988).

The Stylomatophora are one of the more problematic groups. Whereas Solem and Yochelson (1979) have recognized a number of living stylommatophoran families in the late Palaeozoic, no other stylommatophoran has been recognized before the latest Jurassic and the status of these Palaeozoic snails is in dispute (Bandel, 1993; Tillier *et al.*, 1996). Although their shells resemble modern pulmonate land snails, this may be the result of convergence. Many present-day families of Stylommatophora can be traced with confidence back to the Late Cretaceous–Paleocene, and molecular evidence also points to a latest Mesozoic radiation of terrestrial snails (Tillier *et al.*, 1996).

11.4.2   Gastropod diversity and species richness

There was a startling increase in diversity of many benthic marine organisms in Cretaceous times. Ross and Skelton (1996) have indicated the punctuated nature of the increase in diversity of rudistid bivalves, largely limited to tropical carbonate platforms, and their dramatic demise towards the end of the Cretaceous, a pattern paralleled by the increasing abundance of nerineoid and acteonellid gastropods, often associated with them. It was in the more temperate areas, however, where the most striking increases of gastropod diversity occurred. The predatory neogastropods showed the most meteoric increase, with a small number of species through the Aptian–Albian, then an exponential rise in diversity (Taylor *et al.*, 1980), tailing off by the end of the Cretaceous but radiating again through to the Miocene. The reality of this diversity increase can be seen by comparing the actual number of species of neogastropods at particular localities. Twelve species are recorded from the British Albian (Taylor *et al.*, 1983). This compares with the 183 species (143 of which are named) from the Late Campanian of the Mississippi embayment in the southern USA (Sohl, 1964).

Was this temporal increase in diversity controlled by global events, or may it have been caused by the the acquistion of some key innovation(s) which allowed an adaptive radiation? The early diversification of the neogastropods and other predatory caenogastropods occurred at a time of record sea-level highs, represented by maximum extension of epicontinental seas and overall warm temperatures. But, while this may have played a role, the favoured scenario is of an evolutionary breakthrough into carnivory, followed by the establishment of a number of trophic guilds with restricted types of prey taken by each of their families (Taylor *et al.*, 1980). The behavioural and anatomical specializations needed by the gastropods to tackle prey such as sea urchins probably precluded feeding on other types of prey. Since many of the putative prey animals of Cretaceous neogastropods are not usually preserved, or evidence of their destruction, when taken as prey, is not easily preserved, we are unable to test this hypothesis fully. An exception is the shell drilling of molluscan shells by species of the Naticoidea and Muricidae where the evidence of predation is preserved as distinctive boreholes in their prey (Kabat, 1990). Although naticoid-like boreholes have been found in Triassic molluscs (Fursich & Jablonski, 1984) there is some uncertainty whether these holes were actually made by naticoideans (Kabat, 1990; Bandel, 1993). The earliest unequivocal holes associated with recognizable naticid gastropods are from the Albian Blackdown Greensand of England (Taylor *et al.*, 1983). Since then, naticid boreholes became increasingly common in fossil assemblages, with an associated diversification of the Naticidae (Kabat, 1990). In earlier faunas many naticid prey were burrowing gastropods, but there was a shift to dominantly burrowing bivalve prey during the Tertiary (Kabat, 1990).

Drilling by muricid gastropods also began in the Albian (Taylor *et al.*, 1983), with these predators taking mainly epifaunal or shallow burrowing prey as they do today. A major radiation of the Muricidae took place in the Early Tertiary and the family is extremely diverse at the present day. The most diverse superfamily of neogastropods, the Conoidea (estimated 4500 living species), today largely feed on polychaete prey (Kohn, 1990; Taylor *et al.*, 1993). Conoideans first appeared in the Late Cretaceous and are quite diverse by the Campanian (Sohl, 1964), with *Conus* itself appearing in the Early Eocene (Kohn, 1990). By comparison of shell shapes with living species the early *Conus* were probably polychaete feeders, with mollusc and fish feeding species probably appearing in the Miocene.

It is possible that the temporal increase in diversity of the neogastropods was concomitant with an increase in prey diversity, with most groups of benthic animals showing evolutionary radiations in the late Mesozoic. There seems to be a coincident timing, with increased incidence of naticid borings matching a diversity increase in deeper burrowing veneroid bivalves, which are commonly their prey (Kabat, 1990).

The neogastropods may also have been more diverse at temperate latitudes than in the Tropics during Cretaceous times. Data from described lower latitude faunas consistently shows a dip in diversity across these latitudes for Late Cretaceous neogastropods (Taylor *et al.*, 1980). In contrast, at the present day and throughout the Tertiary, the highest diversities of neogastropods are found in the Tropics (Taylor & Taylor, 1977). It is still uncertain whether the relatively low diversity of neogastropods at tropical latitudes in the Late Cretaceous was a real phenomenon. If it was, one suggested cause depends on competitive exclusion. The shallow marine environments of the Late Cretaceous Tethys supported a gastropod fauna that was dominated by a small number of species of the heterobranch gastropod families Acteonellidae and Nerineidae that were extremely abundant and which possibly occupied the potential niches of neogastropods. This scenario is unlikely considering the diversity of feeding adaptations and behaviours found within the neogastropods. Another hypothesis, suggested by Johnson *et al.* (1996), is that Late Cretaceous water temperatures were so high on the carbonate platforms of the Tethys that only highly specialized benthic taxa were adapted to survive. These apparently included the rudists, but the hypothesis could equally well be applied to the acteonellid and the nerineid gastropods. This hypothesis is open to testing by isotopic methods of palaeotemperature estimation.

An alternative view is that the diversity dip at low latitudes is not a real phenomenon but an artefact related to sampling problems. Sampling of the Upper Albian sands of south-west England (Taylor *et al.*, 1983) entailed the examination of approximately 10 000 benthic molluscs. There are virtually no equivalent-sized faunas described from rocks of the Tethyan Late Cretaceous. Studies of the Late Campanian and Maastrichtian of eastern Arabia (Smith *et al.*, 1995a) have revealed a very variable preservation of aragonitic shells and although there are extensive exposures, fossils now made of calcite are difficult to extract and taxa preserved as internal moulds are difficult to identify. This apparently produces a low count of ornately-shelled members of the Muricoidea/Buccinoidea. It is likely that estimates of Late Cretaceous diversity of neogastropods from tropical limestones will rise with increased taxonomic study, but whether this will completely obliterate the apparent diversity dip remains to be seen.

11.4.4   Latitudinal control on distribution

Throughout the Mesozoic, the climate seems to have been more equable than it is today and there is no firm evidence of permanent sea level ice through the Cretaceous (Gale, this volume). In spite of this, there are clear differences between the biota of the area close to the Cretaceous palaeoequator and that occurring closer to the poles, and this situation continues to the present day where temperature and sunlight are clearly among the controlling factors. The broad patterns of gastropod distribution during the Mesozoic have been documented by Sohl (1987). The warm marine shallow waters of the 'tropical' areas throughout the Mesozoic included a biota described as Tethyan. Some taxa, particularly gastropods but other invertebrate groups as well, are virtually restricted to the Tethys. Sohl (1987) recognized a core Tethyan Biota which included colonial corals, megalodont and rudistid bivalves, and nerineoid and acteonellid gastropods. The concept is imprecise and subject to circularity of argument. However, a number of Mesozoic gastropod families or superfamilies, in addition to many lower taxonomic groups, were virtually restricted to the carbonate platform deposits of the 'Tethys' or their inshore facies equivalents. These same families are not known to occur in contemporaneous circumpolar areas. During the Cretaceous, they included the heterobranchs Nerineoidea, Acteonellidae and Pseudomelaniidae *sensu stricto*, the caenogastropods Cassiopidae, Campanilidae, Strombidae and Cypraeidae and the vetigastropood and neritopsine groups including the neritomorph 'limpet', *Pileolus*, the trochoidean *Discotectu*s, Proconulinae and Chilodontidae in addition to a number of genera with shell morphologies similar to modern *Turbo*, *Astraea* and *Angaria*. With the exception of the vetigastropods, only three of the seven family group taxa mentioned here survived into the Tertiary. The large, heavy shelled heterobranchs all disappeared.

The presence of exceptionally large forms of Nerineioida and Acteonellidae (both probably heterobranchs) and *Campanile* (a caenogastropod) in the Late Cretaceous shallow platform carbonate environment remains unexplained. These giant Late Cretaceous gastropods were matched in size by contemporaneous rudists. Their large size may be the result of continuing morphological change in a relatively unchanging environment and, to some extent, a measure of the richness and continuity of primary productivity in tropical shallow waters. In today's gastropod biota, the host of large trochids and turbinids and the large number of cerithioideans in the shallow environments of the Tropics are dependent on highly productive algal turfs. The feeding habits of the Late Cretaceous forms remains unknown. Symbiotic algae have been claimed for some (Skelton & Wright, 1987) or many rudists (Kauffmann & Sohl, 1974; Vogel, 1975). Many of the later Cretaceous acteonellids and some nerineids occurred in abundance in very low diversity associations, in inshore habitats relative to the coral/algal/rudist/larger foraminiferan assemblages, suggesting that they might have had unusual salinity tolerances. The often high abundance in these habitats might suggest they were herbivorous, yet some Recent carnivorous shelled opisthobranchs do sometimes occur in enormous numbers (Taylor & Jensen, 1992).

11.5   CRETACEOUS/TERTIARY BOUNDARY EXTINCTIONS

Extinction and extinction rates were discussed in a detailed summary by Jablonski (1995) (and see MacLeod *et al.*, 1997). He concluded that a background extinction pattern can be linked to biological features of marine invertebrates, such as their larval type and hence dispersal. Major extinctions such as the end-Cretaceous event, however, cuts right across these biological features and are more random. Jablonski's (1986*a*) data for both bivalves and gastropods suggest that end Cretaceous extinctions were not latitudinally controlled when the rudists are removed from the dataset. However, taking into account also the associated acteonellid and nerineid gastropods, together with the dramatic drop in diversity in apparently hermatypic corals (Rosen, this volume), it seems to be the shallow carbonate platform biota of the low latitudes that was most affected at this time.

The widely held view (Jablonski & Raup, 1995) that the rudists may well have been in decline before the end of the Cretaceous, has considerable implications on the importance of the end Cretaceous event to evolutionary history. Their demise might well have been paralleled by the decline in global water temperatures (Gale, this volume) lessening the significance of the impact of a bolide impact collision on extinction. However, Smith *et al.* (1995*a*) and Morris and Skelton (1995) have shown that in Arabia rudists continued well into the late Maastrichtian, disappearing only when there was a change to deeper water facies towards the end of that period. What can be learnt from the stratigraphical distributions of the Nerineoidea and the Acteonellidae (Fig. 11.5), which to some extent parallel both the stratigraphical and geographical distribution of the rudists?

Taking the Acteonellidae first, in eastern Arabia (Smith *et al.*, 1995*a*), a few species occur commonly up to the approximate mid-point of the Maastrichtian sequences. These include *Acteonella borneensis* which has a widespread distribution including Borneo, Arabia and the Caribbean. One species of acteonellid, *Acteonella crassa*, commonly occurs, but in restricted facies, in eastern Arabia in the lower part of the upper Maastrichtian succession. The higher part is represented by thin-bedded, deeper water limestones, apparently devoid of benthic fossils. In-shore facies of late Maastrichtian age are represented in Pakistan. No acteonellids have been described from the sections of the Mari Hills where shallow water upper Maastrichtian deposits are known. In this case, however, there is an ammonite fauna present, indicating open marine conditions, not the low diversity, sand biota marginal to the carbonate platform that was characterized by acteonellids. A second comparable fauna, but again without acteonellids, has been described from Luristan in western Iran by Douvillé (1904). Here, the highest strata of Cretaceous age, known to be Maastrichtian, include a rich fauna of the gastropod superfamily Cerithoidea in rubbly, inshore limestones. Of particular note are the large Campanilidae (Fig. 11.5) which, together with some of the other cerithioids, did sometimes co-occur with acteonellids in eastern Arabia. There is clearly some overlap of distribution of dead shells of Campanilidae and Acteonellidae, but it is also apparent that some accumulations of acteonellids occurred without cerithioids or

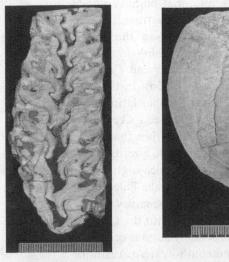

Fig. 11.5. Nerineoid, acteonellid and campanilid gastropods from Tethyan environments; only *Campanile* survived into the Tertiary.

with different cerithioids that apparently belonged to the Potamididae. We are not certain that the right habitat for the acteonellids was preserved in every one of these upper Maastrichtian sections and therefore we do not know whether the Acteonellidae died out gradually through the Maastrichtian or whether their extinction was instantaneous and the result of the end Cretaceous event.

As far as the Nerineoidea are concerned, regional differences come into play. There are no authenticated old world members of the Nerineoidea in the Maastrichtian (personal observation and H. Kollmann, personal communication). The youngest taxa all seem to be of Late Campanian age. However, they do occur in the Maastrichtian of the Caribbean region (Sohl, 1987). Although it is possible that the old world Maastrichtian facies were unsuitable, it seems more likely that the nerineoids became extinct in this region by the end of the Campanian. Thus the Nerineoidea were probably declining before the Cretaceous–Tertiary boundary.

The vetigastropods and neritopsines are even more difficult to evaluate because it is not easy to be certain to which living groups the Mesozoic taxa belong. The marine neritopsines are today largely restricted to warm waters and are often intertidal (e.g. *Nerita*), while some brackish and fresh water neritinids are more tolerant of lower temperatures (e.g. *Theodoxus*). During the Mesozoic the genus *Neritopsis* and its near relatives were the most abundant and most widespread, at least up to the end of the Early Cretaceous. The pan-Tethyan limpet-like *Pileolus* was almost restricted to the Tethys, ranging from western Europe to the western Pacific by Late Jurassic times. In the Caribbean region *Pileolus* continues into the Maastrichtian, but species of this age have not been found in the Old World.

Extinctions were not confined to the equatorial biota. Volutidae formed a higher proportion of species of the carnivorous neogastropods during the latest Cretaceous than they do at the present day. Among the Late Cretaceous Volutidae, large forms such as *Volutomorpha* and *Volutoderma* were common away from the lower latitudes whereas *Lyria*-like species and *Caricella* dominated some areas of the Tethys (Sohl, 1964; Smith *et al.*, 1995a). In this case it was the Tethyan taxa that survived into the Tertiary, but the higher latitude taxa did not.

We support Jablonski's (1986a) view that the end Cretaceous extinction among gastropods, gradual or sudden, was random rather than related to some biological lack of ability to cope with a changing world. Size itself does not seem to have been an important factor because the Nerineoidea and the Acteonellidae, which include 'giant' species, and were all extinct before the Paleocene, occurred at the same time as the giant caenogastropod, *Campanile*. *Campanile* clearly survived the changes, with giant forms up to 900 mm long well into the Late Eocene (Jung, 1987) and is represented today by a single, moderately sized species, *C. symbolicum* (240 mm), which lives in shallow water around south-western Australia.

11.6   SOME MAJOR CHANGES IN GASTROPOD FAUNAS DURING THE TERTIARY

The migration of climatic belts in response to long-term temperature changes continued to affect gastropod distribution throughout the Cenozoic. Thus much of the change in the molluscan fauna at the end of the Eocene in north west Europe represents a southerly migration of cooler water forms paralleling a general global cooling event (S. Young, unpublished PhD Thesis, University of Birmingham). Real extinctions on a global scale, did, however, occur at this time.

For the later Tertiary there are many examples of the response of molluscan faunas to changes in global climate. One of the most interesting concerns how global temperature change in conjunction with the positions of the continents after the Miocene isolated closely related temperate species in the North Atlantic and North Pacific oceans. Following the opening of the Bering Strait in the Pliocene, coupled with warmer sea temperatures, there was a major migration of Pacific marine animals into the Atlantic via the Arctic seaway, usually assumed to have been via Arctic Canada (Durham & McNeil, 1967; Vermeij, 1991). This trans-Arctic interchange of marine animals caused massive changes in the composition of the North Atlantic marine fauna, the earliest taxa arriving around 3–3.5 Ma. For molluscs alone, a total of 295 species took part in the interchange or are descended from species that did so (Vermeij, 1991). With deteriorating temperatures and periodic closure of the Bering Straits by lowered sea level, the interchange ceased, and the Atlantic and North Pacific faunas became isolated again. Many common animals on modern northern Atlantic shores have north Pacific affinities; a particularly well-documented example is the genus *Littorina* (Fig. 11.6). A combination of anatomical and molecular phylogenetic studies have demonstrated that two separate lineages of littorinids entered the Atlantic from the north Pacific (Reid, 1996). The well-known Atlantic edible winkle, *Littorina littorea,* which has a planktonic larval stage, is a sister species of the north-west Pacific *L. squalida.* The other five living species of Atlantic *Littorina,* which do not have a planktonic larval stage (*L. saxatilis, L.*

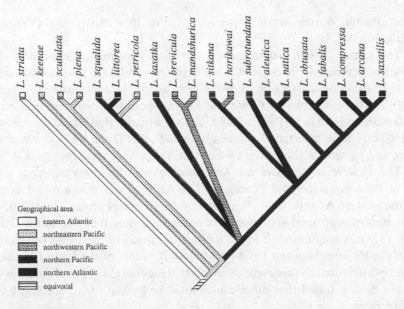

Fig. 11.6. The trans-Arctic interchange. Area cladogram of living *Littorina* species obtained by superimposing areas of endemicity on a phylogeny. *Littorina littorea* and *L. obtusata-L. saxatilis* occur in the Atlantic, but both have northern Pacific sister species. Morphological, molecular and fossil evidence supports the hypothesis that there have been two separate invasions of the Atlantic by *Littorina* species (from Reid, 1996).

obtustata, etc.), again have a sister-group in the north Pacific and form another clade which has speciated since invading the Atlantic.

The common Atlantic dogwhelk *Nucella lapillus* is also similar in shell and anatomical characters to six *Nucella* species inhabiting shores of the North Pacific. Recent molecular and morphological analysis has clarified the relationships of *Nucella* species (Collins *et al.*, 1996) and demonstrated that the sister species of *N. lapillus* is the Japanese *N. freycinetti,* but also suggests that a trans-Eurasian migration route to the Atlantic was as likely as a trans-Canadian route. The earliest *Nucella* in the Atlantic is the form *N. lapillus incrassata* recorded from Upper Pliocene Red Crag deposits. Another Red Crag species, '*Nucella*' *tetragona*, has previously been regarded as the earliest Atlantic member of the clade, but is now regarded as belonging to a distinct genus *Spinucella* that neither descended from or gave rise to *Nucella* (Vermeij, 1993*b*). Other common North Atlantic molluscs with north Pacific affinities, but where the evolutionary and biogeographical histories are not so well worked out, are *Mytilus*, *Tectura* and *Neptunea* species.

Changes in global climate and the disposition of the continents has had further major repercussions. The approach of the Afro-Arabian plate to Eurasia led to the physical division of Tethys during the Miocene. Although similar species occur today in the Caribbean area and the Indo-west Pacific they may have been isolated since the Late Miocene by this closure of Tethys. However, around 20 species with long-lived larval stages (e.g. Ranellidae) are thought to have dispersed to the Atlantic and hence the Caribbean via the southern coast of Africa from the

Indo-West Pacific during warmer periods of the Plio–Pleistocene (Vermeij & Rosenberg, 1993).

The same collision led to the eventual and continuing dessication of the Mediterranean during the Messinian (Late Miocene). When a near-normal marine fauna was re-established, it came from the west as the Atlantic broke through the straights of Gibraltar. Thus there are few Recent species in common between the Mediterranean and the Indo-Pacific fauna. Conversely, the Red Sea, once connected to the Mediterranean, reverted to a basically Indo-Pacific fauna when it was isolated near Suez and opened in the region of the Afar Triangle (Coleman, 1993).

The closure of the Central American seaway by the emergence of the Isthmus of Panama in the Mid Pliocene (3.1 to 3.6 Ma) was also a major event. It isolated the western Atlantic fauna from that of the Eastern Pacific. As a result, pairs of similar Recent gastropod species may be recognized on either side of the Isthmus (Vermeij, 1978). A major change in the molluscan faunas of the western Atlantic at the end of the Pliocene between 2 and 1.5 Ma (60–70% loss of species in Florida) has been directly attributed to this event (Vermeij & Rosenberg, 1993) but Jackson *et al.* (1993) have demonstrated that other causes must be looked for to account for this faunal turnover.

These few examples demonstrate the dynamic nature of the marine faunal provinces during the Cenozoic. Detailed studies of fossil gastropod lineages combined with phylogenetic and molecular studies are showing that the distributional limits of species change over time and that far from being a cohesive unit the provincial faunas are composed of species often with widely different geographical origins and histories (Vermeij & Rosenberg, 1993).

11.7   CONCLUSIONS

The apparent coincidence of global events during the Cretaceous and Cenozoic with major changes in the evolution and distribution of gastropods is documented. Major falls in sea level across the Jurassic–Cretaceous boundary and the concomitant development of extensive low-lying areas of non-marine sedimentation may have encouraged the development of several groups of freshwater snails. Similarly, another sea-level fall across the Cretaceous–Paleocene boundary coincided with a major radiation of land snails, demonstrated by both fossil and molecular evidence. During the intervening period, long-term high sea levels in the mid Cretaceous correlate with increasing diversity of many gastropod groups.

In the later Tertiary, the opening of the Bering Strait allowed Pliocene trans-Arctic exchange of fauna from the north Pacific to cause profound changes in the composition of the intertidal communities of the North Atlantic. Conversely, the closure of the Isthmus of Panama in the Mid Pliocene led to a divergence of the gastropod faunas to the east and west.

Although the mid Cretaceous diversification coincided with a major rise in sea level, it probably owed much to the development of predation by, and upon, the gastropods. It was marked, in particular, by massive diversification of the Neogastropoda, along with the Naticoidea and Tonnoidea, which together prey on most marine phyla. This diversification, along with predators in other phyla, caused profound changes in the structure of benthic communities. Gastropods were themselves subject to predation and during the Mesozoic there were three major changes in shell morphology involving the evolution of predator-resistant morphologies.

Finally, the well-documented extinctions at the Cretaceous–Tertiary boundary (MacLeod *et al.*, 1997) took considerable toll on the gastropods, but only a few families disappeared, and some of those such as the Nerineoida and Acteonellidae may well have been already in decline.

12

Algal symbiosis, and the collapse and recovery of reef communities: Lazarus corals across the K–T boundary

BRIAN R. ROSEN

12.1 INTRODUCTION

It has long been customary to consider the ecology of corals in terms of reef and non-reef groups. Yonge's (1940) landmark review also emphasized the critical role of algal symbiosis in the ecology of modern reefs and reef corals, and this view has since been reinforced by numerous other authors. This in turn has influenced perceptions of ancient reefs, and more recently it has generated a series of important reviews specifically relating patterns of reef-building through geological time to the history of algal symbiosis in corals and other reef organisms (e.g. Cowen, 1988; Talent, 1988; Copper, 1989; Stanley, 1992; Wood, 1993; 1995; Stanley & Swart, 1995). As part of this, intervals of reef absence in the geological record are attributed by some to collapse (*sic*) of algal symbiosis. Alongside this, most of these authors have also accepted and furthered the idea first expressed by Newell (1971) that the pattern of reef communities through time has consisted of relatively stable 'packages', punctuated by phases of short-term, rapid change, mediated by global events (see also Boucot, 1983; Heckel, 1974; James, 1983; Sheehan, 1985; Fagerstrom, 1987; Jackson, 1992; 1994; Kauffman & Fagerstrom, 1993).

Hallock and Schlager (1986) added an extra strand to these arguments by suggesting that, since modern reef corals, reefs and algal symbiosis are adversely affected by nutrient-rich (eutrophic) waters, fluctuating patterns of reef occurrence on various geological timescales might also be controlled by regional to global fluctuations in nutrient levels. In their model, eutrophic conditions lead directly or otherwise to demise of both algal symbiosis and reef-building (see also Hottinger, 1987; Cowen, 1988; Wood, 1993; 1995; Done *et al.*, 1996). Since nutrient flux is linked to climate through factors such as terrestrial run-off and oceanic circulation (Hallock, 1987), global change would appear to be the underlying factor affecting distribution patterns of corals and reefs in the geological column. Thus the 'collapse and recovery' model links global change, nutrients, algal symbiosis, reef-building in general, and reef corals in particular, as the reigning orthodox explanation for historical patterns of reefs and reef communities, as summarized by Talent (1988):

> The episodic pattern of reef building would seem to be readily explicable if it is interpreted in terms of major patterns of global climate and global sedimentary–tectonic events and, apparently more important, the sequence of global life crises –

including extinctions and delayed substitutions in vacated niches – and the, doubtless often related, changing fabric of symbiotic relationships.

And yet there remains the fundamental drawback that the algal symbionts themselves are not preserved in their fossil host organisms. Since reef occurrence itself is used as the main evidence for symbiosis, this lays the model open to the criticism of being a self-fulfilling 'prediction'. One way to evaluate it therefore, is to examine the corals themselves for evidence of algal symbiosis, independently of their reef occurrence, especially over a geological interval with mass extinctions. This chapter reviews such evidence after summarising the broad global evolutionary pattern for scleractinians in general, and uses the K–T interval and Paleocene for a case history re-appraisal of the 'collapse-and-recovery' model.

12.2   OUTLINE HISTORY OF POST-JURASSIC SCLERACTINIANS: TAXONOMIC RICHNESS, EXTINCTIONS AND RADIATIONS

The fossil record of the order Scleractinia began in the Mid Triassic. Subjective phylogenies (e.g. Wells, 1956) give no clear indication of common ancestry for the major subordinal groups and no detailed skeleton-based cladistic analysis at this level is yet available. Recent molecular phylogenies (Veron *et al.*, 1996; Romano & Palumbi, 1996), however, suggest that scleractinians as a whole are monophyletic but that there is a deep phylogenetic split predating the first recorded fossil examples. Assuming that these earliest records reasonably reflect the date of the emergence of skeletonized scleractinians in the Mid Triassic, the clade itself appears to have originated much earlier as an initially unskeletonized group within which two distinct subordinal groups appeared which acquired skeletons by convergent evolution.

Many of the still widely accepted family–level groupings of genera go back to Vaughan and Wells (1943) or before, and were characterized by shared internal and external details of septa. The same applies to Roniewicz and Morycowa's (1993) microstructural groups. Molecular studies of extant examples of these familial groups mostly show closeness of relationships between confamilial genera, implying that these morphologically defined families may be paraphyletic or monophyletic, but rarely polyphyletic. However, relationships of extinct familial groups to each other and to molecularly determined phylogenies remain unsolved. Cladistic analysis will be needed to study this, the most likely basis for which appears to be the 'internal micro-skeletal characters' sensu Roniewicz and Morycowa (1993).

Diversity data (taxonomic richness) for Cretaceous to Recent scleractinians are shown in Fig. 12.1. The present compilation, however, is secondarily assembled from previous authors' compilations and so is dependent on the methods and ground rules adopted by those authors. On phylogenetic grounds (above), generic patterns are probably more reliable than suprageneric patterns, for which there will be considerable paraphyletic noise. Offsetting this, different coral works are more consistent at familial level, becoming progressively less so at lower levels. Approaches to Mesozoic and Tertiary coral taxonomy have often been very different, and there have also been longstanding differences in the taxonomic treatment of living and fossil corals, though Budd and coworkers have been greatly improving on this for

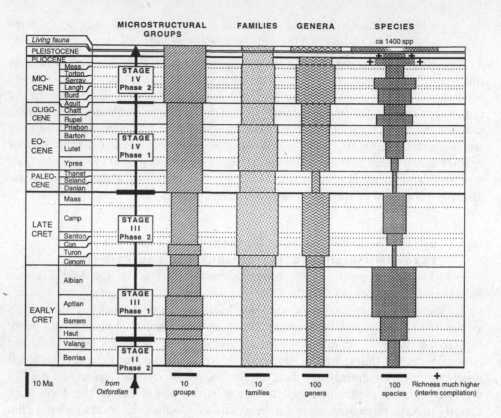

Fig. 12.1 Taxonomic richness of scleractinian corals from Cretaceous to present. Compilations are not always available for each separate stratigraphical division. Information which has been compiled by source authors for more than one of the stratigraphical divisions used here is shown as a single block spanning the respective divisions. All data are simple totals without statistical adjustment for geological duration. Gaps indicate absence of available compilation. For further discussion of data, see text. *Microstructural groups*: Sequence and numerical data from Roniewicz and Morycowa (1993). *Families*: Data from Newell (1971) based on Wells (1956). *Genera*: Cretaceous and Paleocene data from Rosen and Turnšek (1989, fig. 1: 'whole system records included') derived from Wells (1956). Other data from Veron (1995). *Species*: Data from Coates and Jackson (1985) except Pliocene to Recent: Pliocene – dearth of well-dated described faunas of Pliocene age outside the Carribean. Data here an approximate summation of Indo-Pacific Era Beds reefal fauna (Veron & Kelley, 1988) plus Caribbean reef coral fauna (Johnson *et al.*, 1995). Total therefore an absolute minimum since it takes no account of any other regions, nor of non-reefal corals. Pleistocene – plot shows basic minimum from the Caribbean compilation by Johnson *et al.* (1995), rounded up to 100 to make notional allowance for non-reef corals and for the Indo-Pacific region. True figure likely to be very much higher. Recent – from Paulay (1997).

Caribbean Neogene–to–Recent corals (Budd *et al.*, 1994 for compilation; Johnson *et al.*, 1995). These authors are also providing rigorous statistical and stratigraphical treatment of specimen-based data for the Caribbean which is rare for other regions. Records culled from the older literature, especially over the Maastrichtian–Eocene interval, are fraught with problems of stratigraphical reinterpretation, and are also therefore a source of compilation noise. All the foregoing problems probably explain why different authors give contrasting numerical data for the same taxonomic level and stratigraphical interval, and why in Fig. 12.1 generic totals sometimes exceed species totals for the same interval (e.g. Paleocene). In the former instances, the largest available figure has usually been chosen. Problems are also highlighted by the contrast in available figures (3 and 14, respectively) for family-level extinctions at the K–T boundary.

For the sake of consistency at family level, the single author source of Newell (1971) after Wells (1956) is preferred here to the multiauthor source of the familial compilation by Veron *et al.* (1996). The quasifamily 'microstructural groups' of Roniewicz and Morycowa (1993) are also compiled in Fig. 12.1. These authors distinguish four major evolutionary 'stages' in corals, each further divided into 'phases'. It is noticeable that the number of groups does not vary much through the last 145 Ma, a pattern which is similar to that of families proper. Intriguingly, however, the K–T boundary is marked by a lack of extinctions of any microstructural groups in these authors' range charts, while the number of groups actually increases from 12 in the Senonian to 16 in the Paleogene. Therefore, K–T extinctions, as with other patterns in Fig. 12.1, seem to become more marked at progressively lower taxonomic levels.

The combined data suggest a pattern of four main 'boom and bust' cycles:

(i)   Valanginian–Hauterivian turnover and radiation, leading to an Aptian–Albian 'high' and then Albian–Cenomanian extinctions;

(ii) Early Late Cretaceous 'low' with a build-up to a late Late Cretaceous high followed by major extinctions around the K–T boundary (see also Rosen & Turnšek, 1989, Rosen in MacLeod *et al.*, 1997);

(iii) Marked 'low' in the Paleocene followed by an Eocene to Early Oligocene radiation and extinctions at, or towards, the end of the Oligocene (Chattian);

(iv) Miocene radiation.

With respect to the K–T boundary in particular, although stratigraphical resolution in the available data are coarse, Rosen and Turnšek (1989) suggested that scleractinian extinctions took place at a progressively increasing rate through the last stages of the Cretaceous. However, recent work in the Oman Mountains has revealed a rich 'Gosau-type' fauna of Maastrichtian age (Smith *et al.*, 1995*b*). The Gosau fauna proper occurs in the Coniacian–Santonian of the eastern Alps in Austria, and few such faunas are found as late as the Maastrichtian (Beauvais & Beauvais, 1974). The new fauna suggests that extinction rates prior to the Maastrichtian may have been lower, that taxonomic richness remained higher, and that extinctions were concentrated more in the later Maastrichtian, than has previously been realized for corals. The pattern from the Miocene onwards is the least clear part of the record, the high quality of Caribbean reef coral data (see above) not being matched by data from elsewhere. The late Cenozoic data should also be contrasted with the living coral fauna, which consists of at least 1400 species (Paulay, 1997) and cannot be explained simply as an entirely Holocene radiation.

If emergences of different microstructural groups are an indication of evolutionary innovations, then Roniewicz and Morycowa's (1993) work suggests that modern faunas mostly have deep-seated origins in the Late Cretaceous, not as might be expected, in the aftermath of K–T extinctions. Origination rate of extant microstructural groups was about the same in the Late Cretaceous and Neogene (*ca.* 0.2 groups per million years) and three times greater than in the Paleogene (0.07). The Late Cretaceous accounts for about 35% of living groups. Implications of the patterns in Fig. 12.1, with respect to global change, are discussed later.

12.3   REEFS OR CORALS?

Reefs and reef corals have featured noticeably in the recent debate about how far communities have existed as stable entities in space and time, partly because their distribution is closely related to climate and climate-related factors (Rosen, 1984; Fraser & Currie, 1996). Palaeontology has contributed substantially to this debate, especially because fossil reef communities are largely preserved *in situ* (Jackson, 1992, 1994), leading to the idea, mentioned already, of reef communities as 'packages' which collapsed or otherwise changed relatively suddenly at times of short-term, rapid change, like mass extinctions. Boucot (1983) and Miller (1991), however, considered that there are several spatiotemporal tiers of such ecological packages. Miller (1991) and Jackson (1991, 1992), pointed out that this leads to confusion in the debate about community change vs. stability. This chapter concentrates on Boucot's third tier ('Ecological Evolutionary Units' or 'EEUs') which are tens of millions of years in duration (see also Sheehan, 1985; Fagerstrom, 1987). According to Boucot, the K–T boundary, discussed later, represents the changeover from his last EEU to the modern one.

A second and previously unremarked source of confusion is that of inadequate distinction between the history of corals and the history of reef communities. The main frame of reference for coral palaeoecology has long been reef-based (e.g. 'reef corals', 'reef-building corals', 'bioconstructional corals', 'hermatypic corals', and their respective antonyms). While concentration on reef-related corals might seem an advantage for considering global change, it also locks all analysis of such coral patterns on to definitions and concepts of reefs. In practice, it is very difficult to give precise definitions of terms like 'reef coral' because (i) authors differ about which facies within a reef or carbonate complex constitute the reef proper; (ii) it is usually unclear whether particular corals are confined exclusively to whatever facies and formations are regarded as reefal; and (iii) the nature of reefs has clearly been very different over large spans of time (Wood, 1993, 1995; Webb, 1996), and to treat them as a single ecological entity can only lead to false conclusions. As in situ biogenic carbonate structures, it is evidently valid, sedimentologically, to group reefs together (James, 1983; Hüssner, 1994), but the great palaeoenvironmental range of reefs, and the boundless variety of organisms that have inhabited or constructed them over the whole of their geological record (*ca.* 3500 Ma: Hüssner, 1994), beg the question of why most palaeoecologists invariably treat them as a continuously changing sequence of a single kind of community. Terrestrial ecologists do not treat all forests and woodlands as a single kind of ecological unit.

The problem is particularly highlighted for the Cretaceous, for which there are substantial unresolved differences in the literature about which organisms were the reef builders, the kinds of reefs or 'bioconstructions' that exist, and whether in fact true reefs even existed at all at this time (Masse, 1977; Tchechmedjieva, 1986; Coates & Jackson, 1987; Kauffman & Fagerstrom, 1993; Wood, 1993; Gili *et al.*, 1995*a*; Scott, 1995; Webb, 1996; Skelton *et al.*, 1997). Part of the problem lies with the general, longstanding question of reef and framework definition, for which two recent papers in particular (Webb, 1996; Insalaco, 1998) offer a useful way forward from a biological standpoint. In the meantime, it is clear that the term 'reef coral' has no consistent palaeoecological or sedimentological meaning that can be used here as evidence for algal symbiosis in fossil corals. A reef-independent approach is needed instead. Although algal symbiosis in fossil corals cannot be observed directly, it is preferable to 'reef corals' as a palaeoecological criterion because the most fundamental distributional patterns of scleractinians are much more precisely related to whether they are symbiotic or not, than to reef-based criteria. There has been no previous review of the history of algal symbiosis in fossil corals based on all the different existing methods of inference. As will be seen, a reef-independent approach also reveals a very different picture of coral history over the K–T boundary, than the customary one derived from reef-based studies.

12.4 TOWARDS A HISTORY OF ALGAL SYMBIOSIS IN SCLERACTINIAN CORALS

12.4.1 Background

Algal symbiosis occurs in about half the known number of living genera (Wilson & Rosen, 1998). The distribution of symbiosis in extant coral taxa cuts across formal scleractinian classifications, suggesting that it has probably been lost in some groups and/or has evolved more than once. Extant symbiotic corals (i.e. zooxanthellate or z-corals) are restricted to warm, shallow waters of the Tropics and tropical margins, whereas non-symbiotic (i.e. azooxanthellate corals or az-corals) are not restricted, though they also overlap in distribution with z-corals (Rosen, 1981; Schuhmacher & Zibrowius, 1985; Stanley & Cairns, 1988). Amongst other things, z-corals derive carbon compounds from their zooxanthellae whose presence also appears to facilitate coral calcification, while the partners recycle their nitrogen and phosphorous. Spencer Davies (1992) has reviewed these metabolic aspects.

Algal symbiosis gives z-corals a novel metabolic capability compared with az-corals. In addition to being carnivorous (like most cnidarians), they are also autotrophic and effectively photosynthetic. Also, the following features are of particular interest in the search for palaeontological clues to symbiosis, although as Wood (1993) points out, the possibility of preadaptation must also be borne in mind. Z-corals exhibit photoadaptive habits, some of which at least are inherited, suggesting coevolution with algal symbiosis (Coates & Oliver, 1973; Porter, 1976; Stearn, 1982; Coates & Jackson, 1987; Wood, 1993; Insalaco, 1996). Symbiosis appears to confer faster skeletal extension rates on z-corals, which also attain a much larger skeleton size and exhibit a broader range of colonial organizations and colony shapes than az-corals. Coloniality is far more common and often more complex amongst z-corals than az-corals.

12.4.2   Evidence of symbiosis

Since the algal symbionts are not preserved in fossil corals, indirect methods of inference have to be used, as summarized in Table 12.1. All methods shown are independent of reef-based evidence. Although stable isotopes and taxonomic uni-formitarianism are probably the most reliable methods, these methods cannot always be applied. Morphological methods, though probably more hazardous, can be applied to almost any reasonably well-preserved coral, and can even be applied to good published figures and descriptions. The scope for morphological methods is derived from the way in which particular morphologies are found only in those modern corals which are symbiotic (Section 12.4.1 above). Examples of this approach, applied to Paleocene corals, are shown in Table 12.2. Since inference of algal symbiosis in extinct corals represents an historical hypothesis for the corals concerned, the terms 'zooxanthellate-like' (z-like) or 'azooxanthellate-like' (az-like), should be used (Rosen & Turnšek, 1989). Ideally, such hypotheses are strengthened by obtaining similar indications for the same corals, using different independent methods, although so far, insufficient work has been done to make this possible in more than a few cases.

12.4.3   Results: (i) Historical overview of algal symbiosis in scleractinians

Works which indicate the historical record of algal symbiosis in fossil corals are shown in Table 12.1. In addition, *Goniopora websteri* samples from the Bracklesham Group (Eocene: Early Lutetian) of southern England have z-like isotopic composition (P. Swart, B. Rosen and J. Darrell, unpublished data). This is consistent both with the morphology of the genus and the z-status of its living congeneric counterparts (Table 12.2). It is also compatible with warm water conditions in southern Britain, notwithstanding a relatively high palaeolatitude of about 45° N at that time (Adams *et al.*, 1990). Since the corals occur in a bedded glauconitic sand (Curry *et al.*, 1977), however, they also exemplify the weakness of relying on reef occurrence as a criterion for inferring algal symbiosis.

A substantial proportion of Eocene genera are extant z-corals, so algal symbiosis can be inferred in these by taxonomic uniformitarianism. On these grounds alone therefore, it seems that algal symbiosis was continuous throughout the Cenozoic. This conclusion is also consistent with evidence from morphology, since nearly all the Paleocene examples of extant z-genera in Table 12.2 have colony organizations found only in living z-taxa.

Whether algal symbiosis was also continuous from the Carnian (Triassic) is still not clear as sampling so far is insufficient to rule out hiatuses ('collapses') especially at times of mass extinctions. However, data assembled here for the K–T boundary and Early Paleocene do shed interesting light on coral history and algal symbiosis over this particularly well-known extinction interval.

Table 12.1. *History of algal symbiosis in fossil scleractinians*

	Stable isotope signatures	Coloniality methods		Ecological functional morphology	Taxonomic uniformitarianism (*sensu* Dodd & Stanton 1990)
		Comparative morphology	Assemblage characteristics		
	Stable isotope signatures characteristic of skeletons of living z-corals indicate symbiosis in fossil corals (Stanley & Swart, 1995).	Empirical, comparative morphology of individual taxa: morphologies found only in living z-corals indicate symbiosis in fossil corals (Rosen, 1977; Rosen & Turnšek, 1989; this chapter).	Empirical comparisons of coloniality characteristics of whole coral assemblages: characteristics of living z-coral assemblages indicate symbiosis in fossil coral assemblages (Coates & Jackson, 1987).	To date, only Schlichter–Insalaco model: lamellar colonies with confluent corallites and pennular septa are a known adaptation to symbiosis in poorly lit waters (Schlichter, 1991) and indicate symbiosis in fossil corals (Insalaco, 1996).	(Species and genera that are zooxanthellate today, were also zooxanthellate in the past.
Neogene					This chapter
Oligocene			Coates & Jackson, 1987		This chapter
Eocene	P.K. Swart *et al.* unpublished data		Coates & Jackson, 1987		This chapter
Palaeocene		Rosen & Turnšek, 1989; this chapter			This chapter
Late Cret.		Rosen & Turnšek 1989	Coates & Jackson, 1987;[a] Skelton *et al.*, 1997	[a]Coates, 1997; [a]Darrell *et al.*, 1995, [a]Tchechmedjieva, 1986; [a]Skelton *et al.*, 1997	[a]Budd & Coates, 1992
Early Cret.				[a]Masse, 1977; [a]Scott, 1990 Insalaco, 1996	
Late Jur.			Coates & Jackson, 1987		
Mid-Jur.					
Early Jur.					
Late Trias	Stanley & Swart, 1995		Coates & Jackson, 1987		
Mid-Trias					

Upper rows show methods and the authors responsible. Stratigraphical part of the table cites works in which z-like corals have been inferred for the age indicated. All works cited are based on published or accessible coral collections. Corals from those parts of the geological record without citations are uninvestigated, though z-like corals are not necessarily absent from them. For further details, see text.
[a]Works in which corals are inferred here to be z-like, using the method shown (i.e. not necessarily the authors' interpretation).

Table 12.2. *Partial compilation of zooxanthellate coral records from selected Paleocene localities*

1	2	3	4	PALEOCENE (65–55 Ma)									
				DANIAN (65–60.5)				THANETIAN (60.5–55)					
				'earlier'		'later'							
GENERA	**GLOBAL STRATIGRAPHICAL RANGES**	**MORPH-OLOGY** (i.e. of these records)	**SYMBIOTIC CONDITION OF GENUS**	Slovenia	Italy	Pakistan	Slovenia	Alabama	Egypt	Italy	Niger	Pakistan	Somalia
Acropora	Paleoc - Rec 1,2	P-PC	z										●
Actinacis	M.Cret - Olig	**P-PC**	**z-like**		△▽			△▽		△▽		+ △▽	
'Astraea' 1		T	z-like 1									(◇)	
Cyathoseris	U.Cret - Mio	(T)	z-like[?] 2									▲▼	
Diploria	U.Cret - Rec	**M**	**z** 2							▲▼	▲▼	+ + ▲▼	
Euphyllia	Paleoc - Rec 1	M	z									+	
Goniopora 2	M.Cret - Rec	**CP-PW**	**z**		▲▼	●	▲▼	▲▼		▲▼	▲▼	+ ▲▼	
'Goniaraea' 2		CP-PW	z 1,3										
'Porites'		C-PW	z 1,3										
Hydnophora 3	Cret - Rec	**MH**	**z**									▲▼	
Leptoria	U.Cret - Rec	**M**	**z**										
Pachygyra 4	U.Cret - Eoc	**M**	**z-like**						△▽			+ ○ +	
Pachyseris	Paleoc - Olig 1	M	z-like 1									+	
Pironastrea		T	z-like										
'Pseudastraea' 1	Paleoc - Olig	P-PC	z-like 1									+	
'Reussastraea' 5		T	z-like 1,4	▲▼								+	
Siderastrea 3	Cret - Rec 3	**A/C-PW**	**z**					●				+ ▲▼	●
'Thamnastraea' 6		T	z-like 1,4										
Stephanocoenia	Cret - Rec	**c-s**	**z**									●	▲▼ ▲▼
Stylophora	Paleoc - Rec 1	p-sc	z										

▲▼ Lazarus z-genus (extant)

△▽ Lazarus z-like genus (extinct)

(◇) z-like genus (extinct) ranges across K–T boundary; present record problematic

● z-genus (extant; not Lazarus)

○ z-like genus (extinct; not Lazarus)

+ z-like coral record; problematic identification

The table emphasizes the presence of z- and z-like Lazarus taxa (*sensu* Jablonski, 1986*b*), shown in bold characters, suggesting that algal symbiosis did not 'collapse' over the K–T boundary. Actual records of z- and z-like-corals also appear earlier than is widely stated (i.e. for 'reef corals') (see text).

Genera

Records are cited from the literature without revision unless otherwise indicated. Problematic usage or identifications shown in inverted commas. Records in alphabetical order except for probable synonyms which are grouped and indented beneath senior name.

1. Change of generic name probably required on nomenclatural grounds.
2. Probably *Goniopora* as listed above.
3. Described by author as *Staminocoenia*, his genus being a synonym of *Hydnophora* according to Wells (1956).
4. May not be this genus.
5. May be *Pavona*, Oligocene-Recent (Wells, 1956).
6. May not be this genus (usually written as *Thamnasteria*). One of the species (*T. balli*) recorded by the author, may be a *Siderastrea* as listed above.

Global stratigraphical ranges

Ranges after Wells (1956) unless otherwise indicated, and shown only for those records whose generic names are not problematic. Lazarus taxa are those known from both Cretaceous and Tertiary, but not recorded from earlier parts of the Paleocene. Information on youngest Cretaceous records is not compiled here, so Cretaceous duration of their K–T hiatuses are 'unknown'.

1. Ranges updated here as a result of this compilation.
2. Record also appears to be first appearance of this genus.
3. Although record is Lower Danian, this coral is not present at the base of cited authors' sequence, so is regarded here as a Lazarus genus.

Morphology

Morphological characters shown are those of the particular records compiled here and confined to features which bear on inference of algal symbiosis. Symbols in capitals denote morphologies which are found exclusively in modern z-corals. This is not an exhaustive list of possible morphological indicators of algal symbiosis.

A/C-PW   Astreoid/cerioid with porous wall
C-PW   Cerioid with porous wall
c-s   Cerioid but with unusual inner wall structure
CP-PW   Ceriod-plocoid with porous wall
M   Meandroid

MH   Meandroid (hydnophoroid)
P-PC   Plocoid with porous coenosteum
p-sc   Plocoid with solid coenosteum
T   Thamnasterioid
(t)   Genus used by author is usually thamnasterioid but apparently not in the form recorded by him.

Symbiotic condition of genus

Relevant morphological evidence is shown in column 3 and unless otherwise indicated applies to both the present records and the concept of the genus as a whole. Inference of algal symbiosis in records with problematic identifications, or where present record has morphological characteristics inconsistent with genus as a whole, is explained in annotations below.

z   present record belongs to extant zooxanthellate genus, i.e. algal symbiosis inferred here through taxonomic uniformitarian evidence (Table 12.1)
z-like   present record belongs to extinct zooxanthellate-like genus, for which algal symbiosis is inferred through morphology (column 3 and Table 12.1). Unless otherwise indicated, inference applies to both present records and concept of genus as a whole.
z-like (?) Generic identification problematic, but this genus is 'z-like' (see column 1).

1. Generic identification problematic but morphology is z-like.
2. Morphological indications for symbiosis for this particular record are uncertain.
3. Generic identification problematic, but probable valid name is extant z-genus (see column 1).
4. Generic identification is problematic, but possible senior synonym is extant z-genus (see column 1).

Sources of records

Alabama   Salt Mountain Limestone, south-western Alabama, USA (Bryan, 1991).
Egypt   Abu Tartur Plateau, Western Desert (Schuster *et al.*, 1993).
Italy   Maiella carbonate platform of Umbria-Marche part of the Appenines, Italy (Moussavian and Vecsei, 1995).
Niger   Close to base of Thanetian at Touhout, near Kaou, collected by Noel Morris, The Natural History Museum, London (B.R. Rosen and T.S. Foster, unpublished data).
Pakistan Ranikot Series, western Pakistan (Duncan, 1980; Gregory, 1930); revised ages from Adams (1970).
Slovenia Dolenja Vas, western Slovenia (Drobne *et al.*, 1988).
Somalia   Auradu Limestone, Berbera-Sheikh area, dated as Thanetian in this area (Carbone *et al.*, 1993).

12.4.4   Results: (ii) Algal symbiosis in corals from the K–T boundary to the Paleocene

The K–T boundary and Paleocene are of particular interest, because of the supposed collapse of algal symbiosis at this time, perhaps in response to changes in global nutrient flux. Symbiosis is also said by some authors to have taken a particularly long time (up to 10 Ma) to recover (Copper, 1989; Veron, 1995). This scenario is based largely on the supposed absence of reefs over this interval (e.g. Talent, 1988; Copper, 1989), an approach which is rejected here as basically circular. In any case, it is now clear that the earliest occurrence of reefs in the Paleocene is sooner than this (see below).

It must be emphasized that there is, as yet, no evidence from the corals themselves that algal symbiosis collapsed, only that z-like corals were significantly more susceptible to extinction than az-like corals (the same appears to have been the case at the Albian–Cenomanian boundary) (Rosen & Turnšek, 1989). Both z-like and az-like corals show parallel but not quantitatively similar changes in taxonomic richness to those of the scleractinian order as whole (Fig. 12.1), with about 70% of z-like corals going extinct compared with about 40% of az-like corals. Z-like corals, however, did not completely disappear and Rosen and Turnšek (1989) estimated that there were nine post-Paleocene Lazarus corals. Four of these corals were z-like: *Actinacis*, *Favia*, *Goniopora*, and *Montastraea*, the last three being extant z-genera (Rosen in MacLeod *et al.*, 1997; see also Budd *et al.*, 1994, Fig. 5).

Further details are shown in Table 12.2. Although this is not a comprehensive compilation of Paleocene z-like corals, it is sufficient to support the following points:

(i)   If there was a breakdown in algal symbiosis during K–T extinctions, the hiatus was much shorter than previously thought, perhaps up to 2–3 my and probably not more than about 5 my.

(ii)   Even though the coral records themselves imply a hiatus, the presence of z-like Lazarus genera implies that algal symbiosis probably continued in corals from the Cretaceous, notwithstanding extinctions.

In addition to the Lazarus taxa mentioned by Rosen and Turnšek (1989), listed above, seven further such genera are shown in Table 12.2 (*Cyathoseris*, *Diploria*, *Hydnophora*, *Leptoria*, *Pachygyra*, *Siderastrea* and *Stephanocoenia*), bringing the total to eleven. All but two of these (*Actinacis* and *Cyathoseris*), moreover, are living z-genera. Moreover, Budd and Coates (1992) have demonstrated the presence of *Montastraea*-like corals throughout the Cretaceous. These are z-like in character on morphological grounds (porous coenosteum).

12.4.5   Remarks on the results

In the present approach, coral taxa must be taken to be az-like unless there is positive evidence that they were z-like. However, there are also positive ways of inferring az-status, mostly using parallel methods to those used in Table 12.1. These can be used positively to corroborate times and places where algal symbiosis was absent. Many Danian coral assemblages consist of corals whose morphology and mode of occurrence points positively to their being az-like and often include the extant az-genus *Dendrophyllia* (e.g. Stanley & Cairns, 1988; Bernecker & Weidlich, 1990). Drobne *et al.* (1988) described a Danian sequence which commenced with az-corals before z-corals appeared. Their earliest z-coral record (*Siderastrea*) appears to be the earliest known record to date for z-(or z-like) corals in the Tertiary (Table 12.2).

While the evidence of the Lazarus z-taxa would seem to be overwhelming in favour of continuity rather than collapse of algal symbiosis in corals over the K–T boundary, it is important to point out the limitations of the above Lazarus evidence. First, the taxonomy and stratigraphical ages of nearly all Cretaceous to Paleocene records of the taxa involved should be properly revised, with the possible exception of *Montastraea* which has been studied by Budd and Coates (1992) and Budd *et al.* (1994, fig. 5). Secondly, although the incidence of algal symbiosis is largely based on genera rather than species, there are a few genera living today comprising both z- and az-species, suggesting that generic-level inferences should not regarded as water-tight. However, all Lazarus z-genera in Table 12.2 also have z-like morphologies.

12.5   DISCUSSION: GLOBAL CHANGE, DIVERSITY AND ALGAL SYMBIOSIS

12.5.1   Diversity and extinction patterns

The environments inhabited by living scleractinians range from the abyssal floors of the oceans and the shelves of cooler high latitudes, to deep water coral banks and tropical carbonate shelves, platforms and reefs. This environmental range for scleractinians was established by the Early Jurassic (Toarcian) (Roniewicz & Morycowa, 1993). It follows that post-Jurassic scleractinians could have occupied the same possible environmental range, and this has to be kept in mind in any attempt to understand the historical patterns treated in Section 12.2. It is difficult to see how most of the events and changes discussed by Gale and Pickering (this volume) might apply to these coral patterns because different kinds of corals are known to be affected in different ways, and because we do not know how they respond on these scales. While shallow water corals would presumably be more affected by sea level changes than deep water corals, the reverse would be true for all but the most extreme anoxic and dysaerobic events. The restricted temperature range of z-corals suggests that they should be more directly affected by climatic fluctuations than az-corals (e.g. Adams *et al.*, 1990); and whereas z-corals are widely thought to be adversely affected by eutrophic conditions (Hallock & Schlager, 1986), eutrophic effects on az-corals are not known. These corals, being non-symbiotic, might even benefit from increased nutrients. It is possible that all the patterns in Fig. 12.1 are dominated by data from shallow-water, reefal or z-corals, or that the general pattern simply mirrors the patterns shown by these kinds of corals, but this cannot be assumed *a priori*.

However, since first-order spatial patterns of species richness are closely related to energy levels (Fraser & Currie, 1996), first order explanations for temporal patterns might also be sought in broadly climatic terms. Even if most az-corals inhabit cool, dark conditions, their diet will presumably depend at least in part on energy levels in the shallowest waters above them. This said, the direct effects of climate will be overprinted by factors which affect nutrient levels, such as land erosion, run-off and oceanic turnover, and these cannot be assumed to respond in concert with climatic change. When the four radiation phases of Table 12.1 are compared with the global temperature curves shown by Gale and Pickering (this volume), we find that only the Hauterivian to Albian radiation took place against a climatic warming trend, and the other three all correspond to phases when climate was cooling or showed a combination of broadly stable conditions and cooling.

Conversely, diversity 'lows' and major extinctions in corals might be expected to correspond to climatic coolings. From Gale's and Pickering's curves, this holds for the K–T boundary (cool), but not for Albian–Cenomanian extinctions (warm). The temperature fall around the Eocene–Oligocene boundary is also more marked than any other Paleogene fall, but this is too early to explain Chattian (Oligocene) extinctions. The severity of climatic cooling during the Quaternary (Pickering, this volume) combined with Caribbean reef-coral turnover and extinctions in particular (Budd *et al.*, 1996) were possibly part of a larger global pattern at this time, but global coral data are insufficient to investigate this. The evidently poor correlation between the coral patterns of Fig. 12.1 and temperature curves suggests that (i) the time scales and resolution of the coral data need to be matched more closely with those of the temperature curves, (ii) removal of corals from the dataset which are not confined to warm shallow tropical waters would make it easier to identify possible correlations (see below), or (iii) that factors other than warm temperatures are exerting a more significant influence on coral patterns. It is also possible, of course, that the coral data are inadequate to draw useful conclusions.

However, the quest for all such 'geoecological' (*sensu* Rosen, 1988a) explanations is simplistic, since it is naïve to attempt to explain temporal diversity patterns entirely in terms of contemporaneous ambient conditions alone, such as temperatures or sea levels. Such factors may well be important in maintaining both abundance and diversity of communities once established, and, conversely, in explaining extinctions, but a complete theory also needs to consider their possible prior historical effect on speciations (e.g. Flessa & Jablonski, 1996). Changing geographies must be considered here too, not only as potential isolating mechanisms (e.g. vicariance), but also to help explain the differential accumulation of species in some areas rather than others (e.g. Wilson & Rosen, 1998). Rosen (1981, 1984, 1988a, b, 1992) has grouped these biogeographical processes under the headings 'maintenance', 'distributional change' and 'originations', emphasizing that biogeographical patterns (e.g. diversity) potentially derive from all three. While maintenance processes act contemporaneously, however, the significant events arising from the other processes are likely to have happened some time previously (on various possible timescales). In short, the longstanding tradition of trying to explain biogeographical patterns in terms of contemporaneous conditions alone is unsatisfactory, but one also has to match temporal scales of likely processes to the temporal scales of the observed patterns.

One possible proxy for isolation and vicariance is endemicity, and it is of interest that both the Cretaceous radiation phases of Fig. 12.1 correspond to episodes of rising endemicity as determined by Coates (1973), his coral data being, albeit, for Tethyan reef dwellers only. Unfortunately, other more recent palaeobiogeographical accounts lack consistent data to examine this further for the Cretaceous, although vicariance, isolation and endemicity seem to have had an important bearing on reef–coral patterns in the Tertiary (Rosen, 1984, 1988*b*; McCall *et al.*, 1994; Wilson & Rosen, 1998).

12.5.2   Global change and the history of algal symbiosis

The patterns of Fig. 12.1 might be more revealing with respect to global change if the coral data were broken down into more obviously environment-related groups. 'Reef–coral' data are already available for some of the post-Jurassic record, but were not compiled here, first because they are not available for the whole time-span required, and secondly, as already explained, there are pitfalls in this approach, especially when considering the Cretaceous and Cenozoic together. Use of z/az-coral categories is therefore preferred.

Present data point to historical continuity of algal symbiosis in corals at least from the Oxfordian onwards, although the Valanginian–Hauterivian; and Albian–Cenomanian extinction/turnover phases should be particularly investigated. The possibility of symbiosis going back as far as the Carnian, whether continuously, or with hiatuses, notwithstanding evolutionary turnover of both hosts and symbionts, might be thought an evolutionary problem. As discussed by Buddemeier and Fautin (1993), however, there is continuous recombination of different living taxa on each side of the partnership, at least on an ecological timescale. This suggests a possible evolutionary model in which the symbiosis is maintained through time by lineages of both algae and coral hosts, and this may be the reason why it has persisted longer in geological time than the likely duration of any single species involved in the partnership.

The most important conclusion here is that, contrary to what is widely believed, algal symbiosis seems to have been continuous across the K–T boundary. None the less, diversity of z- and z-like corals appears to have been greatly reduced in the Paleocene. Although coral reefs are now known from much earlier Paleocene intervals than previously realized, there still remains an initial phase of up to 2–3 my, possibly 5 my, in which they were absent (Bryan, 1991; Moussavian & Vecsei, 1995). This suggests severe disruption of z-coral habitats, and this could have resulted from both cooler conditions and reduction of shelf areas through lowered sea levels at around this time (Gale and Pickering, this volume). Darkness resulting from high dust levels generated by meteoritic impact and/or volcanic activity could also have contributed (Rosen & Turnšek, 1989). On the other hand, continuity of algal symbiosis in the form of Lazarus z-taxa implies that z-corals did survive these various changes, so this must have happened in reduced communities and populations in various refuges, too few or too inaccessible to have yet been discovered. Rosen and Turnšek (1989) proposed that one likely region was that of Pacific coral islands and sea-mounts whose Paleogene sequences are almost entirely subsurface (Winterer, 1991).

It might be argued that Lazarus z-genera are poor evidence for z-continuity since there is no way of knowing if genera or even species could have lost their zooxanthellae but nevertheless continued to exist without significant morphological changes, effectively as az-corals with z-like 'morphological façades'. Isotopic methods would be the only way to check such corals, if and when unaltered aragonitic specimens are found. Veron (1995, p.118) has suggested that such corals might also have migrated into typical az-coral environments (cooler and deeper waters) where their chances of preservation would be poor. If so, evidence is going to be difficult to find for this ingenious scenario, which is reminiscent of naval submarine tactics, with corals effectively diving to escape adverse surface conditions (meteorite flak?), and then waiting for the 'all clear' to resume their symbiosis later in the Paleocene.

12.5.3   Algal symbiosis, reefs and global nutrients

Having considered coral history independently from reefs, it is now possible to return to reef history with this added perspective. The most recent detailed reviews (Bryan, 1991, Moussavian & Vecsei, 1995) of reef distributions in the early Paleogene indicate that the well-known Paleocene hiatus in reef building was shorter than the figure of 10 my or so supposed by previous authors (cited above). Apparently, such hiatuses, whatever their duration, are 'best regarded as intervals during which the decimated populations [of frame-builders] took several million years to recover or for re-establishment of symbiotic relationships appropriate for luxuriant growth and grand-scale frame-building.' (Talent, 1988, p.341). The second of these explanations, however, is circular, as it implicitly assumes that reef corals are synonymous with z-corals, and that z-corals occur only on reefs, hence absence of reefs signifies absence of z-corals. Reef-independent evidence of z-like Lazarus genera (Table 12.2), however, shows that they probably persisted either in the absence of reefs or in reef formations not yet discovered. Even if the tropical shallow water reef ecosystem 'collapsed', it seems that algal symbiosis did not, since z-taxa, albeit impoverished, appear to have persisted, even though there are no direct records until later in the Paleocene. 'Collapse' and 'recovery' are attractively dramatic metaphors, but can mislead if not adequately qualified.

As a rider to the collapse model Kauffman and Fagerstrom (1993, p.321) stated that 'Reef communities with their complex biological interactions evolve over long time intervals; they are not inherited from surviving component lineages.' There is no evidence to support the second part of this statement for the K–T extinction. Veron (1995) has pointed out the continuity of reef-building faviids from the Cretaceous, and every one of the above Lazarus corals either became an important reef contributor at some time during the Cenozoic (e.g. *Actinacis*, *Diploria*, *Favia*, *Goniopora*, *Montastraea*), and/or has close sister-group relationships with other typical Cenozoic reef-building corals (*Cyathoseris*, *Hydnophora*, *Leptoria*, *Plerogyra*, *Siderastrea*, *Stephanocoenia*) (Wells, 1956; Pandolfi, 1992; Veron *et al.*, 1996; Romano & Palumbi, 1996; Johnson, 1998). Moreover, about 35% of living microstructural lineages of corals began in the Late Cretaceous (see above).

There are also doubts about the nutrient argument. First, Atkinson *et al.* (1995) have shown from experimental evidence that statements about z-corals being unable to grow in high-nutrient conditions are oversimplifications. Secondly, Done *et al.* (1996) pointed out that standing levels of nutrients on reefs may be low, but flux is high. Thirdly, $\delta^{13}C$ curves (Gale and Pickering, this volume; Martin, 1996) show a strong fall at the K–T boundary indicating decreased burial of organic carbon and perhaps therefore lowered global productivity and nutrient levels (i.e. global oligotrophy). This is in direct conflict with the nutrient excess hypothesis for explaining the K–T reef collapse.

So long as algal symbiosis is thought to have collapsed, it has been possible to argue that contemporaneous disappearance of reefs was linked with nutrients, perhaps through global eutrophication. Reef-independent evidence for persistence of symbiosis weakens this argument, but if algal symbiosis did not collapse, why were reefs absent in the Early Paleocene? An alternative model is needed to that of eutrophy-led reef demise, but there is space here to consider only a few of the possible factors and their synergistic interactions (Wood, 1993, 1995; Done *et al.*, 1996). Disruption of suitable environments remains the simplest proximal explanation for the combined coral and reef patterns. Perhaps the effects of eutrophication were more regional than global: eutrophic conditions might have intensified close to land masses, especially during the lowered sea-level phase, but not around open ocean settings like sea-mounts. The case of *Goniopora websteri* from the British Eocene (above), however, shows that some Tertiary z-corals could occur far from any reefs, in sandy, glauconitic inshore (and therefore nutrient-rich?) conditions. Worldwide reduction of suitable habitats would have resulted in greater isolation of communities, perhaps sufficiently to have significantly reduced larval exchange between their benthic metapopulations. It is notable that coral reef formations continued to be rare in south-east Asia, not just in the Paleocene, but also until the Late Oligocene, probably because of regional tectonic factors (Wilson & Rosen, 1998).

Turning to ecological function, and extrapolating from late Plio–Pleistocene Caribbean extinctions (Johnson *et al.*, 1995), initial extinctions at the K–T boundary may have been sufficiently drastic to have removed not only the small-colony species, but also the large-colony, clonal, 'keystone' species. Reefs would then not have reappeared until large-colony species returned – though in fact, such species are also scarce in the Cretaceous (see below). Alternatively or additionally, if as Webb (1996) has suggested, rigid reef growth-fabrics are due largely to early microbially mediated lithification and cementation, not to dense growth of large skeletal frame builders, then the relevant organisms, or conditions favouring them, were possibly absent in the Early Paleocene.

Discussion so far has concentrated on the Paleocene, but the Cretaceous itself is perhaps the key. Confusion about the nature of Cretaceous reefs has already been mentioned, and this may be symptomatic, since there seem to be very few instances of large-scale, large-colony coral growth-fabrics. Platy coral facies are perhaps one of the main exceptions, though in other respects these are not comparable with modern coral reefs either. Corals are otherwise typically modest in size (< 0.5 m) (Rosen in MacLeod *et al.*, 1997) and occur in well-stratified deposits. Yet, paradoxically, they are often extremely abundant and diverse (Fig. 12.1) and exhibit a wide range of forms typical of z-corals. It seems then that all the coral ingredients were present for reef-building except possibly large colony size. As with the Paleocene, however, the crucial factor seems to be cementation since Webb (1996) has concluded that true reefs, coral-built or otherwise, are absent from much of the Cretaceous, though global reasons for this are not yet clear. If Webb is right, the supposed post-K–T 'recovery' of reefs during the Paleocene actually marks the first appearance of microbially cemented reefs since the Jurassic.

If reefs as we know them today did not exist in the Cretaceous, however, we then cannot speak of their demise at the K–T boundary, and the role of algal symbiosis (or its supposed collapse) is irrelevant to this. This also throws doubt on the whole model of a succession of reef-EEUs through geological time punctuated by relatively brief collapses. Instead, for the later Mesozoic onwards at least, there was a continuous sequence of different kinds of z-like coral-dominated ecological assemblages, but apparently few if any reef communities, through the Cretaceous until the earlier Paleocene. While coral reefs appear to have commenced sooner in the Paleocene than many authors have appreciated, they actually mark the advent of a new kind of reef-coral community, not the recovery of an erstwhile one (see also Wood, 1993). This then brought to a close a remarkable reef hiatus that had actually lasted not just for the earliest Paleocene, but for most of the Cretaceous. By extension from the present study, the combined evidence of algal symbiosis history in fossil corals and Webb's cementation model of reefs, demands a critical re-evaluation of the idea that algal symbiosis holds the key to the distribution of reefs in space and time.

12.6   CONCLUSIONS

Diversity data point to three major 'boom and bust' cycles over the last 145 Ma, broadly Early Cretaceous, Late Cretaceous, Paleogene, and a fourth radiation in the Neogene. These cycles do not consistently correlate with any particular climatic trends, though the coral data are perhaps not adequate to investigate this properly. It is likely that historical factors like vicariance and endemicity were important but these have been discussed in detail elsewhere.

Coral history has, for too long, been partitioned into reef and non-reef categories, but this approach is flawed, ecologically. Notwithstanding practical difficulties of inferring algal symbiosis in fossil corals, partitioning into symbiotic (i.e. zooxanthellate-like) and non-symbiotic (i.e. azooxanthellate-like) groups is more informative. Z-like corals go back to the Late Triassic (Carnian), since when their record appears to have been continuous or nearly so until the present time. Continuity is especially likely for the Late Cretaceous onwards, notwithstanding great reduction in the number of z-like corals during the K–T extinction event, since at least eleven z-like Lazarus genera survived.

Thus algal symbiosis cannot have collapsed at this time, whether through global eutrophication or other causes, as some have argued, and these ideas, therefore, cannot be used to explain the paucity of reefs in the earliest Paleocene. Global eutrophication is also incompatible with the buried organic carbon record. Extent and variety of habitats for z-like corals must, however, have been severely disrupted at this time, but suitable refuge regions must have existed, perhaps in the western Pacific. Moreover, the popular idea of 'reef recovery' in the Early Paleogene is misleading. Not only is there a complete lack of any consensus about what constituted a coral reef during most of the Cretaceous, but recent evidence from cementation studies suggests that reefs did not exist at all then. Paleocene coral reefs must therefore mark the emergence of modern coral reef communities in their own right, not a recovery from a eutrophically driven collapse. Critical re-evaluation is due of the long-held idea that algal symbiosis holds the key to the distribution of reefs in space and time.

13

Changes in the diversity, taxic composition and life-history patterns of echinoids over the past 145 million years

ANDREW B. SMITH AND CHARLOTTE H. JEFFERY

13.1 INTRODUCTION

Echinoids have been one of the success stories of the Mesozoic and Tertiary. Since the Triassic, echinoids have been increasing in diversity, expanding to occupy new habitats and exploit new sources of food. Today there are approximately 900 extant species, more or less equally divided between regular (pentaradiate forms with the periproct enclosed by the apical system and lying at the opposite pole to the mouth) and irregular (secondarily bilateral forms with the periproct displaced outside the apical system) taxa. Echinoids are found in all marine habitats, from the Poles to the Equator and from the intertidal zone to more than 5000 m depth (Smith, 1984).

Many important changes have taken place in the evolution of echinoids over the past 145 million years. Diversities of echinoid clades have waxed and waned relative to one another, and the range of niches occupied by echinoids in marine benthic communities has expanded significantly. Specialist feeding strategies have been evolved and life history strategies have also been profoundly altered in certain groups. Unfortunately, in most cases insufficient work has been done to establish whether extrinsic, environmental factors are involved in driving these changes. However, in a few cases we do seem to have an environmental correlate. Major changes in preservational potential and perceived diversity probably correlate to rates of sea-level change, and a marked shift in favour of lecithotrophic (i.e. planktic, non-feeding) or brooded development was possibly driven by increasing climatic seasonality.

13.2 CHANGES IN DIVERSITY

Although sampling biases create huge problems for estimating how diversity has changed over time, there seems little doubt that echinoid diversity has increased over the past 145 million years. Kier (1975a, fig. 75) showed this graphically by plotting numbers of described species through time, and a similar pattern emerges from Sepkoski's (1981) factor analysis of familial diversity. However, raw diversity plots do not show a steady increase over time, but fluctuate wildly, forming distinct peaks and troughs which are much more pronounced at species level than family level. This pattern is almost certainly artefactual, caused by sampling or preservational biases that, to a large extent, mirror global or regional sea-level fluctuations. Preservational bias arises because during slow and progressive marine transgression onshore facies have an excellent chance of being preserved, whereas regression results in condensed sedimentary successions and loss of onshore facies. Furthermore, a sharp and rapid rise in sea level, such as occurred in the Late Cenomanian and Early Paleocene, results in platform drowning and a greatly restricted distribution of onshore facies. This is important because not only are echinoids most diverse and abundant within the photic zone in onshore settings, but their preservation potential is also highest in onshore environments (Kidwell & Baumiller, 1990). The net result of these two factors is clearly reflected in the diversity curve for the British Cretaceous fauna (Fig. 13.1). The Aptian, Cenomanian and Early Santonian represent times of progressively rising sea level, and are periods when high-diversity, more onshore echinoid faunas are recorded in abundance. In marked contrast the Late Cenomanian marks a rapid increase in water depth with highstand in the Early Turonian (Gale, this volume) and both have particularly poorly known faunas.

Rates of sea-level change thus directly influence preservational potential and our perception of taxonomic diversity. On a global scale McKinney *et al.* (1992) attributed the poor fossil record of the Early Oligocene to extinction driven by climatic cooling. Yet the alternative, that it represents a sampling artefact, was not even considered. Only through taking a phylogenetic approach will it be possible to recognize preservational biases and distinguish between sampling artefacts and genuine extinction patterns.

The number of fossil species recorded from any local region is to a large extent dependent upon intensity of sampling, the range of environments represented at outcrop, taphonomic biases affecting preservation, and the taxonomy employed by monographers. To examine how echinoid diversity has changed, it is therefore essential to compare faunas from similar habitats over time. Ideally, these should also have been intensively studied and recently monographed to minimize taxonomic problems. When well-studied faunas from low- to mid-latitude shallow-water carbonate platforms are compared, diversity is at a maximum during the Late Cretaceous and shows a drop after the Cretaceous and again after the Early Miocene (Table 13.1). Echinoids in shallow carbonate shelf environments appear to have been more species rich during the Late Cretaceous due largely to the high diversity of regular echinoids at this time. Whereas Late Cretaceous faunas include almost 50% regular echinoid species, post-Cretaceous faunas are dominated both in terms of species and numbers of specimens, by irregular echinoids. The loss of taxonomic diversity at the end of the Cretaceous is probably related to habitat loss through platform drowning, as discussed below.

Diversity in present-day shallow-water environments can be high, but a fairer comparison is with species that actually end up entombed in present-day sediment samples. Only Nebelsick (1992) has carried out this sort of large-scale actuopalaeontological investigation of echinoids in a modern carbonate platform environment with diverse facies. He found that, in his Red Sea study area, only 13 out of 23 species entered the fossil record as dead tests, although a further seven species could be distinguished on the basis of fragments. The diversity he records

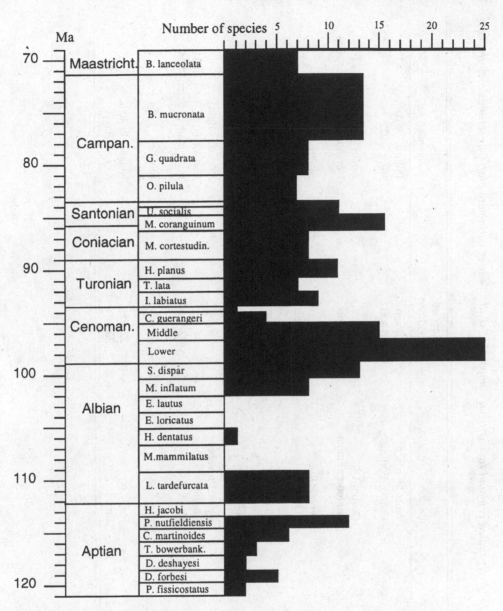

Fig. 13.1. Diversity of regular echinoid species recorded from the Aptian to Maastrichtian of the British Isles (data from Smith & Wright, 1989–1996). The major peaks coincide with times of progressive transgression. Timescale from Jenkins *et al.* (1994).

seems rather low in comparison with fossil assemblages, but Kier (1975*a*) found similar levels of diversity in the carbonate platform sediments off Belize. There appears, therefore, to have been a genuine drop in species richness since the Miocene. This decline in faunal diversity coincides with a marked deterioration in global climate and general dwindling of carbonate shelf habitat.

Table 13.1. *Species diversities for well-studied echinoid faunas from comparable shallow-water carbonate platform settings over the last 145 Ma*

Age	Region	Species recorded	Reference
Late Valanginian –	West Morocco	28 spp.	Petitot, 1961
Early Hauterivian	Lisbon, Portugal	36 spp.	Reys, 1972
Early-mid-Cenomanian	Devon, UK	44 spp.	Smith *et al.*, 1988
	Charente, France	45 spp.	Néraudeau & Moreau, 1989
Maastrichtian (Simsima Fm)	Oman-UAE	45 spp.	Smith, 1995*b*
Maastrichtian (Maastricht Fm)	Maastricht	46 spp.	Ham *et al.*, 1987
Eocene: Ypresian	Egypt	30 spp.	Roman & Strougo, 1994
Lutetian	Carolinas, USA	27 spp.	Kier, 1980
Lutetian (Khirthar Fm)	Pakistan	ca. 32 spp.	Duncan & Sladen, 1884
Late Oligocene –	Basse Provence	25 spp.	Negretti, 1984
Early Miocene	Anguilla	34 spp.	Poddubiuk & Rose, 1984
	Malta	32 spp.	Poddubiuk & Rose, 1984
Pliocene	Red Sea	23 spp.	Ali, 1985
Recent	Red Sea	22 spp.	Nebelsick, 1992
	Belize	20 spp.	Kier, 1975*b*

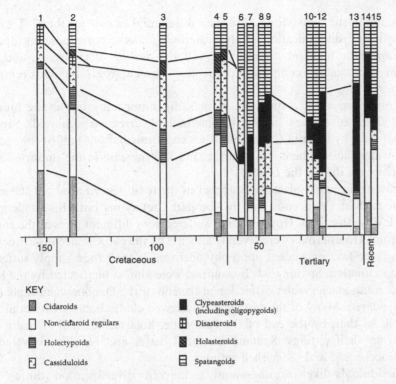

Fig. 13.2. Changes in the taxonomic composition of shallow-water carbonate platform echinoid faunas over time. Columns 1–13 represent well-studied faunas and show the relative proportions of major taxa as a percentage of the total. Faunas are as follows:- 1, Sequenian/Kimmeridgian of Morocco (Petitot, 1961); 2, Valanginian/Lower Hauterivian of Portugal (Reys, 1972); 3, Lower-Middle Cenomanian of Charente Maritime, France (Néraudeau and Moreau, 1989); 4, Maastricht Formation, Upper Maastrichtian of the Maastricht district, Netherlands (Ham *et al.*, 1987); 5, Simsima Formation, Upper Maastrichtian, United Arab Emirates – Oman borders region (Smith, 1995); 6, Upper Ranikot Limestone, highest Thanetian of Sind, Pakistan (revised and modified from Duncan & Sladen, 1882); 7, Libyan Formation, Ypresian, Egypt (Roman & Strougo, 1994); 8, Yellow Limestone Formation, uppermost Lower to lowermost Middle Eocene (modified from Donovan, 1993); 9, Santee and Castle Hayne Limestones, Middle Eocene of North and South Carolina (Kier, 1980); 10, Late Chattian – Early Burdigalian, Malta (Podubiuk & Rose, 1984); 11, Chattian – Early Burdigalian, Antigua and Anguilla (Podubiuk & Rose, 1984); 12, Burdigalian of Nerthe, southern France (Negretti 1984); 13, Pliocene, Red Sea (Ali, 1985); 14, Recent, Red Sea (identified from sediment samples) (Nebelsick, 1992); 15, Recent, Belize (Kier, 1975*b*). Note that the increased proportion of regular echinoids in Recent samples is at least partially due to taphonomic biases.

13.3   CHANGES IN TAXONOMIC COMPOSITION

The last 145 million years have seen dramatic changes in the taxonomic composition of echinoids in both shallow- and deep-water environments, much of which took place during the first 10 million years of the Cenozoic (Fig. 13.2). The most significant changes that have taken place in echinoids are: (i) the origin and diversification of crown-group atelostomates (Spatangoida and Holasteroida) from the paraphyletic disasteroids, and their differential fates after the K–T event; (ii) the origin and diversification of clypeasteroids and the corresponding decline of the paraphyletic remnant cassiduloids; (iii) the decline of cidaroids; and (iv) changes in taxonomic composition of the non-cidaroid regular echinoid component.

From representing a small but numerically important group in the Jurassic (i.e. about 25% of species in the Kimmeridgian), irregular echinoids rapidly expanded so that, by the mid Cretaceous, they constituted about half of the species in shallow-water facies. There was a further marked increase in the importance of irregular echinoids during the Eocene.

Spatangoids and holasteroids appear in the fossil record at about the same time, in the earliest Cretaceous, and by the mid Cretaceous both had undergone considerable diversification. However, they suffered very different fates at the end of the Cretaceous. Holasteroids were severely affected by the end Cretaceous extinction event, with those taxa dependent upon phytodetritus as their food supply suffering the greatest extinction. Spatangoids by contrast were almost unaffected by the K–T event, undergoing much greater extinction at the end of the Danian when chalk seas finally disappeared. Many of the surviving holasteroid clades became extinct at this time as well, so that, by the end of the Paleocene, holasteroids had virtually disappeared from shelf settings. Spatangoids fared better and rapidly recovered to dominate Eocene mid and outer shelf settings.

Cassiduloids likewise underwent a marked diversification during the Cretaceous and were badly hit at the end of the Cretaceous. Approximately 50% of lineages became extinct, predominantly from shallow carbonate platform settings. In the western Tethys cassiduloids remained important in shallow-water carbonates up to the end of the Thanetian, after which clypeasteroids rapidly diversified to become as diverse as cassiduloids and both groups formed an important component of shallow water benthos during the Eocene. Subsequently, cassiduloids went into decline as climate deteriorated and shallow water carbonate settings contracted, reaching their present low diversity by the Plio–Pleistocene.

Holectypoids, though not a major component at any time, were never important after the Cretaceous and virtually disappear from shallow carbonate settings by the Plio–Pleistocene.

Regular echinoids made up around 50% of the carbonate shelf echinoid fauna throughout the Upper Cretaceous, though usually less than 20% of the specimens (Smith, 1995). Cidaroids were initially diverse, forming up to 50% of the regular species diversity and about 30% of the total species numbers in the early Cretaceous. However, by the Cenomanian they had become numerically less important, forming approximately 20–25% of the regular echinoid fauna and 5–10% of the total fauna, and had declined in importance further by the Maastrichtian. However, in modern shallow carbonate platform settings, such as the one studied by Nebelsick (1992) in the Red Sea, regular echinoids once again comprise about 50% of the fauna but occur in very low abundance and are rarely preserved as whole tests. Part of the apparent decline of regular echinoids may therefore be taphonomic and due to increasing levels of bioturbation.

13.4   CHANGES IN LIFE-HISTORY STRATEGIES

Reproductive strategy within living echinoids can be divided into three types: planktotrophic (i.e. planktic, feeding), lecithotrophic (i.e. planktic, non-feeding) and brooded development. Several lines of evidence suggest that planktotrophy is the primitive developmental mode amongst the echinoids (Strathmann, 1978; Wray, 1996) and 145 million years ago all echinoids appear to have been planktotrophs. Since then it is estimated that a lecithotrophic or brooded mode of larval development has been independently evolved a minimum of 14 times (Emlet, 1990) and today, eight of the fourteen orders of echinoids have at least one species with non-feeding larval development (Emlet, 1990).

In the fossil record, lecithotrophy can be inferred from the orientations of crystallographic axes in the calcite of apical disc plates or from sexual dimorphism in gonopore size in adults (Emlet, 1989). The earliest known lecithotroph, based on gonopore dimorphism, is *Cyclaster platornatus* from the lower Maastrichtian of the Netherlands (Jagt & Michels, 1990). The late Maastrichtian *Cyclaster vilanovae* from Spain, is also inferred to have lecithotrophic development, based on analysis of apical disc c-axes (Jeffery, 1998). There is an earlier report of lecithotrophy based on sexual dimorphism in a Cenomanian hemiasterid spatangoid (Néraudeau, 1993). However, analysis of the crystallographic axes in the apical system does not corroborate this interpretation (R.B. Emlet, personal communication).

Brooded development may be recognized in the fossil record by the presence of depressed areas, or marsupia, on the adult test where offspring are sheltered (Kier, 1967, 1969). Earliest brooders amongst regular echinoids are *Thylechinus*, said to be from the Upper Cretaceous (probably Maastrichtian) of North Africa, and the Maastrichtian *Almucidaris* from the Antarctic and Spain. Amongst irregular sea urchins, the earliest undoubted brooder is the Danian *Abatus pseudoviviparus* from Madagascar, some specimens of which still have juveniles preserved in the sunken petals (Lambert, 1933).

Table 13.2 gives the first occurrences of non-feeding larval development in those lineages where it is known to occur, and Fig. 13.3 shows these first occurrences plotted against the geological timescale. Non-feeding larval development was independently evolved in eight lineages between the beginning of the Maastrichtian and the late Thanetian (a period of approximately 16 Ma). Although closely spaced temporally, lecithotrophy and brooding evolved over a wide latitudinal range from low-latitude carbonate platforms to polar waters. This is in marked contrast to today's distribution of marsupiate echinoids where, of the 28 extant brooding species, 25 are known from polar waters (Jablonski & Lutz, 1983). Past workers (e.g. Kier, 1969; Smith, 1984) have postulated that lecithotrophy and brooding are associated with cold water conditions and a lack of sufficient plankton for food for the larvae (Kier, 1968). It has been demonstrated for recent taxa (Emlet *et al.*, 1987; Emlet, 1990) that there is a significant correlation between decreasing planktotrophy and increasing latitude and depth. McNamara (1994) interprets this as the result of increased seasonality in nutrient supplies, rather than as a decrease in absolute temperature, in high latitudes. Lecithotrophic and brooded juveniles rely only on nutrient reserves in the large yolky eggs from which they hatch and are able to

Table 13.2. *First occurrences of lecithotrophy or brooding in major higher taxa of echinoids*

Lineage	First genera showing non-planktotrophic development	Age	Locality	First non-planktotrophic developmental mode
Cidaroida	Almucidaris[a]	Maastrichtian	Antarctic	Brooding
			Spain[g]	Lecithotrophy
Echinothurioida		Recent		Lecithotrophy
Arbacioida	Goniopygus[b]	Danian	Netherlands	Brooding
Phymosomatoida	Thylechinus[c]	Maastrichtian	North Africa	Brooding
Temnopleuroida	Arbacina?[b]	Danian	Netherlands	Brooding
Echinoida		Recent		Lecithotrophy
				Brooding
Holectypoida	Neoglobator	Thanetian	Northern Europe	Lecithotrophy
Cassiduloida	Oolopygus	Maastrichtian	Western Europe	Lecithotrophy
Holasteroida	Plexechinus[d]	Recent	Antarctic	Brooding
	Urechinus[d]			
Spatangoida	Cyclaster[e]	Maastrichtian	Western Europe	Lecithotrophy
	Diplodetus	Maastrichtian	Madagascar and USA	Lecithotrophy
	Mauritaniaster	Maastrichtian	Algeria	Lecithotrophy
	Tripylus[f]	Danian	Madagascar	Brooding

First occurrence of non-feeding larval development in lineages where non-planktotrophy is known.
[a]Blake & Zinsmeister, 1991; [b]Jagt & van der Ham, 1994; [c]Kier, 1969; [d]David & Mooi, 1990; [e]Jagt & Michels, 1990; [f]Lambert, 1933; [g]Gallemí, personal communication.

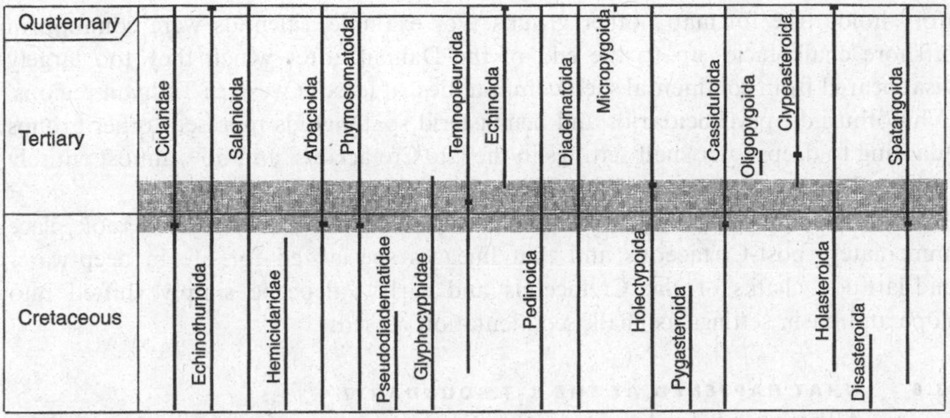

Fig. 13.3. Stratigraphical range chart for major post-Jurassic echinoid clades. First appearance of non-planktotrophic development in clades is indicated by solid rectangles. Note how many independent lineages evolved non-planktotrophic development in the latest Cretaceous–earliest Tertiary (stippled band).

develop in regions where other sources of nutrients are scarce. Planktotrophs, on the other hand, produce larvae capable of feeding and hence of surviving in the plankton for prolonged periods. Echinoids with feeding larvae thus have the highest dispersal capability of the three developmental types. Although this is certainly an advantage if adequate nutrient supplies exist, it is a definite handicap in areas of low or seasonal nutrient supply such as polar waters.

The near-simultaneous adoption of lecithotrophic and brooded development at the end of the Cretaceous and in the early Tertiary in many independent lineages over a wide geographical range is very striking. If McNamara's assertions about nutrient supplies and lecithotrophy and brooding are correct, this sudden move to a non-planktotrophic mode of larval development implies an increased seasonality in global climate starting in the Late Cretaceous and ties in with a marked fall in global temperature (Gale, this volume).

13.5   CHANGES IN HABITAT OCCUPANCY

The fossil record of echinoids from deep-sea environments is virtually non-existent, but it is clear that a number of major groups have moved from the continental shelf into the deep-sea over the last 145 million years. Again, much of this change appears to have taken place during the Paleocene. Today, the deep-sea environment is dominated by echinothurioids and holasteroids, with cidaroids, saleniids, spatangoids and a variety of other regular echinoids more common on the shelf slope. The great majority of recent deep-sea holasteroids form a monophyletic group whose origins can be traced back to Late Cretaceous and Danian taxa living in offshore chalk facies (i.e. their closest sister groups are Maastrichtian and Paleocene holasteroids from deep-water chalk facies). Holasteroids basically disappear from the continental shelf sedimentary record after the Paleocene, although a few taxa continued in onshore shelf environments into the Eocene in Australia. A similar story holds true for many other groups. For example, saleniids were common in offshore chalk facies up to the end of the Danian, after which they too largely disappeared from continental shelf communities, at least in western Tethyan regions. Echinothuriids, psychocidarids and hemiasterid spatangoids represent other groups adapting to deep-water shelf settings in the late Cretaceous, and now almost entirely restricted to the deep-sea environment.

It would appear then that much of the deep-sea colonization took place immediately post-Cretaceous and that lineages specialized for life in deep-water mid-latitude chalks of the Cretaceous and early Paleocene simply shifted into slope and basin settings as chalk sedimentation ceased.

13.6   WHAT HAPPENED AT THE K–T BOUNDARY?

Although major changes in faunal composition have occurred at various periods over the last 145 million years, such as the demise of cassiduloids after the Miocene, the most profound changes took place between the end of the Cretaceous and the late Thanetian (Fig. 13.2), possibly over several million years.

13.6.1   The size of the extinction event

Previous estimates for the number of echinoid taxa disappearing at the K–T boundary have been as high as 70% at the generic level (Roman, 1984). However, our recent phylogenetic-based assessment found only about a third of late Maastrichtian echinoid clades at approximately generic level have no post-Maastrichtian record and thus presumably became extinct (Jeffery & Smith, 1998). Previous estimates had been misled by taxonomic errors that inflated the number of genera present in Maastrichtian deposits, by pseudoextinction arising from inaccurate taxonomy and from errors of dating. Thus, although echinoids were undoubtedly affected by events at the end of the Cretaceous, they were not as catastrophically decimated as had previously been supposed. Estimates of about 50% extinction in other major groups (summarized in Jablonski, 1995) are probably inflated to an equal degree.

13.6.2   Extended extinction

There are marked differences in the extinction patterns of higher taxonomic groups. Certain taxa, such as the cardiasterid and stegasterid holasteroids, extended to the latest Maastrichtian and then largely disappeared. Holasteroids suffered more extinction of generic level clades than any other higher taxon at this boundary (> 50% of clades disappeared at the end of the Cretaceous). By contrast, the majority of spatangoids passed into the Danian, with less than 20% extinction of clades at generic level. However, a number of these spatangoid clades did not survive beyond the Danian, when mid-latitude chalk deposition ceased, and a similar pattern is shown by saleniids and phymosomatids. Thus the major extinction event for spatangoids appears to have occurred later than for the majority of holasteroids. Extinction levels were only slightly lower in the Danian (25%) compared to the Maastrichtian (34%) (Jeffery & Smith, 1998). This suggests that the drop in diversity in echinoids at the end of the Cretaceous was not a simple one-step catastrophic event, but the result of a complex and extended interplay of events.

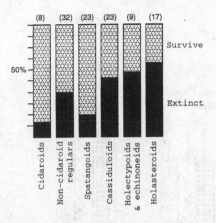

Fig. 13.4. Relative levels of extinction calculated for major higher taxa of echinoids at the end of the Cretaceous (unpublished data). Numbers in brackets = total number of clades.

13.6.3   Selectivity of extinction

Simply looking at the relative extinction levels suffered by different major taxonomic groupings it is clear that some groups suffered much less extinction at the K–T boundary than others (Fig. 13.4). Amongst regular echinoids stomopneustids and arbaciids fared rather badly in comparison to saleniids and phymosomatids, while spatangoids fared much better than either holasteroids or cassiduloids. There was some sort of selectivity in determining which taxa survived and which went extinct.

Yet, within individual major clades there is little evidence that morphological attributes played any significant role in determining survivorship. For example, cassiduloids suffered 50% loss of generic-level clades at the K–T boundary, yet morphological disparity for the group remained unaltered pre- and post-Cretaceous (Fig. 13.5). As Foote (1993) has argued, this finding implies that extinction was effectively random with respect to the morphological characteristics that differ amongst taxa. So the fine morphological adaptations that distinguish the various clades of cassiduloids seem to have played little part in determining which survived or became extinct.

The coarse pattern of survivorship seen at high taxonomic level could therefore be explained either by major differences in the lifestyle of the various clades, or by broad differences in habitat or geographical distribution. Our preliminary analyses indicate that species diversity within a clade makes no difference to survivorship: clades with more Maastrichtian species stood no significantly better chance of survival than species-poor clades (Fig. 13.6). This contrasts with the pattern seen in molluscs (Raup & Jablonski, 1993). Another of our findings directly linked to this observation is that geographically widespread taxa did not fare better at the K–T boundary. Since most echinoid species are geographically highly restricted, these two findings are not surprisingly correlated. There is, however, a strong pattern associated with latitudinal, but not longitudinal gradient. Taxa at high latitude show a higher probability of surviving than do low latitude taxa. Although this might be

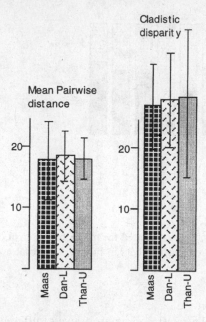

Fig. 13.5. Histogram of morphological disparity calculated from 57 discrete characters scored for Late Maastrichtian (25 taxa), 12 Early Danian (12 taxa) and Late Thanetian (22 taxa) cassiduloid genera. Error bars represent two standard errors. Morphological disparity was calculated in two ways: (i) from pairwise comparisons of the raw distance matrix for taxa present at each time interval, and (ii) from pairwise comparisons of total branch-length separation derived from the most parsimonious cladogram for taxa present at each time interval.

directly related to climate and temperature, a more likely explanation has to do with habitat destruction.

13.6.4   Loss of habitat?

The diversity within individual local communities shows no change across the K–T boundary, at least in the Maastricht district where the single documented Danian community (Ham, 1988) has just as diverse an echinoid fauna as any single Maastrichtian community (Ham et al., 1987, Table 2). A similar picture emerges for the Danish fauna (Gravesen, 1979; Asgaard, 1979), where changes are largely confined to species level. Despite the stability in species richness within these individual environments, the Paleocene fauna, taken as a whole, clearly shows a marked decline in numbers of species. This decline is, therefore, likely to reflect a loss of habitat type rather than a general decline in diversity across all habitats.

The end of the Maastrichtian marked a major sea-level low-stand and was followed by a rapid sea-level rise of maybe as much as 130 m (Keller et al., 1993). Low-latitude shallow carbonate shelf environments, which were extensively developed in the Maastrichtian of the Tethyan region, were drowned and virtually disappeared during the Danian, as did many of the clades associated with these habitats. By contrast, mid-latitude chalk facies, as in Kazachstan and the

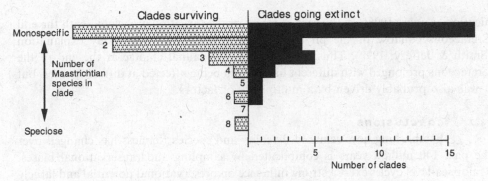

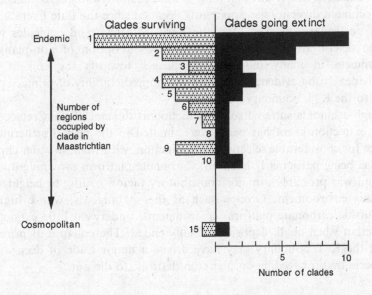

Fig. 13.6. Plots of survivorship vs. extinction for 59 Maastrichtian clades against (i) number of geographical regions occupied by the clade in the Maastrichtian, and (ii) number of included Maastrichtian species in each clade.

Denmark basin, continued well into the Danian along with most of their characteristic echinoid clades (e.g. Jeffery, 1997). Thus habitat loss associated with platform drowning might explain why the predominantly low-latitude, shallow, inshore cassiduloids were so badly affected whereas spatangoids, which have their greatest diversity in more offshore or deeper-water settings, were less affected. Platform drowning has also emerged as the most likely driving force behind extinction amongst Jurassic echinoids. Thierry and Néraudeau (1994) have shown very clearly how major extinction events affecting Jurassic echinoids correlate very strongly with the onset of major transgressive phases.

However, habitat loss cannot explain all disappearances, since we also find deeper-water taxa such as the holasteroid *Stegaster* going extinct from virtually continuous shelf marginal settings in the Biscay region (P. Ward, personal communication, August 1996). In the case of stegasterids starvation associated with the end Cretaceous collapse of the phytoplankton, appears the most likely explanation (Smith & Jeffery, 1998). Thus, not only was the faunal change at the end of the Cretaceous prolonged with different higher taxa being affected at different times, but it was also probably driven by a multiplicity of factors.

13.7   CONCLUSIONS

Establishing how echinoid diversity and species richness has changed over the past 140 million years is complicated by sampling and preservational biases. Major sea-level cycles exert a strong influence on preservational potential and largely control perceived taxic diversity through time. Within shallow-water carbonate shelf environments, echinoid diversity has apparently declined since the Late Cretaceous. This decline is particularly marked after the Early Miocene, and coincides with general climatic deterioration. The near synchronous adoption of non-planktotrophic development in many independent lineages towards the end of the Cretaceous provides strong evidence for increasing unpredictability of primary productivity prior to the K–T boundary.

Although echinoids suffered some extinction at the end of the Cretaceous, the size of the extinction event has been overestimated in the past. Furthermore, there is evidence for considerable selectivity of extinction, with shallow-water carbonate shelf faunas being particularly hard-hit. Carbonate platform drowning during the early Danian was probably a major contributory factor leading to heightened extinction in this environment. Groups such as the spatangoids, whose highest diversity lies outside carbonate platform environments, underwent little extinction until the Thanetian when chalk deposition finally ended. The collapse of primary productivity at the K–T boundary may have driven a major clade of deep-water holasteroids specialized for feeding on plankton detritus, to die out.

14

Origin of the modern bryozoan fauna

PAUL D. TAYLOR

14.1 INTRODUCTION

The great majority of bryozoans secrete calcareous skeletons and the phylum consequently enjoys a rich fossil record extending back to the Early Ordovician. Modern bryozoans are colony-forming, typically sessile, suspension feeders which can be key components of epibenthic environments (for summaries of bryozoan biology, see Ryland, 1970; McKinney & Jackson, 1989). Although marine bryozoans are distributed throughout the world from intertidal to abyssal depths, they peak in relative abundance on non-tropical continental shelves. Here, they may be the dominant carbonate producers, and fossil bryozoans are important components of many Cenozoic temperate limestones (Nelson *et al.*, 1988). Two essential factors controlling local bryozoan distribution are the existence of suitable hard (e.g. rocks and shells) or firm (e.g. algae) substrata for attachment, and an adequate supply of phytoplankton for food. Conversely, high levels of sedimentation and/or disturbance, and stagnant conditions on the sea-bed are disfavourable factors.

The first aim of this chapter is to undertake a short survey of the Cretaceous and Cenozoic fossil record of bryozoans in order to trace the origin of the modern bryozoan fauna. Temporal rather than the biogeographical or phylogenetic origin will be investigated, and evidence will be sought for possible effects of global change on broad-scale evolutionary patterns. Specifically, are there any changes in standing diversity, origination or extinction rates which appear to correlate with important global events? To answer this question, a general survey of the bryozoan fossil record from the Cretaceous to Recent will be first undertaken. The second part of the chapter focuses on the Neogene and Quaternary of the Mediterranean Basin where the bryozoan fossil record is more complete and species ranges are better established than elsewhere. It should be emphasized that local changes, notably the Messinian salinity crisis (see Pickering, this volume), not necessarily of importance on a global scale, may have profoundly influenced the evolution of the Mediterranean bryozoan fauna.

Given the deep limitations in our knowledge of the bryozoan fossil record, and of how specific environmental factors control the distribution of bryozoans living at the present day, it is difficult to draw any well-supported conclusions about the effects of global change on bryozoans through geological time, although areas for further study can be highlighted. Furthermore, bryozoan phylogeny is too poorly known to adopt anything other than a taxic approach with the inherent sampling biases discussed by Smith (1994).

14.2 CRETACEOUS-RECENT PATTERN OF BRYOZOAN EVOLUTION

14.2.1 Taxic diversity patterns

Cretaceous to Recent marine bryozoan faunas comprise species belonging to three orders: Ctenostomata, Cheilostomata and Cyclostomata. Cheilostomes dominate in both diversity and abundance at the present day. For example, Horowitz and Pachut's (1994) database contains 295 ctenostome species (6%), 4268 cheilostomes (81%) and 705 cyclostomes (13%). The fossilization potential of the three orders is not equal: ctenostomes lack mineralized skeletons and are therefore poorly represented as fossils; however, boring species can leave trace fossils, and encrusting species may be preserved by bioimmuration when overgrown by organisms with mineralized skeletons. Cheilostomes and cyclostomes both have rich fossil records, but whereas cyclostomes always have well-mineralized calcite skeletons, the occurrence of entirely aragonitic or bimineralic skeletons and also of weakly mineralized skeletons in some cheilostomes may have depleted cheilostome representation in the fossil record relative to cyclostomes (see below).

In view of their calcareous skeletons, the fossil record of bryozoans is disappointing, with only 11.8% of Recent cheilostome species and 12.3% of cyclostome species recorded as fossils in the database of Horowitz and Pachut (1996). These low figures were compared by Horowitz and Pachut (1994, 1996) with Valentine's (1989) value of 77% for Recent Californian Province mollusc species having a fossil record. One explanation of the poorer fossil record of bryozoans is non-preservation of surface characters hindering taxonomic identification (Horowitz & Pachut, 1994). However, the paucity of systematists actively studying Cenozoic bryozoans may be even more important. For example, the rich Cenozoic bryozoan faunas of the Gulf and Atlantic Coastal Plain of the USA have remained virtually unstudied since the major monographs of Canu and Bassler (1920, 1923), while those of Australia and New Zealand have not been revised since the pre-SEM works of Brown (1952, 1958) which dealt only with cheilostomes from a few faunas. In this context, the considerably better fossil record of Mediterranean bryozoans is significant.

Ctenostomes and cyclostomes both date back to the Ordovician. The fossil record shows that cyclostomes were subordinate to other extinct orders of stenolaemate bryozoans (trepostomes, fenestrates, cryptostomes and cystoporates) until the Early Jurassic when they began a protracted but modest evolutionary radiation lasting into the Cretaceous (Taylor & Larwood, 1990). Diversity had levelled off at a little under 20 families by the mid Cretaceous (Fig. 14.1). Cheilostomes, which undoubtedly evolved their calcareous skeletons independently of cyclostomes, do not appear in the fossil record until the Late Jurassic (Taylor, 1994a). They maintained a low diversity (1–2 families) for most of the Early Cretaceous before commencing an explosive radiation in the Late Albian–Cenomanian (Taylor, 1988). Cheilostomes have continued to diversify ever since (Fig. 14.1), although at a diminishing rate in the late Paleogene when their diversity levelled off at 70–80 families (Lidgard et al., 1993).

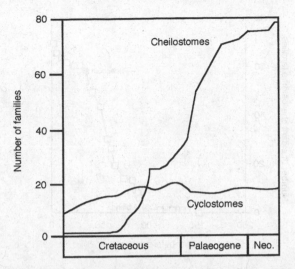

Fig. 14.1. Family diversity of cyclostome and cheilostome bryozoans from the Cretaceous to the present day (based on Lidgard *et al.*, 1993, fig. 6). Note the spectacular increase in cheilostome families since the mid Cretaceous, modest diversification of cyclostomes through the Cretaceous, and failure of either group to show a pronounced family extinction at the end of the Cretaceous.

Considering only extant cheilostome families having a fossil record, a cumulative plot of family diversity shows three phases with different rates of family addition (Fig. 14.2). From their first appearance until the Albian, new families were added at a rate of fewer than one per 100 my. The rate accelerated to approximately one family per my from the Albian to Priabonian, before declining to an average rate of one family per 3 my for the Priabonian to Recent interval but with some indication of an increasing rate of family addition towards the present day.

The general patterns described above are derived from compilations of global, family-level data (Taylor, 1993). The extent to which such higher level patterns can be used as proxies for species richness is debatable (Signor, 1985). As yet, there is no reliable genus- or species-level data for the Cretaceous, but Horowitz and Pachut (1994, 1996) have recently published species-level data for the Cenozoic based on an extensive survey of names in the literature. Their most recent analysis (Fig. 14.3) shows a species diversity peak in the Priabonian (817 spp.) followed by more minor peaks in the Burdigalian (365 spp.), Tortonian (418 spp.) and Piacenzian (306 spp.). However, stage-level dating was unavailable for many of the Mid Miocene occurrences (641 spp.), which were therefore omitted. This quite probably hides a Mid Miocene diversity peak. Horowitz and Pachut (1994) also looked at the times of origin of extant bryozoan species from Cretaceous to Recent. The number of species added per my to the modern fauna has increased continually through time (Table 14.1).

With regard to the K–T transition, Viskova (1994) recorded 178 cyclostome and 172 cheilostome genera in the Maastrichtian, and found that more than half of the genera in each group had become extinct by the Late Paleocene.

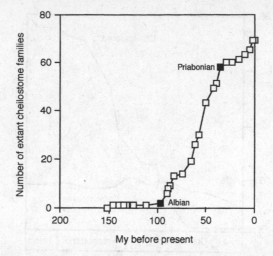

Fig. 14.2. Cumulative origin of the 70 extant cheilostome bryozoan families with fossil records (constructed using data in Taylor, 1993). Rates of family addition are low prior to the Albian, high between the Albian and Priabonian, and moderate but increasing between the Priabonian and present day.

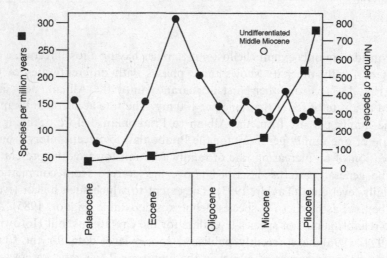

Fig. 14.3. Diversity of species names for stages of the Cenozoic, and numbers of species names per million years for Cenozoic series, from the bryozoan literature database of Horowitz and Pachut (1996). The open circle represents Mid Miocene species not allocated to stage level. Priabonian and likely Mid Miocene peaks in species diversity are evident. Species per million years exhibits a sustained increase towards the present day. Based on Horowitz and Pachut (1996, figs. 3 and 4).

Another measure of long-term diversity changes can be obtained from trends in assemblage (community) species richness. Lidgard *et al.* (1993) compiled a database of 676 Late Triassic to Holocene fossil bryozoan assemblages. Although variances in this database were found to be very high, both mean and median

Table 14.1. *Number and rate of species addition to the modern bryozoan fauna through geological time, using the data given by Horowitz and Pachut (1994, table 4) but recalculating rates according to the Berggren et al. (1995) timescale for the Cenozoic*

Epoch/Period	Number of first appearances of extant species	First appearances of extant species per my
Pleistocene	145	80.6
Pliocene	174	49.7
Miocene	235	12.4
Oligocene	48	4.8
Eocene	41	2.1
Palaeocene	0	0
Cretaceous	6	0.1

assemblage diversities have fluctuated through geological time (Fig. 14.4). Assemblages older than mid Cretaceous typically contain 10–15 species and rarely exceed 20 species, whereas mean diversity is 18–24 species for younger assemblages, with many assemblages having more than 40 species and a few in excess of 100 species. Most of the Cretaceous diversity climb is attributable to the addition of ever increasing numbers of cheilostome species. Average cyclostome diversity within assemblages shows a general decline through the Cenozoic, including a sharp drop in the late Neogene (Fig. 14.4). The rich bryozoan faunas of the latest Cretaceous in north-west Europe (e.g. Håkansson & Thomsen, 1979) are undoubtedly responsible for the high values of assemblage diversities prevalent at this time.

14.2.2 Ecological patterns

Jablonski and Bottjer (1990) examined the origin of evolutionary novelties in cheilostomes along an environmental gradient from onshore to offshore. Although they found that cheilostomes as a clade seem to have originated in an onshore setting, specific novelties showed no apparent pattern of environmental origination. A similar lack of pattern in novelty origination has since been found in cyclostomes which acquired their novelties (many of them convergent with those of cheilostomes) at a much slower rate (Jablonski et al., 1997). Perturbations in the abiotic environment through time appear to play little role in shaping such patterns (Jablonski & Bottjer, 1990).

Whereas data on global generic diversity and within-assemblage species diversity across the K–T boundary shows that relative proportions of cheilostome and cyclostome bryozoans did not change appreciably, a recent analysis of biomass has revealed an unexpected ecological perturbation. McKinney et al. (1998) found that the relative skeletal mass of cyclostomes increased abruptly after the K–T boundary from about 28% in the Maastrichtian to 72% in the Danian. This post-K–T interval of cyclostome ecological dominance lasted somewhere between about 4 and 25 million years; more data is required from the Paleogene before its extent can be established.

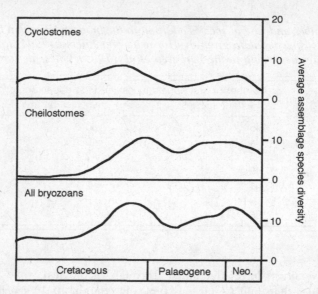

Fig. 14.4. Average species richness in bryozoan assemblages from the Cretaceous to the present day, shown for cyclostomes, cheilostomes and combined. Variances are high, but a general increase in assemblage richness can be seen through the Late Cretaceous, followed by a trough in the Early Palaeogene and a peak in the Early Neogene. Based on Lidgard *et al.* (1993, figs. 8–10).

Very few evolutionary histories have been investigated of specific ecological associations of bryozoans. Diverse bryozoan communities live epiphytically on algae and seagrasses at the present day. Algal associations are poorly represented in the fossil record, but seagrass associations can be traced back to the Late Cretaceous (Voigt, 1981), with some examples described from the Cenozoic (e.g. Ivany *et al.*, 1990). Most of the bryozoan species involved in modern seagrass communities are not specific to seagrasses (e.g. Hayward, 1975), and the variation through geological time in seagrass associations presumably reflects the pool of source species available at the time and cannot be studied as an independent system.

An association with a better fossil record is that between bryozoans and hermit crabs in which the bryozoan colonies construct distinctive helicospiral tubes extending from the apertures of gastropod shells inhabited by the hermit crabs (Taylor, 1994*b*). Few of these symbiotic bryozoans are obligate associates of hermit crabs but many are apparently facultative associates. Tube-building symbioses can be traced back to the Mid Jurassic and are represented by one species in the Jurassic and one in the Cretaceous, becoming commoner in Paleogene deposits. However, the main pulse of diversification occurred in the Early Miocene when bryozoans from many different clades formed tube-building symbioses. Data from New Zealand, where more than ten tube-building bryozoan species can be found in a single assemblage, contributes greatly to this pattern. Unfortunately, the fossil record of hermit crabs is too poor to test whether bryozoan diversification is simply tracking an increase in hermit crab diversity or abundance. It is conceivable that the important

Miocene changes in oceanic circulation patterns and productivity in the Southern Hemisphere could have played a role.

Lunulitiform cheilostomes possess a distinctive bryozoan colony-form adapted to a free-living mode of life. Their small, cap-shaped colonies have convex upper surfaces, on which the feeding zooids and specialized zooids with hair-like appendages open, and concave undersides, sometimes incorporating a sand grain or shell fragment as a tiny substratum at the centre. They live on the surface of sediments ranging in grade from mud to coarse sand, sometimes in high population densities and under strong current regimes. Two main cheilostome families – Lunulitidae and Cupuladriidae – have convergently evolved lunulitiform colonies. Cook and Chimonides (1983) have reviewed the fossil record and palaeobiogeography of lunulitiforms. Whereas the Lunulitidae first occurs in the fossil record in the Santonian of France, the Cupuladriidae date from the Paleocene of West Africa. Both groups attained an almost worldwide distribution and considerable diversity during the Cenozoic. The restriction of Lunulitidae to Australasia at the present day, and the progressive contraction in the distribution of Cupuladriidae, were inferred by Cook and Chimonides to have resulted from a combination of global environmental changes, including drops in sea temperature after the establishment of a circum-Antarctic current in the late Oligocene. A detailed study by Lagaaij (1963) of *Cupuladria canariensis* is important in the latter context. This abundant species occurs at the present day in the Atlantic and East Pacific between the 14 °C surface isotherms, consequently extending no further north than southern Spain along the east Atlantic continental shelf. However, during the warmer Miocene and Pliocene epochs, *C. canariensis* lived in the southern part of the North Sea Basin. Similar contractions in geographical range are not uncommon when Neogene and Recent distributions of species or higher taxa are compared (e.g. Cheetham, 1967; Vávra, 1980), attesting to the palaeobiogeographical effect of climatic cooling (see also Buge, 1982).

14.2.3 Interpretation

The prevailing pattern evident in the bryozoan fossil record since the beginning of the Cretaceous is one of an appreciable and largely uninterrupted rise in bryozoan diversity, both in terms of global diversity and local diversity within individual assemblages. At this stage in our knowledge it is difficult to eliminate the possibility that, at least following the explosive radiation of cheilostomes in the mid Cretaceous, the pattern observed owes more to sampling artefacts associated with proximity to the Recent, such as volume of fossiliferous rock available for study and the 'pull of the recent' in terms of intensity of study (e.g. see Benton, 1990), than it does to real diversity changes. Monographic bias may perhaps be responsible for the Priabonian species diversity peak (Fig. 14.3), a stage with well-studied bryozoan faunas in both Europe and North America.

Intrinsic biotic factors may have been largely responsible for the spectacular radiation of the cheilostomes (Taylor, 1988). The onset of this radiation in the Late Albian coincides with the first evidence for larval brooding, thought to signify the origin of short-lived, non-feeding larvae from the primitive cheilostome condition of having long-lived, feeding, non-brooded larvae. Such a switch is predicted to have decreased gene flow within and between populations, thereby increasing the like-lihood of speciation and triggering the explosive radiation. There is no obvious connection between environmental changes and the invention of a novel larval type. However, the Late Albian was also a time of major sea-level rise (Gale, this volume), and presumably greater areas suitable for bryozoan habitation.

The apparent post-Priabonian deceleration in cheilostome radiation evident in the family-level data for extant taxa (Fig. 14.2) coincides with the change from greenhouse to icehouse conditions (Fischer, 1984). If there is a causal link here, it is difficult to envisage the mechanism/s involved, especially in view of the lack of a pronounced latitudinal diversity gradient in bryozoans at the present day (Schopf, 1970), which might otherwise explain why the onset of global cooling would put a brake on diversification.

Mass extinctions are not conspicuous features of standing diversity trends (e.g. Fig. 14.1) in the Mesozoic and Cenozoic, in contrast to the striking effects of the end Permian extinction (e.g. Taylor & Larwood, 1990, fig. 10.1). As currently under-stood, the K–T event did not have a large effect on the evolution of bryozoans, with only 6 of 48 families (13%) becoming extinct, and bryozoans rapidly re-establishing themselves in the abundant and diverse Danian communities seen in Denmark (MacLeod et al., 1997). The effects, if any, of Cenozoic mass extinctions, including the well-known Eocene–Oligocene event (Prothero, 1994), have yet to be demon-strated in the bryozoan fossil record. Nevertheless, the geographical distributions of bryozoan taxa have apparently responded to global changes, in particular cooling in the Neogene–Recent, and one might expect this to influence temporal trends dis-cernible in the fossil record, whether or not these geographical changes have been translated into macroevolutionary patterns.

14.3   EVOLUTION OF MEDITERRANEAN BRYOZOAN FAUNAS

14.3.1   Recent Mediterranean bryozoan fauna

Bryozoans are among the most diverse sessile groups in the Mediterranean (Harmelin, 1992), with more than 350 living species. Zabala and Maluquer (1988) summarized the Mediterranean bryozoan fauna, listing 277 species of cheilostomes and 48 cyclostomes, giving a total of 325 species with potentially fossilizable miner-alized skeletons. Subsequent revisions (e.g. Harmelin et al., 1989) have added a few more species to this fauna. However, saturation diversity has almost certainly not yet been reached. Harmelin (1992, table 2) has summarized the biogeographical affinities of modern Mediterranean bryozoan species, nearly 34% of which are endemics.

14.3.2   Neogene and Quaternary species ranges

A database has been compiled from the literature giving stratigraphical ranges of Neogene and Quaternary bryozoan species recorded from the Mediterranean Basin. The principal references consulted for the database were Barrier et al. (1987), El Hajjaji (1992), Li (1990), Moissette (1988), Moissette and Spjeldnaes (1995), Moissette et al. (1993), and Pouyet (1991). The full database is to stage level, except for the Pliocene where the Piacenzian and Gelasian have been combined into a single stage representing the upper Pliocene. A total of 333 species are included in the database of which 133 are extant in the Mediterranean and an

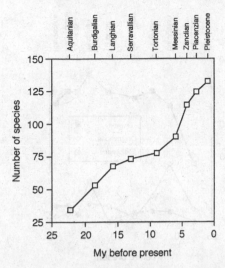

Fig. 14.5. Cumulative origin of the 133 cheilostome bryozoan species which live in the Mediterranean Sea at the present day and have fossil records. Rate of species addition is almost constant from the Aquitainian to Pleistocene, with the exception of the Tortonian and Messinian during which fewer species were added to the modern fauna than expected. Figure constructed using the database mentioned in the text.

additional 15 species do not live in the Mediterranean today but are extant elsewhere in the world.

About 41% of Recent Mediterranean bryozoan species with calcified skeletons are represented in the fossil record. As predicted above, cyclostomes, with 65% of extant species known as fossils, have a better record fossil record than cheilostomes, with 37% of extant species known as fossils. These figures are substantially greater than the *ca.* 12% quoted by Horowitz and Pachut (1996) for the fossil record of extant bryozoans world-wide, presumably reflecting the greater intensity of study of fossil bryozoans in the Mediterranean area. Ranges of individual species vary: some range from the Paleogene to the Recent (26 spp. = 8%) while others (39 spp. = 12%) are confined to a single stage. Species with very uncertain ranges were excluded from the database, and all ranges were extrapolated between known occurrences to give 'range-through' data.

14.3.3   Interpretation

In all databases of this type, incorrect taxonomy and dating are potential (and probable) sources of error, but there is no *a priori* reason to believe that these will have biased temporal patterns. Considerable scope exists, nevertheless, for a more refined database in the future.

Considering only the species still living in the Mediterranean today, Fig. 14.5 is a cumulative plot summarizing their times of origin in the fossil record. The modern Mediterranean bryozoan fauna can be seen to have accumulated at a fairly constant rate of about five species per million years since the beginning of the Neogene. The number of extant species originations was lower only for the two stages of the upper

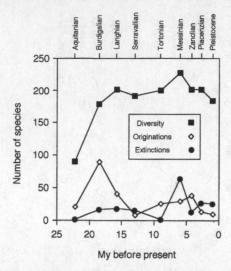

Fig. 14.6. Species diversity, originations and extinctions of bryozoans in the Neogene and Quaternary fossil record of the Mediterranean basin. Figure constructed from the database of 333 species mentioned in the text.

Miocene – the points for the Tortonian and Messinian fall beneath the straight line that could reasonably be drawn through all the other stages. Could this low number of extant species originating in the Late Miocene be a function of an overall low origination and/or low diversity at the time? This appears not to be so. Fig. 14.6 shows that levels of species origination for both stages were fairly typical, and that species diversity actually peaked in the Messinian and was high in the Tortonian. Diversity dropped into the Early Pliocene because of a large number of species becoming extinct during or at the end of the Messinian. The high turnover of species in the Messinian is also reflected by the fact that the proportion of species confined to this stage (5%) is higher than for any other stage except the Burdigalian (9%).

Moissette and Pouyet (1987) have previously investigated the effects of the Messinian salinity crisis on bryozoan evolution. They found that 83 of 122 (68%) of cheilostome species which occurred in the Messinian survived into the Pliocene and, of these survivors, almost 17% were Mediterranean endemics. This led them to conclude that some species persisted in refuges in the Mediterranean during the crisis. However, it is never possible to be sure that particular species were truly endemic in the geological past – endemism can be falsified but never proven – and this comparatively small number of surviving species may actually have had wider distributions outside the Mediterranean during the Messinian and subsequently recolonized the Mediterranean in the Pliocene. Moissette and Pouyet's data do nevertheless show a significantly greater Messinian extinction of apparent endemic species (65%) than of non-endemic species (14%).

Global cooling is reflected by the decreasing numbers of tropical species in Mediterranean bryozoan faunas from the Messinian to Recent. Moissette and Pouyet (1987) recorded 21 species belonging to tropical cheilostome genera in the Messinian, 10 in the Pliocene, 5 in the Pleistocene, decreasing to only 2 in the Recent.

14.4   CONCLUSIONS

An important, though often intractible, issue in any analysis of patterns in the fossil record is to what extent they are driven by: (i) changes in the physical environment; (ii) changes in the extrinsic biotic environment; or (iii) internally through evolutionary innovations (e.g. Allmon & Ross, 1990). Diversity changes are a result of the interplay between origination and extinction rates. Explanations of extinctions involving physical, and to a lesser extent, biotic, environmental changes are attractive and relatively easy to apply, as exemplified by the bolide impact theory for the end Cretaceous extinction. In contrast, it has been more common to look towards innovations to explain changes in origination rates such as evolutionary radiations (e.g. Heard & Hauser, 1995). This is not surprising as innovations can provide direct explanations that do not depend on the often complex chains of hypothesized organism–environment interactions necessary for theories involving environmental changes. Nevertheless, it would be naïve to neglect the possible role of environmental changes in evolutionary radiations. There are at least three ways in which environmental changes might be effective: (i) by providing new kinds of habitat; (ii) by increasing the area of favourable habitat; (iii) by forming geographical barriers promoting speciation through vicariance. Unfortunately, none of these has been adequately explored for post-Palaeozoic bryozoans, although Ross and Ross (1996) have claimed that sea-level cycles had an important influence on bryozoan evolution and dispersal patterns in the Palaeozoic, with lowstands resulting in significant species extinctions and highstands corresponding with times of increased diversity in the fossil record. Likewise, the mid Cretaceous marine transgression may have played a role in bryozoan radiation.

In summary, the Cretaceous–Recent global fossil record of the Bryozoa does not bear a strong signature of environmental change. A steep rise in bryozoan family diversity beginning in the mid Cretaceous largely resulted from the explosive radiation of the Cheilostomata which may have had an intrinsic rather than an environmental trigger. Average species richness in bryozoan assemblages increased in parallel with global family diversity through the Cretaceous. The end Cretaceous extinction had little longlasting effect, and subsequent Cenozoic global extinctions have yet to be clearly demonstrated in the bryozoan fossil record. A new database of Neogene–Recent species ranges in the Mediterranean region shows that species were added to the modern Mediterranean fauna at a fairly constant rate from the Burdigalian to the Pleistocene, except for the Tortonian and Messinian when rates of addition dropped. Species extinction rate was high in the Messinian, correlating with the well-known salinity crisis, and Mediterranean endemics were harder hit than non-endemics. Rebound from the Messinian extinction was relatively rapid, presumably as a result of transfer of species from the Atlantic as is still happening today (see Harmelin & d'Hondt, 1993). Such rapid rebound, in geological terms, may be a more general feature of recovery from Mesozoic and Cenozoic extinctions in the phylum: diversity and other trends appear to be quickly re-established, perhaps suggesting the common existence of refuges. Biogeographical distributions of several warm-water groups have contracted with climatic cooling in the late Cenozoic but any evolutionary effect of such temperature changes has yet to be demonstrated.

A great deal of 'alpha taxonomy' remains to be done on Mesozoic and Cenozoic bryozoans before we can gain an adequate understanding of the distributions of taxa through geological time. This should be combined with phylogenetic analysis to enable ghost lineages to be inferred. It will then become possible, with far greater assurance, to seek correlations between evolutionary patterns and environmental changes on the one hand and intrinsic biological changes on the other.

15

Angiosperm diversification and Cretaceous environmental change

RICHARD LUPIA, PETER R. CRANE AND SCOTT LIDGARD

15.1 INTRODUCTION

Flowering plants (angiosperms) comprise 250 000–300 000 living species and are more diverse than all other groups of extant plants combined. They dominate the vegetation over most of the Earth's land surface and therefore account for most of the primary productivity in terrestrial ecosystems. Despite their present-day importance, flowering plants were the last major group of land plants to appear in the fossil record. While all other major groups have an extensive fossil record that extends back to the Paleozoic or earliest Mesozoic, angiosperms have no well-established fossil record prior to the Cretaceous (Crane *et al.*, 1995). The origin of flowering plants, and especially their rapid diversification during the Cretaceous and Tertiary, is therefore of great evolutionary interest and had profound consequences for the evolution of terrestrial ecosystems and the animals that inhabit them (Friis *et al.*, 1987; Wing *et al.*, 1992).

In this chapter we briefly introduce recent studies of angiosperm diversification, review the changing diversity of major plant groups through the Cretaceous, present new data on changing patterns of abundance, and briefly discuss the possible relationship between vegetational patterns and other aspects of Cretaceous global environmental change. While there is currently no clear evidence that the diversification of flowering plants was triggered by changes in the global environment, several patterns of vegetational change during the mid Cretaceous correlate broadly with major environmental perturbations including oceanic anoxia, increased tectonic activity and rapid sea floor spreading. Future studies of whether or not these disparate factors may be implicated as determinants of vegetational changes will require improved temporal correlation and new interdisciplinary approaches.

15.2 BACKGROUND

The angiosperm fossil record has been studied for over 100 years but recent interest was stimulated by developments in palaeopalynology during the late 1950s and 1960s (e.g. Couper, 1958; Brenner, 1963; Doyle, 1969). Subsequently, studies of fossil pollen, leaves and reproductive structures (Muller, 1970, 1985; Wolfe *et al.*, 1975; Doyle & Hickey, 1976; Hughes, 1976; Doyle & Robbins, 1977; Hickey & Doyle, 1977; Dilcher, 1979) conclusively documented a major systematic and ecolo-

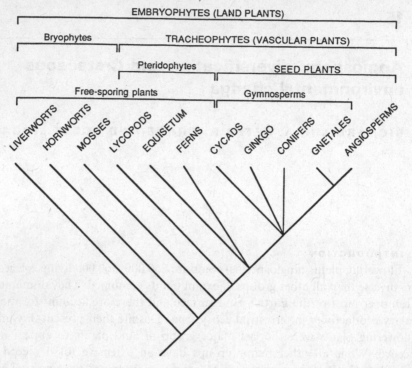

Fig. 15.1. Cladistic relationships among major clades of extant land plants illustrating several of the groups discussed in the text. Upper case indicates probable monophyletic groups; lower case indicates probable paraphyletic groups.

gical diversification of angiosperms during the mid Cretaceous, between 130 and 90 million years ago. Comparative studies of extant angiosperms, also undertaken at about this time (Cronquist, 1968, 1981; Takhtajan, 1969; Walker & Doyle, 1975; Walker, 1976), identified the extant subclass Magnoliidae (*Magnolia*, water lillies and their allies) as a phylogenetically basal assemblage of flowering plants. Many of the earliest angiosperm fossils were also shown to have relationships to extant taxa at the magnoliid grade (Doyle, 1969; Doyle & Hickey, 1976; Upchurch, 1984; Walker & Walker, 1984).

Over the last decade, studies of early angiosperm evolution and diversification have gained further impetus from detailed anatomical and morphological studies of extant magnoliids (e.g. Endress, 1987; 1989), explicit phylogenetic analyses of relationships among seed plants (Fig. 15.1; Crane, 1985; Doyle & Donoghue, 1986; Loconte & Stevenson, 1990; Nixon *et al.*, 1994), and studies of relationships among magnoliid angiosperms based on both morphological and molecular sequence data (Chase *et al.*, 1993; Doyle *et al.*, 1994). In addition, the potential contribution of the palaeobotanical record to understanding early angiosperm evolution has been further increased by bulk sieving techniques that have produced abundant and exquisitely preserved fossil flowers (both mummified and charcoalified) from the Cretaceous of eastern North America and Europe (Friis, 1984; Crane *et al.*, 1995). These discoveries have revolutionized our understanding of the systematic relationships of Cretaceous angiosperms and identified the sources of important dispersed Cretaceous angiosperm pollen.

Angiosperms are first recognized in the Valanginian and Hauterivian based on fossil pollen grains, but so far these lack corroborating evidence from associated macrofossils (Hughes, 1994; Brenner, 1996). Although there are several reports of pre-Cretaceous angiosperms, and the occurrence of Triassic and Jurassic stem-group angiosperms is predicted by current hypotheses of seed–plant relationships (Crane, 1985; Doyle & Donoghue, 1986, 1993), none of these early fossils possesses definitive angiosperm features (Crane *et al.*, 1995). Triaperturate pollen, diagnostic of a major clade of dicotyledons (eudicots) comprising *ca.* 75% of all living angiosperm species, first appeared around the Barremian–Aptian boundary or slightly earlier. At about this time, other early angiosperm fossils (e.g. leaves, flowers) begin to appear in the fossil record. The earliest evidence of monocotyledons is also of about this age (Herendeen & Crane, 1995). Subsequently, the diversity and abundance of angiosperm fossils increases rapidly and angiosperms soon come to dominate many palynological and macrofossil assemblages.

Evidence from fossil flowers, leaves and pollen documents that by the early Cenomanian several extant families of magnoliid angiosperms were already differentiated (Crane *et al.*, 1995). Monocots and several groups of early eudicots were also present (Crane *et al.*, 1995). By the Turonian to Campanian, many extant families in several major angiosperm groups had differentiated, and rapidly accumulating information on fossil flowers is continuing to extend the fossil history of many extant taxa into the Late Cretaceous (Crane *et al.*, 1995).

Numerous hypotheses have been presented to account for the rapid diversification of angiosperms and their extraordinary diversity in the Recent (e.g. Stebbins, 1974, 1981; Raven, 1977; Regal, 1977; Burger, 1981; Doyle *et al.*, 1982; Knoll, 1984; Doyle & Donoghue, 1986; Wing & Tiffney, 1987; Bond, 1989). Most have focused on angiosperms themselves, with little consideration of the broader ecological context in which the initial diversification of the group may have taken place.

15.3 ESTABLISHING THE PATTERN OF VEGETATIONAL CHANGE

Over the last 15 years there have been several attempts to develop a more detailed understanding of the pattern of Cretaceous vegetational change. Initial studies, undertaken as part of a broader analysis of changing patterns of plant diversity through the Phanerozoic (Niklas *et al.*, 1980, 1983, 1985), greatly underestimated the rate and magnitude of the angiosperm radiation as well as declines in other groups (Knoll, 1986; Crane, 1987; Lidgard & Crane, 1988, 1990). More detailed studies, focusing on Cretaceous patterns, examined changing patterns of total diversity (summed within geological time intervals) and within-flora relative diversity (percentage) based on macrofossil floras ranging in age from Late Jurassic to Paleocene (Lidgard & Crane, 1988). Within-flora patterns based on macrofossils were then compared with palynological data to provide parallel, and largely independent, documentation of congruent trends in plant diversity (Lidgard & Crane, 1990). The extensive geographical distribution of fossil pollen and spores has also been used to examine temporal trends in the context of palaeolatitudinal differences in floristic composition (Crane and Lidgard, 1989, 1990). This work has recently

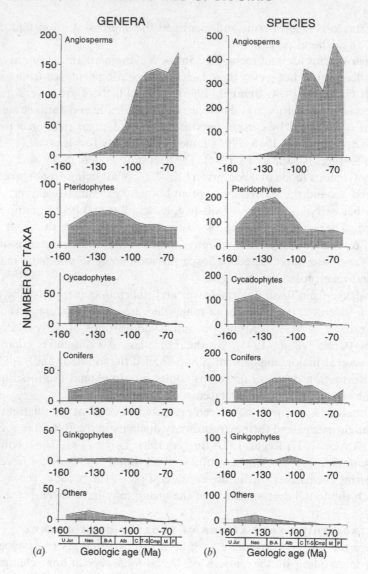

(a)   Geologic age (Ma)   (b)   Geologic age (Ma)

Fig. 15.2. Trends in summed genus and species diversity ('richness') of six major tracheophyte groups between the Upper Jurassic and Paleocene based on fossil leaf taxa primarily from the Northern Hemisphere. 'Pteridophytes' are a paraphyletic assemblage of non-seed plant tracheophytes that includes three major monophyletic groups – lycopsids, sphenopsids and ferns (Fig. 15.1). 'Cycadophytes' include all plants with pinnately compound 'cycad-like' leaves – predominantly true cycads and Bennettitales (both monophyletic), and some 'seed ferns' of probable diverse relationships. Ginkgophytes include *Ginkgo* and related plants (probably monophyletic) plus Czekanowskiales. 'Others' refers to other seed plants not included as angiosperms, conifers, cycadophytes or ginkgophytes (predominantly Mesozoic seed ferns). Diversity totals summed for successive geologic intervals are based on 197 Late Jurassic to Paleocene macrofossil floras (see Lidgard & Crane, 1988). Uneven temporal distribution of floras and fifteenfold differences in stage durations present significant biases for direct comparison of 'summed' diversities among Cretaceous stages. Therefore we combined occurrences in the Late Jurassic, Neocomian, Barremian–Aptian and Turonian–Santonian to

been extended through more extensive sampling and detailed evaluation of the Cretaceous palynological record from North America (Lupia, 1997). These new studies incorporate abundance data and seek to provide a quantitative basis for previous ideas of changing floristic provinciality during the Cretaceous.

15.3.1 The diversity and abundance of angiosperms through the Cretaceous (mid-palaeolatitudes)

Compilations of plant diversity, based on 197 macrofossil floras ranging in age from Late Jurassic to Paleocene, show a dramatic increase in the absolute number (summed diversity) of genera and species of angiosperm leaves through the mid Cretaceous (Fig. 15.2; Lidgard & Crane, 1988). The most pronounced increase occured between the Barremian–Aptian and the Cenomanian. Within-flora percentages of angiosperms, based on the same macrofossil data, also increased rapidly through this interval, rising from near zero to about 70% (Lidgard & Crane, 1988 – see also Knoll, 1986). Trends in within-flora diversity, based on 860 Cretaceous palynofloras from middle palaeolatitudes (25° N – 65° N palaeolatitude) in the Northern Hemisphere, show a broadly similar pattern, but comparison of macrofloras and palynofloras from the same area shows that the increase in angiosperm species diversity is slower in palynofloras (Lidgard & Crane, 1990). In the Cenomanian angiosperms account for a mean of about 30% of the species in palynofloras, rising to levels of 40–50% or more by the late Campanian and Maastrichtian. In macrofloras the equivalent figures are about 60% and 80%. Similar results for palynofloras are also obtained based on an expanded data set of 982 samples from the Cretaceous of North America (Figs. 15.3, 15.4(*a*)).

Discrepancies between estimates of the rate and magnitude of the Cretaceous angiosperm diversification based on macrofossil and palynological data probably reflect factors that differentially affect the representation and recognition of angiosperm leaves and pollen grains in the fossil record (Lidgard & Crane, 1990). Representational effects include the possibility that angiosperms were particularly diverse near environments that favour the preservation of macrofossils, or that they are underrepresented in palynofloras because of low pollen production. For example, direct comparison of pollen *in situ* within fossil flowers, with the associated assemblages of dispersed angiosperm pollen, shows that many presumed insect-pollinated taxa are often not represented in dispersed palynofloras. Pollen floras also substantially underrepresent the angiosperm magnoliid family Lauraceae, which has pollen grains that rarely survive palynological preparation because of the near absence of sporopollenin in the pollen wall. Lauraceae are one of the most common

Fig. 15.2 (*cont.*)
attain more uniform interval lengths and numbers of floras sampled per 'interval', as well as to incorporate floras whose stratigraphical resolution overlaps stage boundaries. (*a*) Temporal trends in diversity at the genus level. (*b*) Temporal trends in diversity at the species level. Scales for the number of taxa are consistent for all major groups at the same systematic level. Note different scales at species and genus levels. U Jur., Upper Jurassic; Neo., Neocomian (Berriasian–Hauterivian); B–A, Barremian–Aptian; Alb., Albian; C, Cenomanian; T–S, Turonian–Santonian; Cmp., Campanian; M, Maastrichtian; P, Paleocene.

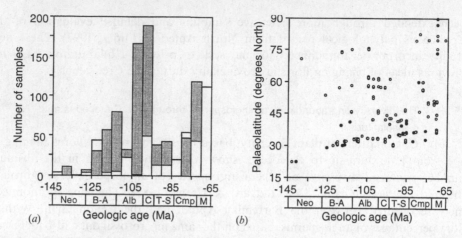

Fig. 15.3. (*a*) Number of palynological samples per successive five million year intervals for North American dataset used in Fig. 15.4. In all intervals, the maximum value shown is the maximum number of palynofloras for which diversity or abundance data were available (whichever was greater). Stippled bars indicate number of samples for calculation of floristic diversity ($n = 982$); open bars indicate number of samples for calculation of abundance ($n = 389$). Values displayed are nested, not additive, and the number of samples used for interval 70–65 Ma is the same for both measures. Paleocene data were not compiled. (*b*) Palaeolatitudinal distribution of samples through time (many samples have the same palaeolatitude and age). Hollow points represent sampling for floristic diversity; filled points represent sampling for ecological abundance. Palaeolatitude calculated using plate rotation algorithm of Dr D. Rowley, University of Chicago. Abbreviations as in Fig. 15.2.

families in Cretaceous macrofloras based on leaves, wood and flowers (Crane *et al.*, 1995).

Recognition problems include the possibility that small or relatively feature-less angiosperm pollen 'species' may have been overlooked in Cretaceous palyno-floras, particularly in light microscope studies. In macrofloras there is also the possibility that intraspecific variability in fossil angiosperm leaves may have been mistakenly reported as multiple species thereby inflating estimates of diversity. Detailed investigations of autochthonous fossil assemblages, and integrated studies of fossil leaves, flowers and pollen from the same fossil deposit, provide the best opportunity to identify representational and recognition biases and assess their potential influence on large scale patterns (Lupia, 1994).

Measures of angiosperm abundance give broadly similar results to those based on within-flora diversity. Data based on 389 palynofloras from North American (Lupia, 1997) indicate a massive increase in the relative abundance of angiosperm pollen between the Barremian–Aptian and Cenomanian (Fig. 15.4(*d*)). In Barremian–Aptian palynofloras from North America, angiosperm pollen rarely accounts for more than 10% of the total pollen grains and spores present, while in Cenomanian and later assemblages, angiosperm pollen is commonly in the range 30–60% (Fig. 15.4(*d*)).

Data compiled from both macrofloras and palynofloras, together with reliable systematic evidence for the early appearance of many extant families of flowering plants (Crane *et al.*, 1995), provides overwhelming evidence of a rapid and substantial mid Cretaceous increase in angiosperm diversity and abundance. However, macrofossils and pollen provide slightly different indications of the prominence of angiosperms in Late Cretaceous floras. Literal interpretation of the palynological data (Figs 15.4(*a*), (*d*)) suggests that angiosperms were less diverse during the Late Cretaceous than is indicated by macrofossils, and also that the abundance of angiosperms may have remained subordinate to 'gymnosperms' and ferns in some habitats through the Late Cretaceous (e.g. Pierce, 1961).

Further evidence relevant to this issue is provided by the discovery of remarkable *in situ* macrofloras preserved beneath Maastrichtian ash falls in Wyoming (Wing *et al.*, 1993). The within-flora species diversity of these *in situ* assemblages conforms to patterns established based on macrofossils from other localities (Lidgard & Crane, 1988, 1990), which show that angiosperms comprised about 68% of Maastrichtian within-flora taxonomic diversity. In terms of abundance, however, dicots in the ash bed floras averaged only 12% of the cover (assessed as percentage cover on bedding planes) and Wing *et al.* (1993) concluded that 'dicots played a subordinate role in these areas of fern dominated vegetation'. These results raise the possibility that angiosperms may not have attained ecological dominance, perhaps in certain types of habitats and over large geographical areas, until the early Tertiary (Wing *et al.*, 1993). Such a scenario, which remains to be tested over a broader geographical area, would have important implications for understanding how angiosperms invaded the Mesozoic landscape, and for interpreting the ecological context in which early angiosperm diversification occurred.

15.3.2   The diversity and abundance of non-angiosperm groups through the Cretaceous (mid-palaeolatitudes)

Changes in the absolute number of genera and species of non-angiosperm leaf species (summed diversity) between the Late Jurassic and Paleocene show that the mid Cretaceous diversification of flowering plants was accompanied by a decline in the diversity of 'cycadophytes' (mainly cycads and extinct Bennettitales) and 'pteridophytes' (mainly ferns), as judged from the fewer 'genera' and 'species' of leaves in the Late Cretaceous (Fig. 15.2; Lidgard & Crane, 1988). In contrast, conifer diversity exhibits less dramatic change through the same interval. Similar patterns for these three groups are also observed in within-flora analyses, although the decline of 'cycadophyte' and 'pteridophyte' leaves is even more marked (Lidgard & Crane, 1988).

For both free-sporing plants (for these purposes more or less equivalent to 'pteridophytes' – mainly ferns) and conifers, temporal patterns of within-flora palynological diversity give results similar to those based on macrofossils (Lidgard & Crane, 1990). Pollen of 'other seed plants', is a consistently low contributor to the diversity of palynological assemblages throughout the Cretaceous and exhibits no marked temporal trend. The low within-flora diversity of this group, which does not track the decline in 'cycadophyte' foliage, is probably attributable to the poor systematic discrimination provided by monosulcate grains. Pollen in this category was

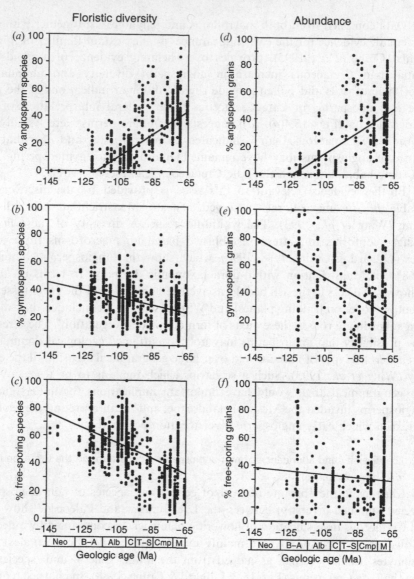

Fig. 15.4. Alternative 'within-flora' measures of the changing composition of Cretaceous vegetation based on palynofloras from North America (see Fig. 15.3) showing linear regressions to indicate major trends. (a–c) Percentage of total 'species' belonging to higher taxon within each of 982 palynological samples (Fig. 15.3; within-flora relative diversity). (d–f) Percentage of total grains belonging to higher taxon within each of 389 palynological samples (Fig. 15.3; within-flora relative abundance). To reduce the effects of small sample size, only complete samples containing 10 or more species were used. Percents were normalized to 100% after removal of aquatic and unassignable palynomorphs. Because proportions of major groups in any one flora must sum to 100%, the power of these analyses depends on the consistency with which many floras independently conform to an overall pattern through time. Linear regressions provide estimates of major trends but do not provide 'tests' of statistical significance in plots of this kind. Monosulcate/monocolpate grains of uncertain affinity were assigned to 'gymnosperms.' Free-sporing plants includes all non-seed plant embryophytes (Fig. 15.1; 'bryophytes', lycopsids, sphenopsids, ferns). (a) Temporal trends in

probably produced by cycads, *Ginkgo* and perhaps certain angiosperms (Lidgard & Crane, 1990). Within-flora analyses of palynological data in which 'other seed plants' and conifers are combined as a single category ('gymnosperms' Fig. 15.4(*b*)) show a clear decline.

Relative abundance data from North America palynofloras (Lupia, 1997) shows considerable variance (Figs. 15.4(*d*), (*e*), (*f*)) and differences from the within-flora diversity patterns (cf. Fig. 15.4(*b*) and (*e*) with Fig. 15.4(*c*), (*f*)) underscore the fact that diverse taxa are not always abundant, and that abundant taxa are not always diverse. However, literal interpretation of these patterns is complicated by many biases of representation and recognition.

15.3.3   Temporal trends and palaeolatitude

Another important factor that contributes to some of the variance in within-flora assessments of changing plant diversity and abundance is geographical, and especially latitudinal, variation. This is seen, for example, in the within-flora abundance data for free-sporing plants (mainly 'pteridophytes'). In the plots based on percentage data combined across about 60° of palaeolatitude (*ca.* 30° N – 90° N) the relative abundance of 'pteridophyte' spores remains more or less constant (Fig. 15.4(*f*)). However, when the data are divided into a northern subset (Fig. 15.5; about 42° N – 90° N) and a southern subset (Fig. 15.5(*a*); about 30° N – 42° N) the within-flora relative abundance of spores is shown to decline markedly in the south, while remaining more or less constant in the north.

There has been considerable discussion of floristic provinces during the Cretaceous and how they changed through time (e.g. Brenner, 1976; Herngreen & Chlonova, 1981; Vakhrameev, 1991), but there have been few attempts to define floristic provinces in quantitative terms, or examine how they change in the context of overall temporal trends in the composition of Cretaceous floras. An alternative approach to clarifying and partially quantifying these temporal and palaeolatitudinal patterns has been to use trend surfaces to summarize within-palynoflora diversity data (Crane & Lidgard, 1989, 1990). Analyses based on percentage data from 1 125 pollen and spore assemblages, mainly from North and South America, Africa and Europe (palaeolatitudes 80° N – 20° S; Crane & Lidgard, 1989, 1990), document the rapid diversification of angiosperm pollen species, across all palaeolatitudes examined, between about 125 and 85 Ma (Fig. 15.6(*b*)). The bulk of this diversification is attributable to increasing diversity of the triaperturate or triaperturate-derived pollen of eudicots (Crane & Lidgard, 1989, 1990). Analyses of within-flora abundance

Fig. 15.4 (*cont.*)

the within-flora relative diversity of angiosperm pollen. (*b*) Temporal trends in the within-flora relative diversity of 'gymnosperm' pollen. (*c*) Temporal trends in the within-flora relative diversity of spores from free-sporing plants (mainly 'pteridophytes'). (*d*) Temporal trends in the within-flora relative abundance of angiosperm pollen. (*e*) Temporal trends in the within-flora relative abundance of 'gymnosperm' pollen. (*f*) Temporal trends in the within-flora relative abundance of spores from free-sporing plants (mainly 'pteridophytes'). Abbreviations as in Fig. 15.2.

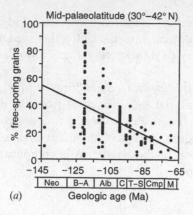
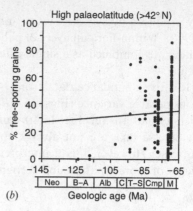

Fig. 15.5. Temporal trends in the relative (percentage) abundance of spores from free-sporing plants in Cretaceous palynofloras based on the data from Fig. 15.3. Linear regressions provide estimates of major trends but do not provide 'tests' of statistical significance in plots of this kind. (*a*) Samples from mid-palaeolatitudes (30–42° N). (*b*) Samples from high palaeolatitudes (> 42° N). Regression omits the three isolated samples older than 115 Ma. Choice of 42° N as limit determined by approximate boundary between Normapolles and *Aquilapollenites* floral provinces in North America.

show a similar pattern of increasing dominance of angiosperm pollen species (Fig. 15.7(*b*)).

Geographical patterns in the increasing within-flora diversity and abundance of angiosperm pollen show two significant features (Figs. 15.6(*b*), 15.7(*a*)). First, angiosperms became important at low palaeolatitudes before they became significant at middle, and then higher, palaeolatitudes. Secondly, even by the end of the Cretaceous, angiosperms were most important as a component of middle and low palaeolatitude palynofloras, where they typically attained relative diversity and abundance values of 70–80%. At higher palaeolatitudes, north of about 40–50° N, they only attained within-flora diversity and abundance levels of 30–40%.

Through the Cretaceous, as judged by decreasing within-flora diversity of pollen and spores, free-sporing plants (mainly 'pteridophytes'), conifers and 'other gymnosperms' all declined. Conifers, as a whole, remained more important at middle and high palaeolatitudes, while 'other gymnosperms' were generally more important at lower palaeolatitudes, especially in the Early Cretaceous. As in the scatter plots of relative diversity (Fig. 15.4), it is free-sporing plants (mainly 'pteridophytes') that declined most precipitously, particularly at low and middle palaeolatitudes. They remained relatively more important (*ca.* 25%) at high palaeolatitudes, where together with conifer pollen (*ca.* 25%) they continue to comprise a major part of the flora.

Among non-angiosperm taxa, the most striking and potentially most significant palaeolatitudinal and temporal pattern in within-flora diversity is shown by distinctive polyplicate pollen ('ephedroids'), which was undoubtedly produced by plants closely related to extant Gnetales (Crane, 1988; Crane & Lidgard, 1989, 1990). This group of three extant genera is often considered to be the closest living relatives of flowering plants (Crane *et al.*, 1995). Ephedroid pollen increased in

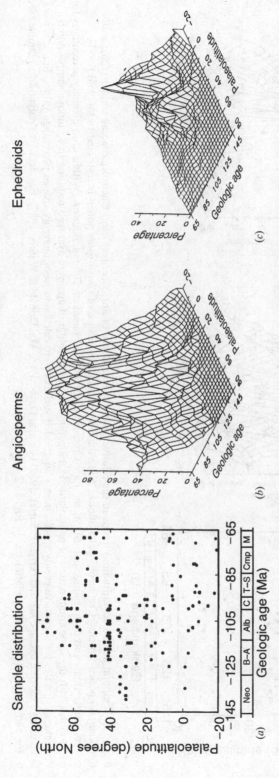

Fig. 15.6. Temporal and palaeolatitudinal trends in the relative (percentage) diversity of angiosperm (b) and gnetalean ('ephedroid') (c) pollen 'species' in Cretaceous palynofloras showing the parallel increase in the relative diversity of angiosperm and ephedroid pollen during the mid Cretaceous at low palaeolatitudes. Note that the greatest diversity of ephedroid pollen is restricted to low palaeolatitude areas, and that ephedroid pollen diversity declines precipitously in the Late Cretaceous. Graphs are based on 1125 palynofloras. Within-flora species diversity percentages for palynofloras with the same combination of palaeolatitude, palaeolongitude and geological age were averaged resulting in 252 unique sample points. Percentage estimates for successive evenly spaced points on gridded surfaces were calculated from unevenly spaced percentage values. Surfaces were fitted using a moving average method (kriging: Surfer, 1987) (see also Davis, 1986; MacDonald & Waters, 1987). For details of palynofloras sampled see Crane and Lidgard (1990). Abbreviations as in Fig. 15.2.

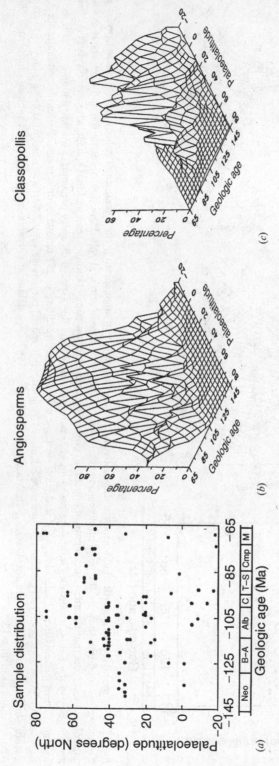

Fig. 15.7. Temporal and palaeolatitudinal trends in the relative (percent) abundance of angiosperm (*b*) and Cheirolepidiaceae conifer (*Classopollis-Corollina*) (*c*) pollen 'species' Cretaceous palynofloras. Note that the greatest abundance of *Classopollis* pollen is restricted to low and middle palaeolatitude areas, and declines precipitously in the Late Cretaceous. Graphs are based on 708 palynofloras. Within-flora diversity percentages for palynofloras with the same combination of palaeolatitude, paleolongitude and geological age were averaged resulting in 95 unique sample points. Percentage estimates for successive evenly spaced points on gridded surfaces were calculated from unevenly spaced percentage values. Surfaces were fitted using a moving average method (kriging: Surfer, 1989) (see also Davis, 1986; MacDonald & Waters, 1987). For details of palynofloras sampled see Crane and Lidgard (1990). Abbreviations as in Fig. 15.2.

diversity and abundance at low palaeolatitudes at about the same time as angiosperm pollen (*ca.* 125–105 Ma) and this contrasts sharply with the declines in most other non-angiosperm groups. However, ephedroid pollen peaked at about 95 Ma and subsequently declined rapidly (Fig. 15.6(*c*)). Unlike that of angiosperms, ephedroid pollen never became diverse or abundant in palynofloras north of about 30° N palaeolatitude.

In addition to the close temporal and geographical parallels between the early phases in the diversification of angiosperm and ephedroid pollen (Crane & Lidgard, 1990; Crane, 1996), there is also evidence that early angiosperms and ephedroids were ecologically associated (Crane & Upchurch, 1987; Crane, 1995). In the Wealden sequence of southern England, Phase I of the angiosperm diversification recognized by Hughes (1994), which began in the late Hauterivian, is marked by the simultaneous appearance of both small monosulcate tectate grains of presumed angiosperm affinity and *Ephedripites*-type pollen. Analyses of mid Cretaceous pollen floras from low palaeolatitudes also show that peaks in angiosperm and ephedroid pollen abundance are positively correlated, while they are negatively correlated with high levels of *Classopollis* pollen that were produced by extinct conifers assigned to the family Cheirolepidiaceae (Doyle *et al.*, 1982).

Classopollis pollen is never an important contributor to the species diversity of Cretaceous palynofloras because the pollen produced by diverse species of Cheirolepidiaceae are virtually identical and therefore difficult to discriminate at the species level (Watson, 1988). The abundance of *Classopollis* in Early Cretaceous palynofloras, south of about 40° N palaeolatitude, is often very high (Vakhrameev, 1991; Spicer *et al.*, 1993). The apparent increase in the relative abundance of *Classopollis* in Early Cretaceous palynofloras (Fig. 15.7(*c*)) may reflect insufficient sampling at this time and in this area (Fig. 15.7(*a*)). However, the very pronounced and rapid decline in the abundance of *Classopollis* at around the Cenomanian–Turonian boundary is not attributable to inadequate sampling (Fig. 15.7(*c*)).

Other temporal trends and palaeolatitudinal patterns in the Cretaceous that remain to be investigated include: (i) the diversification of the very distinctive *Aquilapollenites*-type pollen (triprojectates), that first appeared in the Late Cretaceous and became important at high palaeolatitudes; (ii) the diversification of Normapolles-type pollen, which became abundant at middle palaeolatitudes through the Late Cretaceous; and (iii) the decline in the abundance and diversity of different subgroups of fern spores (e.g. Gleicheniaceae, Schizaeaceae). Detailed studies of these and other patterns are necessary to develop a more thorough understanding of the systematic, ecological and geographical components of Cretaceous vegetational change. They are also crucial to establishing whether the patterns of diversification and extinction in different groups coincide temporally and thus potentially reflect shared responses to global environmental change.

15.4   CAUSAL FACTORS IN ANGIOSPERM DIVERSIFICATION

15.4.1   Angiosperm biology

Numerous causal explanations have been offered to account for the angiosperm radiation, and most have focused on potential 'key innovations' (Crane &

Lidgard, 1990). Among vegetative features, vessels in angiosperm woods may have facilitated the development of laminar, reticulately veined leaves that were perhaps competitively advantageous compared to the more stereotyped leaves of other seed plants. Vessels may also have conferred resistance to high evapo-transpiration rates, particularly in regionally arid Cretaceous low latitude areas. Among reproductive features, the origin of flowers and insect pollination is widely cited as a key innovation that facilitated angiosperm diversification through providing new possibilities for reproductive isolation and effective out-crossing at low population densities. Developmentally, the origin of the flower through reduction, simplification and aggregation of sporophylls, as well as other features (e.g. enclosure of the ovules by a carpel, truncation of the gametophyte phase of the life-cycle) may have been a secondary effect of selection for a weedy life-history or precocious reproduction (progenesis) in open, perhaps seasonally arid (Takhtajan, 1969; Stebbins, 1974, 1981) or disturbed (Wing & Tiffney, 1987), environments. Similar progenetic effects may also account for some of the vegetative features of angiosperms including the characteristic laminar, reticulate-veined, angiosperm leaf (Hickey & Doyle, 1977) and scalariform pitting in the secondary xylem (Doyle, 1978; Doyle & Donoghue, 1986).

The great variety of plausible explanations to account for angiosperm diversification makes it difficult to resolve which features of angiosperms were especially significant during the mid Cretaceous. There is also the difficulty of disentangling which features may have promoted diversification or enhanced local competitive abilities. Basically, superior competitive abilities is one explanation for angiosperm success (Knoll, 1984), but another is that dramatically accelerated speciation rates (Doyle & Donoghue, 1986) simply generated an overwhelming diversity of adaptive types. In either case it is also important to recognize that the angiosperm diversification took place against a backdrop of constantly changing biotic, edaphic and climatic factors that would have continuously modified the competitive landscape.

15.4.2   Palynological patterns and mid Cretaceous global events

While the mid Cretaceous increase in angiosperm diversity and abundance was undoubtedly mediated by biological factors, changes in the physical environment may have influenced the dynamics of Cretaceous vegetational change. The best evidence of such effects is provided by parallel patterns in the history of different plant groups that are broadly correlative with exceptional geologic events.

The parallel increase in the diversity and abundance of angiosperm and ephedroid pollen between about 125 and 105 Ma suggests that the diversification of both groups may have been influenced by similar environmental factors. The latitudinal component of the pattern also suggests the influence of climatic factors. Gnetalean fossils from the Triassic (Crane, 1996) indicate that the angiosperm and gnetalean lineages had been distinct for at least 80 million years before both began to diversify at about the same time (Crane, 1985; Doyle & Donoghue, 1993). The onset of diversification broadly coincides with the onset of a major increase in the production of oceanic crust, both at the mid-ocean ridges and as plateaus in the Pacific Ocean (Gale, this volume). These tectonic events have also been linked to the beginning of the long mid Cretaceous geomagnetic normal, the beginning of an 'almost continuous overall sea-level rise which continued into the Late Cenomanian' (Gale, this volume), high levels of atmospheric CO_2, and possibly oceanic anoxic events (Larson, 1991; Kauffman & Hart, 1996; Gale, this volume).

While changes in atmospheric CO_2 and associated climatic trends may have had a significant impact on terrestrial vegetation (Field *et al.* 1992), there is currently no clear mechanism that would favour the diversification of angiosperms and Gnetales over other groups. It is interesting, however, that extant Gnetales and angiosperms share several biological features (e.g. presence of vessels, reticulate veined leaves, insect pollination), that may account for their apparent palaeoecological similarities. Detailed evidence of mid Cretaceous Gnetales from low palaeolatitudes is lacking, but such ecological similarities between Early Cretaceous Gnetales and angiosperms may have predisposed them either to similar responses to environmental change or to the origin of comparable innovations in response to similar environmental, biotic or other selective pressures.

In the Late Cretaceous, the most pronounced pattern in the palynological data is the abrupt, and more or less coordinated, decline in the abundance of ephedroid and *Classopollis* pollen at around the Cenomanian–Turonian boundary. There is also evidence of a concordant decline in certain groups of fern spores (Crane & Lidgard, 1990). Again, the most dramatic effects are at low palaeolatitudes implying the influence of climatic factors. While the precise timing of these palynological events is unclear, they are broadly coincident with the Cenomanian–Turonian mass extinction (Sepkoski, 1986; Kauffman & Hart, 1996), as well as the climatic cooling inferred to have resulted from the drawdown of atmospheric CO_2 (from the burial of large amounts of organic material in marine sediments; Gale, this volume). Climatic effects might be expected to have had a particular impact on two fundamentally xeromorphic (dry-adapted) groups (Gnetales, Cheirolepidiaceae) that both appeared in the Late Triassic and subsequently became important in semiarid environments, but it is uncertain why angiosperms were not similarly affected. If shared ecological characteristics stimulated the parallel diversification of angiosperms and ephedroids, then these must have been less influential after the Cenomanian when the two groups show very different evolutionary histories (Fig. 15.6).

15.5   CONCLUSIONS

Attempts to quantify temporal changes in the diversity and abundance of different groups of plants confront many difficulties, especially concerning biases in the representation and recognition of different plant groups (e.g. Niklas *et al.*, 1980; Signor, 1990). However, some of these difficulties can be overcome and biasing factors alone are unable to account for the obvious and concordant trends in the changing composition of mid Cretaceous fossil floras. Quantitative analyses are also necessary to provide an explicit repeatable framework in which detailed examination of local stratigraphical sequences can be undertaken (e.g. Wing *et al.*, 1993) and the broader context in which causal explanations of angiosperm diversification can be evaluated.

The initial phase of angiosperm diversification between the Barremian and Cenomanian broadly correlates with an increase in the diversity and abundance of ephedroid (gnetalean) pollen grains. While most attempts to explain the rapid radiation of angiosperms during the Cretaceous have focused on 'key innovations', this parallel diversification of two closely related, but distinct, groups of plants suggests that environmental factors may have influenced the dynamics of mid Cretaceous vegetational change. Subsequent more or less synchronous declines in ephedroids (Gnetales), cheirolepidiaceous conifers and other plant groups at around the Cenomanian–Turonian boundary, also imply the influence of external factors. Improved stratigraphical resolution of these changes in Cretaceous palynofloras, to clarify how they correlate with global climatic and marine events (e.g. Arthur *et al.*, 1985; Reyment & Bengston, 1986; Leckie, 1989; Larson, 1991; Gale, this volume), could be very instructive in understanding the linkages among mid Cretaceous biotic and environmental changes. Because stratigraphical 'smearing' may result from the reworking of dispersed palynomorphs, it will also be important to check the palynological record against macrofossil reports of typical Mesozoic groups (e.g. Bennettitales, Czekanowskiales, Caytoniales, Eucommiales and other groups) that are presumed to have gone extinct during the Late Cretaceous or Early Tertiary. Further clarification of temporal and geographical vegetational patterns will be indispensable in starting to unravel the complex processes and feedbacks inherent in Cretaceous global change.

16

Cenozoic evolution of modern plant communities and vegetation

MARGARET E. COLLINSON

16.1 INTRODUCTION

In the Cenozoic, the modernization and continued diversification of flowering plants (Figs. 16.1 and 16.2) was one of the major controls on global change in land biomes. Therefore, it is necessary to understand how this group of plants differed from those which had dominated previous vegetation and hence the impacts which these differences might have had on community evolution and on our ability to relate this to physical global change. In addition, if the evolution of modern vegetation is to be considered, it is necessary to establish how the record of plant fossils can be used to reconstruct ancient communities and to compare them with those existing today.

This chapter considers the nature and application of the plant fossil record, reviews the plant groups which typify latest Cretaceous and Cenozoic floras, addresses the significance of flowering plants and documents the changing distributions of selected dominant plants in response to global change through the Cenozoic. An interpretation of plant communities and their response to global change uses examples of reconstructed communities at their maximum poleward extent during the Eocene thermal maximum. Patterns of community change and species diversity are documented through the Cenozoic.

16.2 NATURE AND APPLICATION OF THE PLANT FOSSIL RECORD

Partially complete 'whole' plants can be reconstructed from plant fossils. This is done in various ways including organic connection (e.g. fruits and leaves found on a single twig; pollen in situ in stamens); biological connection (e.g. pollen grains on stigma remnants of fruits); anatomical similarity between organs (e.g. similar epidermal hairs or cuticle anatomy on leaf and fruit cuticles; similar twig and trunk wood anatomy; fruits developing from flowers) and recurrent association (i.e. exclusive co-occurrence or repeated co-occurrence of isolated organs at several localities). The best examples are those of the Platanaceae (Pigg & Stockey, 1991; Crane et al., 1993) Cercidiphyllaceae/Trochodendraceae, Betulaceae, Ulmaceae, Juglandaceae etc (for reviews see Crane & Blackmore, 1989; Collinson, 1990) and the aquatic herb Limnobiophyllum (Stockey et al., 1997). These reconstructions enable community reconstruction to be based on whole plants and they provide conclusive proof of the nature of the plant.

However, plants are typically represented in the fossil record by isolated organs which are shed from the parent plant during the natural life cycle. Those with the highest chances of preservation also possess resistant macromolecules such as cutan in leaf cuticles, lignin in wood and fruits and seeds and sporopollenin in the walls of pollen and spores (Van Bergen *et al.*, 1995). The fossil record of each of these organs can give valuable information on the environmental factors affecting plant growth and distribution as well as on plant communities and vegetation.

Pollen, being small, easily transported and highly resistant, tend to represent the more regional vegetation not represented in macrofloras. Pollen and fruits and seeds are also important in providing a record of herbaceous plants whose leaves are not shed but die back on the plant. Leaves, fruits and seeds and woods, provide evidence of mammal/plant or insect/plant interactions; the latter are especially well documented by trace fossils of insect herbivore damage on fossil leaves (Collinson, 1990; Collinson & Hooker, 1991; Lang *et al.*, 1995; Scott *et al.*, 1994).

Flowers are especially valued in phylogenetic studies and floral architecture reveals pollination biology. For example, flowers lacking a perianth and in catkins reflect wind pollination whilst a complex perianth and hidden nectar indicates insect pollination. Architecture of pollen also indicates wind (small and smooth) vs. insect (larger and ornamented) pollination. Wind-pollinated plants dominate in open vegetation of cooler climates with low insect numbers, while biotic pollination dominates in closed vegetation in warm climates which support large and diverse insect populations.

Fruits and seeds provide evidence of dispersal conditions, animal interaction and source plant ecology. Architecture of fossil fruits and seeds indicates their dispersal biology and potential role in animal diet. Winged forms are wind dispersed whilst thick-walled nuts or fleshy fruits are consumed and dispersed by mammals; large numbers of fleshy fruits characterize warm and equable climates while many hard dry nuts or winged nutlets characterize cooler temperate forests. Seed size reflects the ecological role of the parent plant (small-seeded 'weedy' colonizers or large-seeded climax forest species). Seed architecture indicates mode of germination, thus showing if germination had occurred to produce a young plant in the ancient environment (Collinson, 1999; Collinson & Hooker, 1991).

Woods and flowering plant leaves are valued for interpretation of growth environment because their morphology (physiognomy) is under climate control (Spicer, this volume). Large leaves with entire margins characterize warm and wet climates, whilst small leaves with serrate margins characterise cooler and drier climates. Percentages of entire margined leaves in a leaf flora are considered to provide a good estimate of Mean Annual Temperature, and a biome type can often be extrapolated from this using the known distribution of vegetation with respect to MAT and mean annual range of temperature e.g. tropical forest above 20 °C MAT in megathermal conditions.

Flowering plant leaf morphology is also indicative of distinctive habitats, e.g. elongate stenophyllous leaves characterize riparian settings while cordate leaves with drip tips typify tropical rainforest understory. Leaf cuticle thickness is indicative of water availability, forms adapted to water deficit having thick cuticles and sunken or protected stomata. Internal anatomy preserved in permineralized leaf or stem fossils etc. can indicate modifications to life in water, e.g. numerous air spaces for storing gases as in the wetland flora of the Eocene Princeton chert (Cevallos-Ferriz *et al.*, 1991).

At even the simplest level, the existence of fossil wood indicates that the source plant required prolonged structural support and thus was a shrub or tree, the larger the obvious original circumference, the larger stature the tree. *In situ* tree stumps (e.g. Basinger *et al.*, in Boulter & Fisher, 1994; Greenwood & Basinger, 1994) confirm the existence of forests and indicate tree density as well as persistence in an area.

Wood is produced by a cylinder of dividing or meristematic cells (vascular cambium). The activity of this meristem is controlled largely by the availability of sufficient light, water and sufficiently warm temperatures. Meristematic activity produces patterns in the wood known as growth rings which indicate growth conditions. In flowering plant woods the distribution and size of vessels (water conducting cells) is also controlled by growth conditions (Wheeler & Baas, 1991,1993). For example, a wood with no growth rings and very large isolated vessels throughout would be characteristic of continuous, very favourable, growth conditions such as might be found in continuously well-lit, warm and wet areas of the Tropics. Conversely, a wood with extremely narrow growth rings, essentially no late wood and with very few tiny vessels only at the start of the early wood, could be produced where conditions favouring growth were very short lived and ceased suddenly, for example at high altitude (mainly temperature control) or latitude (temperature and light control). False rings (i.e. rings which do not completely encircle the stem and/or which exhibit a gradual rather than sudden restart of growth) indicate temporary interuptions during otherwise favourable conditions.

16.3   THE CENOZOIC PLANT RECORD
16.3.1   Cenozoic plant groups
Flowering plants

The mid Cretaceous marks the emplacement of flowering plants (Magnoliophyta, angiosperms) in Earth's vegetation (Fig. 16.1(*a*)) while in a short interval of the mid to Late Cretaceous they diversified and rose to dominance in the rock record (Fig. 16.1(*b*); Lupia, *et al.*, this volume; Wing & Boucher, 1998). The latest Cretaceous and early Cenozoic (Paleogene) marks the second radiation of flowering plants and the commencement of a modernization (Collinson, 1990), which continued with almost exponential diversification through the entire Cenozoic (Figs 16.1(*a*) and (*c*)) and would probably be continuing to the present were it not for the activities of humans (Niklas *et al.*, 1983, Knoll, 1986). This Cenozoic diversification is best examined in the context of the subdivision of flowering plants, by many authors, into two monophyletic clades – Monocots (some 22% of modern species) and Eudicots (75%) – embedded in a grade of Magnoliid Dicots (3%) based upon cladistic analyses of morphological and molecular data (see discussion in Drinnan *et al.*, in Endress & Friis, 1994). Only those groups for which the

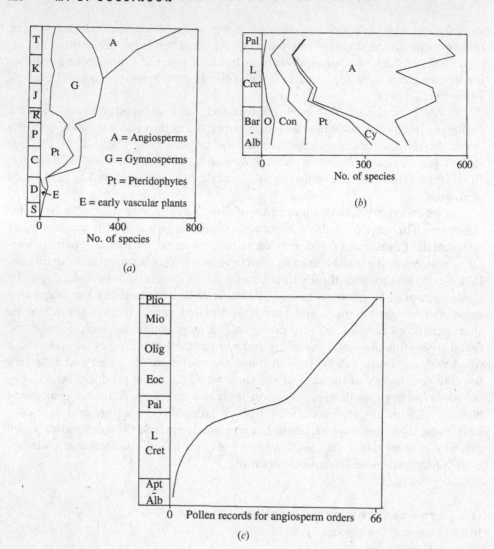

Fig. 16.1. Diversity of land plants and flowering plants through geological time. (*a*) Diversity of vascular plant groups showing general distribution since the evolution of land plants, modified from Niklas *et al.* (1983). S (Silurian) through T (Tertiary) are Periods and represent geological time. (*b*) Rise to dominance of the flowering plants (A) during the Late Cretaceous also showing the low impact of this on conifers (Con) but impact on cycads (Cy), other seed plants (O) and pteridophytes (Pt), modified from Lidgard and Crane (1988) – see also Lupia *et al.* (this volume). Bar–Alb = Barremian, Aptian and Albian ages, L Cret = Late Cretaceous (Cenomanian to Maastrichtian ages), Pal = Paleocene (Early Tertiary of the Cenozoic). (*c*) Diversification and modernization of flowering plants through the Tertiary as expressed by records of modern flowering plant Orders through time, as recognized from fossil pollen (Muller, 1981, 1985), modified from Muller (1985).The overall pattern remains broadly correct although the entire curve would be shifted down column by new records (see Collinson *et al.*, 1993; Herendeen & Crane, 1995; Endress & Friis, 1994; Crane *et al.*, 1995). Apt–Alb = Aptian and Albian ages, L Cret = Late Cretaceous, Pal–Plio = epochs of the Tertiary; Pal = Paleocene, Eoc = Eocene, Olig = Oligocene (together the Paleogene), Mio = Miocene, Plio = Pliocene (together the Neogene).

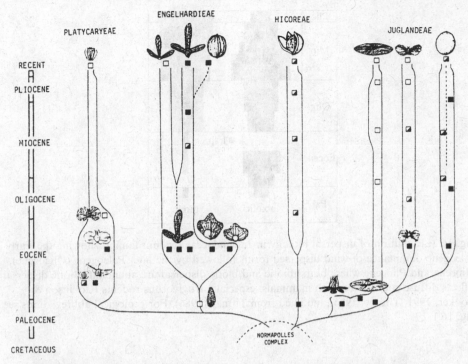

Fig. 16.2. Modernization and diversification of Juglandaceae showing the Cretaceous 'Normapolles complex' and the origins and radiations of modern tribes in the Paleogene of the Cenozoic. Important data points are based on fossil fruits of which key examples are figured. Occurrences – Dark square North America; open square Eurasia; half darkened square both regions. Reproduced with permission of the author and publishers from Manchester (1987).

fossil record (Collinson *et al.*, 1993) indicates a significant role in Cenozoic vegetation are considered here. All families cited are modern.

Magnoliid dicot trees are very well represented in Paleogene floras by fruits and seeds of Magnoliaceae (Magnolia trees), Lauraceae (Bay trees) and Annonaceae (custard apple trees). Wetland herbs of the Nymphaeaceae and Cabombaceae s.s. (water lilies) and Ceratophyllaceae (hornworts) are also common (Collinson, 1990; Collinson, 1999; Collinson *et al.*, 1993). Several of these families first appear in the early Late Cretaceous, including the Magnoliaceae, Chloranthaceae, Lauraceae and Winteraceae (the latter based on pollen only) (Collinson *et al.*, 1993; Herendeen & Crane, 1995, p.15; Endress & Friis, 1994; Friis *et al.*, 1997). The Chloranthaceae, so important amongst early flowering plants, have not apparently been recorded in Cenozoic assemblages, a striking Lazarus effect. The most common ranunculid fossils are endocarps of the Menispermaceae (moon seed lianas), which are known in the Cretaceous and are abundant in the Paleogene.

Trees of the Subclass Hamamelidae are amongst the earliest Eudicots both to appear and radiate. In the Paleogene the families Betulaceae (birch trees), Trochodendraceae/ Cercidiphyllaceae (Kadsura trees), Platanaceae (plane trees = sycamores in North America), Ulmaceae (elm trees), Fagaceae (beech trees), Juglandaceae (Fig. 16.2; walnut and pecan trees) and Casuarinaceae (casuarina

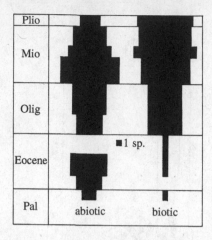

Fig. 16.3. Evolution of dispersal biology in Juglandaceae (walnut family) showing the Early Paleogene dominance of wind dispersed forms followed by the later Paleogene (Oligocene), Miocene and Pliocene where both abiotic and biotic dispersal are abundant. Biotic dispersal reflects diffuse coevolution with mammals, especially frugivorous rodents (Collinson & Hooker, 1991; Tiffney, 1986), modified from Tiffney (1986). For geological abbreviations see Fig. 16.1(c).

trees of the Southern Hemisphere) are diverse and well represented, many by partially reconstructed whole plants as well as isolated organ fossils (Crane & Blackmore, 1989; Collinson, 1990; Collinson et al., 1993; Collinson, 1999; Pigg & Stockey, 1991; Manchester, 1987,1989; Crane et al., 1993). Hamamelidaceae (witch hazel trees), Moraceae (mulberries) and Myricaceae (myrtle shrubs) also occur. Platanaceae are also well-represented in the Cretaceous (extending down to the Albian), Fagaceae and Hamamelidaceae occur in the Late Cretaceous, whilst Juglandaceae (Fig. 16.2) and Trochodendraceae have close extinct Cretaceous relatives (Herendeen et al., 1995; Crane et al., 1993; Endress & Friis, 1994). Modernization occurred in the latest Cretaceous or Early Cenozoic. For example, in Platanaceae the earliest achenes with dispersal hairs like those of modern *Platanus* are from the Eocene (Manchester, 1994), in Fagaceae the earliest representatives of the modern genera are fruits and leaves of *Quercus* and fruits of *Castanopsis* from the Mid Eocene (Manchester, 1994; Kvacek & Walther, 1989). In Juglandaceae the major radiation of modern tribes occured in the Paleogene (Fig. 16.2; Manchester, 1987) and the subsequent diversifications reflect diffuse coevolution with mammals evidenced by changes in dominant dispersal biology (Fig. 16.3; Tiffney, 1986; Collinson & Hooker, 1991).

Rosidae are well represented in Paleogene floras especially by Fabales (legumes and Acacia trees) (Herendeen & Dilcher, 1992) and also by Cornales (dogwood trees) and Sapindales (many families including Aceraceae [maple trees, sycamore trees in Europe] Anacardiaceae, Sapindaceae, Rutaceae and Burseraceae). Icacinaceae (Icacina lianas and trees) and Vitaceae (grape vines) are also common. Buxaceae, Aceraceae and Santalales (?loranthaceous) and Sapindaceae appear in the Cretaceous (Collinson et al., 1993; Endress & Friis, 1994). Within the Dilleniidae significant Paleogene representatives include Theaceae (tea shrubs and trees),Tiliaceae (lime trees) and Salicaceae (willow and poplar trees). Dilleniidae and Rosidae also have Late Cretaceous representatives, such as Clusiaceae (Crepet & Nixon, 1998).

In contrast, the subclasses Caryophyllidae and Asteridae, and some orders of Dilleniidae, which include many herbaceous representatives today, have a more recent earliest occurrence with any evident diversification only late in the Neogene (Collinson et al., 1993). Amongst this category are plants of considerable importance in modern herb-dominated vegetation and forms prolific as weeds today such as the Brassicaceae (cabbages and cresses – Pliocene record), Asteraceae (daisies – one possible Eocene record otherwise Miocene onwards) and Apiaceae (umbellifers – possible Miocene record). The earliest confirmed example of a flowering plant epiphyte is *Viscoxylon* from within Late Miocene pine wood (see Viscaceae and Loranthaceae in Collinson et al., 1993).

The Monocot (Class Liliopsida) fossil record has been reviewed by Collinson et al., (1993) and Herendeen and Crane (1995). The Subclass Alismatidae is very well represented in the Paleogene by herbaceous water plants of the family Hydrocharitaceae (*Stratiotes*, water soldier) and order Najadales (Najads, pond weeds and ditch grasses). Within the Arecidae the Arecales (palms) are widespread and numerous in the Paleogene including the mangrove palm *Nypa*. Palms certainly also existed in the Late Cretaceous (e.g. leaves in Wing et al., 1993; Crabtree, 1987). The Arales (aroids or arums) are also well represented from the Eocene onwards, including whole plants (Stockey et al., 1997). In Zingiberidae the Zingiberaceae (gingers) and Musaceae (bananas) appear in the Late Cretaceous with records of modern genera of Musaceae in the Eocene (Collinson et al. 1993; Manchester & Kress, 1993). Several 'intermediate' fossils also exist (Herendeen & Crane, 1995).

In the Commelinidae only the Cyperales (sedges) and the Typhales (bullrushes or cattails), both emergent wetland plants, have an extensive record which for the latter extends back into the latest Cretaceous. Poaceae (grasses) are known from whole plants in the Paleogene but modern grassland genera are not recorded until the Miocene (Collinson et al., 1993; Herendeen & Crane, 1995). The Liliidae and many families of Commelinidae, which include many herbaceous and bulbous plants such as lilies, agaves, irises, orchids, have a minimal to absent Cenozoic fossil record (Collinson et al., 1993).

In summary, flowering plant trees and lianas, including many whose nearest living relatives are tropical, subtropical or deciduous temperate, dominate the Paleogene whilst herb-dominated flowering plant vegetation, including grasslands, has a largely Neogene (or younger) origin. The latter is a response to opening habitats with continued climatic cooling trends commencing at the Eocene–Oligocene transition.

Non-flowering plants

As can be seen in Fig. 16.1(*a*) and (*b*) other major plant groups such as pteridophytes (including lycophytes [club mosses] and ferns) and gymnosperms (including conifers, cycads and *Ginkgos*) have a much longer history on Earth.

The evolution and diversification of the flowering plants displaced ferns and allies and some seed plant groups (e.g. cycads and bennettites) from former dominance and drastically reduced the abundance of 'fern prairies' (Collinson, 1996a; Lupia *et al.*, this volume). However, almost all the plant groups (with some exceptions such as bennetites) survived to play a more minor role in Cenozoic and Recent vegetation (Fig. 16.1(*b*); Lupia *et al.*, this volume).

Many modern fern families which were widespread and abundant in the Mesozoic were very rare in the Cenozoic due to range restriction and displacement by flowering plants. Other modern fern families lack a Mesozoic fossil record but are represented by modern genera in the Cenozoic whilst modern genera of the the water ferns (Salviniales) are common from the Late Cretaceous onwards (Collinson, 1996a).

Amongst non-angiospermous seed plants, the cycads, widespread in the Mesozoic, have a relatively poor Cenozoic record but modern genera are recorded, for example, in the Tertiary of Australia (Hill, 1994). The genus *Ginkgo* (monotypic Ginkgophyta) is known through the Mesozoic and Cenozoic.

Amongst the conifers many modern families were present in the Mesozoic. As with flowering plants, modernization occurred through the Tertiary at generic level. For example, *Metasequoia* (dawn redwood) is known from an almost complete reconstructed whole plant in the Early Eocene and this is essentially indistinguishable from the modern species (for review see Collinson, 1990). Modern genera of Taxodiaceae (redwoods and swamp cypresses), Araucariaceae (monkey puzzle and kauri) and Pinaceae (pines, spruce, larch, etc.) occur in Paleogene floras. There is little evidence of any major impact on conifer diversity following the evolution of the flowering plants (Fig. 16.1(*b*); Lupia *et al.*, this volume) but conifer forests (apart from taxodiaceous swamps) are largely lacking during warm intervals (Wolfe, 1985, 1995; Basinger *et al.* in Boulter & Fisher, 1994).

16.3.2 The significance of flowering plants: intrinsic factors

Many factors have been invoked to account for the rapid rise to dominance, subsequent radiations and maintenance of dominance of flowering plants. Intrinsic aspects of flowering plant functional biology are a major causal component of the biotic global change which led to the evolution of modern land biomes. These intrinsic factors are outlined here; for further discussion the reader is referred to Collinson (1990); Friis *et al.* (1987); Collinson and Hooker (1991) and Behrensmeyer *et al.*, (1992).

Species diversity and floral biology

The flowering plant egg is enclosed in an ovule within the carpel. Male gametes do not have direct access to the egg but must be carried through the carpel wall. The carpel carries a receptive stigma on which pollen containing the male gamete is recognised and pollen germination is controlled. Male gametes can be rejected at any stage in this process and cross-fertilization is favoured. Animals visiting plants can be exploited in gamete transfer due to the precise position at which pollen are to be received (the stigma). This process of pollination in many flowering plants involves specialist biotic vectors including many insects (and less commonly birds, mammals and other groups). Vectors may receive a reward (e.g. nectar) or may be duped by the plant through mimicry into a visit with no reward. Diverse, faithful biotic pollination interactions are exclusive to angiosperm-dominated vegetation. The resulting mutual dependence tends to ensure survival of even rare genomes and species. An example could be an orchid represented only by two plants growing as epiphytes on trees in widely separated areas of a forest. Specialist, faithful pollination would mean that one insect would seek out the two orchid plants attracted by any combination of flower colour, shape, aroma and nectar. In orchids, pollen are carried in pollinia (pollen packets) which may contain numerous grains. Orchid carpels contain numerous ovules housing eggs. Just one flight from one plant to the other by one insect could result in many seeds, hence many new plants thus ensuring survival.

In summary, cross-fertilization combined with specialist and faithful pollination maintains genetic variation and ensures high species survival, even of isolates. This results in the characteristic high species diversity in angiosperm dominated vegetation. It follows that, once evolved, flowering plants would be expected to diversify as seen in the fossil record (Figs. 16.1 and 2) and Knoll (1986) suggested that an essentially exponential diversification might still be continuing were it not for the intervention of humans.

Vegetation diversity

The high diversity of flowering plants is expressed in Recent vegetation, for example, in the layering of tree canopies in tropical forests, the extensive shrub and herb layers in woodland and the variety of herbs in meadowlands and downlands. By comparison, coniferous forests may be almost monotypic stands, dominated by a single tree species with a single canopy and essentially no ground cover or understorey. There are no herbaceous conifers. Ferns and their allies may form an important part of the herbaceous layers and may dominate certain areas as they did in the widespread Mesozoic fern prairies. Tree ferns may be very evident in moist forests especially along water courses. Otherwise, flowering plants have colonized and come to dominate almost all areas of land with the exception of moss–lichen tundra and tracts of conifer forest or Taiga in the Northern Hemisphere. Flowering plants occur widely as epiphytes on other plants, they dominate amongst freshwater macrophytes and the seagrasses excell in shallow seas. Flowering plants range from high altitude and latitude tundra to the low latitude and altitude mangroves and tropical rain forests. This widespread and diverse flowering plant dominated vegetation provides the numerous niches for the diversity of other organisms including other plants and insects, mammals and birds thus resulting in the highly diverse land biomes of the present Earth. As a consequence of flowering plant evolution, land biome diversity should be expected to increase along with flowering plant species diversity (see above).

Fruit and seeds

Many parts of the flower (in particular, the ovule and carpel walls) can be transformed in numerous ways following fertilization. Combined with species diversity (see above), this results in another major distinction of flowering plants, their huge variety of fruits and seeds which are exploited by animals (for food) and exploit animals (for plant dispersal) (see Section 16.2). Furthermore, the food reserve (endosperm) for the embryo inside the seed results, in flowering plants, from double fertilization (synapomorphy of flowering plants). This contrasts with other seed plants where the food reserve forms prior to fertilisation, slowing down the life cycle and often being wasted. The food reserve supports the early growth of the young flowering plant and, as discussed above, must be able to do this in conditions ranging from an open meadow to a dense forest. The range of seed size and food reserve reflects an ability to colonize and dominate newly available areas with great rapidity (small-seeded colonizers or R-strategists 'weeds') or to maintain a population in a stable climax forest (large-seeded K-strategists). Extremes are the minute dust seeds such as heathers and the huge *Lodoicea*, the Seychelles coconut. This wide variety of size of seed food reserve is exploited by many seed-eating fauna, which consume everything from large dry nuts (e.g. squirrels and walnuts) to the tiny seeds of meadowland herbs fed upon by birds.

Vegetative biology and physiognomy – leaves and woods

Flowering plant woods are specialized in that a division of labour occurs between cells providing structural support (fibres) and those conducting water (vessels) (cf. for example, the dual function compromise in tracheids of conifers). The widely laminate leaves of flowering plants provide a large area for light interception, gaseous exchange and temperature regulation (through water loss) maximizing photosynthesis, hence metabolite production and growth (cf. for example the needle leaves of many conifers). These specializations mean that flowering plant leaves and woods also serve as active sensors of growth conditions, especially climate (Section 16.2; Spicer, this volume).

16.4   PALAEOBIOGEOGRAPHY - GLOBAL PLANT DISTRIBUTIONS

The overall palaeobiogeographical pattern is one of widespread ranges in the early Cenozoic which subsequently contracted to a much reduced present distribution. All these range contractions are consistent with dominantly climate control with vegetation responding to the overall cooling climate of the Cenozoic following the Eocene thermal maximum (Fig. 16.6).

Flowering plants

Classic examples are the many flowering plants that were widespread in western Europe and North America in the Paleogene but are represented today only in south-east Asia and environs (Collinson, 1983, 1990; Manchester, 1994, 1999; Tiffney in Boulter & Fisher, 1994). These examples include *Nypa* the mangrove palm (Fig. 16.4), other Arecaceae, and genera of Sabiaceae, Alangiaceae, Mastixiaceae, Lauraceae, Menispermaceae, Icacinaceae, Staphyleaceae and members of the Juglandaceae such as *Platycarya* (Fig. 16.5) and *Engelhardia*. Three distinct examples are given in more detail here.

In the case of *Nypa* (a 'stemless' palm living in mangroves) the fossil record is based mainly on fruits and *Spinizonocolpites* pollen (Fig. 16.4). In Icacinaceae tribe Iodeae (tropical lianas) the evidence comes from endocarps (e.g. like those of modern

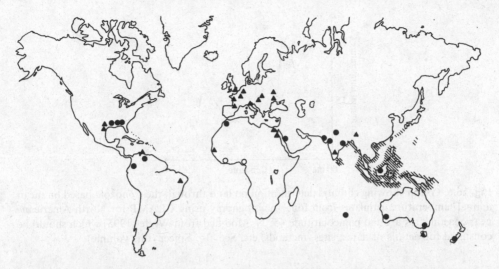

Fig. 16.4. Modern (cross hatching) and fossil distribution of the mangrove palm *Nypa* in the Paleogene. Triangles indicate co-occurrence of fruit and pollen, circles indicate pollen alone. Open circles indicate records dated as Maastrichtian only, not recorded as extending into the Paleogene. Compiled from references cited in Collinson (1993, 1996*b*) with Pole and Macphail (1996).

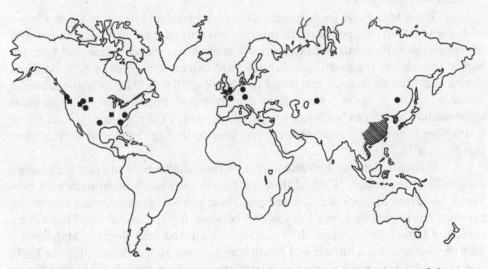

Fig. 16.5. Modern (cross-hatching) and fossil distribution of *Platycarya* (a tree of the Juglandaceae – walnut family) and related genera in tribe Platycaryeae. Solid squares represent Eocene fossil fruits, solid circles represent fossil pollen. Modified from Manchester (1987) in which full references to sources can be found.

Hosiea, Iodes and *Natsiatum*) and *Compositoipollenites rhizophorus* pollen together with rarer leaves (Manchester, 1994; Collinson, 1983; Kvacek & Buzek, 1995). *C. rhizophorus* pollen plotted in the Eocene Paratropical forest association of Boulter and Hubbard (1982). Icacinaceae are exclusively tropical today (Collinson, 1983). Juglandaceae tribe Platycaryeae (Fig. 16.5) has a single modern species (*Platycarya*

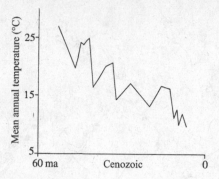

Fig. 16.6. Graph showing climatic deterioration on land through the Cenozoic based on mean annual temperature estimates from foliar physiognomy using CLAMP for North American near coastal sites around palaeolatitude 45° N. Modified from Wolfe (1995) which should be consulted for details of dates, sites, methods, etc. See also Spicer (this volume).

strobilacea), a tree confined to the sclerophyllous broad-leaved and mesophytic forests of eastern China and Japan. Fossil evidence comes from macrofossils (leaves, catkins, infructescences, etc. the latter named *Platycarya*, *Pterocaryopsis*, *Palaeoplatycarya* and *Hooleya*) and pollen named *Platycaryapollenites* (Manchester, 1987).

　　These taxa range from mangroves to forest trees and lianas and their former widespread occurrence points to the presence not only of these plant groups but also of tropical to subtropical vegetation over a much wider global extent and to much higher latitudes in the past. Maximal poleward extent of vegetation belts occurred during the Eocene thermal maximum (see Section 16.5.1). This was the maximal global warming to occur following the evolution and emplacement of these plant communities, which can be closely compared to those of the present day. Latitudinal restrictions follow the climate changes with cooling during the Cenozoic (Fig. 16.6) (see Section 16.5.2).

　　Changing land bridges and migration routes influenced interchange causing geographical restrictions. Critical thresholds occurred when land bridges were broken or lost from appropriate climate belts. These phenomena controlled migrations between North America and Europe and between Europe and Asia. Thus strong similarity (about 24% genera, 10% species in fruit and seed floras – Manchester, 1994) between western Europe and North America (see above) occurred in the Early and early Mid Eocene as a result of migrations over land bridges across the North Atlantic and possibly also via the Bering Straits. The latter is less likely as it was probably too far north for evergreen thermophiles which would have been restricted by light and warmth and also because the Turgai Straits might have limited exchange between Asia and Europe at this time (Tiffney, 1985; Tiffney in Boulter & Fisher, 1994).

　　According to evidence from fossil mammals land interchange in the latest Paleocene/Early Eocene via the North Atlantic was only likely to have occurred between about 55.5 and 53 Ma (Hooker, 1998; this volume) and this leads one to look for evidence in fossil floras of a short-term warming enabling tropical to subtropical vegetation to interchange at this time. However, Paleocene and Eocene floras from Greenland, Spitsbergen, Mull and Canada are rather similar to one another and lack any evidence of strongly thermophilic elements such as Menispermaceae or Icacinaceae. Instead they are dominated by low diversity deciduous elements (Kvacek ; Manum; Basinger *et al.*, all in Boulter & Fisher, 1994). However, it is uncertain if any of these floras correlates exactly with the time period inferred above. In southern England macrofloras across this short interval show little or no change (Collinson & Hooker, 1987). However, palynological assemblages show a short-term increase in the proportions of tropical elements in both the North Sea and the London Basin (Schröder, 1992; Allen, 1982). North American pollen floras show evidence of migrations from Europe but exact datings are uncertain (e.g. the *Platycarya* pollen appearance in North America is now considered to be Late Paleocene not earliest Eocene) and no strongly thermophilic pollen types are involved (Frederiksen, 1991, 1994).

The flowering plant trees and shrubs of the genus *Nothofagus* (southern beech), usually included in Fagaceae, have been regarded as the key genus in the study of Southern Hemisphere palaeobiogeography. The fruits are small nutlets with a very poor dispersal potential so that the distribution of the genus must have resulted from spread across land connections (Hill in Hill, 1994). Earliest occurrences based on 'ancestral' pollen across Gondwana are followed by Late Cretaceous occurrences of modern subgenera in the Antarctic Peninsula and South America followed by later occurences of these subgenera in Australia and New Zealand. The presence of members of all four of the modern *Nothofagus* subgenera in the Oligocene of Tasmania (based on macrofossils) shows the former widespread distribution of subgenera restricted at the present day to southern South America (subgenus *Nothofagus*) and to New Guineau and New Caledonia (subgenus *Brassopora*). Extinctions resulting in these relict modern distributions must have occurred more recently. These extinctions (at least in Tasmania) are attributed to climatic deterioration later in the Tertiary (Hill in Hill, 1994).

In the Northern Hemisphere it seems that land bridges control the initial distribution of Cenozoic plants but climate changes control their subsequent persistence in any one area. In the Southern Hemisphere the Gondwanan continental land mass was responsible for the initial widespread distribution of *Nothofagus* which persisted after Gondwana breakup but for which restricted present-day distributions are most likely due to subsequent climate change.

Non-flowering plants

Ginkgo was widespread in the early Cenozoic of North America (Basinger *et al.* in Boulter & Fisher, 1994) but is today restricted to one small area in China. Amongst conifers, *Metasequoia* shows a similar pattern whilst *Taxodium*, the dominant tree of Paleogene and Neogene coal-forming environments in North America and Europe, is today restricted to the subtropics of Florida (Collinson, 1990; Mosbrugger, 1995).

Many fern families were formerly widespread in the Mesozoic but underwent range restriction during the Cenozoic, e.g. to south-east Asia (Matoniaceae and Dipteridaceae), to tropical and subtropical zones (Gleicheniaceae and Marattiaceae), or to the Tropics and the Southern Hemisphere (Dicksoniaceae). The Cenozoic macrofossil record of these families is poor, probably largely as a result of displacement by angiosperms combined with displacement from disturbed sites of high preservation potential (Collinson, 1996a). Studies using well-represented and highly diagnostic spores of the fern genera *Anemia* (Schizaeaceae), *Ceratopteris* and *Cnemedaria* (Cyatheaceae) also show range contraction during the Cenozoic with present ranges established in the Neogene. *Cnemedaria* reached maximum southern extent during the Eocene thermal maximum, being a thermophilic fern this parallels the patterns shown by tropical and paratropical flowering plants (see references in Collinson, 1996a).

Patterns of distribution in non-flowering plants thus also reflect response to climate deterioration as was the case in flowering plants. However, the competetive exclusion of ferns by flowering plants, particularly from disturbed sites where flowering plant weeds largely took over from fern prairies, was probably also an important factor.

16.5   CENOZOIC PLANT COMMUNITIES

This section first demonstrates how plant communities can be reconstructed, using those of the Eocene thermal maximum as a case history. It then documents the patterns of change through the Cenozoic relating these to several other parameters of global change.

16.5.1   Reconstruction of plant communities and their distribution in response to the Eocene thermal maximum

Given the nature of the plant fossil record, a variety of approaches are necessary to deduce the nature of an ancient plant community from the fossil record. Three examples are given below. The major approaches involve (i) *in situ* plant fossils (roots, tree stumps, litter layers etc.), (ii) reconstructed whole or partial plants, (iii) functional morphology (see Sections 16.2 and 16.3.2), (iv) physiognomy of leaves and woods (see Section 16.2; Spicer this volume), (v) taphonomic evidence, sedimentary context and facies associations and (vi) extrapolation of plant form and ecology from nearest living relatives.

Mangroves

A *Nypa*-dominated mangrove with subordinate Rhizophoraceae (Ceriops) and other rarer taxa is reconstructed as the Early and Mid Eocene coastal vegetation of the southern North Sea area (present England, France and Belgium). This reconstruction is based upon functional morphology, taphonomic behaviour, nearest living relatives of pollen and fruits, facies association and co-occurrence of fruits with pollen. *Nypa* fruits occur as unfertilized, aborted and fully mature embryo-containing examples both on the strandlines in areas of modern vegetation and at the fossil sites. The range of size represented in modern fruiting heads, modern strandlines and fossil assemblages is closely comparable. The most autochthonous *Nypa* fossils occur in bioturbated silty sands with *Ostrea* deposited in a restricted embayment of an estuarine or lagoonal complex. *Nypa* fruits co-occur with *Spinizonocolpites* pollen and the pollen occurs at many additional sites. The pollen and fruits are essentially identical with those of modern *Nypa*, in the case of the pollen this applies at LM, SEM and TEM level. The tips of the viviparous embryo or sea pencil of *Ceriops* occur with the *Nypa* in the London Clay flora. The embryo morphology is directly related to the functional biology of the mangrove habit and the anatomy is identical to that of living *Ceriops*. (For further details on mangroves see Collinson, 1993, 1996b, 1999.) *Nypa* fruits, pollen and leaves co-occur at palaeolatitudes estimated at 65° S in the Eocene of Tasmania (Pole & Macphail, 1996) confirming a Southern Hemisphere *Nypa* mangrove previously based largely only on pollen.

The *Nypa*-dominated mangrove may have been in place in the latest Cretaceous as fruits and pollen are known from that time though no detailed taphonomic, facies association or co-occurrence work has been done on these. However, the Cretaceous occurrences are at lower latitudes and the maximum poleward expansion of the mangrove clearly reflects temperature maxima during the Eocene (Collinson, 1993, 1996b; Pole & Macphail, 1996). The distribution of *Nypa* (Fig. 16.4) is typical for many flowering plants which in the Paleogene were very widespread and which today are restricted, in this case to south-east Asia and environs. (See also Fig. 16.5.)

Tropical aspect forest

Vegetation to landward of the mangrove discussed above is reconstructed as a tropical to paratropical rainforest (Collinson, 1983; Collinson & Hooker, 1991). This is based almost entirely on the nearest living relative approach (but see supporting evidence below). The fossil fruit and seed flora is highly diverse (some 350 named species). Many of the nearest living relatives grow today in tropical or paratropical rainforests with many of those in south east Asia. Palms (Arecaceae) are both abundant and diverse. Other important families with entirely, or dominantly, tropical to subtropical affinities are Anacardiaceae, Anonaceae, Burseraceae, mastixioid Cornaceae, Dilleniaceae, Icacinaceae, Lauraceae, Menispermaceae, Sabiaceae, Sapindaceae, Vitaceae. In those cases where a few members of the family do extend into temperate zones today (e.g. Vitaceae, Menispermaceae, Anacardiaceae) a high diversity and the particular genera represented in the London Clay flora argue convincingly for these to be non-temperate representatives. In assessing the 143 species with close-living relatives Collinson (1983) determined that 92% of them grew today in tropical areas. The similarity with south-east Asia floras was strengthened by the discovery (Poole, 1993) of a twig of Dipterocarpaceae from the London Clay flora. This family, so widespread in south east Asia today, has an almost non-existent fossil record outside the Neogene of south-east Asia (Collinson *et al.*, 1993) so the fact that only a single twig was found is hardly suprising.

Whilst it is true that arguments based almost entirely on nearest living relatives are least satisfactory, there is considerable support for the existence of megathermal vegetation at these palaeolatitudes from other evidence. First of all, the large woods from the London Clay flora lack growth rings. Growth rings are absent in about 70% of the twigs (Collinson, 1983; Poole, 1992) so very little seasonality is indicated and that which did exist may have related to water availability not temperature. Secondly, leaf and wood floras and physiognomy from contemporaneous sites at similar palaeolatitudes indicate the presence of megathermal or warm mesothermal vegetation, i.e. tropical to subtropical in temperature regime (Wolfe, 1985, 1995; Wilde, 1989) and equable continental interiors (Wing & Greenwood, 1993). Thirdly, the London Clay flora contains a wide variety of different flowering plant groups, and their habits vary from vines to lianas to large trees. Whilst some of these might have changed their tolerances through time, it would be a very complicated argument to claim that each had changed in exactly the same way and thus remained in association. The abundance and diversity of palms argues strongly for a frost-free climate as the single apex of palms is vulnerable to freezing and if that single apex is killed the plant will die. Thus, although one or two palm species reach higher latitudes today, most are confined to low latitudes and mega- to mesothermal climates. Finally, it must be emphasized that the Eocene rainforests are not considered to be identical to those in south east Asia today, merely that the rainforest vegetation of that area is the closest modern analogue with many (but not all) homologous features. A detailed consideration of the origin of tropical rainforests is given by Morley (1999).

Taking into account all lines of evidence (fruit and seed floras, leaf and wood floras and their physiognomy, etc.) it is thought that this tropical to paratropical aspect vegetation (with fringing mangrove on suitably tranquil coastlines) extended to palaeolatitudes of at least 55° (ranging up to 60° or 65°) North and South during the Eocene thermal maximum (Collinson, 1990; Wolfe, 1985, 1995; Hill, 1994; Pole & Macphail, 1996). This vegetation is generally accepted to have been emplaced following increased humidity after the Cretaceous/Tertiary transition (see Section 16.5.2) but at that time was restricted to low latitudes (Wolfe & Upchurch, 1987; Upchurch & Wolfe in Friis et al., 1987; Wheeler & Baas, 1991, 1993). At least three subclasses of dicotyledons were in fact represented by trees in the Late Cretaceous, and large trees are known (Wheeler et al., 1994), based on fossil wood from Texas which implies a tree height of over 30 m. Their wood physiognomy implies a lack of pronounced seasonality but an overall warm and dry climate rather than a rainforest setting.

Polar deciduous forests

Eocene vegetation from the Canadian Arctic Archipelago (Axel Heiberg and Ellesmere Islands) (Basinger et al. in Boulter & Fisher, 1994; Greenwood & Basinger, 1994) and from the Paleocene and Eocene throughout the high northern palaeolatitudes has been termed polar broad-leaved deciduous forest by Wolfe (1985). There is no modern homologue as no tree-dominated vegetation now exists at such high latitudes under polar light regimes. This biome is thus extinct though many of the floral elements survive in hardwood and mixed hardwood/coniferous deciduous forests in various parts of the world. At Axel Heiberg direct evidence for polar deciduous forest comes from in situ trees with litter mats at palaeolatitudes of 75–80° N with growth rings indicating prolonged favourable tree growth and seasonality, and with deciduousness due to light availability (winter darkness) rather than temperature control. The community was dominated by wind-pollinated, dry-fruited, deciduous trees including conifers of the Taxodiaceae, Ginkgos and flowering plants of the Cercidiphyllaceae/Trochodendraceae, Betulaceae, Ulmaceae, Juglandaceae and Platanaceae all with serrate-margined leaves. Taxodiaceae dominated swamp forests, whilst hardwoods dominated better-drained habitats. The climate is deduced to have been free from severe frost (Taxodiaceae and crocodiles are present) with mild, moist summers (Basinger *et al.* in Boulter & Fisher, 1994; Greenwood & Basinger, 1994).

In the Late Paleocene similar forests grew slightly further south and one well-understood example comes from Alberta, Canada. Here, although *Metasequoia* (Taxodiaceae) is present, deciduous hardwood trees dominated and both are known from reconstructed 'whole' plants, i.e. a *Cercidiphyllum*-like tree and a tree of the Platanaceae. Fossil seeds and seedlings of both these plants show that they represent a colonizing 'r-selective' deciduous forest which facies association analysis indicates grew on open disturbed areas of floodplains (Pigg & Stockey, 1991). In the Mid Eocene of similar palaeolatitude, the Princeton flora from British Columbia is very different and includes the expected warmer-loving elements (Vitaceae, Sapindaceae, Magnoliaceae) in addition to the conifer *Metasequoia* (Cevallos-Ferriz *et al.*, 1991).

16.5.2   Patterns of community change

Five critical phases may be recognized in the evolution of modern plant communities. Major controlling factors are the timing of the establishment of communities followed by changing distributions largely due to overall cooling climate (Fig. 16.6) through both the Paleogene and Neogene (Wolfe, 1995; Mosbrugger, *et al.*, 1995; Knobloch *et al.*, 1993; Prothero & Berggren, 1992). The earliest critical phase is the Cretaceous emplacement of the flowering plants which is described by Lupia *et al.* (this volume). The other four are discussed here.

Cretaceous–Tertiary (K–T) transition

Studies of leaf floras in mid-palaeolatitudes of North America do not indicate any major modernization immediately associated with the K–T boundary in either leaf morphotype, architecture or taxonomic category (Johnson, 1992; Johnson & Hickey, in Sharpton & Ward, 1990). Fruit and seed floras in Europe also show no major modernization between Late Cretaceous and early Tertiary times (Knobloch *et al.*, 1993) though exact location of the boundary is not possible in the areas from which these fossils are derived. Cretaceous relatives of the Juglandaceae belong to the 'Normapolles' complex (Fig. 16.2). In North America extinction of 'Normapolles' types had occurred through the Late Cretaceous whilst nine species survived the K–T event into the Paleocene and three survived on into the Eocene. Two Juglandaceae pollen occurred in the Late Cretaceous and Paleocene, while diversity rose to eight species in the Late Paleocene and Eocene (Frederiksen, 1994). Overall, the patterns evident in the Late Cretaceous and early Cenozoic modernization and diversification of flowering plants seem to be part of a continuum resulting from the inherent functional biology of flowering plants themselves (see Section 16.3.2) and are not significantly affected by the K–T event.

In middle palaeolatitudes in North America vegetation was modified following the K–T event (Wheeler & Baas, 1991, 1993; Wheeler, *et al.*, 1994; Wolfe & Upchurch, 1987; Wolfe, 1987,1990; Johnson, 1992; Sweet & Braman, 1992; papers in Friis *et al.*, 1987 and Sharpton & Ward, 1990). 'Abrupt' ecological disruption is evidenced by low diversity opportunistic floras including the 'spike' of high numbers of fern spores in palynological preparations with rarer occurrences of fern macro-fossils, major turnover (up to 80%) in pollen and leaf morphotypes (but see caveats in Section 16.5.3) and by changes in wetland and marginal plant communities. In the longer term, vegetational change resulted in the initial emplacement of multistratal tropical rainforest, and selective extinctions of evergreen taxa led to selection for deciduous flowering plants altering the future aspect of tropical and Northern Hemisphere vegetation. Differential extinctions characterize different latitudes and vegetational belts. Lower palaeolatitudes, hence evergreen, animal pollinated flower-ing plants, were most affected. Overall, both megafloral evidence (leaf and wood physiognomy) and lithofacies (coals and pond and swamp settings) indicate increased wetness and humidity post-boundary (four-fold precipitation increase). Increased warmth (up to 10 °C increase in MAT) is also suggested by Wolfe (1990) from leaf phsyiognomy. Mammalian radiations post K–T in North America (Hooker, this volume) may partly reflect a response to the emplacement of rainforests, e.g. in the retention of relatively small body size and expansion of frugivory. It must be emphasised that there is no evidence for ecological trauma in the high Arctic and there is only minimal evidence of vegetational change outside North America (Spicer *et al.* in Boulter & Fisher, 1994). It is not known to what extent the North American events influenced future global land biomes elsewhere.

Eocene thermal maximum

The Eocene thermal maximum marked the maximum north and south pole-ward extent of megathermal vegetation including rainforests and mangroves and is documented world-wide. All forest vegetation belts were pushed poleward with polar broad-leaved deciduous forests at the poles. Details are given in Section 16.5.1. There followed gradual range restrictions with cooling climate culminating in the terminal Eocene event.

Eocene–Oligocene transition

Major effects are due to the sharpness of the cooling seen at this time. Dominantly deciduous or semi-evergreen forests were emplaced where former ever-green subtropical to tropical forest belts had existed. More arid floral elements appeared in continental interiors and forest canopies were lost with opening of habitats (Collinson, Wolfe, Hooker, Leopold *et al.*, all in Prothero & Berggren, 1992; Knobloch *et al.*, 1993; Hill, 1994). These changes may have influenced food resources and caused mammalian extinctions (Hooker, this volume). *Nypa* became extinct in North America, the Caribbean and Europe (Frederiksen 1991), while Icacinaceae persisted to the Early Oligocene in Europe (Knobloch *et al.*, 1993; Kvacek & Buzek, 1995). Eocene–Oligocene climatic decline in the northern high latitudes may be evidenced by an increase in Pinaceae which begins in Mid Eocene floras (Basinger *et al.*, in Boulter & Fisher, 1994) and indicates the shift of conifers to lowland settings.

Vegetational changes, which began in the Late Eocene, intensified at the Eocene–Oligocene transition and continued to some extent in the Oligocene with further climate deterioration. They resulted (especially in mid-latitudes) in more open habitats, lower diversity and lower stature vegetation, increase in proportions of smaller and dry fruits produced from wind-pollinated flowers and an increase in proportions of smaller and more xeromorphic foliage. Mammalian communities responded with increasing proportions of larger mammals, decreased proportions of arboreal mammals, increased proportions of herbivores with increasing crown height and a shift from fleshy to dry fruit frugivory (Collinson & Hooker, 1987, 1991; Hooker, this volume).

Miocene and Pliocene

Forest vegetation belts during the Neogene generally reflected conditions imposed by the Eocene–Oligocene deterioration. Herb dominated communities were insignificant (Wolfe, 1985; Wing, 1998). For example, floras of the Canadian Arctic Archipelago show that mixed evergreen coniferous and broad-leaved deciduous forests persisted in high northern palaeolatitudes throughout the Neogene until the Pleistocene glaciations (Basinger *et al.*, in Boulter & Fisher, 1994). Coniferous forests spread in central and western northern Europe, while mixed mesophytic to broad-leaved deciduous forests persisted at mid-latitudes of Europe, North America and Asia (Mosbrugger, 1995; Knobloch *et al.*, 1993; Wolfe, 1985, 1987; Kovar-Eder *et al.*, 1996; Boulter & Fisher, 1994; Wing, 1998). Peat-forming vegetation in Europe during the Miocene comprises vegetation units which can be closely linked to those in the south-eastern USA today, i.e. *Taxodium* swamp, *Nyssa* swamp, *Cyrilla* shrub and a pine/Ericaceae woodland (Mosbrugger, 1995). The swamp forests of Europe were much reduced during the Late Miocene to Pliocene climate deterioration (Mosbrugger, 1995) resulting eventually in the restricted distribution to the southern USA today. Local tectonic effects and sea-level changes (Mosbrugger, 1995), combining to produce base-level changes in water table, may have contributed to this restriction in the distribution of tree-dominated wetlands.

Climate deterioration (cooling and local aridity especially in continental interiors) ultimately led to the origin of low biomass, low stature vegetation including, eventually, grasslands (Wolfe, 1985; Wing, 1998). Unequivocal grass fossils are known from the Early Paleogene but grasses only became floristically significant much later (Collinson *et al.*, 1993). Early suggestions that the three-toed horse *Hipparion* lived in savanna grasslands at around 11.5 Ma (latest Mid Miocene) in Europe were corrected by Bernor *et al.* (1988), who documented their occurrence in warm–temperate mixed mesophytic forests.

Evidence for the earliest savanna grasslands comes from the High Plains continental interior of North America. There grass pollen and macrofossils of grassland grass groups (e.g. *Stipa*) attain significant representation in Mid to Late Miocene fossil floras. Grazing mammals also occured at this time (Collinson & Hooker 1991).

Isotopic evidence (tooth enamel, soil carbonates) shows an increase in C4 plants (assumed to be grasses) at about 7 million years ago in low latitudes of Asia, America and Africa. Other areas, including higher latitudes, show this increase later, around 2–3 million years ago (Koch, 1998).

True open grasslands were probably never in place in the Neogene (Leopold *et al.*, in Prothero & Berggren, 1992). Low biomass vegetation in the Oligocene of the central USA and Asia is considered to have largely or entirely lacked grasses, having been dominated by low growing, arid-adapted woody plants and scattered trees (based on evidence from modern pollen rain, soils and macrofossils; Leopold *et al.*, in Prothero & Berggren, 1992). Decline in microthermal rainforests (*Nothofagus*-dominated) to be replaced by arid-loving elements is also a prominent feature of the Miocene and Pliocene in Australia (Kerhsaw *et al.*, in Hill, 1994).

16.5.3   Patterns of species diversity and floral turnover

Large-scale patterns of floral and vegetational change are recognized relatively easily (see, for example, Fig 16.1; Sections 16.4, 16.5.1, 16.5.2; Lupia *et al.*, this volume). However, the nature of the plant fossil record (Section 16.2) means that detailed species diversity patterns and floral turnover can only be based on single organs. These must have been reliably and consistently identified and any taphonomic bias must have been eliminated.

A close approach to the necessary ideal is found in the studies of Wing *et al.* (1995), where one group of specialists has undertaken a long-term study of leaf floras through the Paleocene and Early Eocene of the Bighorn Basin, Wyoming, and has taken account of facies associations and other variables. Frederiksen (e.g. 1991, 1994) has addressed species diversity and floral turnover in the Late Cretaceous and Cenozoic of North America using pollen fossils which only he himself has identified. In these studies various factors including migrations across land bridges, niche availability in changing flowering plant dominated vegetation and climate change were all deemed important controlling factors. Both Wing *et al.* (1995) and Frederiksen (1994) noted a sharp decline in diversity at the Paleocene–Eocene transition which might be explained by slow adaptation to rapid climatic shifts (see Hooker, this volume). Frederiksen (1991) noted coincidence in species diversity decline and the Eocene–Oligocene cooling. In another group study, of the Neogene of the Lower Rhine Embayment, a diversity decline was recorded in leaf floras from the Miocene to Pliocene coincident with cooling climate (Mosbrugger, 1995).

Thus Cenozoic plant species diversity patterns, like patterns of community change, seem to reflect climate change. However, it must be emphasized that this interpretation requires confirmation because most fossil leaf and pollen morphotypes have not yet been placed in phylogenetic (cladistic) lineages. This is necessary to ensure that pseudo-extinctions (of named morphotypes but which were in fact part of one lineage) and pseudooriginations (e.g. migrations from elsewhere) can be distinguished from true extinctions and evolutionary originations (of species).

16.6   CONCLUSIONS

Four main phenomena have been recognized controlling Cenozoic evolution of modern plant communities. First, the evolution of the plant groups themselves, especially various flowering plants, and their distinctive functional biology; secondly, the emplacement of communities dominated by those plants and their establishment over extensive ranges; thirdly, the existence of land bridges and land masses which enabled floral interchange, and fourthly climate change which has acted mainly to cause extinctions from large parts of former widespread and high latitude ranges, towards recent relict and restricted, often tropical to subtropical settings. The first and second factors strongly controlled biotic global change such as coevolution of flowering plants with insects and mammals and in so doing controlled the evolution of land biomes. The second, third and fourth factors reflect biotic response to physical global change where land connections enabled migration and spread, while climate change caused extinctions and range restriction.

17

Leaf physiognomy and climate change

ROBERT A. SPICER

17.1 INTRODUCTION

A principal driving force for evolutionary change through selection is environmental change. This is borne out by the widespread recognition that the 'Red Queen World' (Van Valen, 1973), where competition between organisms is divorced from change in the physical surroundings, is largely a theoretical (but useful) abstraction because the effect of environmental change in the real world is all pervasive. Of all the possible mechanisms for environmental change, those due to climate perturbations are the most universal and persistent. Understanding the pattern and process of climate change is therefore of extreme importance to an understanding of the pattern and process of biotic responses to that change.

Climate change is not only an expression of atmospheric phenomena: it is intimately linked to variations in sea level, ocean circulation and tectonics. The bulk of the ocean realm, however, is to a large extent buffered from the short-term (days to weeks) high magnitude fluctuations in temperature, pressure, and fluid flow regimes experienced by non-marine environments that are directly exposed to atmospheric conditions. In terms of magnitude and frequency, climate change is, therefore, most strongly expressed in the terrestrial realm. Here, climatic conditions are recorded at high temporal and spatial resolutions by a variety of features possessed by the plants, animals and sediments present in that environment.

For many years a purely qualitative interpretation of past climate was regarded as sufficient to provide a context for geological or biological phenomena. Relative wetness/dryness could be mapped from climatically sensitive sediments such as coals and evaporites, while relative temperature could often be determined from the biota based on the presumed climatic tolerance of characteristic species (e.g. Vakhrameev, 1991). These qualitative interpretations are still useful and can be collated to produce graphs showing relative climate change through time (e.g. Frakes, 1979). One of the most significant observations derived from such graphs is that viewed over geological time the present climate that we as a species regard as 'normal' is an aberration: it is abnormally cold.

The possibility of anthropogenically induced climate change occurring over hundreds, rather than thousands, of years stimulated the development of computer based climate models to forecast the pattern of possible change.

Testing such Atmospheric General Circulation Models (AGCMs) can only be done in the context of their ability to retrodict ancient climates that were manifestly different to those of the present. In this context qualitative palaeoclimate proxies are distinctly inferior to those providing quantitative climate estimates with measurable uncertainties.

Perhaps the best-known quantitative temperature proxy is that derived from oxygen isotopes preserved in the remains of shelly marine organisms. However, because many atmospheric circulation climate models require sea surface temperature to be specified, these measurements are often used to define the boundary conditions used by the model. When used in this way they are invalid for testing the model. The only legitimate means of testing these simulations is by using land-based palaeoclimate proxies.

For Quaternary studies the relatively small amounts of evolutionary change involved permit the extrapolation of the observed climatic tolerances of living species back to their fossil ancestors. This Nearest Living Relative (NLR) approach relies on the accurate identification of the fossil specimen to a living species with a known, and preferably limited, distribution under a discrete climatic regime. This technique may be applied to both animals and plants. To dilute the effect of erroneous identification or saltational change in environmental tolerance, species aggregations are often used. This has the added advantage that anomalies between species can be identified and might even improve the precision of climatic data. With pollen and foraminiferal data, the concept of transfer functions has been developed (Imbrie & Kipp, 1971; Huntley, 1994). In the case of pollen, a response surface is constructed for each pollen taxon in relation to environmental variables. Statistical manipulation of these responses results in transfer functions for assemblages. However the inclusion of exotic or extinct taxa inherently degrades the precision of the transfer function and most palynologists have been reluctant to apply transfer functions for assemblages older than the Quaternary (Wolfe, 1995).

For pre-Quaternary studies the NLR technique, however conducted, becomes more and more inappropriate the further back in time one goes. This is simply a function of evolutionary change because NLR relies on evolutionary stasis with respect to morphology and climatic and other environmental tolerances (Spicer, 1990). Even if one assumes that the climatic tolerance of all individuals within a species has remained the same, extinct species can only at best be assigned to an extant genus and genera invariably have a wider climatic tolerance than any single constituent species, living or extinct.

For tracking and mapping climate change over long timescales, a more robust approach, largely independent of taxonomic attribution, is required. Animals whose physiology is poorly understood, who annually migrate between environments or over long distances, or who have effective intrinsic (endothermy) or extrinsic (burrowing/hibernating) homeostatic mechanisms inherently make poor environmental proxies, but plants suffer none of these problems. This contribution outlines the development of a new and extremely powerful quantitative climate proxy based on leaf physiognomy and compares initial results to current model predictions.

17.2   PLANT PHYSIOGNOMY AS A CLIMATE INDICATOR

The process of land plant photosynthesis is responsible for a large portion of the atmospheric CO_2 regulation, oxygen levels (and thereby indirectly the ozone concentration and UV flux at the Earth's surface), and the flux of water through the atmosphere (as a consequence of evapotranspiration). This intimate relationship has modified both plants and atmosphere significantly and continues to do so. This has resulted in a highly tuned adaptive legacy that we can use to determine past environmental conditions.

After a land plant has developed from a spore or seed, it is spatially fixed; living at the interface between the geosphere and the atmosphere. It is therefore constrained by the conditions it is exposed to in these two realms, but to survive in competition with other organisms it has to function as efficiently as possible. 'Function' here is taken to mean the production of carbohydrates through photosynthesis for without this primary process all other activities such as growth and reproduction are impossible.

In the course of evolution, therefore, there has been a selective premium on those features which confer the maximum functional advantage under a variety of environmental conditions. Broadly speaking, such features centre on maximizing photosynthetic capacity (for example by maximizing leaf surface area and stomatal size/numbers) while minimizing water loss (for example, by developing coverings largely impervious to water, reducing stomatal size/numbers, minimizing leaf surface area), and simultaneously minimizing the energy invested in photosynthetic and structural components. For any given environmental condition there tends to be only a limited number of 'engineering solutions' that come close to satisfying these conflicting constraints and it is not surprising, therefore, that land plants growing under similar environmental conditions, separated either in time or space, or both, tend to develop similar architectures. Moreover, because these architectures (physiognomic traits) are constrained by the laws of physics, which we assume have remained constant since the advent of land plants, it is safe to assume (within limits imposed by responses to changes in vascular efficiency and responses to atmospheric concentrations of carbon dioxide) that we can interpret palaeoenvironments using suitable plant fossil remains calibrated against the architecture of modern land plants.

While the above constraints affect all land plants, some plant groups appear to have evolved more sensitive physiognomic expressions of climate than others. In long-lived woody angiosperms, for example, we see higher levels of foliar diversity and phenotypic plasticity in leaf architecture, even on a single individual, than in gymnosperms (e.g. cycads and conifers) or ferns. This confers on angiosperms a particularly close relationship with climate that can be precisely calibrated. It would be wrong to dismiss non- or pre-angiosperm vegetation in this context as it also displays a physiognomic relationship with climate, but to date that relationship has been less well quantified.

In the modern world, vegetation can be described and mapped independant of its taxonomic composition (e.g. Holdridge, 1947; Walter, 1979; Box, 1981; Olsen et al., 1983; Leemans, 1989; Prentice, 1990; Prentice et al., 1992). Furthermore the resulting patterns of plant distributions have a clear relationship with climate. The physiognomy (architecture) of individual organs, whole plants, and vegetational units up to the scale of global biomes is, therefore, strongly influenced by climate and can thus be used to infer past climatic conditions. Moreover, adaptive relations between environmental factors and some physiognomic characteristics have been shown to have a theoretical basis (Givnish, 1979, 1986; Givnish & Vermeij, 1976; Parkhurst & Loucks, 1972).

Because physiognomic adaptations to environment are relatively robust in an evolutionary context we can, within certain limits, quantify the technique for determining pre-Quaternary climates. The methodology that has been in use longest is that of leaf margin analysis. Bailey and Sinnott (1915) noted that the leaves of modern woody dicotyledonous flowering plants tend to have smooth leaf margins in warm climates and toothed margins in cool climates. Wolfe (1979) revisited that methodology and by using species growing in drought-free environments in southeast Asia plotted the percentage of entire (non-toothed) margined leaf species against Mean Annual Temperature (MAT) and obtained Fig. 17.1(*a*). This relationship degrades when water is limiting to growth and so can only be applied to palaeobotanical data when the original vegetation can be inferred to have grown in drought-free regimes.

This technique of leaf margin analysis, when applied to Cretaceous leaves in North America (Fig. 17.1(*b*)), demonstrates clearly that: (i) a relationship existed between palaeolatitude (a proxy for MAT) and the ratio of toothed to entire-margined dicot leaves as early as the Cenomanian, (ii) this relationship does not appear to be disrupted by the light regime at high latitudes, and (iii) the relationship reflects the overall global cooling (expressed as an increase in the Equator to Pole thermal gradient) that took place between the Cenomanian and Campanian.

Climate change tends to be most strongly expressed at the Poles (Goody, 1980) and analyses of high-latitude climate proxies, therefore, is of particular importance. During the Late Cretaceous, Arctic forests flourished at latitudes as high as 85° N (Spicer & Chapman, 1990) and fossil floras were deposited throughout the circum-Arctic region (e.g. Herman, 1994; Herman & Shczepetov, 1992; Samylina, 1963, 1973, 1974; Scott & Smiley, 1979; Smiley, 1966, 1967, 1969*a, b*; Spicer *et al.*, 1992). This region, therefore, provides an ideal natural laboratory for investigating terrestrial biotic and environmental change and its relevance to global greenhouse climates.

We know that in Cenomanian time northern Alaska rarely experienced drought because there are a large number of thick coal seams (Spicer *et al.*, 1992) and although charcoal occurs frequently it is present in only small amounts (Spicer & Parrish, 1990). In addition, tree rings show that growth was uniform throughout the summers and therefore water, to name but one factor, was not limiting (Parrish & Spicer, 1988a). The sediments have yielded a large number of leaf fossils and, of all the woody dicot leaf forms that have been distinguished, approximately one-third are entire-margined, suggesting a MAT of 10 °C (Parrish & Spicer, 1988b). By Coniacian times this proportion of entire-margined species had risen slightly, translating to a MAT of 13 °C (Spicer and Parrish, 1990).

A Cenomanian MAT of 10 °C is similar to that of London today but, bearing in mind the polar light regime, it is likely that the Mean Annual Range of

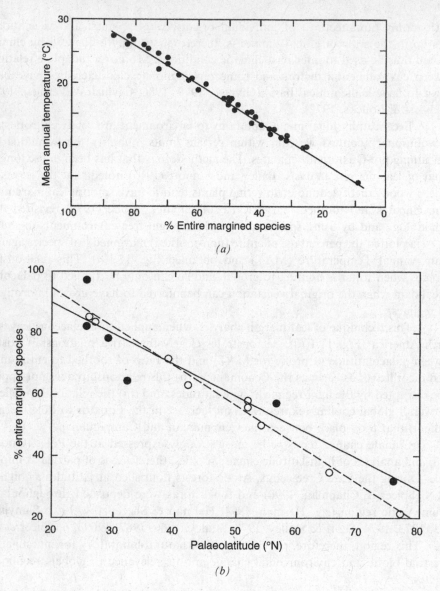

Fig. 17.1. (*a*) Graph of the percentage of entire margined woody dicotyledonous angiosperm leaves in humid modern south-east Asian forests plotted against the observed mean annual temperature (from Wolfe, 1979). (*b*) Graph of the palaeolatitude of fossil floras in North America of Campanian (open circles and dashed line) and Cenomanian (closed circles and unbroken line) ages plotted against the percentage of entire margined woody dicot leaves in those floras. Data from Wolfe (personal communication).

Temperature (MART) was much greater. Simple leaf margin analysis cannot provide MART data. Even in the context of MAT we are making a number of assumptions when we use leaf margin analysis. First we are assuming that there is no calibration drift over time, i.e. that the modern slope of the graph has remained the same over time. Secondly, we are assuming that the relationship between margin characteristics and temperature was not different due to possible effects of the polar light regime. In part, these concerns are addressed by the observation that, by the Campanian, there was established a relationship between leaf margin type and palaeolatitude that can be used as a proxy for temperature despite the shallower (than today) Equator to Pole thermal gradient and the absence of a quantitative measure of any drift that might have occurred. Similarly, there is no obvious break in the slope of that relationship above or below palaeolatitude 66°; the likely palaeoarctic circle (see Spicer & Parrish, 1990, for discussion of why obliquity changes are unlikely to have exceeded those recognized in Milankovich–Croll cycles).

17.2.1   Plants in hyperspace

Clearly, there is a relationship between the leaf margins of woody angiosperms and at least one climatic characteristic, i.e. temperature. Nevertheless there are other leaf characteristics which may correlate with other climatic variables. One such character is leaf size which is likely to be related to precipitation or water availability. As an example of this, consider the contrast between desert and rainforest vegetation. In deserts, leaf size is small; large trees with large leaf loads are absent while some plants (e.g. euphorbs and cacti) have dispensed with leaves altogether and photosynthesize using their stems; stems which are also adapted for water storage. Rainforests, on the other hand, are characterized by a large range of leaf sizes (some more than $1m^2$), large trees with large leaf loads, and a complex structure made up of several bush and tree layers. Moreover, rainforests or deserts have similar appearances in different parts of the world even though they may have few taxa in common.

It is possible to extend this physiognomic approach considerably beyond leaf margin and size characteristics. In an analysis of more than 20 species of woody dicot leaves from each of over a hundred vegetation samples, Wolfe was able to demonstrate that initially 18 (Wolfe, 1990), then 29 (Wolfe, 1993) and subsequently 31 (Wolfe, personal communication) leaf characteristics, including margin type, apex and base types, leaf shapes, leaf sizes etc. are correlated with at least eight climate parameters. Wolfe (1990) termed this technique 'Climate Leaf Multivariate Program' or 'CLAMP'. A typical completed CLAMP scoresheet for the 31 character states is shown in Table 17.1.

Wolfe's (1993) scoring method is summarized here because it is of critical importance to the outcome of such an analysis and is somewhat unusual. The data sheet (Table 17.1) records the 31 leaf characters for each leaf and the characters are divided into groups within which scores sum to unity. For example, size categories record the full size range exhibited by a given taxon at a given site, so the total score for various size categories combined should be unity. Therefore, if a leaf taxon exhibits Microphyll III, Mesophyll I and Mesophyll II size categories, each category is scored as one-third. If all leaves of a taxon fall within a single category, that category is scored as one for that taxon. Other grouped characters that also score to unity are apices 'round', 'acute' and 'attenuate', bases 'cordate', 'round' and 'acute', the five length-to-width categories, and the three shape categories. Margin characters are divided into 'lobed', 'no teeth', and several tooth types. Each of the margin categories may score up to one, but may be scored only as 0.25 or 0.5

Table 17.1. *A typical score sheet for a fossil assemblage: the bottom line of the sheet is added to the modern vegetation matrix*

Species	Lobed	No teeth	Teeth regular	Teeth close	Teeth round	Teeth acute	Tth compound	Nanophyll	Leptophyll I	Leptophyll II	Microphyll I	Microphyll II	Microphyll III	Mesophyll I	Mesophyll II	Mesophyll III	Apex emarg.	Apex round	Apex acute	Apex attenuate	Base cordate	Base round	Base acute	L:W < 1:1	L:W 1 to 2:1	L:W 2 to 3:1	L:W 3 to 4:1	L:W > 4:1	Shape obovate	Shape elliptic	Shape ovate
Menispermites septentrionalis			1	1	1							0.5	0.5								1				1					1	
Paraprotophyllum cf. *cordatum*			1	1		1									1			1			1				1						1
Platanus	0.5		1	1		1					0.25	0.25	0.25	0.25				1					1	0.5	0.5					0.5	0.5
lauro-vetanica			1	1		1					0.25	0.25	0.25	0.25					1			0.5	0.5		1				0.5	0.5	
Araliaephyllum parvidens	1		1	1	0.5	0.5					0.33	0.33	0.33						1			0.5	0.5		1				0.5	0.5	
Araliaephyllum subitum	1		1	1		1	1				0.33	0.33	0.33						1		1				1					1	
Arthollia pacifica			1	1		1						0.5	0.5					1			0.5	0.5		0.5	0.5					1	
Dalembia vachrameevii	0.5		1	1	1		0.5					0.5	0.5						1				1	0.5	0.5					0.5	0.5
Trochodendroides sp.2		1										1						1				1			1					1	
Zizyphus smilacifolia			1	1		1					0.5	0.5							1						0.5	0.5				0.5	0.5
Cissites cordatus	1		1	1	0.5	0.5					0.5	0.5							1		1			1						0.5	0.5
Terechovia intermedia	1		1	1	0.5	0.5					1								1					1						0.5	0.5
'Leguminosites acuminata'		1									1									1			1			1					1

														Total	Percentage
Zizyphus microphylla	0.5			0.5			0.5	0.5			1			2.5	10
Pseudoprotophyllum boreale	0.5		0.5		0.5					1				15.5	62
Magnoliaephyllum sp.	1	1		1		1	1					1			20
Menispermites kryshtofovichii	0.5		0.5	0.5											
Celastrophyllum latifolium	1		1	1			1					1			
Dalbergites sp.															
Araliaephyllum montanum	1		0.5	0.5	0.5		1					1			
Celastrophyllum orientalis	1			1		1	1	0.33	0.33	0.33					
Schefflieraephyllum venustum	0.5	0.5	0.5	0.5	0.5	0.5	0.5	0.33	0.33	0.33					
Cissites bidentatus	1	1	1	1	1	1	1	0.33	0.33	0.33					
Trochodendroides sp. 1															
Trochodendroides armanensis	0.5	0.5									1				
Araliaephyllum sp. 3											1				

Arman River, May 12 1996.

depending on the nature of the character on all leaves of a given taxon, or its frequency of occurrence in the population of leaves within that taxon at that site.

All characters are defined in Wolfe (1993). Length-to-width categories are self explanatory, and shape categories are conventional ones. Leaf size categories are defined differently from those of Raunkiaer (1934) and Webb (1959). The template for sizing leaves differs to that given in Wolfe (1993) only in the addition of the nanophyll and mesophyll III categories.

Leaf margin categories and their scoring require some explanation. A leaf is considered pinnately lobed (as distinct from toothed) if a line joining all the sinuses is approximately parallel to the midvein. If the line is parallel to the margin, the leaf is toothed rather than lobed. Lobes on a palmate leaf are entered by major veins (termed pectinal veins in Spicer, 1986) originating at, or near, the base of the leaf. If all leaves in a taxon are lobed, a score of unity is assigned to that taxon. If some leaves are lobed and others not, a score of 0.5 is given, otherwise the taxon is given no score. The 'no teeth' category is essentially the same as 'entire' of Bailey and Sinnott (1915) but also includes spinose teeth where a fimbrial vein forms abmedial projections beyond the leaf margin. If all leaves in a taxon lack teeth a score of unity is given, if some are toothed and others not 0.5 is awarded, and a score of 0 is given if all leaves are toothed. Teeth are considered 'regular' if the lengths of the basal flanks of two adjacent teeth differ by less than one-third. If the teeth are both regular and irregular and some leaves are toothed and others not, a score of 0.25 is given, but if the teeth are only regular and some leaves have teeth and others do not a score of 0.5 is given. A score of 0.5 is also assigned if the teeth are both regular and irregular and all leaves are toothed, but a score of 1.0 if all teeth are regular and all are toothed.

The statistical engine that Wolfe used initially for CLAMP was Correspondence Analysis (CA) (Wolfe, 1990, 1993); a multivariate ordination technique originally devised for use in phytosociological studies (Hill, 1973, 1974) but which subsequently found widespread application including palaeoecological studies (Spicer & Hill, 1981). In the 1995 version of CLAMP, Wolfe switched to Canonical Correspondence Analaysis (CCA). CCA is termed a direct ordination method (Kent & Coker, 1992) because it explicitly correlates environmental data with attributes (e.g. leaf character states) that are ordered based on observed distribution (e.g. within vegetation samples). This is in contrast to indirect ordination methods such as principal components analysis (PCA) or conventional CA which can only be related to possible causal parameters subjectively (ter Braak, 1986, 1987; Kovach & Spicer, 1995).

CCA has other advantages over other commonly used statistical methods. Unlike with PCA compositional or closed data sets (e.g. percentages) pose no problem for CA or CCA. Moreover, CA and CCA are particularly robust when dealing with variables that covary (i.e. that are not independent of each other) and display the degree of covariance as the proximity of the variables to each other. Such covariance is inevitable in environmental data and, in particular, in data relating to foliar physiognomy. Leaf architecture always represents a compromise solution to conflicting constraints (e.g. size is a function of the need to intercept as much light as possible while minimizing water loss and structural cost), and a climate parameter such as MAT is always a function of both Cold Month Mean Temperature (CMMT) and Warm Month Mean Temperature (WMMT) which, in turn, is linked to the duration of the growing season, humidity, etc. Statistical methods such as PCA and multiple regression analysis that are sensitive to covariance should never be used in this context.

This highly integrated set of correlated variables effectively prevents naïve statistical partialling out of individual leaf and climate characters. Beyond the two primary leaf features (margin type and size) the particular suites of other characters present in populations of leaves from a range of taxa taken from a single site, relate to an array of related climate parameters. Within this melange of hybrid characters and correlated climate parameters it is highly unlikely there is hidden a simple correlation between individual character states and individual climate parameters. Even the relationship between leaf margin type and MAT breaks down when water is growth limiting (Wolfe, 1979). For this reason, direct ordination methods that align climate vectors within distributions of samples based on aggregate characters are the only ones suitable for CLAMP analyses. Fig. 17.2 illustrates the current CLAMP methodology.

Fig. 17.3(a) is a CLAMP (CA) plot and represents 31 dimensional space collapsed to two dimensions. Axes 1 and 2 represent the two axes of greatest variation in the data so the plot is the least distorted projection from 31–dimensional space. The points represent each of the 103 vegetation samples positioned relative to its neighbours. The position of each point is based on the characters that are possessed by the leaves that are on the trees in that vegetation. Points that plot close to each other have similar foliar compositions.

The points are coded to demonstrate that they are arranged according to their mean annual temperature even though MAT information was not part of the analysis; the leaf characters alone determine the relative positions of the sites.

Figs. 17.3(b) and (c) are complementary CCA plots. Fig. 17.3(b) is a plot of vegetation sites and 17.3(c) is derived from the same data but displays selected leaf characters. They both have eight environmental variables explicitly assigned by the program. Note the coincidence of the visually perceived MAT vector in Fig. 17.3(a) with the mathematically assigned vector in Fig. 17.3(b). Other climate vectors such as those relating to precipitation can also be defined mathematically. Fig. 17.3(c) is the complementary plot showing the relationship between a selection of the leaf characters (only those relating to size and margin type are illustrated for simplicity) to the climate vectors. The margin characters run subparallel to the MAT vector, while the leaf size characters are broadly aligned with the precipitation vectors.

Because these vectors are positioned using modern vegetation and climate we can calibrate them using the known climatic parameters measured at the sites where the vegetation was sampled. The small open square marks the position of a fossil site from the Cretaceous (Coniacian) Arctic positioned on the basis of the characters displayed by the leaf assemblage. The fossil site has been added to the analysis as a 'passive' sample so that it does not, by its presence, influence the structure of the modern vegetation physiognomic space. The position of the fossil assemblage projected on to the mean annual temperature vector calibrated by the modern vegetation can be used to determine the temperature under which the fossil

CLAMP FLOWCHART

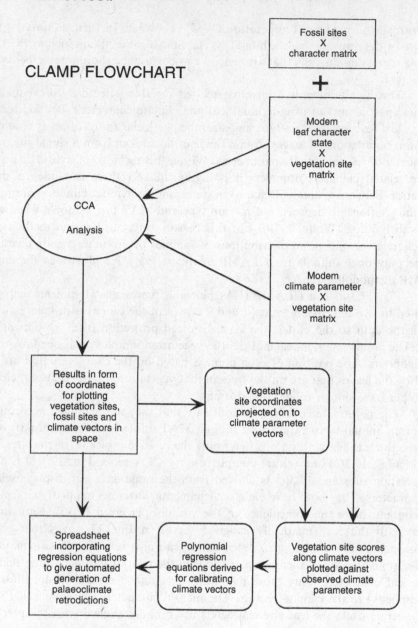

Fig. 17.2. A flow chart showing schematically how to undertake a CLAMP analysis.

vegetation grew within a statistical measured range of uncertainty (Figs. 17.4(*a*) and (*b*). Other palaeoclimatic parameters can be retrodicted in a similar manner.

In Fig. 17.3(*b*) the fossil sample, that from the Coniacian of the North Slope of Alaska (Spicer & Herman, 1997; Herman & Spicer, 1996), sits in a space that is devoid of modern vegetation. Interestingly, this assemblage represents an example of an extinct ecosystem; a polar forest growing at 75° N when the average global temperature was perhaps about 5 °C warmer than now. There is no modern analo-

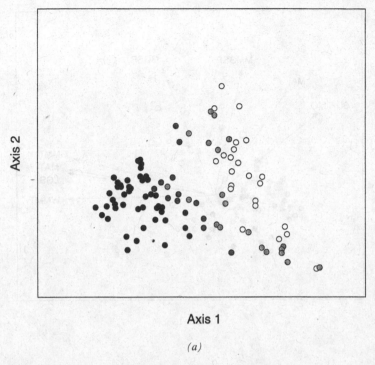

Axis 1

(a)

Fig. 17.3. (*a*) A scatter plot on the two major axes of variation in a CA analysis of modern vegetation sites scored using the 31 leaf character states as in figure. The sites have been coded based on the mean annual temperature observed at the sites where the vegetation was growing. Open circles represent sites with the highest mean annual temperatures while solid circles are those with the coolest mean annual temperatures. (*b*) A scatter plot of modern vegetation sites derived as in Fig. 17.3(*a*) but with climate vectors positioned explicitly using CCA. MAT – mean annual temperature; WMMT – warm month mean temperature; CMMT – cold month mean temperature; LGS – length of the growing season; MAP – mean annual precipitation; MGSP – mean growing season precipitation; MMGSP – mean monthly growing season precipitation; 3DRIMO – precipitation during the three consecutive driest months. (*c*) A scatter plot showing the relationship between selected leaf character states and the climate vectors. Vector designations as in Fig. 17.3(*b*). Leaf character labels are as follows: Tth Cmpd – teeth compound (most characters relating to teeth plot in this area), No Teeth – entire margins. Leaf size characters in ascending size order are Nanophl – nanophyll; Lepto1 – leptophyll 1; Lepto2 – leptophyll 2; Micro1 – microphyll 1; Micro2 – microphyll 2; Micro3 – microphyll 3; Meso1 – mesophyll 1; Meso2 – mesophyll 2; Meso3 – mesophyll 3. With respect to Fig. 17.3(*b*) the plot is slightly rotated about axis 2 to display better the relationships.

gue for this polar broad-leaved deciduous vegetation and it occupies its own space at the edge of the modern vegetation fields. However, its projected position along the MAT vector score provides an estimate of the palaeo-MAT when plotted on to the second-order polynomial regression line produced by comparing the MAT vector scores of the modern vegetation with the observed MAT at those sites (Fig. 17.4(*a*)). The estimated Coniacian MAT is 12.5 °C (compared with 13 °C from simple leaf margin analysis). The dispersion of the residuals about this regression line provide an estimate of the statistical uncertainty associated with the data set (Fig. 17.4(*b*)). In

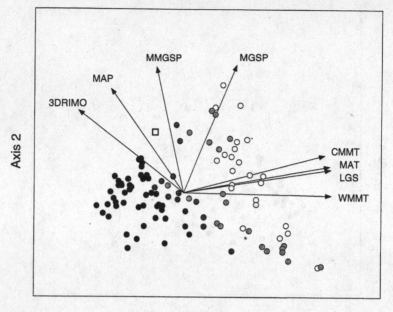

(b)

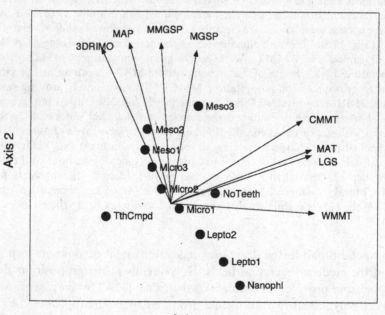

(c)

Fig. 17.3 (*cont.*)

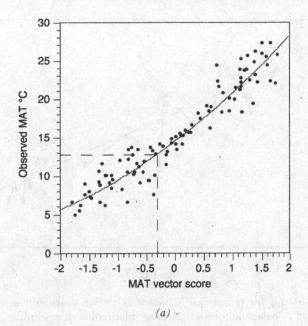

(a)

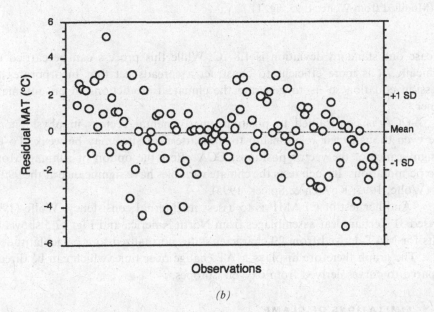

(b)

Fig. 17.4. (*a*) Graph of the calibrated MAT vector score derived from CCA plotted against the observed MAT for modern vegetation sites. The curve is a best fit second-order polynomial regression line. The dotted line represents the projected MAT vector score for a Coniacian leaf assemblage from the North Slope of Alaska. (*b*) Scatter of residuals about the fitted second-order polynomial regression line shown in Fig. 17.4(*a*).

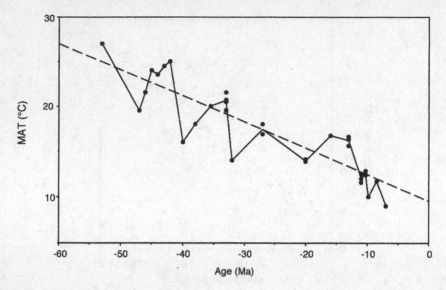

Fig. 17.5. Graph showing the retrodicted MAT for Tertiary leaf assemblages in North America normalised to palaeolatitude 45° N using a latitudinal temperature gradient of 0.3 °C/degree latitude for assemblages > 33 Ma and 0.4 °C/degree latitude for assemblages < 33 Ma. (Modified from Wolfe, 1995, fig. 11.)

this case one standard deviation is 1.8 °C. While this process can be carried out graphically, it is more efficient to construct a spreadsheet that incorporates the regression equation so as to generate the climate retrodictions in an automated manner.

CCA is constrained to position environmental vectors in physiognomic space even though their relationship to the pattern of sites may be weak. To test the significance of the vector positions, CCA offers the option of running Monte Carlo permutations. In such tests the climate variables have significance at the 99.9% level (Wolfe, 1995; Kovach & Spicer, 1995).

Another test of CLAMP is to assess its internal consistency. Wolfe (1995) analysed 31 Tertiary leaf assemblages from North America and Fig. 17.5 shows the results for MAT derived from 29 character states normalized to a palaeolatitude of 45° N. The graph therefore displays MAT change over time, which can be directly compared to curves derived from oxygen isotopes.

17.3   LIMITATIONS OF CLAMP

Powerful as this method is, it does have some limitations. Some are inherent in the present statistical methodology while others are a function of taphonomy and evolution.

17.3.1   Limitations imposed by the methodology

Although the principal axes are plotted as straight lines in CCA for convenience their actual definition in multidimensional space may be curved. Currently, because the climate vectors are defined by the origin of the plot and only one other

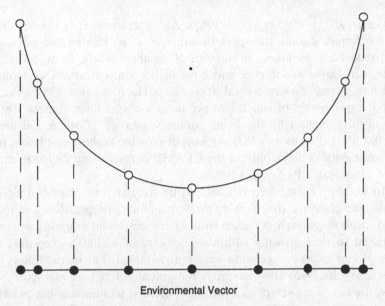

Environmental Vector

Fig. 17.6. Illustration of the calibration effect of projecting a curved foliar physiognomic/ climate relationship on to a linear climate vector. The vector scaling varies with the position along the vector.

set of coordinates, they can only be straight lines. In reality, however, the real climate/vegetation relationship may be better described as a curve. A good example of this is the three-dimensional representation of leaf size with respect to the MAP vector. The MAP vector is, by definition, a straight line but the path described by joining ordered leaf size categories in three-dimensional space is a helix. This can partly be seen in the projection on to axes 1 and 2 shown in Fig. 17.3(c) where the leaf size character states, when joined together in two-dimensional space, describe a curve. This problem is, in part, overcome by recognizing the fact that the calibration scale along the vectors is not uniform: rather it may be stretched in one part and compressed in others as illustrated in Fig. 17.6. Naturally any technique that assumes a linear relationship between leaf characters and climate (e.g. multiple regression analysis) is inappropriate in this context.

Therefore, where the curvature is complex and/or severe, the use of linear climate vectors is limited. A better approach to estimating the climate of an unknown (fossil) site would be either to use curved climate relationships or to determine its position in physiognomic space in relation to its nearest neighbours with known climates (Stranks & England, 1997). The distances to the nearest ten or so neighbours with known climates would then define the climate of the unknown fossil site. Because only local sites are being considered, spatial complexities that exist within the total data set are simplified.

17.3.2   Limitations imposed by taphonomy

It has been long recognized that taphonomic processes have the potential to bias the leaf physiognomic signal as recovered in the fossil record (Roth & Dilcher,

1979; Spicer, 1980, 1981; Ferguson, 1985). As an example it is known that wind favours the longer distant transport of sun leaves, so biasing the physiognomic record, particularly in lakes, in favour of smaller leaves from taller species. Moreover, sun leaves are thicker and have higher concentrations of tannins and other substances that slow microbial degradation (Lichtenhaler, 1985). The overall preferential survivorship of sun leaves produces a cooler drier climate signal than was actually experienced by the living source vegetation (Roth & Dilcher, 1978; Spicer, 1980, 1981; Ferguson, 1985). Aspects of possible taphonomic biases in lacustrine environments as they apply to the CLAMP dataset using 29 foliar characters have been investigated by Stranks (1996).

In building his modern dataset, Wolfe simulated the taphonomic capture that would take place in a river flowing through a patch of vegetation: a range of leaf sizes and morphologies were collected from each species, and in most cases sampling was restricted to trees growing within a kilometre of a climate-recording station. Successional and 'climax' vegetation were both sampled. The methodology of scoring leaf size (Wolfe,1995) also minimizes the potential size bias that exists in fossil sites. Nevertheless, when sampling fossil sites potential taphonomic biases have to be recognized. Ideally, assemblages must come from a suite of sedimentary environments that together closely simulate the range of communities and depositional settings likely to have existed in the local regional palaeovegetation. This is the case with all the sites referred to in this chapter. Unreliable climate results are inevitable if only single communities with specialized taxa (e.g. streamside or mire) are represented. Wolfe regarded 20 as the minimum number of woody angiosperm taxa that were required to yield a useful result, and the greater the number above this minimum the better. This minimum has been shown by experiment (Povey, 1995) to be justified.

17.3.3   Limitations imposed by biotic and environmental evolution

While the observed leaf characters in the fossil assemblage define accurately the position of the fossil sample with respect to the modern swarm, the calibration relationship between key features such as margin type and MAT or size and precipitation may have changed over time. This is likely to be most acute in the context of size and water relations because water relations are strongly influenced by stomatal features, which in turn are a function of atmospheric CO_2 concentrations (Woodward, 1987; McElwain & Chaloner, 1995; Kürschner, 1996).

To attempt to investigate some of these problems, Herman and Spicer (1997) compared analyses from a modern flora and one that was likely to have existed and evolved under elevated atmospheric CO_2. A large (84 species) latest Albian – Cenomanian flora from the Grebenka River, north-east Russia (Shczepetov et al., 1992), was scored and a CCA-based CLAMP analysis was used to obtain temperature and precipitation predictions from the whole flora. Forty-four species were selected randomly from the 84 species for subsequent analysis. This sample size (44) matched one from modern vegetation in Honshu Japan which has an observed MAT similar to that predicted for the Cretaceous of the Grebenka River. From each of the Grebenka and Honshu samples of 44 species, 23 species (the size of a typical fossil assemblage) were selected randomly 20 times and analysed for each of the eight climate variables using CCA. The F test was then used to compare the variances of the 20 random subsets from the Grebenka and Honshu samples. There were no significant differences at the 5% level in the estimates for mean annual temperature, cold month mean temperature, mean annual precipitation and length of the growing season. Warm month mean temperature and the other precipitation estimates did show some difference and, apart from the duration of the growing season (strongly correlated with MAT), all are most relevant to periods of active growth. While encouraging for the major climate variables, these results are inconclusive and more experiments need to be done to examine, for example, the effect of sample size and the missing data on variance differences and their significance.

17.3.4   Phytogeographic factors

The current data sets are highly biased towards North America and it is legitimate to question the use of CLAMP for areas that have a different phytogeographic history. However, if the arguments advanced above in favour of the foliar physiognomic approach are valid, phytogeographic history should have little effect because convergent evolution will ensure a similar range of physiognomic traits in relation to climate will evolve under similar climatic regimes. This assumes that physiognomic/climate equilibrium exists in all areas which may not be the case. For most of the Northern Hemisphere, the North American dataset is likely to be valid because there have been few barriers to plant migration. A far more serious problem is finding mature vegetation systems from which to collect calibration material. However, in the Southern Hemisphere significant problems may exist, most notably in Australia and New Zealand due to their long periods of isolation which has resulted in their unique floras. The suitability of the current data sets for that area and the effects of incorporating Australian and New Zealand data into CLAMP is currently under investigation (Stranks, 1996; Kennedy, 1996).

17.3.5   Comparability of results

The development of CLAMP has seen the expansion and tuning of the modern vegetation reference set such that not all CLAMP results in the literature are now comparable. This is inevitable during the development of any methodology but unfortunately authors have not always been explicit about the details of the dataset or its analysis. The initial 18 leaf character states has been through several experimental phases which included, for a brief period, an inflation to 52 states. Since 1995, however, continued exploration of the structure of the data set has resulted in a stabilization on the 31 states given in Table 17.1.

The total number of reference sites has also grown as opportunities for additional sampling have arisen; sites from Mexico and Japan have recently been added. The current experimental set stands at over 150 sites. Furthermore within the reference set, some sites are from what might be termed 'specialized environments'. For example, sites that are at high altitude comprise a well-defined 'sub alpine nest' (SAN) that consistently plot together because the leaf morphology is uniquely adapted to freeze-induced drought, snow cover, and short growing seasons. Because of the very small leaf size, many taxa have lost marginal teeth (small leaves

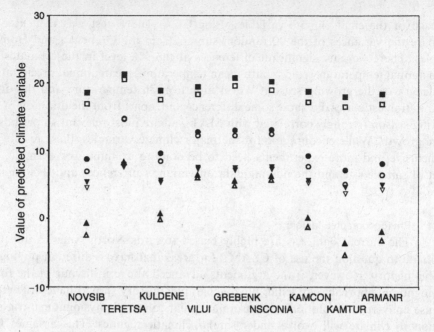

(a)

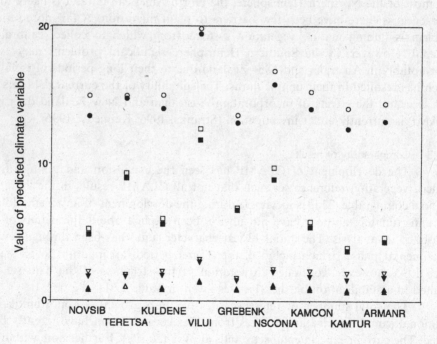

(b)

don't have 'room' for teeth) that would otherwise indicate cool conditions and as a consequence this features of the SAN consistently suggest warmer conditions than are observed. For the majority of fossil floras which are preserved in lowland settings, it is legitimate to exclude such exotic vegetation from the reference set, but for studies of Tertiary elevation change (e.g. Povey *et al.*, 1994) the inclusion of the SAN sites is necessary.

The inclusion of any new sites inevitably slightly alters the structure of physiognomic space, the position of the climate vectors, and therefore the climate predictions. For example, Fig. 17.7(*a*) and (*b*) shows the effect of excluding and including the SAN (103 and 123 sites, respectively). In respect of the temperature related variables, the 103 dataset uniformly gave slightly higher predictions than the 123 set, while the opposite was true for the precipitation-related variables.

The results for the Asian and Alaskan fossil sites analysed using the 103 dataset and shown in Fig. 17.7(*a*) and (*b*) have been compared to climate model results (Spicer *et al.*, 1996). There is reasonably good agreement between the model retrodictions and the CLAMP results in coastal regions but in more continental interior positions CLAMP showed a more equable temperature regime and overall a wetter climate than that suggested by the model. The inability of climate models to simulate equable continental interiors as indicated by a wide array of palaeontological data has been noted for the Eocene (Wing & Greenwood, 1993) while the similarity in the results from coastal sites shows

Fig. 17.7. (*a*) Scatter graph showing the differences in temperature related climate retrodictions using a data set that includes vegetation growing in subalpine conditions (123) and one that excludes subalpine vegetation (103). NOVSIB – Novaya Sibir' Island, north-east Russia, Turonian, 82° N palaeolatitude; TERETSA – Teretky–Sai (Kazakhstan), Cenomanian, 40° N palaeolatitude; KULDENE – Kuldenen–Temir (Kazakhstan), late Albian, 40° N palaeolatitude; VILUI – Vilui Basin, Cenomanian, 65° N palaeolatitude; GREBENK – Grebenka River, north-east Russia, late Albian–Cenomanian, 70° N palaeolatitude; NSCONIA – North Slope Alaska, Coniacian, 75° N palaeolatitude; KAMCON – north-west Kamchatka, Coniacian, 72° N palaeolatitude; KAMTUR – north-west Kamchatka, Turonian, 72° N palaeolatitude; ARMANR – Arman River, north-east Russia, Late Albian–Cenomanian, 68° N palaeolatitude. Squares represent warm month mean estimates, circles – mean annual temperature estimates, triangles – cold month mean temperatures and inverted triangles – length of the growing season. Closed symbols relate to results using the 103 modern vegetation sample reference set while the open symbols are those for the 123 sample set that includes the subalpine nest. (*b*) Scatter graph showing the differences in precipitation-related climate retrodictions using a data set that includes vegetation growing in subalpine conditions (123) and one that excludes sub-alpine vegetation (103). Floral locality abbreviations as for Fig. 17.7(*a*). Circles represent mean annual precipitation estimates, squares – mean growing season precipitation, triangles – mean monthly growing season precipitation, and inverted triangles – precipitation during the three consecutive driest months. Closed symbols represent estimates using the 103 modern vegetation sample reference set and open circles the 123 reference set that includes the subalpine nest.

that CLAMP is again providing temperature data that are consistent with those obtained by other means (the model sea surface temperature being defined by oxygen isotope data).

17.4   CONCLUSIONS

In our present relatively cool world with a strongly defined Equator to Pole temperature gradient practically all climate regimes that existed in the more equable past are represented. In vegetation terms only high latitude palaeofloras represent communities for which no modern analogues exist. The various modern data sets that have been used inevitably result in slightly different structures in physiognomic space and therefore different regression equations for calibration purposes. It follows that any analyses should state explicitly which reference set was used. As more sites are added, the framework of physiognomic space will become more complete and stable.

CLAMP is still under development but initial results such as those reviewed here show that the technique is remarkably robust and yet capable of useful precision (repeatability). The accuracy (validity) is supported by comparisons with model results constrained by temperatures derived from oxygen isotopes and climate trends in space and time determined by a variety of other methods. Being based on the foliar physiognomy of woody dicotyledonous angiosperms, the CLAMP technique is inherently limited to the last 100 million years of climate history. Questions remain as to the calibration drift that might have occurred as the angiosperms diversified and atmospheric composition changed. Nevertheless, preliminary investigations into these issues suggest a remarkable degree of stability.

If CLAMP does prove to be a precise, accurate, and robust tool, it will provide the most powerful means so far developed of determining past climate change in non-marine environments. Used in conjunction with other techniques such as oxygen isotope analysis, it is likely that climate models can be engineered to work successfully in the context of a wide range of boundary conditions and therefore provide new insights into both past and the future environmental conditions.

18

Biotic response to Late Quaternary global change – the pollen record: a case study from the Upper Thames Valley, England

ADRIAN G. PARKER

18.1   INTRODUCTION

Palynology (pollen analysis) is concerned with seed-bearing plants (Phanerograms) which produce pollen in the male gamete (anther), or with spores which are asexual reproductive cells of Cryptogram plants. Primarily, it is the study of fossil pollen grain and spore assemblages, which have been isolated from their sedimented deposit in the recent past or as far back as the Palaeozoic era. Pollen analysis is the most widely adopted and perhaps the most successful technique used in the reconstruction of terrestrial palaeoenvironments, especially for the Quaternary (Lowe & Walker, 1984).

The identification of pollen and spores enables a picture of the past vegetational history to be made through time. Pollen and spores are normally preserved in three major site types: lakes, peats and soils. The reconstruction of former vegetation by means of pollen analysis provides a picture of past landscape and environment (Faegri & Iverson, 1989; Moore *et al.*, 1991). In many cases, changes in the vegetation of an area may lead to strong inferences about the former climate of an area. However, not all changes in vegetation are necessarily due to change in climate. For example, fire, insect infestation, disease, plant successional changes, human interference and factors leading to the accumulation and preservation of the material often make the interpretation of the pollen and spore record complex (Bradley, 1985).

Pollen grains range in size from 5 to 200 µm. Pollen is protected by a chemically resistant outer layer, the exine, which is made of sporopollenin, an inert and resistant natural organic compound. Thus, pollen and spore grains may be found in deposits in which other types of fossils have been diagenetically destroyed (usually under anaerobic conditions), although this casing can be destroyed by oxidation processes.

It is the morphology and chemical properties of the exine which lie at the foundation of pollen analysis. Pollen and spore grains of many plant families are different morphologically and can be recognized by their distinct shape, size, sculpture and apertures. In some genera (e.g. *Plantago*) it is possible to identify to species level; in other cases identification can only be made to generic (e.g. *Ulmus*) or family (e.g. Chenopodiaceae) level. The amount of pollen produced varies from species to species, being dependent on the pollination mechanism. Identification is made by comparison with prepared reference slide material whose specific identity is known with certainty. With pre-Quaternary pollen and spore assemblages, where relationships of fossil to living plants is uncertain, names founded on morphology are needed for fossil taxa (Traverse, 1988).

   This chapter briefly summarizes the fossil record of pollen over the past 145 Ma and then concentrates on the major environmental changes that have taken place in the Upper Thames region of England during the Late Glacial (Late Devensian (15 000–10 000 yr BP)) and Holocene (Flandrian (10 000 yr BP to present)). Natural and anthropogenic changes in the landscape and their relationships to biotic response and global change are highlighted throughout, illustrated by the use of pollen analysis.

18.2   THE CRETACEOUS-RECENT RECORD

   Over the past 145 million years there has been great diversity in palynomorphs with the evolution and extinctions of a large number of taxa. During Late Jurassic times there was great diversification of ferns, cycads, cycadeoids, conifers and ginkgophytes all of which reached their maximum abundance during this period. In addition, tectate pollen first appeared, while glossopterids became extinct (Stewart & Rothwell, 1993). During the Early Cretaceous evidence for monosulcate angiosperm pollen (dispersed by wind and insects) is evident, with monocotyledonous and dicotyledonous plants present. Palynomorphs from cycads, cycadaceoids and ginkgophytes rapidly decline from the record, with pteridosperms becoming extinct. The Late Cretaceous record reveals a great diversification of angiosperm pollen types (many of which are characteristic of extant families in addition to those with no modern counterparts), displacing gymnosperms (Traverse, 1988; Stewart & Rothwell, 1993). This remarkable takeover may have been encouraged by the increasing seasonal climate of this period and the ensuing Cenozoic, calling for new reproductive strategies (Brasier, 1980). Cycadeoids became extinct during the Late Cretaceous and evidence for cycads and gingkophytes dramatically declines. The Late Cretaceous flora was characterized by broad-leaved woodlands, though grasslands did not seem to develop until the early Cenozoic Era (Brasier, 1980). During the Tertiary steady evolution of new angiosperm types occurred with a decline in the diversity of conifers. The Quaternary pollen record provides the most remarkable response to climatic and geographical change.

18.3   THE STUDY AREA

   The Upper Thames Basin lies in the heart of central southern England (Fig. 18.1) and covers an area of approximately 4 500 km^2. Its extent is not determined by any single geographical parameter, and thus it traverses a variety of landscapes dominated by a scarp and vale topography with considerable variations in relief.

   The juxtaposition of the Mesozoic clays and limestones across the Upper Thames region gives rise to numerous calcium-rich springs and seepages which can lead to the development of peats. These peats can provide invaluable information on the nature of environmental changes during the Late Devensian (Late Glacial) and

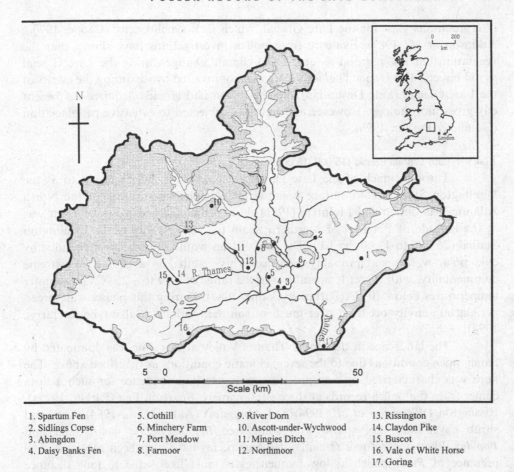

Fig. 18.1. Pollen sites referred to in the text from Upper Thames Valley.

1. Spartum Fen
2. Sidlings Copse
3. Abingdon
4. Daisy Banks Fen
5. Cothill
6. Minchery Farm
7. Port Meadow
8. Farmoor
9. River Dorn
10. Ascott-under-Wychwood
11. Mingies Ditch
12. Northmoor
13. Rissington
14. Claydon Pike
15. Buscot
16. Vale of White Horse
17. Goring

Flandrian (Holocene) especially in relation to changes in the vegetation caused by natural and anthropogenic factors on both a local and regional scale.

Information on changes in the environment come from a large number of proxy sources. This evidence for climatic change, human activity and vegetational change will be given in a temporal sequence to achieve a 'broad' synthesis. The biotic responses to environmental change in the Upper Thames Valley over the 15 000 years is illustrated through the use of pollen analysis. The pollen diagrams are divided into zones on the basis of pollen stratigraphical changes. These boundaries are of local chronstratigraphical value as species arrivals are time transgressive over large areas.

18.4   THE LATE GLACIAL ENVIRONMENTAL HISTORY (15 000–10 000 YR BP)

Much debate surrounds the definition and delimitation of the Late Glacial and the boundaries of its associated substages within the British Isles. The detailed subdivisions advocated by Walker (1995) are not clearly defined within the Upper Thames region, with only three key subdivisions recognized. Conflicting climatic interpretations exist for the Late Glacial, which lack synchroneity (Coope, 1970*a*; Atkinson *et al.*, 1987). Evidence from pollen investigations have shown that the vegetation did not respond as quickly to climate change during the Late Glacial as did insects. The Upper Thames Valley was not invaded by ice during the events of the Late Glacial (Late Devensian) and the region had already acquired its present day gross morphology. However, the area was subjected to extensive periglaciation (Goudie & Parker, 1996).

18.4.1   Late Glacial Period (15 000–13 000 yr BP)

Through much of the Late Devensian cold-stage, here referred to as the Dimlington Stadial, cold surface polar water occupied large areas of the North Atlantic. Ruddiman and McIntyre (1981) showed that the oceanic polar front was at the latitude of Portugal or Northern Spain (*ca.* 40° N) during the Dimlington Stadial (26 000 to 13 000 yr BP). The British Isles would have been surrounded by cold polar waters and probably pack ice. This resulted in a climate of extreme continentality, with winter temperatures in the range of −20 to −25 °C and summer temperatures below 10 °C (Barber & Coope, 1987). During this phase, widespread periglacial activity occurred over much of southern Britain (Ballantyne & Harris, 1994).

The landscape in the Upper Thames Valley at this time was dominated by harsh, open conditions due to the severe climatic conditions as described above. The flora was characterized by a steppe, tundra vegetation. Evidence for such a flora comes from the pollen records at three sites, namely, Spartum Fen (Parker, 1995a), Rissington (Wilkinson *et al.*, 1994) and Abingdon (Aalto *et al.*, 1984). Tree and shrub cover was very sparse with some tree *Betula*, *Betula nana, Salix* and *Populus*. Plant macrofossil remains of these species have also been recorded. The presence of *Pinus* pollen at low frequencies is most likely due to long distance transport. The flora was dominated by herbaceous taxa, tolerant of the extreme conditions, these included *Artemisia, Rumex, Plantago, Filipendula*, Gramineae and Cyperaceae. The flora was restricted by temperature (often subzero for much of the year), nutrient and water availability, and poor soil development. The Thames and its tributaries at this time was characterized by a braided river regime, with shallow eutrophic pools or slow moving water. These were colonized at all three sites by *Myriophyllum, Sparganium, Typha* and *Equisetum*. The organic deposits at Abingdon yielded radiocarbon dates of 13 260 ± 120 (Hel–1092) yr BP and 13 580 ± 120 (Q–2017) yr BP (Aalto *et al.*, 1984). Similar conditions existed elsewhere in southern England (see Godwin, 1964; Gibbard, 1977; Gibbard & Hall, 1982).

18.4.2   Late Glacial Interstadial (13 000 to 11 000 yr BP)

At the end of the Dimlington stadial the rapid migration of the North Atlantic polar front to the north of the British Isles, possibly to the latitude of Iceland (Bard *et al.*, 1987), has been recognized as causing the rapid climatic warming in Britain known as the Late Glacial Interstadial (also known as the Windermere Interstadial or Allerød). The Late Glacial Interstadial displays an extremely rapid climatic amelioration, which permitted thermophilous insect taxa to migrate into many parts of the British Isles (Osborne, 1972; Coope & Brophy, 1972). Atkinson *et al.* (1987) estimated that by about 12 500 yr BP the climate of England was as warm as it is today, though more continental in nature, with mean monthly temperature ranges between 0–17 °C. During the Late Glacial Interstadial the rivers in southern England adopted meandering forms and locally cut into the gravel in response to the change in climate (Briggs *et al.*, 1985). Barber and Coope (1987) suggested that this period of incision may explain why organic sediments of this age are almost unknown in central Britain. Despite the sudden change to temperate conditions during the Late Glacial Interstadial, this event did not last long enough to permit soils to develop fully enough for temperate thermophilous tree taxa to migrate into southern England from their refugia in southern Europe. The only tree taxon which did expand during this interlude was tree *Betula*. It is thought that tree birch was present in localized populations during the Late Glacial in Britain (Birks, 1989). Thus it was able to rapidly colonize from these populations during the interstadial (due to its relatively short lifespan, low nutrient requirements, its fruits are readily dispersed by wind and thus can travel far and it produces viable fruits at a fairly young age).

In the Upper Thames Valley there are, however, a number of sites which do record the floristic response to the sudden change in climate at this time. Pollen evidence (along with plant and snail macrofossil remains) is available from four known sites, namely, Spartum Fen (Parker, 1995a), Rissington (Wilkinson *et al.*, 1994), Northmoor (Sandford, 1965) and the Vale of the White Horse (Paterson, 1976).

The flora at Spartum Fen during the Late Glacial Interstadial can be seen in the upper part of zone SPF 1 in Fig. 18.2. Two radiocarbon dates of Late Glacial interstadial age were measured, 12 900 ± 90 yr BP (AA–11606) and 12 660 ± 95 yr BP (AA–11607), (though reversed). The pollen record revealed damp tundra vegetation characterized by tree *Betula, Salix* and Cyperaceae. Macrofossil remains of the frost sensitive species *Cladium mariscus* were also found, indicating an amelioration in conditions. In the dry habitats pollen evidence revealed heliophilous communities with *Juniperus, Helianthemum, Artemisia* and *Sanguisorba*.

At Rissington, *Betula* and *Salix* woodland was more pronounced than at Spartum Fen perhaps due to fewer constraints on its local expansion (Wilkinson *et al.*, 1994). Peat lenses from the Floodplain Terrace at Northmoor were studied by Sparks and West (in Sandford, 1965) for their floral and molluscan faunal assemblages. The pollen showed a high proportion of non-arboreal pollen, suggesting open, relatively treeless conditions at the time of deposition. *Salix* may have been present as a carr vegetation. The molluscan fauna contained climatically tolerant species, except *Carychium minimum*, which is usually found in warmer conditions. The peats were radiocarbon dated to 11 250 ± 100 (Birm–105), 11 012 ± 70 (IGS–161) and 10 931 ± 70 yr BP (IGS–162) (Shotton *et al.*, 1970) yielding Late Glacial Interstadial ages, marking the end of warmer conditions and the onset of climatic deterioration.

Late Glacial Interstadial palaeosols in the Vale of the White Horse provide additional evidence for milder conditions. An extensive phase of physical weathering was represented by chalk muds and fine gravels with varying proportions of flint and Greensand rubble followed by a phase of relative stability during which fossil soils

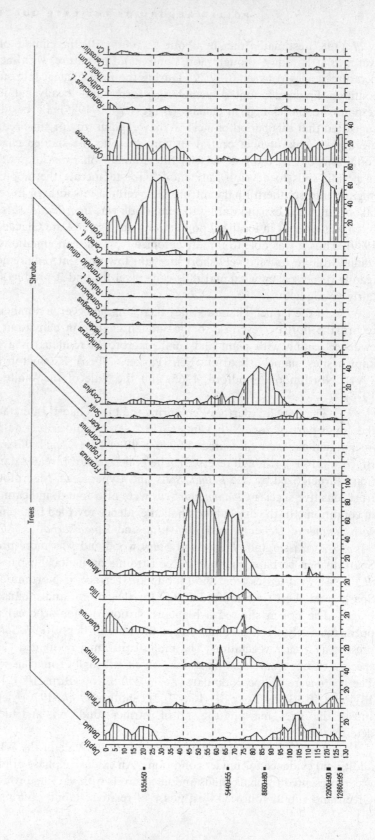

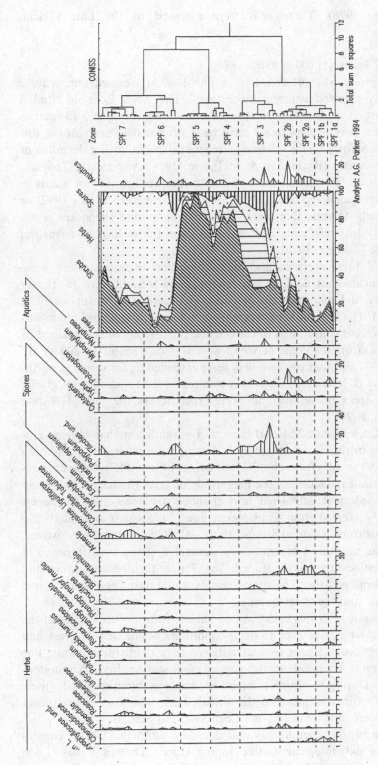

Fig. 18.2. Percentage pollen diagram for Spartum Fen. Depth shown in cm. Pollen sum is of total identifiable pollen and spores of vascular plants, excluding aquatics (based on 300 grains). Sum for aquatics is main sum + sum of group. (Parker, 1995a.)

developed (Paterson, 1976). These soils were assigned to the Late Glacial Interstadial.

18.4.3   Loch Lomond Stadial (11 000 to 10 000 yr BP)

This warm phase was succeeded by a phase of accelerated temperature decline which mainly affected winter temperatures. The Loch Lomond Stadial (Younger Dryas event *ca.* 11 000 to 10 000 yr BP) was the last major cold spell to affect the British Isles. Several hypotheses have been proposed to account for this event (Harvey, 1989) and include the following: iceberg rafting involving the influx of tabular ice from the Arctic (Ruddiman & McIntyre, 1981; Andrews & Tedesco, 1992), lower atmospheric CO_2 concentrations due to the rapid growth of plants in the Late Glacial Interstadial (Kudrass *et al.*, 1992; Lauenberger *et al.*, 1992), a shutdown of the North Atlantic Deepwater (NADW) formation (Broecker *et al.*, 1989) and low-salinity lens refreezing of freshwater from the melting of terrestrial ice-sheets (Johnson & Maclure, 1976).

Whichever of these mechanisms, or combination of mechanisms, were responsible, they indicate that cooling was triggered by the influx of meltwater from the Laurentide and Fennoscandian ice-sheets (Ruddiman & McIntyre, 1981; Broecker *et al.*, 1985). The hypotheses suggest that the onset of this cold phase was caused by rapid movement southwards of the oceanic polar front, which reached its most southerly position off the coast of south-west Ireland around 10 200 yr BP (Ruddiman *et al.*, 1977) and that pack ice may have extended as far south as northern Scotland (Johnson & Maclure, 1976). This resulted in the re-advance of glacier ice in some upland areas of Britain and permafrost across the rest of Britain (Ballantyne & Harris, 1994).

During the Loch Lomond Stadial there was a significant change from erosion to sedimentation in the fluvial systems of Britain, with extensive aggradation about 11 500 to 10 250 yr BP (Rose *et al.*, 1980; Briggs *et al.*, 1985). Such changes have been attributed to a reduction in the magnitude of peak discharge, and to an increased supply of sediment geliflucted into channels from adjacent hillslopes. Atkinson *et al.* (1987) inferred from Coleopteran assemblages that around 10 500 yr BP the mean temperatures stood at around 10 to −17 °C for summer and winter, respectively. The mean monthly temperature for central England was suggested at about −5 °C for the coldest part of the stadial. The climate then became Arctic and the rivers followed a braided form. The annual runoff would have been concentrated during the spring thaw and the short summer.

The biotic response to this event can be shown from pollen studies in the British Isles, which show a revertance to open conditions dominated by grass and sedges and other herbaceous plants characteristic of bare or disturbed soils. For example, the xerophyte *Artemisia* would suggest regional aridity (Pennington, 1977; Parker, 1995a). During the Loch Lomond Stadial, tree birch declined widely and possibly became extinct in much of the British Isles. The floral assemblages reflect a prevailing severe climate during the Loch Lomond Stadial.

The response in the vegetation is well documented in the Upper Thames through a number of palynological studies. In the Upper Thames Valley, Loch Lomond Stadial sediments at Farmoor (Coope, 1976; Lambrick & Robinson,

1979), Mingies Ditch (Allen & Robinson, 1993) and near Great Rissington in the Windrush Valley (Wilkinson *et al.*, 1994) have revealed floras indicative of a park tundra. At Farmoor (Coope, 1976) and Mingies Ditch (Allen & Robinson, 1993) the deposits contained similar coleopteran assemblages and were radiocarbon dated between about 10 900 and 10 500 yr BP. Wood remains identified as *Cornus*, were radiocarbon dated to 10 710 ± 100 yr BP from Rissington (Wilkinson *et al.*, 1994).

The pollen diagrams from Cothill Fen (Fig. 18.3, zone CF 1) (Day, 1991) and Spartum Fen (Parker, 1995*a*, zone SPF 2) show that the vegetation responded to the sudden cold phase by reverting from a more 'temperate' flora back to a xerophytic, arctic assemblage. The Loch Lomond Stadial flora was dominated by Gramineae, Cyperaceae and herbs including *Artemisia* with sparse tree and shrub cover. At Goring, Oxfordshire, an undated pollen profile from a palaeochannel yielded a cold stage flora characteristic of Loch Lomond Stadial conditions (Parker, 1995*b*). The flora was dominated by grass and sedges with a large herbaceous component including *Artemisia* and Chenopodiaceae.

Little is known about the extent of human occupation during the Late Devensian. Briggs *et al.* (1985) have suggested that the environment was suitable only for brief and infrequent hunting visits. Firm evidence of human activity in the Upper Thames Valley during the Late Glacial is provided by Palaeolithic tools from at least two of the sites, namely Goring (Allen, 1995) and Mingies Ditch (Allen & Robinson, 1993).

18.5 THE FLANDRIAN ENVIRONMENTAL HISTORY

18.5.1 Early Flandrian (10 000–8000 yr BP)

The Loch Lomond Stadial concluded with an abrupt change to milder conditions as shown in the Greenland ice cores (Dansgaard *et al.*, 1989; Alley *et al.*, 1993). This period of rapid warming was associated with the northward movement of the North Atlantic oceanic polar front (Ruddiman & McIntyre, 1981), which, according to Bard *et al.* (1987), migrated between 35 and 55° N, almost immediately at the end of the stadial with a strong influx of warm Atlantic waters. Koç *et al.* (1992) and Koç and Jansen (1994) have suggested that within 50 years the sea surface temperatures (SST) within the Nordic Sea rose 9 °C from very cold Younger Dryas (Loch Lomond Stadial) to Holocene (Flandrian) values. During this drastic change of climatic conditions in the North Atlantic and Nordic Sea, summer insolation reached its maximum between 10 000 and 9000 yr BP. In response to this warming, there was rapid wasting of the Laurentide and Fennoscandian icesheets, which is reflected at the 2/1 boundary in the isotopic traces found in ocean cores from the North Atlantic (Ruddiman & Duplessy, 1985). Proxy climatic data, notably ^{18}O concentrations from polar ice cores, reflect changing temperatures. At Camp Century, Greenland ^{18}O values increased about 18 000 yr BP and rose rapidly until about 9800 yr BP (Dansgaard *et al.*, 1984). Alley *et al.* (1993) have suggested that the Younger Dryas–Holocene transition took less than 20 years. According to recent estimates based on fossil coleopteran assemblages (Atkinson *et al.*, 1987), mean annual temperatures in Britain may have risen by 1.7 °C per century, and possibly by as much as 7 °C in 50 years (Dansgaard *et al.*, 1989). The effects on the landscape were dramatic. The rivers again adjusted to a meandering form around

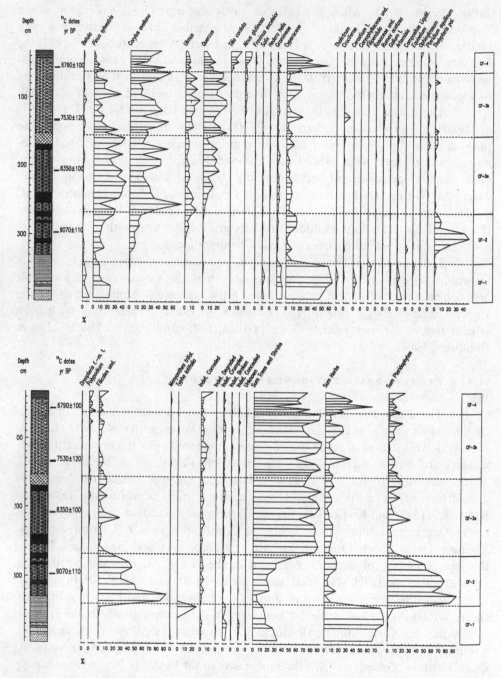

Fig. 18.3. Percentage pollen diagram for Cothill Fen. Depth shown in cm. Pollen sum is of total identifiable pollen and spores of vascular plants, excluding aquatics (based on 300 grains). Sum for aquatics is main sum + sum of group. (Day, 1991.)

9000 yr BP. Valley aggradation commenced and the channels became infilled with silts, peats and marls before further incision occurred. Such deposition reflects the river size resulting from hydrological change influenced by climatic modification. Lamb (1977) suggested pre-Boreal mean summer temperatures were 16.3 °C, while Atkinson *et al.* (1987) proposed a figure of 15.8 °C based on Coleopteran data. Lamb (1977) suggested that around 8500 yr BP anticyclonic weather dominated Europe. This resulted in warmer air reaching Britain from more southerly latitudes, culminating in a drier climate. The existing thermal gradient became reduced. East–west circulation was enhanced by the Laurentide Ice sheet melt at about 8000 yr BP. The predicted temperatures at this time are 2 °C greater than those of today, with mild winters, prevailing westerlies, a warm ocean and increased precipitation. The development of woodland led to an abrupt decline in the loading of fluvial systems, a cessation of alluviation and the development of a meandering system. The rivers formed stable channels with small discharge variation and limited bedload transport. Gradual backswamp sedimentation and extension of the floodplains were characteristic features of well-vegetated low relief environments (Rose *et al.*, 1980).

Despite the rapid amelioration in temperature the landscape initially remained open with a lag in the response of the vegetation from their refugia during the full glacial episode. The lags were due to the time required for soil maturation following glaciation and periglaciation, soil instability and frost drought (Birks, 1989).

In the Upper Thames the expansion of pioneer arboreal vegetation at about 10 000 yr BP was by *Juniperus* and *Betula* as shown in the pollen diagrams from Spartum Fen, Cothill Fen and Sidlings Copse (Fig. 18.4; Day, 1991) with *Betula* and *Pinus* woodland establishing a little later. This early Flandrian woodland comprised shade-intolerant, short-lived species.

Due to increasing soil development and fertility these were later replaced by *Corylus, Ulmus* and *Quercus* woodland as shown at Mingies Ditch, Cothill Fen, Sidlings Copse and Spartum Fen, the River Dorn (Dury, 1965) and Lechlade (Briggs *et al.*, 1985). At Cotill Fen, Sidlings Copse (Day, 1991) and Spartum Fen, *Corylus* began to expand at about 9400 yr BP, followed by *Ulmus* dated at about 9100 yr BP and *Quercus* at about 8800 BP. At Cothill, *Pinus* declined as populations of *Quercus* and *Ulmus* expanded, although it persisted until about 7700 yr BP. Day (1991) inferred a prolonged presence may have been connected to human disturbance of the vegetation. However, it is more likely that *Pinus* persisted on the nutrient poor acidic soils of the Corallian sands and the Plateau Drifts within the region due to less competition from the thermophilous species (Parker, 1995a).

Similar patterns of early Flandrian vegetational development are illustrated by a number of sites outside the Upper Thames. The arrival and expansion of different taxa were dependent upon a variety of factors including: competition, dispersal, distance of their journey from refugia, edaphic development, temperature, and water availability. Scaife (1987) has stated that 'Changes in the vegetation were rapid, reflecting a complex reaction of communities to variations in external factors and are problems of dynamic phytosociology.'

Huntley (1993b) discussed a number of hypotheses which may explain the rapid migration and expansion of *Corylus* during the early Flandrian. Smith (1970) argued that human disturbance using fire may have promoted its spread, accounting

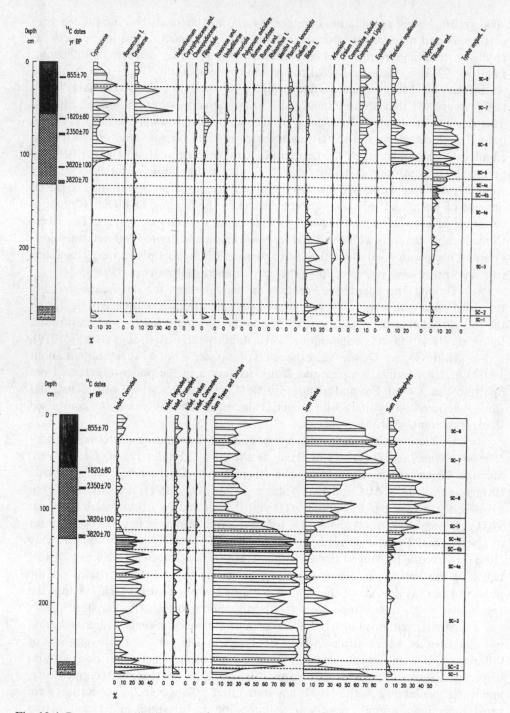

Fig. 18.4. Percentage pollen diagram for Sidlings Copse. Depth shown in cm. Pollen sum is of total identifiable pollen and spores of vascular plants, excluding aquatics (based on 300 grains). Sum for aquatics is main sum + sum of group. (Day, 1991.)

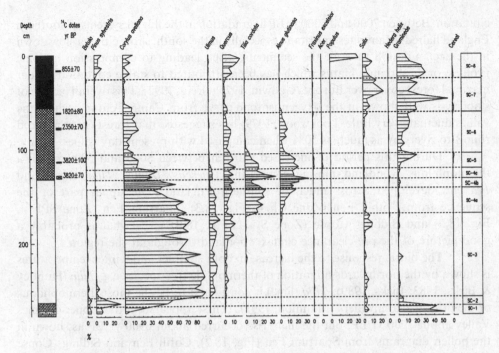

Fig. 18.4 (cont.)

for the unusually high abundance of *Corylus* from a large number sites. However, his hypothesis was based on observations of a North American species, and not *Corylus avellana*. During the hazel maxima at Spartum Fen charcoal levels were low (Parker, 1995a). There is very little indication from the pollen curves at this site to infer human disturbance. This view is supported by evidence from a number of other workers (Edwards, 1982; Edwards & Ralston, 1984). Thus, there is no support for the idea that hazel represents a fire-climax community. It is apparent from archaeological finds of hazel nuts that they were probably used as a food source by Mesolithic people, e.g. charred hazelnut fragments were found at Ascott-under-Wychwood (Dimbleby & Evans, 1974). However, there is no direct evidence that yields were deliberately increased by humans. The reduction in *Corylus* pollen frequencies was probably due to the suppression of flowering by shading as the high forest canopy developed.

18.5.2 Mid Flandrian (8000–5000 yr BP)

The period from 8000 to 5000 yr BP is generally acknowledged to be the period of greatest post-glacial warmth and is generally referred to as the 'climatic optimum'. The Nordic Seas were at their warmest between 8000 and 5000 yr BP, based on marine diatom records (Koç & Jansen, 1994) and waters off Greenland were much less influenced by sea ice than they are at the present day. Koç and Jansen (1994) have suggested that sea surface temperatures were as much as 6 °C warmer than present ones. Lamb (1977) has suggested that summer mean temperatures reached 17.8 °C and mean winter temperatures 5.2 °C, with an increase in precipitation. Between 7000 and 5000 yr BP aggradation in the fluvial systems of southern England halted. Many sites in Britain, especially in the south, show either a slow down in the accumulation rates of the sediments, often leading to compression in some profiles, or an Atlantic hiatus which has been attributed to stable conditions under maximal forest cover over Europe (Godwin, 1975; Greig, 1992). This overall picture of stability is well shown in the oxygen isotope curve from Camp Century, where the values fluctuate very little. Lockwood (1979) has suggested that forest cover reduced runoff to rivers by as much as 50% when compared with present-day values.

During this period of increased warmth between 8000 and 5000 yr BP a thermophilous woodland mosaic comprising *Quercus, Ulmus, Tilia, Alnus* and *Fraxinus* developed across much of southern England. The upper part of the sequence from Cothill Fen (Zone CF 4, Fig. 18.3), Spartum Fen (Zone SPF 4, Fig. 18.2) and Sidlings Copse (Zone SC 4, Fig. 18.4) gives what is probably a good picture of the pre-clearance climax woodland throughout the region.

The biotic response to the increase in both summer and winter temperatures is shown by the northwards migration of thermophile tree species e.g. *Tilia* (Huntley & Birks, 1983; Birks, 1989). *Tilia* (lime) is considered to be the most thermophilous of the post-glacial invaders (Mannion, 1991). It first appeared in the Upper Thames Valley about 7500 yr BP, but the main rise occurred about 6900 yr BP as shown in the pollen diagrams from Spartum Fen (Fig. 18.2), Cotill Fen and Sidlings Copse (Figs. 18.3 and 18.4). The high values for *Tilia*, when its poor pollen dispersal is taken into account, would suggest that lime was probably the dominant forest tree in base rich areas as well as drained soils. This is reflected in the pollen diagrams from Spartum Fen (Parker, 1995a) and those from Buscot Lock (Dimbleby, in Robinson, 1981), Ascott-under-Wychwood (Dimbleby & Evans, 1974), Cothill Fen and Sidlings Copse (Day, 1991). Piggot (1969) showed that pollen sites from this age reflect a more northerly distribution of *Tilia* than at the present day. For the seeds of *Tilia* to germinate a mean July temperature of at least 17 °C is required (Moore, 1977).

The early expansion of *Alnus* into some areas of southern England has been recorded at about 8200 yr BP in the Lower Thames estuary (Devoy, 1979), the Isle of Wight (Scaife, 1982) and Poole Harbour (Haskins, 1978). The pollen diagrams (Figs. 18.2, 18.3, 18.4) from the Upper Thames show that *Alnus* persisted at low frequencies prior to the main expansion of *Alnus* about 6800 BP. The expansion was most probably in response to local changes in the water table, which allowed alder to expand into habitats which had been either too dry or too wet due to local, site-dependent events such as hydrological thresholds, topography and climate. By 6540 ± 80 (Har–8355) yr BP, closed dominant *Alnus* woodland had developed on the Windrush floodplain at Mingies Ditch (Allen & Robinson, 1993), with a similar picture recorded at Spartum Fen (Parker, 1995a), Cothill and Sidling Copse (Day, 1991). Dense alder woodland was also recorded at Buscot Lock and an *Alnus glutinosa* log from the lower layer of peat was radiocarbon-dated to 4010 ± 90 yr BP (Dimbleby, in Robinson, 1981). The results from these sites show a strong alder component in the vegetation across the Upper Thames, perhaps along the floodplain and river and stream corridors. This view is supported from the record in the basal sediments from Daisy Banks Fen (Fig. 18.5) where there was very little alder pollen recorded on the drier, fertile soils of the terrace gravels. At this site, the small scale spread of *Alnus* took place when clearance occurred leading to a localized rise in the water table (Parker, 1995*a*).

Fire has been suggested as a plausible mechanism for the spread of *Alnus* during the Flandrian (McVean, 1956; Smith, 1984; Chambers & Price, 1985; Smith & Cloutman, 1988). The widespread use of fire by hunter–gatherers and its hypothesized use in pre-history as part of a hunting and browse-creating regime has been well attested (Edwards, 1988). Mesolithic disturbance and burning are seen as promoting catchment runoff and flushing on low ground (Moore, 1986). At Spartum Fen the rise in *Alnus* occurred at around the same time as a rise in charcoal. It would be difficult to determine whether there was a causal relationship between burning and the expansion of *Alnus*; however, it cannot be discounted. A similar picture was shown at Sidlings Copse (Day, 1991).

Fraxinus appears in the pollen record with *Fagus* and *Carpinus* in low frequencies a little later in the Flandrian. Wood remains of *Fagus* and *Carpinus* from Abingdon have been dated to about 5000 BP, perhaps the earliest firm evidence for their presence in England (Miles, 1986).

Cereal-type pollen of pre-elm decline age was recorded at Spartum Fen, Daisy Banks Fen and Sidlings Copse, implying that early agriculture may have occurred. Cereal pollen from pre-elm decline contexts have been noted from an increasing number of sites (Edwards & Hirons, 1984; Edwards, 1989; Day, 1991), though the possibility that the grains came from a wild grass species cannot be excluded.

The diverse lithological nature of the Upper Thames Valley bedrocks, and associated hydrological and edaphic variability would have led to a complex mosaic of vegetation, especially woodland, with many species contributing to the canopy. The results from the sites described, yield additional information on the regional woodland types as suggested by Robinson and Wilson (1987, p. 29) for the Upper Thames Valley and are as follows:

(i) Lime would have been favoured on well-drained, base-rich, calcareous soils, e.g. Jurassic limestones.

(ii) The floodplains and river corridors of the Upper Thames Valley would have favoured alder with willow and possible small numbers of birch.

(iii) On areas of clay and gleyed soils, a mixture of mainly oak with elm, lime and ash with an understorey of hazel would have been favoured.

(iv) On the base-poor, acidic sand and gravels across the area (e.g. the Lower Greensand, Corallian sands and Plateau Drift) oak and hazel would have been favoured. It is possible that small numbers of birch and pine may have persisted.

Behre (1986) stated that human impact has been the most important factor affecting vegetation change in Europe during the last 7000 years. This has led many workers to note increased difficulty in separating human and natural factors (especially climatic) when reconstructing patterns of vegetation change (Birks, 1986). For example, at Hampstead Heath (Greig, 1982) and Daisy Banks Fen (Parker, 1995*a*) the sediment deposition seems to have started only when forest clearance commenced causing soil erosion (cf. Rybnicek & Rybnickova, 1987).

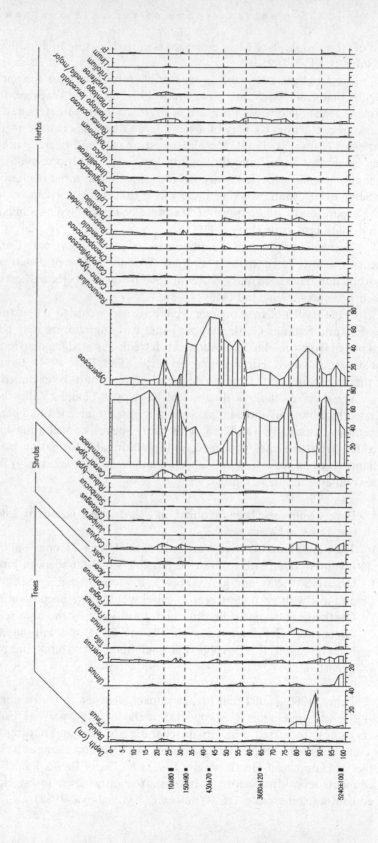

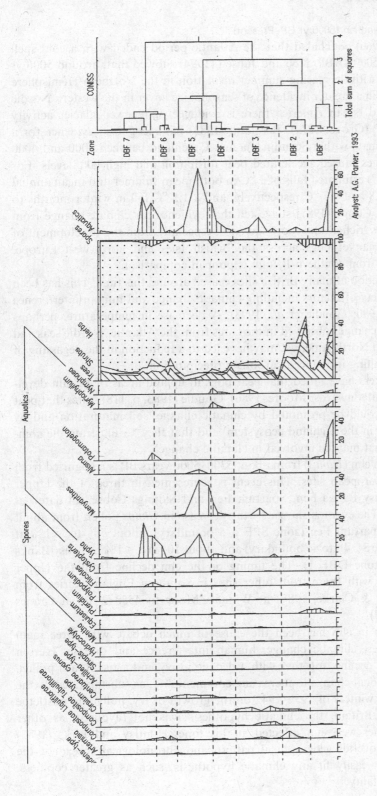

Fig. 18.5. Percentage pollen diagram for Daisy Banks Fen. Depth shown in cm. Pollen sum is of total identifiable pollen and spores of vascular plants, excluding aquatics (based on 300 grains). Sum for aquatics is main sum + sum of group. (Parker, 1995a.)

18.5.3   The Late Flandrian (5000 yr BP–Present)

Frenzel (1966) concluded that the Atlantic period ended with a cold spell between 5400 and 5000 yr BP. Koç and Jansen (1994) showed that, around 5000 yr BP, a cooling step with decreasing summer insolation in the Northern Hemisphere occurred, resulting in a greater incidence of seasonal ice cover in the western Nordic Sea. In Europe and North America there is evidence of renewed glacier activity around 5500 yr BP (Grove, 1979). Berglund (1983) has recognized evidence for a deterioration in climate within Scandinavia and Greenland between 5000 and 4000 yr BP. These changes caused accelerated peat formation and high lake levels. For Britain, Lamb (1977) estimated falls of 1 °C in both mean summer and mean annual temperatures to 16.8 and 9.7 °C respectively, and a 1.5 °C fall in winter months to 3.7 °C. Barber and Coope (1987) suggested that evidence for climate change from 5000 yr BP becomes 'richer' and more 'diverse', especially from the development of raised mires and blanket bogs across the British Isles and north-west Europe. However, evidence from southern Britain is poor in comparison.

Tufa deposition at many British sites ceased around this time. This has been attributed to a variety of factors including climatic change and human interference (Viles & Goudie, 1990; Goudie et al., 1993). A decrease in temperature, perhaps associated with the changes observed in the cores from the Atlantic, Nordic seas and onshore Europe and North America, may have led to a reduction in the degassing of carbon dioxide resulting in reduced carbonate precipitation.

Some workers have noted that peat bogs in upland areas of Britain developed from 5000 yr BP onwards (Moore, 1986). Goudie (1983, p. 118) states that peat bog development 'could be promoted by climatic change, soil maturation and by human interference in the highland ecosystem' and that they '… illustrate the complexity of factors that may be involved in climatic change'.

A decline in elm (*Ulmus*) from about 5300–5000 years BP is recognized from many north-west European sites. This event is represented in three of the Upper Thames sites – Daisy Banks Fen, Spartum Fen and Sidlings Copse, all dated at about 5200 yr BP. The event is shown by the decrease in *Ulmus* pollen from about 15% to <1% at Spartum Fen (zone SPF 4–5 boundary), about 7% to <1% at Sidlings Copse (zone SC 4b to 4c boundary) and about 10% to <1% at Daisy Banks Fen (mid-way up zone DBF 1). The timing of the elm decline from the Upper Thames sites along with those from other sites in southern England varies from 5630 ± 90 (Barber & Clarke, 1987) at Winnals Moor to 4550 ± 60 at Cranes Moor (Waton, 1982).

The decline in elm has been the cause of much debate with three main contested hypotheses, climatic change, human interference and disease. Iversen (1941) used the elm decline, together with a decline in ivy and a rise in ash pollen, to attribute vegetational change to increasing continentality, as substantiated by the effects of the severe winters of 1939–1942 on the growth of ivy, holly and mistletoe (Iversen, 1949). In Britain, the climatic hypothesis was not favoured, as other thermophilous species were not affected at this time. Huntley and Birks (1983) suggested that the overall geographical pattern and chronological spread of the elm decline did not easily fit any climatic hypothesis, such as greater coolness, wetness or continentality.

Traditional views tend only to have focused on percentage pollen data. However, some sites in the British Isles show drastic reductions in total pollen influx and concentrations between 5300 and 4800 yr BP; at Waun-Fignen-Felen, south Wales, Smith and Cloutman (1988) suggested that this indicated a decline in elm possibly due to climate change. Coleopteran evidence from the Midlands of a deterioration in the climate between 5500 and 4800 yr BP (Osborne, 1976, 1982) would support the theory. If there was a deterioration in climate which affected the thermophile pollen concentrations, an increased incidence of late frosts may have been a contributing factor accounting for this hypothesis.

As the elm decline in many places marks the Mesolithic–Neolithic transition, in which there was a general change from hunter–gatherers to agriculturalists, woodland clearance via burning is another possible explanation. Hence charcoal records are of great interest during this period of cultural and economic change. The Upper Thames sites show low frequencies of charcoal at this time, perhaps due to a wetter climate, thus contradicting the view that clearance via burning was a major factor. Instead, there may have been a greater frequency of fires (both natural and human) prior to the elm decline (Edwards, 1988).

The cause of the elm decline has also been attributed to the advent of the new Neolithic economy in which elm shoots were collected for animal fodder. Pollarding every two or three years would have prevented flowering, and thus reduce pollen production. The problem with this hypothesis as the cause for the decline is that it does not take account of the scale of the event. How did such small communities affect the elm population to such a great extent, and almost simultaneously across the country?

Dutch elm disease has also been suggested as a plausible mechanism for the decline. The ascomycete fungus *Ceratocystis ulmi* is carried by the host vectors, the elm bark beetles *Scolytus scolytus* and *S. multistriatus*. Fragments of *S. scolytus* have been found in horizons immediately below the elm decline at Hampstead Heath (Greig, 1992). On the basis of the evidence from many sites across southern England, it is most likely that all three hypotheses contributed to the decline of *Ulmus* across the region.

Many views on the clearance of land for agriculture and cereal production at or immediately following the elm decline have come from pollen work often far from known archaeological sites. Moffett *et al.* (1989) reviewed the evidence of charred plant remains from a large number of Neolithic sites in England and Wales and showed that there is a general paucity of cereal remains relative to that derived from collected food resources. They suggested that arable cultivation was widely adopted (most Neolithic sites with charred food plant remains yield a few cereal grains), but that it did not fully supplant the collection of wild food plants until well into the Bronze Age. Most of the sites examined were within southern England and the Upper Thames Basin. Here, cereals were produced, but there was not an agricultural revolution, merely a broadening (rather than narrowing) of resources. Clearance episodes were often short and abandoned after about 100 years. They were usually followed by the regeneration of woodland or the development of grassland (M.A. Robinson, personal communication). Isolation from continental Europe and a low population level may explain the slow transition to a fully agricultural, settled society (Moffett *et al.*, 1989).

The Daisy Banks Fen pollen diagram supports these finds with relatively high frequencies of cereal pollen during the Neolithic and a substantially open environment. In contrast, the Spartum Fen profile shows very low frequencies of cereal pollen and a substantially wooded environment, despite the sites being relatively close to one another.

In Britain, there are marked regional differences in the extent and timing of early woodland clearance, especially across the Upper Thames region. At Daisy Banks Fen the local woodland was cleared by 5000 BP. Major clearance at Sidlings Copse started about 3800 BP and at Spartum Fen about 3500 BP as shown by the reduction in tree and shrub pollen and the rise in herbaceous pollen types, e.g. Gramineae. Evidence from archaeological sites (e.g. Drayton and Mingies Ditch) have revealed a number of tree-throw pits (Robinson, 1992), suggesting ring-barking of mature trees, which were pulled over once the roots had rotted and then burnt. Mature trees would not have been felled by axes. At Sidlings Copse there was a second decline in elm associated with the first occurrence of cereal pollen, followed by a slightly later decline of lime (Day, 1991). After about 3100 yr BP both trees appear to have been absent from the local vegetation.

The distribution of lime has been discussed by Moore (1977) and Greig (1982). Its decline is recognized in both British and European pollen diagrams, and was traditionally used to delimit the Atlantic/Sub-Boreal transition. Early views assumed synchroneity between sites, although dating has shown declines from the Neolithic (e.g. Morden Carr, Yorkshire) to Anglo-Saxon (e.g. Epping Forest). However, the majority of dates occur around the late Bronze Age, e.g. about 3800 yr BP from Sidlings Copse, and have been closely associated with land clearance for agriculture. This view is based on the enantiomorphic relationship observed at many sites across Britain between *Tilia* and the abundance of herbaceous and ruderal pollen.

Lamb (1977) argued that during the Bronze Age, drier conditions prevailed. This view is supported by the general absence of Bronze Age waterlogged organic materials from the Upper Thames Valley (Robinson & Wilson, 1987; Parker, 1995*a*) and from elsewhere in the country, e.g. Willow Garth (Bush, 1993). The pollen diagrams from Daisy Banks Fen and Spartum Fen both record a hiatus at about 3300 yr BP, probably associated with a warmer and drier climate, resulting in lowered water tables. There are only a few sites in the upper Thames which cover this period, e.g. Sidlings Copse (Day, 1991), and Minchery Farm (Parker & Anderson, 1996). Other sites with preserved pollen of this age come from archaeological sites, e.g. Radley (Parker, 1996) and the Vineyard, Abingdon (A.G. Parker, unpublished data).

The onset of renewed flooding, alluviation and raised water tables can generally be correlated with accelerated forest clearance and agricultural intensification across the Upper Thames. There is good evidence for alluviation starting around 2500 yr BP (Robinson & Lambrick, 1984; Robinson & Wilson, 1987). Evidence for clearance from the pollen record shows the main clearance episodes were initiated between 3300 yr BP and 2500 yr BP (Day, 1991; Parker, 1995*a*).

From 3500 yr BP, but especially from about 2500 yr BP, temperatures apparently fell or the climate became wetter, culminating in the sub-Atlantic period about 2500 yr BP. Grove (1979) has shown that there was a period of Alpine glacier expansion from 3300 to 2400 yr BP. This signal is also shown in the ice cores from Crête and Camp Century, Greenland. This led to a cooling in Arctic latitudes which was probably responsible for the emergence of more powerful atmospheric circulation at this time. Evidence at about 3300 yr BP for a fall in the upper limit of pine forest has been inferred as a deterioration in the climate (Bridge *et al.*, 1990). Deuterium isotope work on pine macrofossils from the Cairngorms suggests that particularly wet conditions prevailed at this time (Dubois & Ferguson, 1985). Lamb (1977) estimated that, between 2800 and 2400 yr BP, mean July temperatures fell to 15.1 °C, winter temperatures became milder (4.7 °C) and there was an increase in precipitation. Osborne (1982) has shown a deteriorating climate from a number of sites in Lincolnshire and the Midlands between 3300 and 2200 yr BP, based on coleoptera. He noticed that at this time northerly beetle faunas had migrated to these more southerly sites. By 2500 yr BP there was a particularly marked deterioration across north-west Europe with a catastrophic decline between 2800 and 2200 yr BP (Aaby, 1976; Barber, 1982). Lamb (1977) suggested that annual temperatures around this time became at least 2 °C below those of the climatic optimum.

In the Thames, channel and floodplain metamorphosis in the late Iron Age and Roman times are thought to have been caused by increased runoff and flood magnitude. This was linked to a shift to a wetter climate, with flow augmented by Iron Age and Roman woodland clearance. Such views are supported in the Upper Thames region by pollen evidence from Sidlings Copse (Day, 1991). In addition a number of sites situated on the floodplains of the Thames and its tributaries at Mingies Ditch (Allen & Robinson, 1993), Farmoor (Robinson & Lambrick, 1979) and Port Meadow (Lambrick & Robinson, 1988) have yielded waterlogged sediments. At these sites, investigations were made from seasonally occupied Iron Age enclosures. The low arboreal pollen (about 3% of the total pollen sum here suggests an open landscape, with a pastural emphasis.

Bell and Walker (1992) suggested that intensive Iron Age and Roman activities rather than climatic change were responsible for alluviation. However, Macklin and Lewin (1991) saw river alluviation in Britain as climatically driven but culturally blurred, thus deciphering climate change through the pollen record becomes increasingly difficult.

The proxy-climate curve from Bolton Fell Moss (Barber, 1981; Barber *et al.*, 1993, 1994) shows that temperatures reached a peak around 1500 yr BP and then record four shifts from fairly dry bog surface to wet lawn conditions between 1350 to 1150 yr BP. Between 1050 and 850 yr BP, pools and wetter conditions prevailed. Unfortunately, sites situated in the limestone regions of Britain do not exhibit such a detailed proxy record for climate change. However, both Spartum Fen and Daisy Banks Fen record a rise in the water table resulting in net sediment accumulation at about 1000 yr BP, which may coincide with this event. Both sites indicate an intensification in agriculture at this time.

Between about AD 700 and 1300, there was a short-lived episode of climatic warming which has become known as the Little Optimum. This was a period of glacial retreat in many mountain regions of the Northern Hemisphere (Grove, 1979). The ice-core data from Greenland indicate temperatures 1–2 °C higher than the present (Dansgaard *et al.*, 1989).

Pollen evidence from the Mediaeval Period in the Upper Thames region comes from a number of sites including Sidlings Copse, Spartum Fen and Daisy Banks Fen. At Daisy Banks there is an almost continuous curve of *Centaurea cyanus* (cornflower). This species, a weed of cereal crops, became more common at this time at sites in many parts of Britain and may suggest increased rye cultivation (Greig, 1991). The pollen evidence indicated regional woodland differences, though, in general, an open grassland landscape dominated the region.

It is difficult to suggest biotic responses to environmental changes from this period onwards, which is a reminder that these are proxy-climatic records with which the sensitivity suffers both temporally and spatially. There are numerous historical sources which cover this period onwards (e.g. Bradley & Jones, 1993; Grove, 1988).

18.6   CONCLUSIONS

Fig. 18.6 shows a proposed framework of biotic response to environmental change in the Upper Thames region over the last 15 000 years. Although fewer Late Glacial and Flandrian sites have been investigated in central-southern England (when compared to those outside of these areas), they form an important record for environmental change. This enables us to understand biotic responses to local, regional and global changes which have taken place over the past 15 000 years. Late Glacial sites are present, thus offering long records of information. Three main phases are recognized, two very cold with arctic floral assemblages and an intervening temperate phase during which birch woodland became established. During the early Flandrian, pioneer colonies of *Salix* and *Betula* with *Pinus* and *Corylus* were replaced with *Corylus* and *Pinus* woodland. Incoming thermophilous trees later replaced these (*Ulmus ca.* 9100 BP and *Quercus ca.* 8800 yr BP). The ultimate dominance by these species reflected changes in conditions within the surrounding area which probably include increasing soil stability, pedogenesis and freer drainage, due to climatic amelioration.

No association between burning and the *Corylus* maxima as suggested by Smith (1970) was observed, indeed the lowest charcoal records were recorded during this phase. The rise of *Alnus* and *Tilia* have been associated with the Boreal–Atlantic transition, the expansion of which has been suggested at around about 6900 BP. Both *Alnus* and *Tilia,* although present at low frequencies before this stage, would have been out-competed during the earlier post glacial phases by *Salix* and *Betula* until favoured by changing edaphic conditions. The expansion of *Alnus* is associated with the presence of suitable damp soils perhaps associated with regional burning; *Tilia* would have expanded and been dominant in the surrounding limestone hills.

The elm decline dated about 5200 yr BP was probably associated with disease, human disturbance and perhaps a deterioration in climate. The Daisy Banks Fen sequence records early phases of clearance and agricultural activity about 5000 BP with relatively high frequencies of cereal pollen. This occurred slightly after first elm decline. At Spartum Fen there are few signs of human disturbance. Extensive regional woodland clearance occurred from 3500 yr BP, and

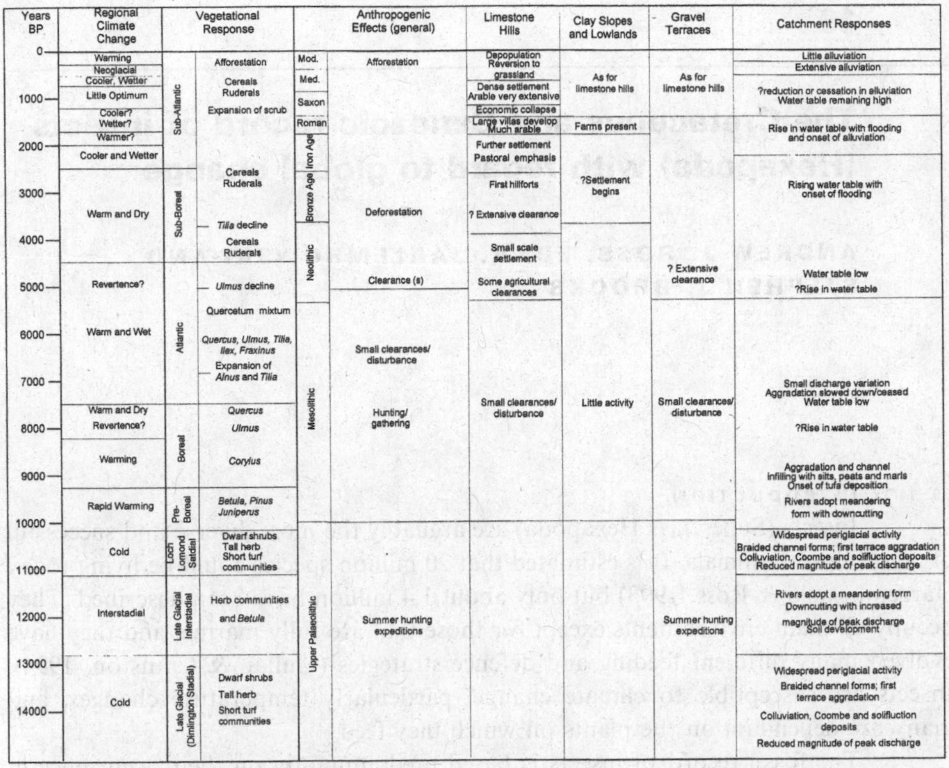

Years BP	Regional Climate Change	(Climate zone)	Vegetational Response		Anthropogenic Effects (general)	(Archaeol. period)	Limestone Hills	Clay Slopes and Lowlands	Gravel Terraces	Catchment Responses
0	Warming	Sub-Atlantic	Afforestation	Mod.	Afforestation	Bronze Age / Iron Age	Depopulation Reversion to grassland	As for limestone hills	As for limestone hills	Little alluviation
	Neoglacial Cooler, Wetter		Cereals Ruderals	Med.			Dense settlement Arable very extensive			Extensive alluviation
1000	Little Optimum		Expansion of scrub	Saxon			Economic collapse			?reduction or cessation in elluviation Water table remaining high
	Cooler? Wetter?			Roman			Large villas develop Much arable	Farms present		Rise in water table with flooding and onset of alluviation
	Warmer?									
2000	Cooler and Wetter						Further settlement Pastoral emphasis			
		Sub-Boreal	Cereals Ruderals							
3000	Warm and Dry				Deforestation		First hillforts	?Settlement begins		Rising water table with onset of flooding
			Tilia decline				? Extensive clearance			
4000			Cereals Ruderals			Neolithic	Small scale settlement		? Extensive clearance	
	Revertence?	Atlantic	Ulmus decline		Clearance (s)		Some agricultural clearances			Water table low ?Rise in water table
5000			Quercetum mixtum							
6000	Warm and Wet		Quercus, Ulmus, Tilia, Ilex, Fraxinus		Small clearances/disturbance					
7000			Expansion of Alnus and Tilia			Mesolithic	Small clearances/disturbance	Little activity	Small clearances/disturbance	Small discharge variation Aggradation slowed down/ceased Water table low
8000	Warm and Dry Revertence?	Boreal	Quercus		Hunting/gathering					?Rise in water table
			Ulmus							
	Warming		Corylus							
9000										Aggradation and channel infilling with silts, peats and marls Onset of tufa deposition
	Rapid Warming	Pre-Boreal	Betula, Pinus Juniperus							Rivers adopt meandering form with downcutting
10000	Cold	Loch Lomond Stadial	Dwarf shrubs Tall herb Short turf communities			Upper Palaeolithic				Widespread periglacial activity Braided channel forms; first terrace aggradation Colluvium, Coombe and solifluction deposits Reduced magnitude of peak discharge
11000										
12000	Interstadial	Late Glacial Interstadial	Herb communities and Betula		Summer hunting expeditions				Summer hunting expeditions	Rivers adopt a meandering form Downcutting with increased magnitude of peak discharge Soil development
13000										Widespread periglacial activity Braided channel forms; first terrace aggradation
14000	Cold	Late Glacial (Dimlington Stadial)	Dwarf shrubs Tall herb Short turf communities							Colluviation, Coombe and solifluction deposits Reduced magnitude of peak discharge

Fig. 18.6. Diagram showing the biotic response to environmental change in the Upper Thames during the last 15 000 years. (Parker, 1995a.)

thus may be associated with the activities of late Bronze and early Iron age human activity in the region.

A hiatus in deposition from about 3300 BP to about 1000 BP is recorded from some sites. The cause of this is most likely due to drier conditions in the mid Bronze Age with a lowered water table. There is a general absence of waterlogged remains of this age from archaeological sites across the region. Extensive woodland clearance continued from the late Bronze Age into Roman times. A rise in the water table during the late Saxon period led to sediment reaccumulation at a number of sites. The pollen record reveals arable farming on the drier land and pastoral practice on the wetter floodplains at this time. The sites have remained important ecotones throughout the Flandrian as witnessed by the dominance of differing species at each site. This allows some differentiation of spatial patterns from one another and indeed from other sites previously explored.

Pollen analysis is undoubtedly a powerful tool in reconstructing floristic changes on local, regional and even global scales. However, its value as a technique is greatly increased in the reconstruction and disentanglement of palaeoenvironments when adopted as part of a multidisciplinary regime. Thus, there has been limited research on the linkages between human activity and climatic change across Britain and the associated biotic response to such events.

19

The Cretaceous and Cenozoic record of insects (Hexapoda) with regard to global change

ANDREW J. ROSS, ED. A. JARZEMBOWSKI AND
STEPHEN J. BROOKS

19.1 INTRODUCTION

Insects (Superclass Hexapoda) are arguably the most diverse and successful of all groups of animals. It is estimated that 20 million species could be living today (Jarzembowski & Ross, 1993) but only about 1.4 million have been described. They occupy all main environments except for those that are fully marine, and they have evolved many different feeding and defence strategies (Gullan & Cranston, 1994). Insects are susceptible to climate change, particularly temperature changes, and many are dependant on the plants on which they feed.

The fossil record of insects is based predominantly on their occurrence in non-marine (lacustrine and fluvial) sediments. They are usually preserved as isolated wings or rarely as complete insects. The wings of different groups of insects have a distinct venation which provides characters to enable their identification and classification. The Cretaceous and Tertiary insect record is supplemented by their remarkable preservation in amber.

The richness of the fossil record of insects has only begun to be realized in the last few years with the publication of three main databases. Two of these are of insect families that occur in the fossil record (Ross & Jarzembowski, 1993: updated in Jarzembowski & Ross, 1996; Labandeira, 1995). The third is a database of genera (Carpenter, 1992), although unfortunately this only contains data up to the end of 1983 and it is now out of date. Since 1983, an additional 500 families and roughly 1 000 genera have been recorded as fossils. Most of these genera occur in Cretaceous deposits. A species database is being developed but it will be many years before it will be complete enough to be useful. The last fossil insect species database was published in 1908 (Handlirsch, 1906-1908). So, at present, only the family data are reliable enough to look at changes in origination, extinction and diversity through time. As yet, little has been done to try to tie in these changes to changes in the physical environment globally.

Up to the end of 1995 there were 1275 families of insect in the fossil record (Jarzembowski & Ross, 1996, plus unpublished data), compared with some 967 families living today (Naumann, 1995), of which 648 (67%) have a fossil record. The oldest fossil hexapod is of Early Devonian age, although the first flying insects did not appear until the Mid Carboniferous (Jarzembowski & Ross, 1996). They

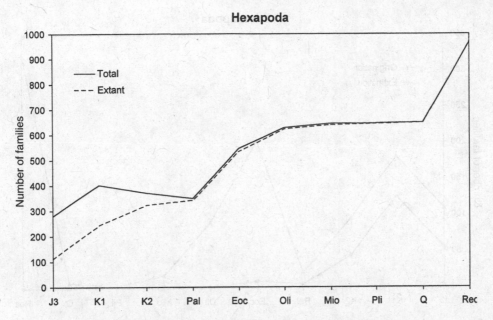

Fig. 19.1. Graph showing the total number of hexapod (insect) families known from the Upper Jurassic (J3) to Recent (Rec), based on data from Ross and Jarzembowski (1993), Jarzembowski and Ross (1996), unpublished MS, and Naumann (1995). The dashed line represents the number of extant families.

underwent high turnover during the Late Carboniferous and Permian, extinction at the end of the Permian, then gradual diversification from the Mid Triassic to the Late Jurassic (Jarzembowski & Ross, 1996). The first extant families appeared in Late Triassic times (Ross & Jarzembowski, 1993). The Cretaceous record is particularly interesting because the radiation of flowering plants had a major impact on the insect fauna, but the insect families were hardly affected at the K–T boundary. The Tertiary saw continued diversification of insects with little extinction. The Quaternary insect fossils belong to extant species, which have enabled in depth studies into the changes in their biogeographical distributions in response to climatic change. This chapter will discuss the Cretaceous, Tertiary and Quaternary record of insects with emphasis on the effects of the angiosperm radiation, the K–T boundary event, climatic cooling through the Tertiary, and Quaternary climate changes.

19.2   THE EARLY CRETACEOUS INSECT RECORD
19.2.1   The general pattern
The Early Cretaceous saw the largest upheaval to the insect fauna in the Mesozoic. This epoch saw the highest origination of new families, most of which are extant (Fig. 19.1), and also the highest extinction of families (31% of all the Early Cretaceous families, Fig. 19.2). The most likely cause for this turnover is the rapid radiation of the flowering plants (angiosperms) replacing the gymnosperm, cycad and fern flora (Lupia *et al.*, this volume). Labandeira and Sepkoski (1993), in contrast, believed that the angiosperms had little affect on the insects, but they considered total

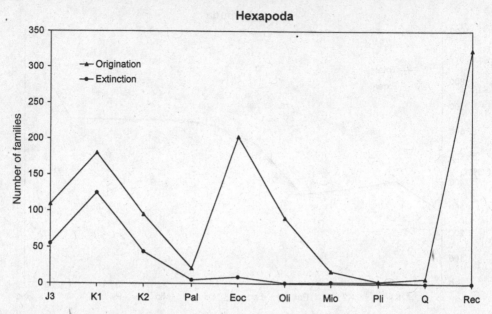

Fig. 19.2. Graph showing episodes of origination and extinction of hexapod families from the Upper Jurassic to Recent. The apparent high origination of Recent families is a reflection of the number of families which as yet do not have a fossil record, such as many specialized vertebrate parasites.

diversity and did not take into consideration the origination and extinction data separately. The rise of angiosperms is likely to have had an initial detrimental effect on phytophagous (plant-feeding) insects resulting in widespread extinction, such as in the orders Orthoptera (grasshoppers, crickets and locusts, Fig. 19.3) and Hemiptera (bugs, Fig. 19.4), although both orders also contain some predatory families. This in turn would have affected predatory insects, such as the Odonata (dragonflies and damselflies, Fig. 19.5) and Neuroptera (lacewings, Fig. 19.6). Many species of the latter order feed on small phytophagous Hemiptera today. At the same time the angiosperms provided a new food source which encouraged the origination of new families (e.g. families of Hemiptera: Aphidoidea (greenfly)), most of which are still living today (see Ross & Jarzembowski, 1993). Most orders of insects were affected in some way, probably due to resultant changes in ecosystems. Unfortunately, precise ages for many of the Early Cretaceous insect localities are not known so it is difficult to say whether the extinction was rapid or gradual, although Zherikhin (1993) suggested it was rapid and occurred in the mid Cretaceous. Hopefully, as stratigraphical resolution is improved, the overall picture will become clearer.

The Early Cretaceous is notable for the first appearances of social insects – the termites (Isoptera) (Martinez-Delclos & Martinell, 1995) and social wasps (Hymenoptera: Vespidae) (Carpenter & Rasnitsyn, 1990). The oldest vertebrate parasites (Siphonaptera) are also recorded, some of which may have parasitized pterosaurs (Ponomarenko, 1976; Jell & Duncan, 1986). The first extant insect genera also appeared, e.g. the dipteran (fly) genus *Olbiogaster* and hemipteran genus *Cixius*.

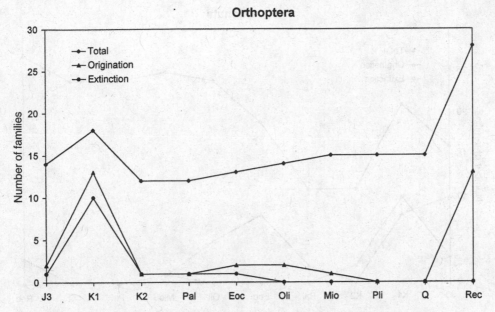

Fig. 19.3. Graph showing the total number, origination and extinction of Orthoptera (grasshopper, cricket and locust) families from the Upper Jurassic to Recent.

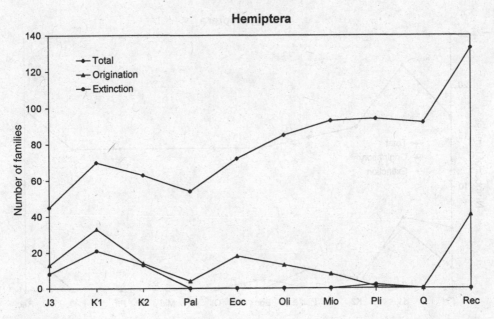

Fig. 19.4. Graph showing the total number, origination and extinction of Hemiptera (bug) families from the Upper Jurassic to Recent.

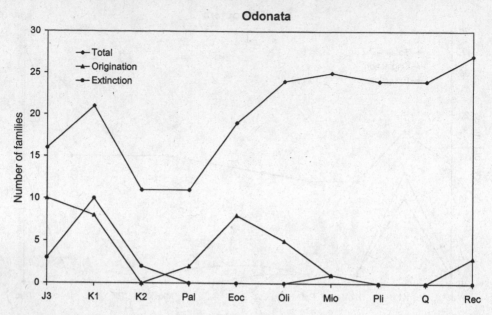

Fig. 19.5. Graph showing the total number, origination and extinction of Odonata (dragonfly and damselfly) families from the Upper Jurassic to Recent.

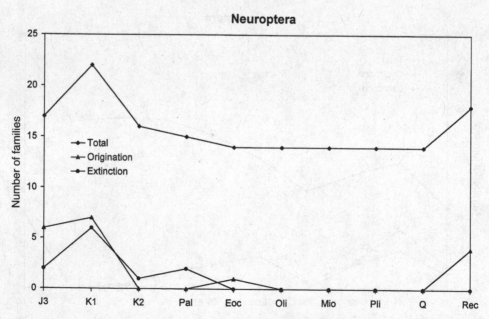

Fig. 19.6. Graph showing the total number, origination and extinction of Neuroptera (lacewing) families from the Upper Jurassic to Recent.

The main Early Cretaceous insect localities occur in England, Spain, Brazil, Australia, Lebanon (amber), Russia, Mongolia and China.

19.2.2   Purbeck–Wealden insects of southern England: a case study

Fossil insect localities in southern England provide potential for studying changes in the Early Cretaceous insect faunas and attempting to correlate these with environmental change. Insects occur in the Purbeck Limestone Group (Berriasian) of Dorset and Wiltshire (Clifford *et al.*, 1994; Coram *et al.*, 1995; Ross & Jarzembowski, 1996) and the Wealden Supergroup (Berriasian–Barremian) of the Weald (Jarzembowski, 1984; Ruffell *et al.*, 1996; Rasnitsyn *et al.*, 1998). The Purbeck and Wealden taxa belong to a mixture of extinct and extant families and may total 500 species, although their description is still at an early stage.

The Purbeck Limestone Group was deposited in predominantly arid conditions, while the Wealden Supergroup accumulated in predominantly humid conditions (see Ruffell & Rawson, 1994). Most of the Wealden insects are from the Lower Weald Clay (Hauterivian) and Upper Weald Clay (Barremian). The Weald Clay insect fauna differs from the Purbeck fauna. Many of the insect groups have not been adequately studied yet but preliminary results can be used for the Isoptera (termites), Blattodea (cockroaches) and Hymenoptera (wasps). Coleoptera (beetles) wing-cases are the most abundant insect fossils but they are difficult to identify. They have not been studied so it is not possible yet to determine whether they can be used to monitor climate change, such as been done using extant Coleoptera in the Quaternary.

Both Isoptera and Blattodea are scavengers, and any changes in the flora were unlikely to have had much of an effect on them. Thus they are more useful than phytophagous insects for trying to correlate faunal changes with changes in the physical environment. Termites (Isoptera) are good environmental indicators and indicate warm temperate to tropical conditions. One species (*Valditermes brenanae* Jarzembowski) occurs in the Weald Clay, but it has not been recorded from the Purbeck Limestone Group. This termite belongs to the family Termopsidae and probably lived in damp, rotting logs (Jarzembowski, 1995), which would have been sparse in the arid Purbeck environment. Blattodea are extremely abundant (second only to Coleoptera) in Purbeck and Wealden sediments which makes them more useful than the rarer insect groups for monitoring changes through time. Eleven species of Blattodea occur in the Purbeck whereas about 20 species (mostly undescribed) occur in the Wealden (five are common to both). Cockroaches can only be used as general environmental indicators, i.e. the higher the diversity, the warmer the climate. However it is unlikely that the Wealden was warmer than the Purbeck environment. Interestingly, the cockroach fauna of the Lulworth Formation is dominated by the extinct genus *Rithma* whereas the Durlston Formation above and the Wealden Supergroup are dominated by the extinct genus *Elisama*. It seems likely that the numerical discrepancy between the faunas is due to the change from arid to humid conditions. Generally insects prefer the latter (see Gullan & Cranston, 1994).

Notable for its absence from the Purbeck and Wealden fauna is the hymenopteran family Xyelidae (Rasnitsyn *et al.*, 1998). This family of sawflies is well known from Asian Lower Cretaceous deposits and today it lives in temperate conditions. The lack of Xyelidae from the Purbeck and Wealden fauna also indicates that the environment was warm. The family Sphecidae (digger wasps) are closely related to the bees and are important pollinators today. There is a higher diversity of sphecids in the Weald Clay Group than in the Purbeck Limestone Group which possibly reflects the angiosperm radiation. Interestingly the Weald Clay has yielded many specimens of a primitive aquatic flowering plant (Hill, 1996) which may have been pollinated by such wasps.

19.3   THE LATE CRETACEOUS INSECT RECORD

The Late Cretaceous insect record is not as good as for the Early Cretaceous. Sea-level rise during the Late Cretaceous flooded a lot of land (Gale, this volume), which would have reduced the size of insect populations. The rise also reduced the amount of terrestrially derived sediments being deposited and as a result insect deposits are few and far between. Most Late Cretaceous insects have come from amber deposits in Canada, New Jersey, Siberia, France and Burma (Poinar, 1992; Ross, 1997). The main non-amber insect deposit was deposited in a crater lake after the eruption of a kimberlite pipe in Botswana (Rayner et al., 1991).

Unfortunately, amber can only be dated by accompanying fossils in the amber-bearing sediments. This only gives a minimum age as there is no way of knowing how long the amber took to be deposited. Burmese amber comes from Eocene deposits, but it contains representatives of two families of insects that are not known later than the Late Cretaceous elsewhere; this implies that it has been reworked. The amber from New Jersey has yielded the earliest representatives of another two social hymenopteran groups – Apoidea (bees) and Formicidae (ants), although the age of the bee is doubtful (see Rasnitsyn & Michener, 1991).

From this more limited record, it appears there was a continuation in the rise of extant families with another 78 appearing (Fig. 19.1). Although there was some extinction at the family level, this does not appear to be as dramatic as for the Early Cretaceous. Forty-four families became extinct by the end of the Cretaceous (Fig. 19.2) which is 14% of the total known diversity. It is uncertain whether this was due to the continued radiation of the flowering plants affecting ecosystems or to the end Cretaceous mass extinction event, although the fact that the predominantly phytophagous order Hemiptera was the most affected (Fig. 19.4), suggests the former. The Hemiptera and some other groups of insects were probably affected by the plant extinction at the end of the Cenomanian (Lupia et al., this volume). It is possible that some of the families that apparently became extinct by the end of the Early Cretaceous actually continued into the Cenomanian but have not been recorded as fossils.

The end Cretaceous extinction apparently had little effect on insects at the family level (Jarzembowski & Ross, 1996). There could be several reasons for this:

(i)   Whalley (1988) suggested that insects in northern latitudes could survive a catastrophe by going into diapause (hibernation), but they would only survive if the event was short lived. He suggested that insects in the Tropics, which do not need to hibernate, would be devastated. But there are no

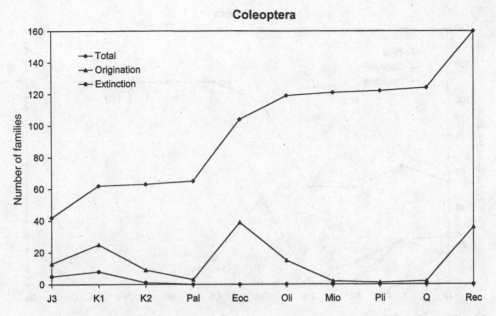

Fig. 19.7. Graph showing the total number, origination and extinction of Coleoptera (beetle) families from the Upper Jurassic to Recent.

known Late Cretaceous tropical insect faunas to check for this. Recent studies of amber from Burma (Myanmar), which would have been closest of all the insect deposits to the Equator at that time, have shown that the insect fauna is not tropical (Rasnitsyn, 1996). Diapause can be triggered by changes in day length, temperature, food quality, moisture, pH and chemicals, and usually lasts from several days to several months (Gullan & Cranston, 1994). Therefore, at the end of the Cretaceous, diapause could have been triggered by a reduction in daylight due to particulate matter being thrown into the atmosphere and a resultant reduction in temperature, caused by a bolide impact and/or increased volcanism (Pickering, this volume).

(ii)   It appears that insects with complete metamorphosis (endopterygotes) were proportionally less susceptible to extinction during this time than those with incomplete metamorphosis (exopterygotes) (compare figs. 10–13 in Jarzembowski & Ross, 1996). For example, two of the most diverse endopterygote orders, the Coleoptera (beetles, Fig. 19.7) and Diptera (flies, Fig. 19.8) were hardly affected with only one family of each becoming extinct. Insects such as these with complete metamorphosis would have had a much better chance of surviving a short-term catastrophe if they were in the pupal stage at the time. The pupa is a tough-walled, closed system in which its only contact with the outside world is through gaseous exchange. Many insects protect their pupae by surrounding them with silk and organic debris or pupating underground or within wood. Most insects pupate for a period of a couple of weeks to several months; however, they can also go into dia-

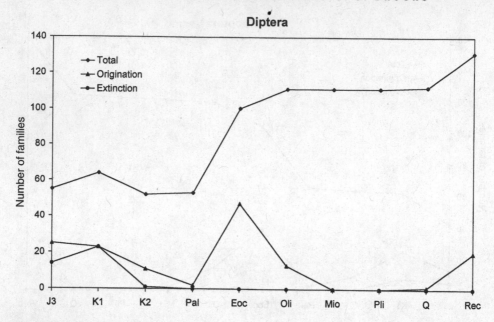

Fig. 19.8. Graph showing the total number, origination and extinction of Diptera (fly) families from the Upper Jurassic to Recent.

pause where the fully formed adult will wait (up to a couple of years in some cases) until conditions are right before emerging.

(iii)   Work on Quaternary Coleoptera has shown that they can respond quickly to rapid environmental change by migrating to other regions (Elias, 1994). If the end Cretaceous extinction event was long term then there may have been refugia in some parts of the world where insects could have survived. If it was short term then it is unlikely that they would have had time to migrate.

Unfortunately, there are hardly any fossil insects known from the Maastrichtian and Danian to test these ideas. However, indirect evidence has recently come to light that shows that some insects were affected by the end Cretaceous extinction event. Examination of insect damage to leaves (leaf-mines, galls, etc.) collected from either side of the K–T boundary in North America has shown that there is a much higher diversity of damage types below the boundary than above it (Peter Lang, personal communication). However the same types of damage reappear later on in the Tertiary. So it is possible that there could have been an insect extinction at the K–T boundary but at a taxonomic level lower than family. Interestingly Pike (1994) indicated that at least 72% of the identified genera in Canadian amber (Campanian) are extinct.

19.4   THE TERTIARY INSECT RECORD

The Tertiary saw continued diversification of insects with another 333 families appearing by the end of the Pliocene. There was hardly any extinction with only 19 families becoming extinct during this period. This is surprising because

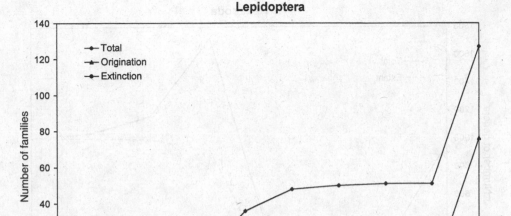

Fig. 19.9. Graph showing the total number, origination and extinction of Lepidoptera (moth and butterfly) families from the Upper Jurassic to Recent.

there was a decline in global temperature (Pickering, this volume) during the Tertiary which could be assumed to have had a detrimental rather than positive effect on insect diversity.

Most of what we know of the Tertiary insect fauna comes from deposits and amber of Eocene, Oligocene and Miocene age (listed in Evenhuis, 1994). Unfortunately, there are very few Paleocene and Pliocene insect deposits, which means that little can be said on the recovery of the fauna immediately after the K–T boundary event and on the origin of the extant fauna during the Pliocene. Certainly, by the end of the Eocene 223 families had appeared since the Cretaceous. However, it is not possible to say for certain whether most of them appeared during the Paleocene or Eocene and hence whether the renewed radiation after the end of the Cretaceous was gradual or rapid. There was continued radiation of extant angiosperm families during the Paleogene which would have facilitated evolution among the phytophagous insects, e.g. Lepidoptera (moths and butterflies) (Fig. 19.9).

Perhaps family level data is not as good an indicator of change during the Tertiary as for the Mesozoic and lower level taxa should be considered instead. The graph of generic data, based on Carpenter (1992), displays a peak for the Oligocene (Fig. 19.10). Caution should be used when interpreting this as most Baltic amber is generally considered to be Late Eocene in age whereas Carpenter used an Oligocene age. Nevertheless there appears to have been a significant generic extinction towards the end of the Paleogene. Global cooling, which started at about the Eocene–Oligocene boundary, could account for this. However, many of the extinct genera are very similar to extant genera, and rather than becoming extinct they may have

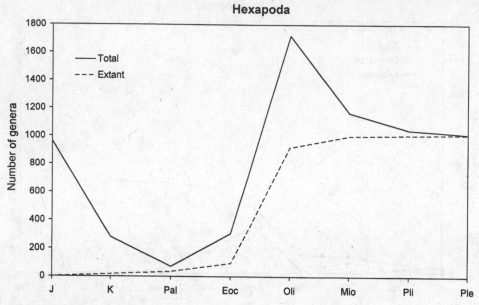

Fig. 19.10. Graph showing the total number of hexapod genera known from the Jurassic to Pleistocene, based on data from Carpenter (1992). The dashed line represents the number of extant genera.

evolved into more modern forms. During the Paleogene many extant genera appeared. Again, due to the lack of good Paleocene localities it is uncertain whether this radiation was gradual or rapid. The continued radiation of the angiosperms is likely to be responsible for many of the new generic occurrences.

The earliest records of extant species of insects come from Baltic amber (Late Eocene–Early Oligocene). Several species have been recorded but recent re-examination of some of them have shown that they are fake, misidentified or pre-served in copal (subfossil resin of Quaternary age). Two of the records are as yet undisputed, a specimen of the ephemeropteran (mayfly) *Heptagenia fuscogrisea* and several specimens of the tiny parasitic hymenopteran (wasp) *Palaeomymar duisburgi* (Grimaldi *et al.*, 1995; Ross, 1997). Both belong to primitive groups that have hardly changed with time. Several more extant species have been recorded from the Oligocene and Miocene. It is likely that most of the extant species appeared during the Neogene; certainly in Quaternary deposits virtually all the species present are extant.

Comparison of some of the Tertiary insect taxa with their extant relatives has displayed some considerable changes in their biogeographical distributions through time. Many entomologists examining the present-day biogeographical dis-tributions of closely related taxa relate their distributions to plate tectonics. For example, the present day insect faunas of eastern North America and Europe contain some species that occur on both sides of the Atlantic. However, the faunas of western and eastern North America are significantly different. This has been explained by

Cretaceous plate configerations (Noonan, 1988) when eastern North America was initially close to Europe prior to the opening of the Atlantic. During the Late Cretaceous the Mid-Continental Seaway separated the eastern and western American faunas.

However, the study of insect fossils shows that this interpretation is not as simple as that. It appears that many modern insects have a more restricted distribution than their Tertiary predecessors. A classic example is the termite genus *Mastotermes*. This genus is only known from one species today and occurs in Australia (it has been accidentally introduced to New Guinea and New Zealand by man). But, during the Tertiary this genus may have had a worldwide distribution. It has been recorded from Eocene deposits in southern England, from Oligocene deposits in Germany, France, Mexico (amber) and the Dominican Republic (amber), and from Miocene deposits in Croatia (Nel & Paicheler, 1993). Another example is the beetle genus *Pactopus* which has been recorded from Eocene deposits in southern England and Baltic amber. Today, this genus lives only in western North America. Other examples include many flies in Baltic amber that are most closely related to species living today in south-east Asia, southern Africa, South America, Australia and New Zealand (e.g. Hennig, 1966, 1967). Other insects that have been found as fossils in the Northern Hemisphere, but only have living relatives in the Southern Hemisphere, are recorded in Eskov (1987, 1992). A genus of scuttle-fly (Diptera: Phoridae) – *Abaristophora*, recorded in Dominican amber is only found today in Nepal and New Zealand (Disney & Ross, 1997). These distributions cannot be explained by plate tectonics alone. The most likely explanation is that these insects had a much wider distribution during the Paleogene but that the decrease of global temperature through the Tertiary has reduced their populations to relict areas.

19.5   THE QUATERNARY INSECT RECORD

The Quaternary was a period of unprecedented climatic change. It was characterized, at least in northern temperate regions, by episodes of rapid climatic oscillations that could swing from warm, temperate interglacials to cold, Arctic phases within a century. Until recently, the effect of these climatic fluctuations on insect faunas was poorly understood. However, following the pioneering work on beetle (Coleoptera) faunas by Coope (e.g. Coope, 1959; 1961), research on Quaternary insects during the last three decades has begun to expand (see Elias, 1994; Hofmann, 1986; Walker, 1995; Wilkinson, 1984; Williams, 1988; Coope, this volume).

Throughout the Northern Hemisphere there is evidence of insect distributions altering in response to climatic change. For example, Nazarov (1984; Nazarov in Elias, 1994) studied Quaternary insect deposits in Byelorussia beginning from 730 000 years ago. At least five Pleistocene glaciations occurred, resulting in almost complete ice cover during three of the stadials. Fossil beetle data suggest that during these glaciations, summer temperatures were about 10–13 °C with winter temperatures falling to −32 °C. The beetle fauna included species associated with tundra, forest-tundra and montane steppe. Cold-adapted Asiatic beetle species were found in the glacial deposits, but these were replaced during the interglacials by thermophilous species which still live in the region today.

Elias (1994) has provided a useful review of the many studies of Quaternary insect assemblages in North America. Insect faunas were significantly affected by the last (Wisconsin) glaciation in North America (about 20 000–15 000 BP) during which the Laurentide ice-sheet covered most of eastern and central Canada and the north-eastern and north-central United States. Cold-adapted insects found refugia from the ice-sheet in Beringia, on the Alaskan–Siberian land-bridge which was dry land at this time, and to the south of the ice-sheet in what is now the northern states of the USA (Morgan et al., 1984). As the climate warmed and the ice-sheets retreated, the cold-adapted Arctic species in the southern refugium became extinct and were replaced by boreal species moving north (Morgan, 1987; Schwert & Ashworth, 1988). However, Beringian populations of thermophobic beetle species were able to recolonize the Arctic regions of North America. Nevertheless, species richness of the modern Arctic beetle fauna is lower in the east than the west. A number of factors may have slowed the easterly migration of species from the Beringian refuge, including the presence of a forestry barrier in the Mackenzie River Valley (Schwert, 1992), the Hudson Bay water barrier (Elias, 1994) and a 3000-year delay in the deglaciation of the central part of the Labrador–Ungava peninsula (Dyke & Prest, 1986).

Chironomidae (Diptera), or non-biting midges, have long been recognized as sensitive indicators of environmental change (for review see Lindegaard, 1995) but it is only relatively recently that their use as indicators of past climate change has been developed. Climatic conditions inferred from chironomid assemblages can provide useful complementary data when compared with the results of pollen or beetle analysis. Chironomid larval head capsules are usually far more abundant in lake sediments than beetle remains; hundreds can be recovered from just a few grammes of sediment, and so can provide an unbroken record with numbers high enough for statistical analysis. A chironomid–temperature transfer function has been developed for eastern Canada (Walker et al., 1991, 1992), and work is now in progress to generate a similar transfer function for European assemblages. Chironomid generated temperature reconstructions can therefore be compared with those generated from beetle assemblages using the mutual climatic range method (Atkinson et al., 1987), or those derived from palynological studies.

The role of climate in governing the distribution of chironomids has been the subject of much debate. Walker and fellow workers (Walker et al., 1992; Walker & Mathewes, 1987, 1989) cite several studies which suggest that temperature has an overriding influence on the composition of chironomid assemblages. However, other workers dispute this (Hann et al., 1992; Warner & Hann, 1987; Warwick, 1989) and advocate sediment composition, turbidity and lake depth as being more influential. Nevertheless, there is a growing body of evidence (e.g. Walker et al., 1991; Levesque et al., 1993; Olander et al., 1997) from studies in North America and Europe, supported by rigorous statistical analysis, that suggests that temperature is the best explanatory variable for chironomid distribution.

One significant development that has resulted from chironomid analysis was the discovery of a hitherto unknown Lateglacial cold oscillation in eastern Canada

(Levesque *et al.*, 1993). Following deglaciation, about 14 500 yr BP, chironomid inferred July surface water temperatures in Killarney Lake, New Brunswick, increased to 18–19 °C. However, at about 11 200 yr BP (uncalibrated conventional radiocarbon date) thermophilic chironomid species were suddenly replaced by cold stenothermic taxa which indicated that temperatures must have fallen by about 4.5 °C. This cold oscillation, named the Killarney Oscillation, lasted about 250 years before another sharp increase in temperatures eliminated the cold-adapted taxa and allowed thermophilic chironomids back into the assemblage. Temperatures appear to have increased to about 17 °C.

The Killarney Oscillation appears to have been synchronous with a cooling event recognized, from stable oxygen isotope analysis, in Switzerland as the Gerzensee Oscillation (Eicher & Siegenthaler, 1976) and other cooling events in Europe identified from pollen (e.g. Riezebos & Slotboom, 1984; Paus, 1989) and beetle (Atkinson *et al.*, 1987) data, and in Greenland from ice core data (Taylor *et al.*, 1993). The Killarney Oscillation therefore appears to have been one manifestation of a general Amphi-Atlantic cooling event (Levesque *et al.*, 1993).

Climatic change may also affect the salinity of inland waters and studies have shown that this is reflected by characteristic changes in the chironomid fauna as the lake becomes dominated by just a few taxa that are tolerant of saline conditions. Walker *et al.* (1995) have developed a chironomid–salinity transfer function in order to reconstruct palaeosalinities.

Climate change during the Holocene in western Europe was less dramatic than in the Lateglacial but nevertheless still had an impact on chironomid faunas. Brooks (1996) showed that there was a marked change in the chironomid assemblage of Lochan Uaine in the Cairngorms about 2500 yr BP when the fauna became dominated by cold-adapted species following the disappearance of thermophilic taxa from the fauna. This faunistic change was in response to the sub-Atlantic cooling event. Subsequently, the chironomid fauna remained largely unaltered until about 100–150 year ago when there was a significant increase in acidophilic species. Similar changes in chironomid faunas in response to anthropogenic-induced acidification have been reported from other lakes in south-west Scotland (Brodin & Gransberg, 1993) and Finland (Olander, 1992).

Chironomid faunas have also been shown to respond to the effects of eutrophication (e.g. Brodin, 1982; Warwick, 1980) which is often brought about by human settlements and agricultural intensification. Typically, lakes so affected have a low diversity of species and become dominated by species of the tribe Chironomini, which can tolerate low levels of dissolved oxygen because of the presence of haemoglobin in their body fluids. Haemoglobin also helps to buffer the effects of acidity so that these species also survive in acidified lakes. Other insect groups have also been used to investigate the impact of organic pollution; for example, Klink (1989) used a variety of freshwater insects, including Ephemeroptera, Plecoptera, Heteroptera, Trichoptera and Diptera in alluvial sediments to chart the pollution history of the River Rhine.

The impact of heavy metal and organophosphate pollutants on chironomids is manifested as deformities affecting the mouthparts and antennae. Attempts have been made to use these abnormalities to indicate and quantify environmental stress

(Warwick, 1985; 1991). However, before a workable index can be formulated, more information is required concerning the cause-and-effect relationships between the frequency and severity of the deformities and the dosages of the specific contaminants.

The effect of climatic and environmental change on insect faunas during the Quaternary is, therefore, well documented in the lake sediments. High resolution analysis of insect remains has allowed a detailed picture of Late Glacial and Holocene climatic change to be pieced together which complements and augments evidence from other proxy-indicators. With the development of transfer functions and the mutual climatic range method, quantitative reconstructions of temperature and salinity changes are now possible. Work on insect faunas in North America and Europe, especially on beetle and midge assemblages, is now mature and the next challenge is to tackle the insect subfossils preserved in lakes in the Tropics and Southern Hemisphere, which will enable the development of a truly global view of the processes and impact of Quaternary climate change.

19.6   CONCLUSIONS

Insects are proving useful for indicating changes in the global environment through time. It is evident from family level data that there was a major turnover in the insect fauna during the Early Cretaceous, probably due to the radiation of the flowering plants affecting ecosystems. Changes in the insect fauna from Lower Cretaceous deposits in southern England reflect a change from arid to humid conditions during the Berriasian. Insect families were hardly affected at the K–T boundary and there are three possible explanations for this.

First, changes in the environment caused the insects to go into diapause (hibernation); secondly, insects with complete metamorphosis would have had a better chance of survival if they were pupating at the time; and thirdly, insects could have migrated to refugia and survived there. An episode of global cooling towards the end of the Paleogene may have caused an extinction at the generic level. Extant species of insect (particularly beetles and chironomid midges) have been used successfully to monitor rapid climatic changes during the Quaternary.

20

The palaeoclimatological significance of Late Cenozoic Coleoptera: familiar species in very unfamiliar circumstances

G. RUSSELL COOPE

20.1   INTRODUCTION

This chapter is concerned with the palaeoenvironmental implications of Quaternary insect faunas, in particular with the ways in which species responded to the large scale climatic events such as the glacial/interglacial oscillations that are such a characteristic feature of this period. These climatic events are probably as near to global changes as any yet documented in the geological timescale, and the manners in which insect species responded to them may well have global significance.

The Quaternary offers a number of unique opportunities for understanding the ways in which organisms responded in the past, and might be expected to respond in the future to such widespread climatic changes. First, this period experienced some of the most numerous, widespread, intense and rapid climatic changes that have yet been recorded, oscillating between prolonged episodes of glacial severity and shorter interludes of temperate conditions (i.e. interglacials, such as the one in which we are living now) some of which were rather warmer than those of today. These varying climatic conditions had to be met by a flora and fauna that was, to a large extent, made up of the same species that are living today with apparently the same environmental preferences and limitations as those of their present-day representatives. Thus the Quaternary (approximately the last two million years) has a good claim to be the most relevant of geological periods in any discussion about current global climatic changes and their possible effects on the modern biosphere. Furthermore, the Quaternary fossil record should be able to provide valuable clues to any future response by these same species to global climatic changes. Finally, this period spans the time when the rise of modern humans was to become one of the most potent global influences both on the biosphere and on the climates of the Earth. Human activities have introduced such novel features into the present-day environment that evidence from the Quaternary past may, in fact, give us few clues as to what the future might hold.

Until recently, Quaternary insects have received much less attention than their contemporary plant and vertebrate fossils. Yet in terrestrial and freshwater fossil assemblages, the insects hold pride of place in both their numbers of species and in the wide variety of habitats to which they have become adapted. They are almost overwhelmingly diverse in their morphology but their disarticulated skeletons are rarely amenable to identification by means of the traditional dichotomous keys used to identify modern species. Almost all of the fossils must therefore be identified by the laborious method of direct comparison with well-authenticated modern material. Their earlier neglect is, therefore, not surprising. They also have an additional problem in that they fall within that no-man's-land between the two scientific disciplines of entomology and classical palaeontology; too old and fragmentary to be of interest to the former and too modern to excite the latter. Yet the study of Quaternary entomology over the last few decades has produced surprises both for the entomologist and for the geologist. Some of these surprises will be discussed here.

20.2   THE NATURE OF THE QUATERNARY INSECT RECORD

Amongst the most abundant and varied fossils of Quaternary insects are the Coleoptera, which often occur in great profusion in any waterlogged sediment that has remained anoxic since its deposition. Thus it is the Coleoptera that have, to date, received the greatest attention and therefore carry the loudest palaeoenvironmental signal.

However, many other insect orders would repay similar attention. These include Hemiptera (true bugs), both Heteroptera and Homoptera, which are often common and highly distinctive. Diptera (true flies) are represented by abundant fragments including numerous larval head capsules of tipulids (crane flies), well-preserved pupal cases and even identifiable simuliid (black flies) heads (Crosskey & Taylor, 1986). The larvae of the Chironomidae (non-biting midges) are abundant in lake sediments and have become a separate area of study indicative of water quality and temperature, closely related to other aspects of palaeolimnology (Walker, 1987; Brooks et al., 1997). Trichoptera (caddis flies) also occur usually as larval sclerites in limnic sediments where, together with chironomid head capsules, they are often abundant (Wilkinson, 1984; Williams, 1988). Even the genitalia of caddis flies have been recovered from interstadial deposits (Coope et al., 1961). Megaloptera (alder flies) are also common as larval sclerites and adult head capsules in deposits laid down in both running and stationary water. Lepidoptera (butterflies and moths) are represented solely by the jaws of caterpillars, which so far cannot be identified further. Distinctive skeletal elements of Hymenoptera–Parasitica (parasitic wasps), are often abundant and very well-preserved but they have not attracted much interest from palaeoentomologists perhaps because of taxonomic difficulties encountered even amongst their modern representatives. Ant fragments are also abundant particularly amongst fossil assemblages that represent debris swept off the nearby land surface and passively incorporated into the deposit. It is curious that the Odonata (dragonflies and damselflies) and Orthoptera (grasshoppers and crickets) are so poorly represented amongst Quaternary fossil insect assemblages, especially considering the frequency with which they have been recorded in classical palaeontology. It would appear that their exoskeletons more readily decompose than some of the other orders of insect. Even extremely frail groups such as the Ephemeroptera (may flies) are occasionally recovered as larval fragments which, no doubt, would repay further investigation. Thus there is a broad spectrum of Quaternary insect fossils available, many of which still await detailed analysis.

Their palaeoecological significance would no doubt equal that of the more adequately studied Coleoptera.

20.3 COLEOPTERA AS PALAEOENVIRONMENTAL INDICATORS

The Coleoptera are invaluable as environmental indicators. Their exoskeletons are particularly robust and can survive largely unchanged in sediments. They are found in all manner of freshwater sediments that have remained in a waterlogged condition since their deposition (not just in peats as is so frequently believed), yet they are not readily redeposited after being eroded from their original sediment. They are complex enough to provide an extensive suite of characters that enable precise identifications to be made at a variety of taxonomic levels, frequently to species level. Surprisingly, they show no morphological difference when compared with their modern counterparts, even down to the intimacies of their male genitalia (Coope, 1970b, 1979). There is thus hardly any evidence of evolutionary change during the Quaternary Period.

Coleopteran species today often exhibit precise environmental preferences and are adapted to almost all terrestrial and freshwater habitats. The fact that Quaternary assemblages, for the most part, resemble communities living today, suggests that they shared the same environmental requirements. Hence they must have maintained physiological constancy as well as morphological constancy, in spite of (or even because of) the large-scale changes in climate (Coope, 1978). The large number of species in Quaternary assemblages spanned a wide spectrum of habitat preferences. Thus a detailed mosaic picture of the local environment of the past can be assembled out of their modern ecological requirements.

The geographical distributions of many species is dependent on factors of the thermal environment. These thermally sensitive species are of greatest value in palaeoclimatic interpretations.

When faced with environmental change, a species has a choice of three options; (i) to evolve out of trouble, (ii) to move out of trouble or (iii) to become extinct. In the Quaternary, the Coleoptera (and many other species of insect) adopted the second option; they responded to the numerous climatic oscillations by tracking acceptable conditions across the continents, thus altering their geographical ranges to maintain more or less constant climatic conditions. Hence, viewed from any particular locality, changes in the assemblages of insect species provide detailed evidence of changing palaeoclimate there.

The great mobility of coleopterans means that they were very prompt in their response to changing climate. Thus they are one of the most valuable indicators of rapid change. Not only could beetle species quickly invade new habitats but, when conditions become intolerable in a particular area, they were readily exterminated locally. Evidence for the rapidity of climatic change can thus be derived from two sources; the synchroneity of both the sudden appearance of new suites of coleopteran species and the sudden disappearance of the old assemblages as the climate switched from one mode to another (Coope & Brophy, 1972).

Past climatic conditions can be quantified using coleopteran assemblages. Reconstructions have been made of the thermal climate in central Britain during the Last (Devensian, Würm, Weichselian) Glaciation (Atkinson et al., 1987) and of the climate along the southern margins of the Laurentide ice-sheet in North America for the period approximately 20 000 to 12 000 years ago (Elias *et al.*, 1996).

Finally, it should be noted that it is only the robust nature of fossil Coleoptera coupled with the precision with which identifications can be made that makes them so valuable as environmental indicators. There is no other aspect that signals them as exceptional, and their observed morphological constancy may similarly be observed among the fossils of other groups of insect which have not yet been investigated to the same degree.

20.4   GEOLOGICAL LONGEVITY OF SPECIES OF COLEOPTERA

Morphological constancy of species of Coleoptera during the last half million years is now well documented for Britain and across much of the northern hemisphere. Beyond the last half million years the record is more scanty, partly because organic deposits become gradually more and more compacted and fossil insects difficult to recover. However, in the permafrost areas of high latitudes, it has been possible to locate fossil assemblages in an extraordinary state of preservation, which extend our knowledge of insect species back at least as far as the Late Miocene; that is to about 5.5 million years ago.

The pioneer work on these faunas has been done by J.V.Matthews Jr (1974) of the Geological Survey of Canada, who has investigated many Quaternary and upper Tertiary sites in Alaska and Arctic Canada, the oldest of these being at Lava Camp mine in Alaska, which is capped by a lava flow well-dated at just over 5.7 Ma. In north-east Asia, A.A.Kiselyov has recorded a small assemblage of Coleoptera from the Kutuyakh suite of strata in the Kolyma Lowlands of Arctic Siberia dating from 1.8 to 2.8 Ma (Sher *et al.*, 1979). Jens Böcher of the Zoological Museum, Copenhagen, has obtained a beautifully preserved, rich and varied fauna (Fig. 20.1), from lower Quaternary deposits at Kap København, Peary Land, in North Greenland (Böcher, 1995). This fossil fauna is far more complex than the present-day fauna of Greenland. This is all the more surprising since the site is situated today about 800 km from the North Pole and must have been located in more or less the same position in Early Pleistocene times. An excellent account of this important site is given by Böcher who also summarizes the Upper Tertiary sites mentioned above.

As an illustration of the geological longevity of insect species, it is convenient to begin with the Kap København assemblage. This fauna and flora dates from the conventional beginning of the Quaternary period namely from just over 2 Ma. By then, Greenland had already experienced its first major glacial period, at about 2.5 Ma, which was followed by a milder episode, during which these insect-bearing fossiliferous deposits were laid down. The total fossil insect assemblage was made up of 275 taxa of which 155 could be named to the species level. Of these 140 (90%) species were beetles (compared with 33 indigenous beetle species in present day Greenland). The rest included Hymenoptera (2 spp.), Diptera (1 sp.), Trichoptera (10 spp.) and Hemiptera (2 spp.). All of these can be referred to species that are still living today, albeit far from northernmost Greenland. Out of all the taxa in this two-million-year-old fossil assemblage, only three species would appear to be extinct. The element of doubt expressed here is because it is difficult to be absolutely certain when a species has become globally extinct, especially amongst small and cryptic animal

Fig. 20.1. Scanning electron microscope photographs of lower Quaternary Coleoptera from Kap København, Peary Land, North Greenland. (*a*) *Grypus equiseti*, right elytron (scale bar 500 μm). (*b*) *Litodactylus leucogaster*, right elytron (scale bar 200 μm). (*c*) Mandible of Cerambicid larva found in *Larix* wood (scale bar 100 μm) (after Böcher, 1995).

species. The Kap København fauna, however, provides objective evidence for specific constancy for a wide variety of insects during the last two million years. On the approximation of one generation per year, this amounts to specific constancy for more than two million generations.

Not surprisingly, the Kap København insect assemblage has no geographical analogue at the present day. This does not mean that the species involved have necessarily changed their environmental requirements during the past two million years. Quaternary entomological investigations have shown that communities of insect do not respond to climatic changes by moving *en bloc*. Each species adjusted its geographic range individually as the necessities and opportunities arose. As time passed, and if the climate remained the same for long enough, old communities reassembled with old associations and old enmities became reestablished. With each climatic change, especially with those that were sudden, the whole process of disruption and reassembly would recur. It is not easy to put a precise figure on the number of these disruptive events because with more and more refined records from both the ocean cores and the deep ice cores, the frequency of recognized climatic oscillations continues to increase dramatically (Ruddiman & Raymo, 1988; Dansgaard *et al.*, 1993). It would, therefore, be surprising indeed had the Kap København fossil insect assemblage, after two million years of this process, been represented today by some similar faunal assemblage of species, given the accidents of space and time that must have intervened as they tracked the numerous Quaternary climatic oscillations.

If the fossil community is looked at from an ecological rather than a geographical point of view, it becomes immediately apparent that the Kap København fauna makes good ecological sense. The palaeoecological reconstruction described by Böcher (1995) is internally consistent in that the present-day requirements of the various species, when put together, enable an ecologically meaningful mosaic picture to be built up. Furthermore, this ecological conformity is supported by other collateral information within the deposit itself. Thus, for instance, larval jaws of the wood-boring cerambycid beetles are found in association with large pieces of larch (*Larix*) wood with galleries and flight holes from which the adult beetles emerged. The patchy forests that grew so close to the North Pole in early Quaternary time, were apparently afflicted by the same pests that attack similar trees much further south at the present day.

Although there is no modern equivalent to these high arctic forests, the plant and insect species which made up their flora and fauna would almost all be familiar to the modern botanist or entomologist. The recentness of the Kap København deposits means that their geographical position at the extreme north of Greenland at the time of deposition cannot have been much different from that of the present day. Nor can there have been much difference in the disposition of land and ocean. Thus the geographical context of Kap Københaven must have been much the same two million years ago as it is today. It is apparent that modern species lived in a modern geographical context where few could survive today. Such 'non-analogue' fossil assemblages show that many present-day species have potentially more ways of exploiting environments than they exhibit today, i.e. not all of nature's goods seem to be currently on view in the shop window. The Kap København precedent shows that, if new environmental opportunities arise in the future, totally unfamiliar ecosystems could develop without the necessity for any evolutionary changes whatsoever. The Kap København fauna and flora demonstrates that one at least of these ecosystems may have already been tried out in the geologically recent past.

Since Quaternary insect species show such a high degree of evolutionary stability, the question arises: how far back in time can this species constancy be traced? Here again, the work of John Matthews and Jens Böcher provide the beginnings of an answer. Matthews (1976) has described complex insect assemblages, in many ways comparable to that from Kap København, from the Beaufort Formation on the Queen Elizabeth Islands and Banks Island in Arctic Canada. These sediments date from the mid Pliocene, that is from between 3 and 4 Ma. Their insect assemblages include rather more extinct species than that from Kap København but the rest of the fossils represent present-day species. The much older Lava Camp Mine insect fauna is of Late Miocene age, namely 5.5 to 6 Ma, and it has only a small proportion of specimens that could be equated with existing species. Much further back in time, the oldest extant species of beetle yet recorded is *Plateumaris nitida* from the Florisant Lake beds of Oligocene age (Askevold, 1990) but species longevity of this sort is probably a rare exception.

It has thus become apparent that many modern insect species had already evolved by Late Tertiary times, but at that time they were accompanied by other species which have subsequently become extinct. With the commencement of the rapidly oscillating climates in the Late Pliocene, leading eventually to the full glacial–interglacial climatic cycles of the Quaternary Period, many thermally sensitive insect species tracked acceptable climates across the continents. Any species incapable of these forced marches, for instance, if photoperiodic adaptations prevented it from changing latitude, or if geographical barriers obstructed its passage, would be at risk of extinction. The present-day insect fauna has thus been filtered by the need and ability of its species to adjust their ranges to the frequent and intense climatic fluctuations. The filter ensured that only those species that were able to adjust their geographical ranges as the climate changed, were able to survive.

Herein may lie an explanation for why there is so little evidence of extinctions amongst Quaternary insect assemblages. Since the amplitude of each climatic oscillation was more or less similar, a species that could avoid the consequences of the first major climatic oscillation could, by using a similar stratagem, avoid the others. By analogy with a hurdle race, if a species could clear the first climatic hurdle, it could similarly clear all the others; if it could not, it would fall at one of the first (Coope, 1995). The discoveries of Matthews (1974) and Böcher (1995) show that the first of these hurdles was already in place by Late Pliocene time.

A further consequence of large-scale changes of geographical ranges forced on many species by the frequent, sudden and intense climatic changes, is that populations are continuously being fragmented and reunited so that any incipient evolutionary changes in isolated populations are undone by numerous opportunities for outbreeding. The gene pools were kept well stirred and sustained evolution rendered almost impossible. The longevity of insect species would thus be one of the expected results of climatic instability rather than a puzzling enigma.

The ability of insect species to track acceptable environments across continents means that the environment in which a species actually lived remained, to a large extent, constant; it was the geography that changed. From this point of view the species would find little incentive to evolve. In different circumstances where changes in geographical range were impossible, such as on remote oceanic islands or isolated mountain tops, where any environmental changes had to be endured on the spot, evolution may have been the sole alternative to extinction.

Finally, the rise in human activity, particularly agriculture and industrialization, has made the option of 'moving out of trouble' almost impossible for a great many species. They have become surrounded by hostile environments across which any movements are now becoming increasingly difficult; they are just as isolated as if they lived on oceanic islands or remote mountain tops. The solution to climatic changes, so successfully employed in the past, of tracking acceptable conditions across the continents may not be available in the future. With the original options now reduced to two, it will be interesting to see which species opt for evolution and which for extinction. Certainly, our knowledge of their past response may be of little value in predicting any future reactions to climatic change, since we have imposed totally new restrictions on their mobility; we have inconveniently moved the goalposts and set up a ball game with totally new rules.

20.5   BRITISH INSECT FAUNAS DURING THE LAST GLACIATION

This period was dominated by prolonged episodes of cold climates with milder interludes, collectively called the Devensian in Britain and the Würm or Weichselian Glaciation on the adjacent continent of Europe and the Wisconsinan in North America. Up to now, plant remains have been the most frequently utilized indicators of past climates with palynology being the most extensively employed method. Climatic conditions were largely inferred from the succession of tree species as they colonized the landscape after periods of climatic amelioration or as they were gradually eliminated as the climates deteriorated. Episodes with low values of tree pollen and thus presumed treeless landscapes have often been taken to indicate quasi-tundra conditions with the implication of arctic climates.

The coleopteran assemblages from some of these treeless periods are dominated by cold-adapted species many of which are typical of the tundras of the present day. Occasionally, however, such treeless periods were characterized by an entirely temperate insect fauna, totally devoid of any cold adapted species and including species which today do not reach as far north as the British Isles. At such times the climate cannot have been of arctic severity and the treeless state of the landscape must have had some other explanation. The coleopteran assemblages show that the inference of tundra conditions based on palynological data alone is at times unsafe.

Some explanation other than extreme coldness is necessary to explain the absence of trees at such times. All of these enigmatic periods have similar climatic contexts. Each of them occur after an episode of sudden climatic amelioration when the flora and fauna, taken as a whole, was temporarily out of equilibrium with the physical environment. As pointed out earlier, the Quaternary fossil record shows that communities of species do not respond to climatic change *en bloc* but each species reacts individually at its own rate. Thus the more mobile species of Coleoptera could evidently take advantage of the new conditions before the trees could migrate far enough northwards from their presumed refuges in the Mediterranean area. The precise reasons for the tardiness in the arrival of trees in Britain at these times may have been complex and not be entirely a product of distance. Factors such as poor soil development, lack of moisture retaining humus, poor availability of nutrients or the failure to develop suitable symbiotic fungal associations, may all have contributed. The list could be prolonged extensively. The fact of the delay in the arrival of trees after the change to an appropriate climate seems inescapable. The Coleoptera indicated very rapid and intense climatic warming that could not have been inferred on the basis of pollen analysis alone.

In the climatic analysis of the beetle faunas it is essential to avoid any risk of beetle distributions being dependent on the distribution of some preferred host plant, thus reducing any climatic inferences to a second order derivative from the palaeo-botanical data. For this reason only carnivorous or general scavenging beetle species were used in the climatic reconstructions. The geographical ranges of these species are not directly dependent upon the macrophyte vegetation. Their foodwebs may often be traced back through various worms and small arthropods, such as collembola, ultimately to the decomposing vegetable matter in general and to the fungi and algae. These Coleoptera thus provide an indication of past climatic conditions which is, to a large extent, independent of palaeobotanical studies whether of pollen or of macroscopic plant fossils

Originally, estimates of past climates were made using the area of maximum overlap of the present day geographic ranges of the species that make up the fossil assemblage (Coope, 1959). This method requires that the geographical range is completely known and that the species occupies its total potential range. Neither of these essential conditions are fulfilled in many cases. Furthermore, species may occur today in geographically separate areas which have similar climatic regimes, such as mountain tops or high latitudes. Thus species living in similar environments but in different locations today could well be found together in fossil assemblages which have no present-day geographical analogues. This could lead to the false inference that the palaeoenvironment had no present-day analogue. With these problems in mind, climatic interpretation using overlapping geographical ranges was abandoned.

As an alternative procedure the total present-day distribution of individual species was plotted on climate space, giving each species its own climatic envelope within which it finds thermal environments tolerable. These envelopes are then easily stored in a database. In practice, these climatic envelopes condensed the distributional data into very much more convenient packages than the original geographical distributions. If the climatic envelopes for all those species in a fossil assemblage that are also in the database are superimposed, the area of overlap is termed their mutual climatic range (Fig. 20.2). In most cases the overlap is between 90 and 100% of the species involved. The climate space delimited by the mutual climatic range is deemed to represent the climate of the times when the fossil assemblage lived. The whole climatic reconstruction programme is termed the Mutual Climatic Range Method (Atkinson et al., 1987).

It is fortunate that the period under consideration here falls within the range of radiocarbon dating techniques because many of the insect faunas come from discrete sites with only crude stratigraphical controls on their ages. Radiocarbon dates provide a temporal precision otherwise unobtainable. However, it should be pointed out that the dates quoted here are given in uncalibrated radiocarbon years which, throughout this period, deviate slightly by variable amounts from calendar years.

Mutual Climatic Range reconstructions have been made from coleopteran assemblages from 27 sites, scattered across central and southern England, that date from the middle of the Devensian glaciation. Reconstructions of the thermal climate

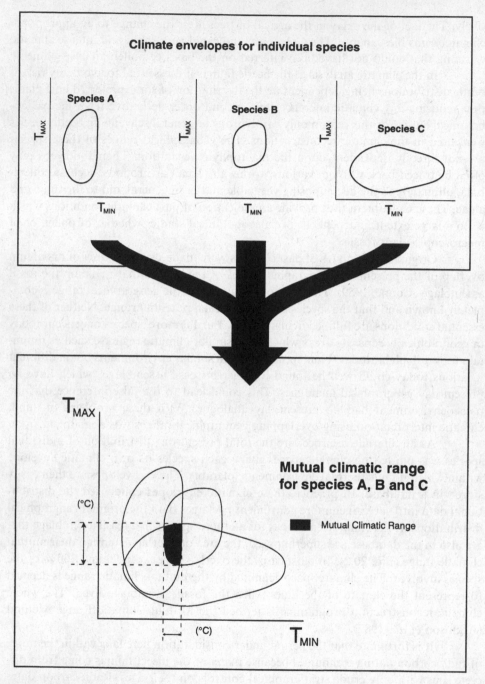

Fig. 20.2. Diagram to illustrate the basic methodology of the Mutual Climatic Range Method. The x axis of the climate space, T_{Max}, is the mean temperature of the warmest month. The y axis of the climate space, T_{Min}, is the mean temperature of the coldest month.

for all sites have been plotted against their age in thousands of radiocarbon years before present (Fig. 20.3). No effort has been made to give the statistical errors of these dates but most should have uncertainties, one standard deviation on either side of the mean, of about one thousand years. At the moment only parameters of the thermal climate have been utilized but there is no reason why any other climatic variable could not be employed. In the reconstruction given in Fig. 20.3, the mean temperature of the warmest month is indicated by TMax and the mean temperature of the coldest month by TMin. The vertical bars indicate that the mean monthly temperature lay somewhere within these limits and not that the temperatures ranged between the extreme ends of the bars; it must be emphasized, therefore, that they are not statistical error bars. Assuming a more or less sinusoidal curve for the variation in the values of mean monthly temperatures throughout the year, it is possible to make a crude estimate of these values. Fig. 20.3 can be used to illustrate the way in which climatic conditions in Britain varied with time and how they differed markedly from those of the present day.

There are a number of significant points that arise from this reconstruction. First, there was an enigmatic episode of considerable warmth between 42 000 and 43 000 radiocarbon years ago. At this time the insect fauna of England was characterized by large, varied and thoroughly temperate insect faunas that lived on a landscape totally devoid of any trees. From a single site at Isleworth in the Thames valley dating from this period, there were 248 named species of Coleoptera (Coope & Angus, 1975). A site of similar age on the Cromwell Road opposite the Natural History Museum in London, yielded 140 named species of Coleoptera (Coope *et al.*, 1997). In these rich and diverse assemblages there was not a single cold-adapted beetle species present and the whole fauna was made up of temperate species, several of which do not range today as far north as the British Isles. The Coleoptera indicate that the thermal climate of the times was at least as warm as that of the present day. If it was not for the suite of temperate Coleoptera, it is likely that the climate of this episode could have been interpreted as similar to that of the present-day tundra. It is evident that the climate immediately before this temperate interlude was of arctic severity but that a sudden amelioration had exterminated the cold-adapted beetles and permitted the incursion of temperate species before the trees had been able to reach these islands.

The second feature of interest in Fig. 20.3 is the prolonged period from 40 000 years ago up to 20 000 years ago when the climate of England was cold, with mean July temperatures at, or below, 10 °C, i.e. too cold for trees to grow. Mean January or February temperatures were of Siberian severity and show that the climate was of extremely continentality. These conditions are indicated by the presence in our fossil faunas of an ecologically diverse group of arctic and eastern Asiatic species, often as common components of our assemblages. For example, the commonest medium sized dung beetle in Britain during this period is now restricted to the Tibetan plateau (Coope, 1973). Inferences of severely continental climatic conditions are supported by independent geological evidence such as casts of thermal contraction cracks (ice wedges) of large size in the same deposits from which the fossil insects were obtained, indicating that mean annual temperatures were of the order of −8 °C.

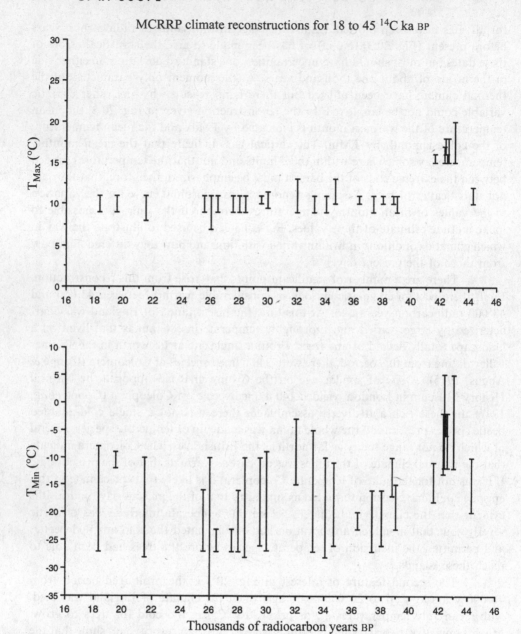

Fig. 20.3. Variations in the mean temperature of the warmest month and the mean temperature of the coldest month for central and southern England during the middle phase of the Last (Devensian, Weichselian) Glacial period reconstructed using the Mutual Climatic Range Method. Time is given in uncalibrated radiocarbon years Before Present (after Coope *et al.*, 1997).

How Britain came to have such a cold continental climate, when its geographic position must have been as oceanic then as it is today, is outside the scope of this chapter, but it was probably related to the changes in the temperatures of the surface water circulation of the adjacent north Atlantic (Ruddiman & McIntyre, 1976). For much of the glacial periods, the Gulf Stream (North Atlantic Drift) did not flow up the west coast of the British Isles as it does today, but turned southwards at about the latitude of central Portugal. During such times, an anticlockwise polar gyre on the ocean surface in the northern part of the North Atlantic would have carried with it masses of pack ice that would eventually have grounded along the west coast of Scotland and Ireland. Whatever the direction of the wind, whether off the ice-covered sea or off the ice-covered land, it would have been cold and dry, and there would have been little ameliorating effect from the adjacent ocean. It is not difficult to visualize Siberian climatic conditions as a consequence of this pattern of North Atlantic surface-water circulation. For at least the last half million years, this glacial circulatory pattern has been more usual than the one to which we have become accustomed today (Ruddiman & McIntyre, 1976). Ultimately, the old glacial insect fauna (or what is left of it after the extinction event of the present day) will return again to Britain and other temperate latitudes when 'normal' conditions are once more resumed.

20.6   CONCLUSIONS

Well-preserved fossil insects are very common in all sorts of waterlogged sediments dating from the Quaternary period. They provide intimate insights into the palaeoecology of terrestrial and freshwater environments and perhaps, more importantly, they provide palaeoclimatic information which has broader, even global, significance.

The fossil record shows that insect species have remained morphologically and probably physiologically stable in spite of, or even because of, the frequent, intense and sudden climatic changes of the Quaternary climate. During episodes of climatic change, many insect species managed to avoid the adverse consequences by changing their geographical ranges to track acceptable climates across the continents. In this way they were able to live in more or less constant environments but in ever changing geographical locations. This dynamic view of insect distributions means that populations were frequently split up or rejoined, so that there was very little chance of speciation. Paradoxically, species stability can thus be seen as one of the consequences of Quaternary climatic instability.

With the widespread appearance of human activities, insect populations are becoming isolated in a hostile landscape, making climatic tracking more and more difficult. Insect species may well find that the stratagem that enabled their ancestors so successfully to avoid the consequences of major climatic changes is no longer available. The response of insect species to global change in the near future, whether man made or natural, may be unpredictable. Under these novel circumstances, their past history may be a poor guide to their future response to such change. Only in the long run, when glacial conditions eventually return, will the old pattern of behaviour reassert itself and the insect faunas once more move in harmony with the global climatic rhythms.

21

Amphibians, reptiles and birds: a biogeographical review

ANGELA C. MILNER, ANDREW R. MILNER AND
SUSAN E. EVANS

21.1   INTRODUCTION

This chapter reviews briefly the apparent responses to global events by the amphibians, reptiles and birds during the past 145 million years. The vast majority of living members of these groups are terrestrial or freshwater (aquatic or amphibious) dwellers. Some 4956 species of amphibians are represented by 28 families of Anura (frogs and toads), ten families of Caudata (newts and salamanders) and five families of Gymnophiona (caecilians) (Cogger & Zweifel, 1998). Over 7427 living species of reptiles are represented by 13 families of Testudinata (turtles and tortoises), 48 families of Lepidosauromorpha (comprising 26 lizard families, *ca.*18 snake families, 4 amphisbaenid (worm lizard) families and a single relict rhynchocephalian genus, the Tuatara (*Sphenodon*)) and the Archosauria (Coggar & Zweifel, 1998). The last group includes the extinct Pterosauria and the wholly terrestrial Dinosauria, and is today represented by three families of Crocodylia (crocodiles and alligators) and by the birds, the most numerous terrestrial vertebrates with over 9700 species in more than 20 orders, more than 50% belonging to the Passeriformes (song birds) (Feduccia, 1996).

The geographical distribution patterns of these groups are intimately linked to the historical distribution of continental areas, and we can attempt to relate the patterns of their past and present distributions to the patterns produced by the global dynamics of plate tectonics. An understanding of the patterns of relationships within these (and other) groups of organisms is fundamental to attempts to interpret their vicariance and dispersal patterns. Modern distribution patterns of the groups can also provide some broad indications of past climatic conditions in areas where related fossil species are recorded.

The fossil record of terrestrial vertebrates is extremely patchy through this span of time and reflects the rarity of conditions and situation for postmortem burial and fossilization for terrestrial organisms and the stochastic effects of taphonomic overlay. The record is rarely so dense as to permit us to relate it to the dynamics of global events as they happen, but a phylogenetic framework allows us to relate evolutionary events to geographical history. Phylogenetic hypotheses of relationships of groups plotted against geological time using the earliest occurrences provide us with the latest possible branching points of stem-lineages,

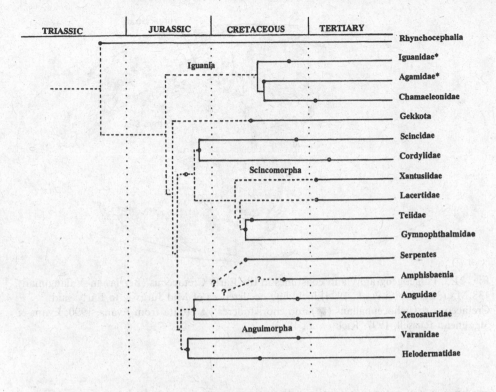

Fig. 21.1. Hypothesis of lepidosaurian relationships plotted against time using earliest occurrences (circles) to give latest possible branching points and lengths of stem lineages. It demonstrates that several major lineages must have evolved by the Mid Jurassic (Evans, 1993) although there are no pre-Cretaceous records of them. Four major lizard groups, Iguania, Gekkota, Scincomorpha and Anguimorpha are recognized together with Serpentes (snakes) and Amphisbaenia (worm lizards) which may be related to Anguimorpha. The common ancestor of all lepidosaurs must have been in the Mid Triassic. The stem lineages of several groups, which have a fossil record only in the Tertiary, originated in Early Cretaceous times. Asterisked taxa are paraphyletic (linked by primitive characters).

which may be much earlier than indicated by literal interpretation of the fossil record (Fig. 21.1).

21.1.1 Early Cretaceous faunas

We summarize here the broad global distribution patterns of amphibians, reptiles and birds in the Early Cretaceous as a starting point to reviewing their subsequent distribution patterns. By 145 Ma, Laurasia and Gondwana were probably substantially separate (Smith *et al.*, 1994), although corridors had existed until the end of the Jurassic (Fig. 21.2).

Frogs, salamanders and caecilians can all be shown, or deduced, to have been diversifying at this time. Cretaceous Laurasian and Gondwanan amphibian faunas were distinct with representatives of different frog and salamander families and stem-lineages, and with caecilians present in Gondwana only. In all, ten lineages

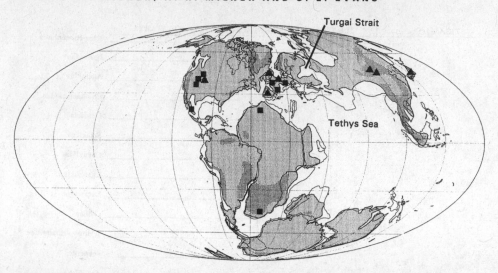

Fig. 21.2. Palaeogeography and coastlines in the Early Cretaceous (Berriasian–Valanginian), 138 Ma (after Smith *et al.*, 1994) including occurrences of Mid Jurassic to Early–mid Cretaceous rhynchocephalians (■) and choristoderes (▲) (data from Evans, 1990; Evans & Sigogneau-Russell, 1997; Rich *et al.*, 1983).

are known as Cretaceous fossils in the two respective areas. In Laurasia, the Turgai Strait had been present since the Callovian and was probably responsible for the vicariant separation of the cryptobranchoid salamanders (Hynobiidae and Cryptobranchidae) in East Asia from the remaining higher salamanders in Euramerica (Milner, 1983) a distribution which largely persists to the present.

Non-archosaurian reptile groups present in the Berriasian include the Testudinata (turtles) and Lepidosauria which survive through to the present; ichthyosaurs which last appeared in the Cenomanian, plesiosaurs and pterosaurs which last appeared at the end of the Maastrichtian and the choristoderes (an enigmatic group of crocodile-like aquatic reptiles including champsosaurs) which last appeared in the Oligocene.

Turtles had already diversified globally into two suborders by the Berriasian and a range of marine and freshwater families were present. Of the lepidosaur groups, rhynchocephalians were declining from a globally diverse peak in the Jurassic, the last record being from the Aptian–Albian of Mexico (Reynoso, 1995). There is then a gap of more than 100 my until subfossil records of the living Tuatara at 40 000 yrs BP in New Zealand. Part of the explanation for this long gap is, undoubtedly, the very limited fossil record of small vertebrates from southern continents, but a Cretaceous extinction of rhynchocephalians in Laurasia seems a reality. There is no obvious correlation with climate change, and a more probable explanation may be competitive displacement by agamid lizards with shearing teeth which were apparently diversifying in East Asia in the Early Cretaceous (Alifanov, 1993). It may be significant that relict rhynchocephalians survive today only in New Zealand, one Old World land area that appears to have separated from Gondwana before agamid lizards reached it. A more complete Cretaceous record from Gondwana (Fig. 21.2) is needed to trace the pattern of extinction and replacement.

Phylogenetic considerations indicate that choristoderes must have been present since the Late Permian (Evans, 1990; Storrs & Gower, 1993) but they first appear in the fossil record in the Late Triassic of Europe and were present in low diversity through the Jurassic and Early Cretaceous of Laurasia as small aquatic predators (Fig. 21.2).

The biogeographical history of archosaurs is one of cosmopolitan distribution across Pangaea in Late Triassic to Early Jurassic times with increasing endemism and dispersal events as Pangaea fragmented, with the ensuing break-up into the isolated Southern Hemisphere continents and the reassembly of a boreal Laurasian supercontinent in the Cretaceous (Fig. 21.3).

Pterosaurs were present in their greatest diversity in the latest Jurassic and Early Cretaceous, but declined in diversity through the Cretaceous. Some families were widespread and there is no evidence for Laurasian–Gondwanan endemism. Crocodiles were present in great diversity in the Berriasian, with a predominance of forms of the 'mesosuchian' grade of organization, plus a few relicts of earlier radiations. Although one family was globally distributed (Buffetaut & Taquet, 1977), Laurasian and Gondwanan (Buffetaut, 1982a) assemblages were evident. Modern crocodiles probably arose from within the Laurasian assemblage and did not appear until later in the Cretaceous. The dinosaur record is comparatively sparse but includes representative of the major divisions; theropods (bipedal carnivores), sauropods (large, long-necked herbivorous quadrupeds), and ornithischians (herbivores including hadrosaurs, horned and armoured dinosaurs); like the crocodile pattern, it reflects the existence of Jurassic connections. East Asia had been isolated from the remainder of Laurasia since the Mid Jurassic by the Turgai Sea. Groups otherwise widely distributed across Laurasia are absent, and endemism occurred in East Asia, including the psittacosaurs (primitive ceratopians or horned dinosaurs), known only from the Aptian–Albian (Sereno, 1990), and the club-tailed armoured dinosaurs (Ankylosauridae: Tumanova, 1983).

The vicariant origin and diversification of a Gondwanan archosaur fauna has been increasingly recognized (Buffetaut & Rage, 1993 and references therein) with diversifying endemic groups of crocodiles, theropod (Bonaparte, 1986) and sauropod (titanosaurs, Hunt et al., 1994) dinosaurs (Fig. 21.3). Sauropods were the dominant herbivores in Gondwana, which was isolated from the spread of advanced hadrosaurs (duck-billed dinosaurs), their ecological counterparts in Late Cretaceous Laurasia. One clade of large theropods, paralleling the large Laurasian theropods, achieved a transcontinental Gondwanan distribution and subsequent isolation before the end of the Early Cretaceous (Sereno et al, 1996).

The dominant birds of the Cretaceous were the enantiornithines, a radiation of diverse forms with well-developed flight apparatus but primitive pelvis and hind limbs. This sister group to modern birds was cosmopolitan in the Early Cretaceous; modern-type birds were also present in Laurasia by that time, but much of the history of birds is poorly represented by the fossil record (Chiappe, 1995).

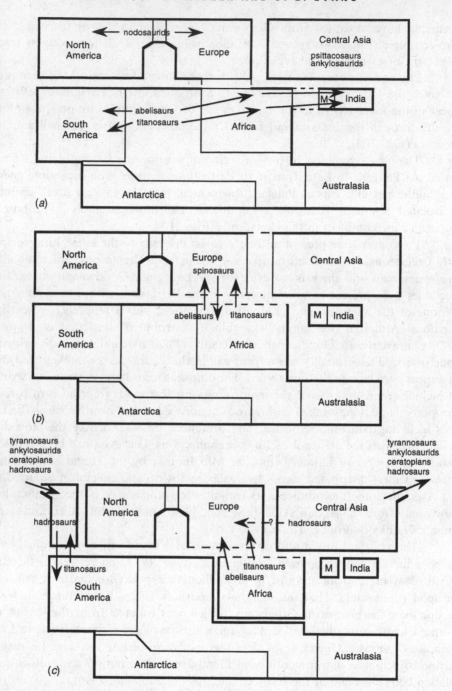

Fig. 21.3. Endemism and dispersal of selected dinosaur groups in Laurasian and Gondwanan continental faunal blocks in (a) Late Jurassic–Early Cretaceous times, with isolated west Laurasia and Central Asia, (b) late Early Cretaceous times, with a Eurasian connection and intermittent corridors between Europe and Africa, and (c) Late Cretaceous times, with the Bering Corridor linking Asia and Western North America. M = Madagascar (modified from Russell, 1993, based on data from Smith et al., 1994).

21.2   CRETACEOUS EVENTS

21.2.1   The fragmentation of Gondwana

The general pattern

In the Southern Hemisphere, the supercontinent of Gondwana was breaking up into component continents which possess characteristic faunas to this day. Some, though not all, of the modern endemism, and hence, for example, modern frog and caecilian diversity was undoubtedly due to the series of isolation events produced by the Cretaceous break-up of Gondwana into New Zealand, Africa, Madagascar, Seychelles, India and the South America + Antarctica + Australia continent. New Zealand (with a much earlier Jurassic separation date) is characterized by a relict primitive frog family and, the remaining components of Gondwana by assemblages of families and subfamilies that are endemic to Africa, Madagascar, the Seychelles, South America and Australasia. Other families show more cosmopolitan distribution, but most can be attributed to one continent of origin with only a few successful genera having a wider range, e.g. ranid frogs and Old World tree frogs (originating in Africa), bufonid toads and hylid tree frogs (originating in South America).

The modern (mostly flightless) ratite birds are restricted to the southern continents and have a poor fossil record. Their distribution is most parsimoniously accounted for by vicariance of an ancestral monophyletic group in which flightlessness arose once (see Cracraft, 1974 and references in Feduccia (1996) for alternative multiple origin and dispersal hypotheses). Primitive flying tinamous and flightless rheas are found in South America, ostriches in Africa, cassowaries and emus in Australasia, kiwis (and formerly moas) in New Zealand, while elephant birds were present in Madagascar until recently. Most, if not all evolved in isolation, as the ancestral biota became separated with the fragmentation of Gondwana.

The separation of the Madagascar–Seychelles–India block

A Madagascar–Seychelles–India block apparently separated from the rest of Gondwana at about 140 Ma. Madagascar became isolated from India by the Coniacian at the latest (Smith *et al.*, 1994). Its Campanian fauna includes Gondwanan snakes (Scanlon, 1993) and dinosaurs (Sampson *et al.*, 1996) including new material of a dome-headed theropod, a previous fragment of which had been confused with an ornithischian group (pachycephalosaurs) known only from Laurasia. This recent discovery removes an hitherto puzzling biogeographical requirement for a prolonged link between Laurasia and Gondwana. The timing and sequence of the ensuing Madagascar–Seychelles–India break-up is a matter of controversy but very important to our understanding of the origin and diversification of Asian herpetofaunas. There seems a general consensus that separation of these continents and fragments occurred in the mid Cretaceous but the original 'slow' hypothesis by which India reached Asia in the Eocene now competes with a 'fast' hypothesis, which argues for an India–Asia suture in the latest Maastrichtian. The presence in India of a Maastrichtian spadefoot toad – otherwise a characteristically Laurasian family – has been used to argue for this Cretaceous contact (Sahni *et al.*, 1982; Jaeger & Rage, 1990) with some criticism (e.g. Thewissen, 1990). Madagascar and the Seychelles both possess endemic frog assemblages, and the

Seychelles have endemic caecilians (curiously Madagascar does not), so it can reasonably be inferred that India carried an endemically modified Cretaceous Gondwana fauna. The Maastrichtian snake (Werner & Rage, 1995), and dinosaur fauna (Hunt *et al.*, 1994; Bonaparte & Novas, 1985) is characteristically Gondwanan. The amphibian fauna interacted with the Asian fauna after contact, perhaps in the latest Cretaceous, but certainly in the Early Cenozoic. Among the India–Asian groups with this attributed history are three caecilian and three frog families (Duellman & Trueb, 1986).

The break-up of the South America–Antarctica–Australia block

Over the Late Cretaceous to Paleocene, the final break-up of Gondwana took place, Australia separating from Antarctica between the Santonian (85 Ma) and the Paleocene (60 Ma) and Antarctica separating from South America at about 60 Ma. There is as yet no evidence of the timing of the development of endemism in the amphibian fauna but it seems likely that the South America and Australian faunas were already distinct in the Late Cretaceous, possibly separated by climatic barriers. The Early Paleogene Australian fauna is unknown but immunological data suggests that the endemic genera were diversifying in the Late Cretaceous (Daugherty & Maxson, 1982).

21.2.2   Corridors

In the Campanian (*ca.* 80 Ma), the Bering Corridor between East Asia and North America became established for the first time, with the result that Central Asia and western North America were effectively one biogeographical area (Russell, 1993) bounded in the east by the Mid-Continental Seaway (Fig. 21.3 and 21.4). The Bering Corridor, combined with the Late Cretaceous cooling, appears to have permitted *Andrias*, a genus of 'giant' Asian salamander, to extend its range into North America. Giant salamanders tangibly appear in the fossil record of North America and East Asia in the Paleocene, but all the other cryptobranchoids are Asian, so the North American material must represent the range extension (Milner, 1983). The group persists in North America to the present day as the genus *Cryptobranchus*. The Bering Corridor also permitted several endemic Asian dinosaur groups (some listed in Fig. 21.3(*c*)) to invade western North America and diversify, reaching their zenith in the Campanian. No dinosaur family is known to have arisen in the Cretaceous of North America, perhaps as a result of the relative sizes of the North American and Asian landmasses (Russell, 1993). There were also similar lizard faunas in the two continents.

The presence of a periodic emergent land link between Africa and Europe (Fig. 21.3), facilitated terrestrial faunal interchange in the Barremian to Cenomanian, via a western Spanish route and an eastern Israel–Croatia–Trieste corridor (Rage, 1988; Tchernov *et al.*, 1996). Gondwanan dinosaurs and crocodiles are present in the Campanian–Maastrichtian archipelago of southern Europe (Le Loeuff & Buffetaut, 1991).

The short-lived Late Cretaceous–Paleocene Panamanian connection permitted a very limited north–south faunal interchange between North and South America. There is no firm evidence of amphibian involvement. A snake

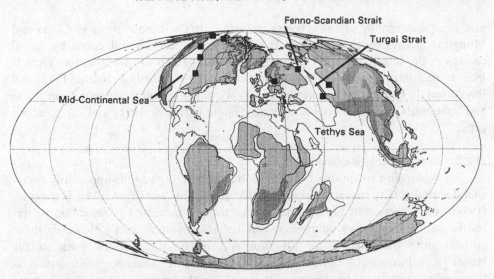

Fig. 21.4. Palaeogeography and coastlines in the Late Cretaceous (Campanian), 80 Ma (after Smith *et al.*, 1994) including occurences of flightless hesperornithform birds around epicontinental coastal margins.

(Coniophis) is present in both North and South America (Estes & Baéz, 1985); two Campanian–Maastrichtian hadrosaurs occur in South America (Bonaparte *et al.*, 1984) and Gondwanan titanosaurid sauropods moved as far north as Wyoming in the Maastrichtian (Lucas & Hunt, 1989). Madtsoiid snakes (as well as similar crocodiles and turtles) in Africa and Brazil (Gayet *et al.*, 1992) provide evidence of at least sporadic east–west migration between major land masses.

21.2.3   Climates

Little can be deduced about the influence of climate change on the vertebrate assemblages although local changes may reflect global events. Changing faunal patterns in the herpetofaunas in Europe may correlate with a change from arid to humid climates in the non-marine basins of Atlantic coastal Europe in the Valanginian to Hauterivian with a reversion to semiarid conditions in the Late Hauterivian to Barremian (Gale, this volume). There are, however, no herpetofaunas recorded from outside Europe at this time.

In Asia, prevailing warm humid Aptian–Albian environments with widespread lacustrine areas may have promoted the radiation of typical gavial-like choristoderans; the development of Cenomanian to Campanian mid-latitude arid zones in Asia might explain their diasappearance. A climate reversal to more humid conditions in the Campanian of the Northern Hemisphere (Gale, this volume) might be correlated with the appearance of a more modern Asian lizard, snake and amphisbaenian assemblage, dominated by small insectivorous/omnivorous forms including gekkos and skinks (Borsuk-Białynicka, 1991: Gao & Hou, 1995).

The global distribution of dinosaurs from coastal floodplains (Berriasian–Barremian, Europe; Campanian–Maastrichtian, western North America) to semiarid or arid internally draining basins in early Late Cretaceous times in China and Mongolia (Jerzykiewicz & Russell, 1991) was not detectably influenced by global cooling in the Campanian and Maastrichtian. Campanian to Maastrichtian records from both polar regions, the North Slope of Alaska (Clemens & Nelms, 1993) and the Antarctic Peninsula (Hooker *et. al.*, 1991), together with evidence from floras and palaeosols suggest a cool temperate climate (Francis, 1991) and no polar ice caps.

21.2.4   Eustatic sea-level changes

According to Haubold (1990), global eustatic highstands apparently correspond with diversity peaks in dinosaur distribution. This was regarded by Hunt *et al.* (1994) as a reflection of taphonomic bias as there is increased preservation of terrestrial sediments during transgression and high stand system tracks. However, those authors note that the model fits sauropod distributions which peak in the Hauterivian–Barremian, Cenomanian and Campanian. Sauropod distribution is temporally disjunct in the western interior of North America, where they became extinct at the end of the Albian, reappearing in the Maastrichtian. This extinction, the 'sauropod hiatus' might be related to a major Late Albian regression (Lucas & Hunt, 1989).

During the Albian, marine transgression produced the North American Mid-Continental Sea separating North America into eastern and western regions and it seems likely that North American higher salamanders underwent diversification by vicariance at this time (Milner, 1983). Marine regression in the latest Cretaceous (65 Ma) resulted in the withdrawal of the Mid-Continental Sea of North America and the reunification of eastern and western North American faunas. Milner (1983) has suggested that some Pacific region salamanders are relict endemics of the western fauna, while many modern eastern salamanders are endemic relicts of the eastern fauna. Other initially eastern salamanders later extended their ranges across North America to become pan-continental taxa.

Four major transgressive–regressive cycles took place in North America in the Late Cretaceous. Faunal studies at the Cenomanian–Turonian boundary, during the Campanian, and at the K–T boundary (references in Archibald, 1996a) point to the conclusion that the transgressions and regressions, resulting from global eustatic sea-level change, may have been forcing factors in dinosaur evolution and extinction. The major late Maastrichtian–Paleocene regression when the Mid-Continental Seaway effectively drained, is implicated by Archibald (1996a and references therein) in the decline and extinction of dinosaurs through the resulting habitat fragmentation. That regression also coincides with the extinction in the Maastrichtian of two modern-type bird groups, the tern-like ichthyornithiforms and the hesperornithiforms, flightless foot-propelled divers. Both were widely distributed along the coasts of the Mid-Continental Sea, hesperornithiforms also extended as far as Kazakhstan, the Turgai and Fenno-Scandian Straits, and southern Sweden (Kurochkin, 1995) (Fig. 21.4). Their extinction is presumably connected with the loss of epicontinental habitats which may also have been a cause of the extinction of sea-going pterosaurs such as *Pteranodon*; only the giant azhdarchids are recorded in the late Maastrichtian and none survived beyond it (Wellnhofer, 1991).

21.2.5   The K–T extinction event

Archibald and Bryant (1990) studied the fate of vertebrates across the K–T boundary in Montana and concluded that extinctions were very selective. The events at the K–T boundary had no detectable effect on the amphibian fauna, as far as the fossil record permits us to determine this. Lizards were affected more substantially and most extinctions were among the Teiidae (whiptails and race-runners) (Gao & Fox, 1991). Apparently teiids became extinct in North America and in Asia at this time, possibly due to a period of cooler temperatures (living teiids being tropical–subtropical animals). However, crocodiles passed though the K–T boundary although large forms disappeared; crocodiles are relatively temperature sensitive, which argues against significant climate deterioration. Turtles, choristoderes and most squamates survived without major effect. There are a number of possible reasons – the ability of all of these groups to cope with a drop in ambient temperature by reducing their metabolic rate; the possibility that some may have been cushioned from environmental change by their freshwater environment; and the potential for small reptiles like lizards to creep into refugia with more tolerable microenvironmental conditions. Apart from the teiids, the only lizard groups to suffer perceptibly were the marine mosasaurs. Sullivan (1987) reported their latest appearance as early Maastrichtian but recent work on Late Cretaceous mosasaurs shows that at least three families persisted into the latest Maastrichtian (G. Bell, personal communication), but none into the Tertiary. Mosasaurs were the top carnivores of marine ecosystems and any event which cut the overall productivity of the lower trophic levels would have affected mosasaurs as well. The Maastrichtian extinction of long-necked plesiosaurs might also be linked to a downturn in marine productivity. That said, it is interesting that large marine turtles were not affected to the same degree (R. Hiragama, personal communication), neither were the marine dyrosaurid crocodiles. Buffetaut (1990) attributed the latter to adaptation to estuarine habitats; dyrosaurids continued to diversify around the proto-North Atlantic and Tethyan coastlines until the Mid to Late Eocene.

The most obvious and dramatic extinction was that of the non-avian dinosaurs but some avians vanished too; enantiornithine birds have no record beyond the Maastrichtian and some modern-type birds (besides the seabird orders mentioned above) also died out. Dinosaur diversity declined steadily from a peak of 45 genera in the Late Campanian to 24 in the late Maastrichtian and 7 at the K–T boundary. The late Maastrichtian record is more consistent with a steep and accelerating decline than with a sudden and catastrophic one (MacLeod et al., 1997, and references therein). Great debate continues as to the causes of the terrestrial K–T extinctions in western North America. Archibald and Bryant (1990) and Archibald (1996a) argue that habitat fragmentation, caused by global sea-level fall, was the primary cause; many others support the Chicxulub bolide impact, Deccan Traps volcanism or a combination of the two (references in Sharpton & Ward, 1990; MacLeod et al., 1997) as agents of sudden climate deterioration, perhaps superimposed on longer-term changes (see also Gale and Pickering, this volume). We do not know if the patterns of dinosaur decline and extinction seen in North America were repeated on a global scale since there are no continuous terrestrial depositional sequences elsewhere in the world.

21.3   PALEOGENE EVENTS
21.3.1   Early Paleogene geographical events
Continental events

After the fragmentation of Gondwana in the Cretaceous, several reconnection events occurred in the Paleogene, forcing pairs of faunas into contact with one another. The exact status and duration of some of these corridors is a matter of some uncertainty and there are alternative explanations for the appearance of some taxa in particular areas. An influx of 'South American' taxa in the Eocene of Europe has been attributed to dispersal via North America or via Africa; likewise the appearance of new taxa in South America at this time has been attributed to both routes. There is similar uncertainty about putative pathways between Europe, Asia and Africa, and the following section should be read with this in mind. Some regions appear to have had relatively restricted herpetofaunas after the K–T event and before Paleocene–Eocene range changes occurred. The depauperate European Paleogene squamate fauna appears to have contained only lacertid and necrosaurid lizards and boiid snakes (Rage & Auge, 1993).

South America and North America

One of the first new corridors was an ephemeral connection between South and North plus Central America (Savage, 1982) which may have been Cretaceous but certainly must have happened by the Paleocene. This would have permitted entry into Central America of South American caecilians and frogs. There is no direct fossil evidence of this, but it is a necessary postulate in order that (i) various molecular divergence dates of North and South American tree frogs could be attained (e.g. Maxson, 1976), (ii) bufonid toads and hylid tree frogs could cross Asia and reach Europe by the Grand Coupure in the Late Oligocene (Blair, 1972; Sanchíz, 1977); and (iii) some frog lineages could reach the West Indies prior to their isolation. The reptile faunas provide little evidence of north–south interchange during this period although iguanid lizards occur in the Paleocene of Brazil and may have entered Central America at this time (Estes & Baéz, 1985).

South America and Africa

The possibility of a late connection between South America and Africa in the Paleogene continues to tantalize. It appears to be a biogeographical necessity although geographically unlikely, but there is possible temporary route via the Walvis and Rio Grande rises which seem to have been emergent at least in the Late Cretaceous (Rage, 1988) (Fig. 21.5). The proposed position of the endemic South American dendrobatid (arrow-poison) frogs within the fundamentally African ranoid complex (Ford & Cannatella, 1993) implies a South American–African link, which has also been invoked for the similarity of fossil pipid frogs in the Paleogene of the two continents (Rage, 1988; Buffetaut & Rage, 1993). Several apparent South American endemic vertebrate groups might conceivably have dispersed from South America to Africa and thence to Europe at this time as they appear in the European Eocene but not in the North American Eocene. These include a leptodactylid frog (Roček & Lamaud, 1995), tropidopheine snakes, zipho-

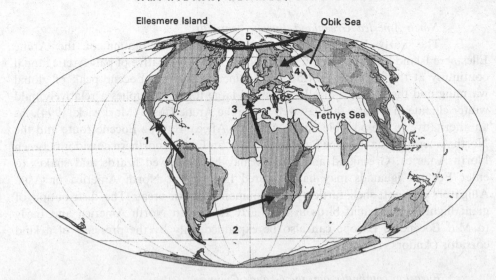

Fig. 21.5. Palaeogeography and coastlines in the Early Eocene (Ypresian), 53 Ma (after Smith *et al.*, 1994) indicating major Palaeogene dispersal corridors. 1, South America – North + Central America; 2, South America – Africa via emergent Rio Grande and Walvis rises; 3, Africa – Europe Gibraltar corridor; 4, Europe – Asia (during periods of withdrawal of the Obik Sea); 5, Pan-Arctic North America – Greenland – Europe link.

dont crocodiles, and flightless giant rails (phorusrhachids) for which the earliest record is Late Paleocene in South America (Mourer-Chauviré, 1981). The occurrence of a Mid Eocene terrestrial ziphodont crocodile in Algeria (Buffetaut, 1982*a*, *b*) provides tangible support that endemic South American groups reached Europe via Africa in a Late Cretaceous or Paleocene exchange (Fig. 21.5).

Europe, Africa and Asia

During the Paleogene, the increasing proximity of Africa to Eurasia brought increasing development of corridor connections but this is only observable in the herpetofaunas of Europe, those of the Paleogene of Africa being very poorly known. Two groups which appear to have evolved in Africa and dispersed to Europe in the Eocene are cordyline lizards and ranid frogs. Ranids make their first appearance in Europe in the Bartonian, Late Eocene (Rage, 1984). Rage attributed their origin to Asian input when a brief withdrawal of the Obik Sea took place in the Paleocene permitting an influx of Asian geckos and agamids into Europe, followed by a second wave in the Early to Mid Eocene (Rage & Auge, 1993). However, because (i) the ranoid complex appears to be an African endemic clade, and (ii) there is evidence (see above) of other vertebrate groups moving between South America and Europe via Africa, an African origin for the European ranids seems likely. The Gibraltar corridor between Europe and North Africa became more firmly established as the Paleogene progressed and there can be assumed to have been a shared fauna between the two regions up to the Late Miocene when the Gibraltar Strait formed (see below).

North America–Greenland–Europe continuity

The Early Eocene (Ypresian equivalent) herpetofauna of the Arctic Ellesmere Island locality (Estes & Hutchison, 1980) is indicative of pan-Arctic faunal continuity at this time (Fig. 21.5). This coincides with an Eocene peak of global warming and the presence of alligators and large tortoises implies a relatively mild winter climate in an area that lay well inside the Arctic Circle (Markwick, 1994). As an alternative to the possible South America–Africa–Europe Eocene route and the North America–Asia–Europe Paleocene route, an Eocene northern corridor between North America, Greenland and Europe may have allowed lizards and snakes to enter Europe; iguanids may have entered from either North America or Asia. Alligators and soft-shell turtles were common to both areas. The distribution of giant flightless diatrymid birds in the Early Eocene in North America and Early to Mid Eocene in Europe can also be explained only by the presence of a land corridor (Andors, 1992).

Eurasian continuity and the Grande Coupure

In the Early Oligocene, the Obik Sea, a late development of the Turgai Strait, dried out and the East Asian and European faunas were placed in sustained contact. A massive interchange took place and this, and the associated extinctions, are known as the Grand Coupure, a very obvious faunal turnover. East to west range extensions marked by sudden appearances in the Oligocene–Miocene of Europe are the cryptobranchid salamander *Andrias* (Böttcher, 1987), bufonid toads (Sanchíz, 1977), hylid frogs (Noble, 1928), and new varanid, agamid and iguanid lizards (Alifanov, 1993). West to east range extensions included two lineages of salamander which reached China and a further salamandrid lineage (the stem of the redbelly–red spotted newt clade) which crossed the still open Bering Corridor to reach North America by the Late Oligocene (Naylor, 1979, summarizing earlier work).

21.3.2   Paleocene–Oligocene climate deterioration

Our knowledge of Paleocene small vertebrate assemblages is still limited with depauperate faunas in Europe and Asia. Alligators and soft-shell turtles, indicative of warm conditions, occur in both Europe and in North America, where the squamate assemblage is more diverse (or better sampled).

In the Eocene, faunal changes in the North American Interior may be correlated with two sharp cooling periods in otherwise wet, tropical conditions. Large choristoderes in both North America and Europe may have been casualties of a temperature decline in the Watsatchian–Bridgerian interval (50 Ma) (Prothero, 1994). Thermally intolerant turtles and crocodiles reportedly also show reduction in diversity (Hutchison, 1982).

The Early to Mid Eocene global high temperatures reflect the wide diversity of repiles restricted to subtropical to tropical conditions. Crocodiles and turtles reached their greatest diversity in North America; marine turtles, aquatic snakes, crocodiles and alligators were present in Europe and North Africa.

The slow climate deterioration from the Mid Eocene onwards involving both cooling and seasonality involved a series of gradual changes in the fauna, most of which did not affect the amphibians at a gross taxonomic level. Two families of salamanders and four families of frogs which were present in the Late Eocene are still present today. However, it has been suggested (Roček, 1996) that Oligocene cooling did produce a developmental response in some amphibian lineages leading to the evolution of neotenous salamander lineages from metamorphosing ones. The fossil evidence is contradictory at a generic level but the broad principal, that there may be a phase of neoteny associated with climate deterioration, notably a shorter growing season, is an interesting and plausible interpretation of the known record.

As far as squamates are concerned, Rage and Auge (1993) have estimated that between 66 and 80% of the lizard and snake species may have been lost in Europe over this period, although this is less obvious at the family level. However, aquatic reptiles present in the Mid to Late Eocene in Europe, and which still live today in mild to subtropical climates are indicative of at least mild conditions; soft-shell and pond turtles, alligatorids and the less cold-tolerant crocodylids are recorded from the European and Chinese Oligocene. On a global scale, the temperature change may have contributed to the virtual extinction of Old World iguanid lizards (already under pressure from diversifying Asian agamid lizards). In the New World, however, iguanids remained successful despite temperature changes, perhaps aided by lack of competion from modern agamids which never reached it.

21.4   NEOGENE EVENTS
21.4.1   Geographical events
The Mediterranean Sea
Between 7.0 and 5.5 Ma, the formation of the Strait of Gibraltar allowed water from the Atlantic Ocean to fill what is now the Mediterranean Sea, and simultaneously to remove a corridor between Western Europe and North Africa. Prior to this, the European herpetofauna had extended south along the Atlas mountains as far as Tunisia and the formation of this barrier has initiated a speciation event. As well as three endemic North African species of frogs and salamanders, there are four trans-Gibraltar Straits species which have been shown to be genetically heterogenous, using electrophoretically obtained allozymic data (Busack, 1986).

The North America–South America link
The first strong evidence of north–south faunal interchange occurs in the Miocene with the appearance of several turtle, crocodile and lizard species common to North and South America. Climatic amelioration may have resulted in the re-entry of teiid lizards to North America (though only the more environmentally tolerant ones) and an increased diversity of snakes. However, there was no land connection through the Central American isthmus in Eocene to Miocene time (Savage, 1982) so 'waif dispersals' may have occurred involving Caribbean and Central American routes (Estes & Baéz, 1985).

In the Pliocene, the Panamanian Isthmus made substantial contact with South America and the 'Great American Interchange' commenced. The amphibian component of this comprised the southward range extension of two genera of lungless salamanders (Wake & Lynch, 1976) and one species of frog (Duellman & Trueb,

1986). Numerous South America frog genera have extended north into Central America, none getting further than the arid regions of northern Mexico. The last record of giant flightless carnivorous rails (phorusrhachids), which diversified in isolation in South America in the absence of mammalian carnivores, coincides with the establishing of the Panama isthmus in the Late Pliocene. Their extinction is presumed to be linked to the invasion of carnivorous mammals from North America. However, a phorusrhachid from the Late Pliocene of Florida provides evidence of limited northwards dispersal (Chandler, 1994). After the Early Miocene, sequential invasions of advanced passerine birds passed across the Bering Corridor from Asia into North America and diversified to produce the New World assemblages, some groups (e.g. tanagers) dispersing on into the Neotropical region across the Panama isthmus (references in Feduccia, 1996). Despite this invasion from the north, primitive passerine diversity has remained high in South America for geographical and climatic reasons; new habitats created by the Andean orogeny, and refugia in the Amazon Basin during period of fluctuating aridity and humidity (Haffer, 1974).

Australian proximity to south-east Asia

The Late Oligocene–Miocene Riversleigh locality provides the first record of lizard diversity in Australia. Around this time, Australia completed its northward movement and made contact with Asia, permitting an exchange of faunas. On the basis of modern and fossil herpetofaunas, Cogger and Heatwole (1981) proposed that the Australian reptilian fauna arrived in two major phases – the first (including archaic skinks, Gondwanan madtsoiid snakes and chelyiid turtles) by a southern route when Australia was still part of Gondwana, and the second (agamids, geckos and monitors) from the north when Australia contacted Asia. Without a Mesozoic record, however, this hypothesis cannot be tested.

21.4.2   Miocene climates

In the Late Oligocene to Early Miocene in North America, the restriction of crocodilians to south-eastern maritime localities (Markwick, 1994) is consistent with a period of relative aridity at that time in the Rocky Mountains–Great Plains, with the fluvial and lacustrine environments of the Early Tertiary replaced by subhumid to dry environments with episodic or seasonal periods of moisture (Hutchison, 1982). This led to the reduction or extinction of aquatic reptiles and amphibians; crocodylids were lost but the more thermally tolerant alligators remained. The mid-continent drying also left several salamander genera represented by disjunct species at the east and west sides of the continent (Wake, 1966).

Crocodile faunas never regained their former diversity either in North America or in Asia and the most northerly extant crocodylians are *Alligator mississippiensis* in North America and *A. sinensis* in China. This combination of cooling and drying also affected Europe and might explain the final extinction of the choristoderes (the last from the Late Oligocene of France, Hecht, 1992).

The Early Miocene saw a return to tropical conditions in Europe and North America marked by a northward re-extension of alligators into the continental interior (Markwick, 1994) and the presence in Europe of birds which today are restricted to sub-Saharan Africa (Ballman, 1969), new monitor lizards (*Varanus*) from Africa or Asia, chamaeleons (from Africa), viperid snakes (? from Asia) and elapid snakes (? from Asia or Africa).

21.4.3   Pliocene cooling and Pleistocene glaciations in northern continents

During the Late Miocene to Pliocene, northern climates began to cool again. In Europe, this led in the Pliocene to the loss or severe southern restriction of many less temperature-tolerant lizard and snake families (Rage & Auge, 1993), as well as crocodiles and most turtles. North America presents a similar picture with a decreased diversity of turtles and crocodiles controlled in part by seasonal aridity but finally in the Pleistocene by critically lower temperatures of the glacial periods. This eliminated large land tortoises and restricted the alligators' range (Hutchison, 1982).

The North American Pleistocene amphibian record is rich and there appear to have been few amphibian extinctions (two frog species only – Holman, 1995), possibly because there are no natural barriers to north–south range alterations and most taxa could move south and north as the climate permitted. Holman has argued that, in the southern region of North America, there was considerable stability with only some northern forms entering and withdrawing during glacial times. It is possible that mountain ranges with a north–south component such as the Appalachians and the Rockies provided habitats and/or refugia at different altitudes such that northern amphibian could extend their ranges into areas where the general lowland climate was not extreme. Climatic deterioration and competition from advanced passerine birds, were factors in the retreat southwards into the tropics of North American primitive passerines (references in Feduccia, 1996). The loss of albatrosses and some shearwaters from the North Atlantic at the end of the Pliocene (Olson, 1985) may also be climate linked. Distribution patterns of other seabirds have been allied to changes in global ocean currents. The shift in puffin and razorbill distribution in the Pacific, from Japan to Oregon in the Eocene–Early Miocene, to southern California in the Pliocene, is correlated with changes in currents associated with the thermal isolation and refrigeration of Antarctica (Warheit, 1992). Rich seabird deposits in the Pliocene of Florida, including a cormorant close to a modern species restricted to a cold-water upwelling in the Pacific, are thought to coincide with a similar upwelling and highstand in the Panamanian region (Elmslie, 1995).

During glacial phases in Europe, the ranges of most amphibians contracted southwards into the Iberian Peninsula, Italy and the Balkans. Aquatic families that were restricted to particular watersheds and less able to alter their range became extinct, notably the palaeobatrachid frogs which last appear in the Mid Pleistocene (Mynarski, 1977) and the European populations of the cryptobranchid salamander *Andrias* which last appeared in the Late Pliocene (Westphal, 1967). Subsequent to the most recent glaciation, some genera (*Rana, Triturus*) have re-extended their ranges across north and central Europe, while others, restricted by barriers such as the Pyrenees and the Italian Alps have remained in southern Europe.

21.5   CONCLUSIONS

Global plate tectonics is one of the major driving force of the changing biogeographical distribution patterns of amphibian, reptile and birds from the

Early Cretaceous to the present. Because of the more complete geographical pattern information available from modern faunas, it is the most susceptible to analysis and some progress has been made. Endemism, dispersals and extinctions of faunas are affected by plate tectonically driven isolation and reconnections of land areas, by land barriers created by orogenic events, and also by global eustatic changes. The sporadic nature of the fossil record allows only limited correlation with global climate change. Nevertheless, modern climatic distribution patterns and thermally sensitive climatic indicators provide a key to interpreting past climates and their influence on distribution at a broad level.

22

Paleogene mammals: crises and ecological change

JEREMY J. HOOKER

22.1   INTRODUCTION

The distribution and evolution of mammals over the past 145 million years show an interesting diversity of patterns. The two main physical parameters that appear to have controlled these patterns are the presence or absence of land bridges between land masses and climate change. A catastrophic event, possibly resulting from either volcanism or extraterrestrial impact, may also have played a part. Emergence of a land bridge may be the result of local tectonic uplift, but more widespread land corridors can be created by eustatic lowering of sea level. Climate change is a critical factor in the latitudinal movement of vegetation zones and in shortening or lengthening the annual period of food availability. Mammals have responded to these parameters of global change by evolving new ecological strategies and/or by shifting their geographical ranges. Long-term climate changes affecting annual ranges of temperature and thus determining seasonality vs. equability may be an important cause of community evolution (Janis, 1989; Collinson & Hooker, 1987, 1991). The absence of land connections may prevent or inhibit geographical range extensions and cause widespread endemism (vicariance). An example of this is the independent evolution of several now extinct mammalian orders in the Tertiary of South America, which until the Late Pliocene was an island continent (Pascual *et al.*, 1985). The effects of tectonics and climate are thus closely linked, but their variable interrelationships may be expected to produce different faunal patterns in the geological record, which are capable of interpretation.

This chapter first reviews the Cretaceous to Paleogene history of the mammals, and then considers the response of Paleogene mammals to climatic fluctuations and sea level changes.

22.2   BACKGROUND: THE FIRST 95 MILLION YEARS

22.2.1   The terrestrial record

Mammalian faunas in the Cretaceous were fundamentally different from modern ones. Although having diversified in the 80 my since their origin in the Late Triassic, mammals remained small, the largest being only the size of a mongoose. The three major modern groups, marsupials, placentals and the more primitive monotremes, differentiated early in the Cretaceous, but no modern orders of placentals are recognized. Thus, by the Albian, the first monotremes had appeared in Australia, evidence of a long, but sparsely documented, history virtually restricted to the Australasian continent (Flannery *et al.*, 1995). The first placentals are recorded from the Aptian/Albian of Uzbekistan and Mongolia (Nessov *et al.*, 1994; Kielan-Jaworowska & Dashzeveg, 1989) and they seem to have been endemic to Asia until quite late in the Cretaceous. The first unquestioned marsupials are from the Cenomanian of North America (Eaton, 1993), there being no unequivocal records of the group outside that continent until the beginning of the Cenozoic (de Muizon, 1994). Alongside these modern groups, several extinct ones flourished, such as the Multituberculata and Triconodonta. They have a stem position with respect to the marsupials and placentals, although their exact relationships within the Mammalia are still far from certain.

Faunal composition at the supraordinal level on the different continents in the latest Cretaceous was very different from today. Asia was dominated by placentals, North America was dominated by marsupials with less diverse and abundant placentals, whereas there is no evidence for any advanced mammals (marsupial or placental) existing at this time on any of the southern (Gondwanan) continents except India, although they may have existed in Africa where there is no Late Cretaceous mammal record. Multituberculates are known from all continents except Australasia and Antarctica and were generally abundant and diverse except in South America. World distributions of mammals in the Cretaceous are still very imperfectly known, but the restriction of some groups to particular continents is likely to be related to the continued fragmentation of Pangaea, enhanced late in the period by globally high sea levels.

By the Eocene, the mammalian world had changed radically. Marsupials had spread to every continent except Asia and placentals to every continent. Multituberculates in contrast had all but disappeared, teetering on until the end of the Eocene only in North America. Thus the Paleocene was a period of major transition. At this time, multituberculates were still abundant in Eurasia and North America, while marsupials and placentals spread successively to South America, Antarctica and Australasia (Woodburne & Case, 1996) and began to diversify into many new ecological types. Antarctica, however, appears to have acted as a filter to land mammals as only marsupials managed to colonize Australasia by this route, and a monotreme escaped only fleetingly to South America in the Paleocene (Woodburne & Case, 1996). No unequivocal representatives of any modern placental order had appeared in the Cretaceous. The Paleocene marks the first records of the Edentata (sloths, anteaters and armadillos) in South America, Carnivora in North America, Lipotyphla (true insectivores) in North America and Asia, Lagomorpha (rabbits and hares) in Asia, Rodentia in Asia and North America (Stucky & McKenna, 1993) and Proboscidea (elephants) in Africa (Gheerbrant *et al.*, 1996). Nearly all the dozen remaining modern orders, however, appeared early in the Eocene in the Northern Hemisphere.

Mammal faunas in the Cretaceous were different from modern ones, not only in their taxonomic composition, but also in their ecology. Apart from being small, most were insectivorous or, in the case of the multituberculates, insectivorous/frugivorous (Krause, 1982). Their niches thus scarcely overlapped with those of the major contemporaneous group of carnivorous and plant-eating land tetrapods: the dinosaurs. Dietary and locomotor diversification took place after dinosaur extinction at the Cretaceous–Tertiary (K–T) boundary (Collinson & Hooker, 1991). Thus, the first undoubted browsing herbivores (pantodonts) were in the Mid Paleocene, the first powered flyers (bats) in the Early Eocene and the first grazing herbivores in the Mid Miocene (where they coincide with the spread of grasslands). More efficient foregut-fermenting herbivores in the form of ruminant artiodactyls gradually replaced the hindgut-fermenting perissodactyls in dominance from the late Paleogene onwards (Janis, 1989). Taxonomic groups as well as communities show evidence of gradual evolution, but several major and widespread faunal turnovers in the Cenozoic increased the pace of both these processes and induced rapid modernization.

22.2.2   The return to the sea

More than 10 million years after the end Cretaceous extinction of the last Mesozoic marine reptiles (plesiosaurs and mosasaurs), several mammal groups invaded the marine realm. Of the modern groups, the first were the whales (Cetacea) and sea cows (Sirenia) in the Early Eocene and later the seals, sealions and walruses (Carnivora) in the Oligocene. The early records of whales and sea cows are from low latitudes in the Tethys seaway. Whales first appeared at higher latitudes (northern Europe) in the Bartonian and had reached polar regions by the end of the Eocene (Fordyce in Prothero & Berggren, 1992).

Today whales and seals are adapted to utilizing the abundant food resources present in the oceans' upwelling zones. This involves both directly by filter feeding on zooplankton as in whalebone whales (mysticetes), and indirectly by active hunting of nektonic animals which themselves depend on the zooplankton, as in the case of toothed whales (odontocetes) and most seals. Upwelling zones occur today especially in high latitudes and on the eastern sides of oceans. They are related to the intensity of atmospheric and oceanic circulation and are enhanced by increase in thermal gradients both latitudinally from the Poles to Equator and vertically in the oceans (Lipps & Mitchell, 1976). Increase in both these gradients occurred during the first major Cenozoic build-up of polar ice, giving rise to cold ocean bottom waters, which originated in the earliest Oligocene (Miller in Prothero & Berggren, 1992). The polar dispersal of cetaceans and subsequent rapid diversification into the types familiar today coincide with this event and is thus considered to represent an important response to global climate change (Fordyce in Prothero & Berggren, 1992). The first spread away from the Tethys in the Bartonian may reflect an earlier, less intense, threshold of cooling as suggested by the initial shift from tectono-eustatic to glacio-eustatic sea-level change at this time (Browning et al., 1996).

Unlike whales and seals, sea cows have been largely restricted throughout their history to lower latitude regions because of their dietary dependance on sea grasses. This dietary adaptation began in the Mid Eocene (Savage et al., 1994). An exception occurs in members of the lineage culminating in the recently extinct Steller's Sea Cow (Hydrodamalis gigas), which until 1768 browsed the North Pacific kelp 'forests'. Loss of much of the protected shallow coastal habitat in the

Pacific suitable for the growth of sea grasses, due to circum-Pacific mountain-building activity combined with major glacio-eustatic sea-level falls from about 15 million years ago, probably provided the selection pressure for the dietary and latitudinal range shift in this unusual sea cow (Domning, 1976).

22.2.3   The Paleogene

This time interval is notable for its climatic fluctuations which turned a greenhouse world into an icehouse world within 15 million years. It is also marked by some major eustatic sea-level falls (Haq *et al.*, 1987). For mammals, it is notable for both sudden faunal turnovers and gradual ecological change leading to community evolution. The patterns and timing of these faunal changes and how they relate to the physical events are discussed below.

22.3   MAJOR PALEOGENE FAUNAL TURNOVERS

Some of the Cenozoic faunal turnovers were intercontinental in scale. The most important are at the K–T boundary, at or near the Paleocene–Eocene and Eocene–Oligocene (Grande Coupure in Europe) boundaries, in the Late Miocene (*'Hipparion'* datum in the Old World) and in the Late Pliocene (Great American Interchange). Short periodicity large-scale changes over the last 1.75 million years can be directly related to the frequent and intense climatic fluctuations of the Pleistocene ice ages. The three major Paleogene turnovers are examined here. Mammal faunas in different continents may be categorized by biochronological units known as Land Mammal Ages (LMA) with prefixes for North America (NA), South America (SA), Asia (A) and Europe (E) (Woodburne and Swisher in Berggren *et al.*, 1995). They are useful when exact correlation to the global timescale is problematic.

When studying faunal turnover (i.e. a phase of increased disappearance and appearance), it is important to be able to distinguish true extinction of a species from 'extinction' of a morphology in the course of evolution along a species lineage (pseudoextinction). It is equally important to distinguish origination by evolution in place from introduction into the local area by dispersal from elsewhere. The most effective way of making the differentiation is by phylogenetic (cladistic) analysis to determine the relationships of the taxa before and after the turnover. Thus, if an outgoing taxon is more distantly related to any of the newcomers than one or other of its contemporaries, then it is judged to have become extinct. If a newcomer has a closest relative in the pre-turnover fauna that is also more primitive in all its character states (metaspecies: see Archibald, 1994), evolution not extinction is judged to have taken place. If the newcomer does not have such a relative, it is judged to have dispersed from elsewhere. Often, such extinctions and immigrations can only be judged on a local basis, but sometimes they can be seen to have continent-wide significance. In many cases, the relevant phylogenetic analyses have yet to be undertaken. In these cases, it is necessary to interpret current ideas of relationships (often non-cladistic) of the groups concerned to assess the minimum true extinction in a fauna.

22.3.1   Cretaceous–Tertiary (K–T) boundary

Although the faunal differences between the Late Cretaceous and Early Paleocene are recognized to be dramatic in all continents where there is a mammal record (none so far in Africa, Australasia or Antarctica), biotic or sedimentary gaps in the vicinity of the boundary, except in Montana, USA, prevent any understanding of the nature and timing of the changes involved. In Mongolia, Cretaceous mammals are represented in strata as late as Maastrichtian, but sediments of Early and Mid Paleocene age are missing (Russell & Zhai, 1987). In China, more of the Paleocene is represented but upper Maastrichtian sediments are missing (Russell et al., 1993). In Uzbekistan and Kazakhstan, the long Cretaceous sequence of mammal faunas terminates within the Campanian (Nessov et al., 1994). In India, the Deccan traps which span the boundary contain mammals only in interbasaltic sediments dated as late Maastrichtian (Sahni et al., 1994). In South America, the latest recorded Cretaceous mammals appear to be from the Campanian, whilst the Early Paleocene Tiupampian fauna probably does not date from the beginning of the epoch (Woodburne & Case, 1996).

In Montana, there is a vertebrate-bearing non-marine sedimentary sequence spanning the boundary. It is well calibrated to marine sequences by means of lateral interfingering with marine sediments of the Western Interior Seaway, by magnetostratigraphy and by the presence of the iridium anomaly in some sections. There are three main mammalian faunal units: the Lancian of undoubted late Maastrichtian age, the Puercan (Zone Pu1) of undoubted early Danian age and an intermediate fauna referred to the Bug Creek Interval (Fig. 22.1). The last of these is known only from channel fills that deeply incise the strata containing the Lancian fauna. Pollen assemblages allow the channel fills to be dated as earliest Paleocene (Lofgren, 1995).

The latest major Lancian fauna is dominated by marsupials (11 species) and multituberculates (10 species), with a lesser representation of placentals (7 species). Of these 28, only two, both multituberculates, have range extensions up into Pu1, where they are accompanied by 20 newcomers. The single new marsupial is closely related to one of the Lancian species and must have evolved in North America since the group is unknown outside that continent before the Paleocene. Two of the Paleocene multituberculates are congeneric with Lancian ones, whilst four more have no close relatives in the North American Cretaceous. The incoming cimolestan could be derived from one of its Lancian relatives and, although not represented until the later Paleocene, leptictids could be derived from a close relative of Lancian Gypsonictops (Kielan-Jaworowska et al., 1979). Cimolestans are also implicated in the later Paleocene radiation of the modern order Carnivora (Fox & Youzwyshyn, 1994) and of several extinct groups (Stucky & McKenna, 1993). The minimum number of true extinctions therefore becomes 22 (79% of the old fauna) and the minimum number of introductions into the area by dispersal, not by evolution in place, becomes 16 (73% of the new fauna).

This is certainly a major faunal turnover. However, it is necessary to consider the sedimentary setting of the channels of the stratigraphically intermediate Bug Creek Interval to try and understand the nature and timing of the turnover (Fastovsky, 1987). The channels have been ordered stratigraphically according to the levels of their bases. In this sequence of channels, many Lancian and Puercan ranges

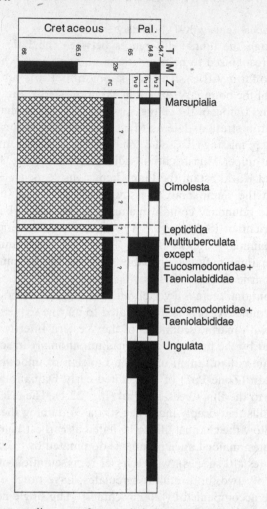

Fig. 22.1. Stratigraphical range diagram of monophyletic mammalian groups with species richness per faunal unit in the latest Cretaceous and earliest Paleocene of western North America, calibrated to the magnetobiochronological scale of Berggren *et al.* (1995). Width of bar proportional to species number; width of Leptictida bar = 1 species. Abbreviations: FC = Flat Creek fauna; M = magnetochron; Pal. = Paleocene; Pu = Puercan NALMA, with numbered zones (see Archibald *et al.* in Woodburne, 1987); T = time in Ma; Z = biozone or other faunal unit. Hatching indicates other Lancian faunas which span the late Maastrichtian Hell Creek and Lance Fms, but whose exact stratigraphical relationships to the Flat Creek fauna are obscure; in total they approximate the diversity of FC. Sources for faunas are Archibald and Clemens (1984), Archibald *et al.* in Woodburne (1987 and references therein) and Lofgren (1995).

overlap and a stepwise pattern of extinction and origination was originally recognized. This suggested a fairly long period of time for the turnover (several 100 ky scale) usually attributed to gradual climate change or habitat modification due to eustatic sea level fall (Archibald, 1987). However, dinosaurs also occur in the channels (giving rise to claims of their survival beyond the K–T boundary – Rigby *et al.*, 1987), but these have been demonstrated to be reworked (Fastovsky, 1987). In some cases they occur in clasts containing Cretaceous pollen and have never been found in floodplain deposits lateral to the channels. The Lancian mammals in the channels are now also judged to have been reworked (Lofgren, 1995), which means that the range overlap of the Lancian and Puercan taxa is superimposed, not real, and suggests a much shorter duration for the turnover. Unfortunately, the top 5 m of Cretaceous strata yield too few mammals (all Lancian species – '?' in Fig. 22.1) to demonstrate whether their decline prior to the Puercan influx was sudden or gradual.

Oddly, the decimation of marsupials in North America was balanced by the appearance of marsupials related to the Lancian ones in the Early Paleocene of South America. As no post-Campanian Cretaceous mammals are known from South America, the dispersal of mammals there from the north could have taken place at any time during the Maastrichtian. Several major eustatic sea-level falls occurred during this interval, which might have meant at least intermittently an emergent archipelago in the present position of Central America.

The earliest Puercan (Zone Pu0 = Bug Creek Interval) newcomers in North America that have no close Cretaceous relatives in that continent comprise 6 species of primitive ungulate and 3 species of multituberculate. Cladistic analysis suggests that the multituberculates of the genus *Catopsalis* originated from a Mongolian Cretaceous species (Archibald, 1993). Moreover, possible Cretaceous ungulates have been reported from Uzbekistan (Archibald, 1996a). Thus, Asia is the most likely source for immigrant North American early Puercan mammals. The exact calibration of a major sea-level fall at about this time is uncertain. It may have coincided with the K–T boundary (Hallam, 1992), providing potential land bridges for dispersal, or have preceded it by 1.5 my (Haq *et al.*, 1987).

22.3.2   Paleocene-Eocene (P-E) boundary

This turnover was less widespread than that between the Cretaceous and Tertiary. However, it was none-the-less dramatic and affected North America, Europe and central Asia. Some changes affected north Africa too, but the scale and timing is less precise (Gheerbrant *et al.*, 1993) and this area will not be discussed here. No major event, however, took place in South America (Pascual *et al.*, 1985) and no faunas spanning the boundary have been discovered in Australasia or Antarctica. The event is characterized by the first and synchronous appearance of the modern orders Perissodactyla (horses, rhinos, tapirs), Artiodactyla (cloven-hooved mammals), Primates and Chiroptera (bats) in Europe and North America. The timing of the turnover in Asia is less precisely known but it is no less major. None of these groups has known close relatives in the Paleocene, either in their area of appearance or elsewhere. Because of this the event is regarded as a dispersal, but the area of origin has not been identified. The turnover in North America is between the Clarkforkian and Wasatchian NALMAs and is best represented in Wyoming, where dense faunal sequences in continental strata are well exposed (Fig. 22.2). Here, using the same criteria as for the K–T boundary, the minimum true extinction is about 40% and the origination at least 29% (sources cited in Hooker, 1998). In north-west Europe (between the Cernaysian and Neustrian ELMAs – Fig. 22.3), the extinction is 53% and the origination 61% (Hooker, 1998). In Mongolia, extinction is 67% and origination 88% (data from Dashzeveg, 1988). Climate and/or land

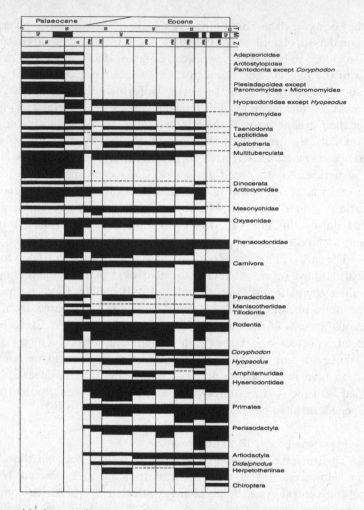

Fig. 22.2. Stratigraphical range diagram as in Fig. 22.1 for well-sampled groups in the latest Paleocene and earliest Eocene of Wyoming, USA. Width of Arctostylopidae bar = 1 species. Abbreviations: Cf, Ti and Wa = Clarkforkian, Tiffanian and Wasatchian NALMAs respectively, with numbered zones (see Archibald *et al.* in Woodburne, 1987; Clyde *et al.*, 1994). Other abbreviations as in Fig. 22.1. Sources for faunas: Uhen and Gingerich (1995) and references in Krishtalka *et al.* in Woodburne (1987) and in Gingerich and Gunnell (1995). Magnetostratigraphical calibration from Clyde *et al.* (1994).

bridges have been invoked to account for the turnover and the evidence for both will be evaluated.

The presence of a major shortlived negative excursion in oceanic carbon and oxygen isotopes at 55.5 Ma (the earliest point within the currently grey area of the Paleocene–Eocene transition) has been interpreted as a sudden global warming event punctuating the overall warming trend that began in the Late Paleocene (Aubry *et al.* in Knox *et al.*, 1996). In the sea it is associated with a major extinction of benthic forams (Thomas & Shackleton in Knox *et al.*, 1996). On land, the carbon excursion

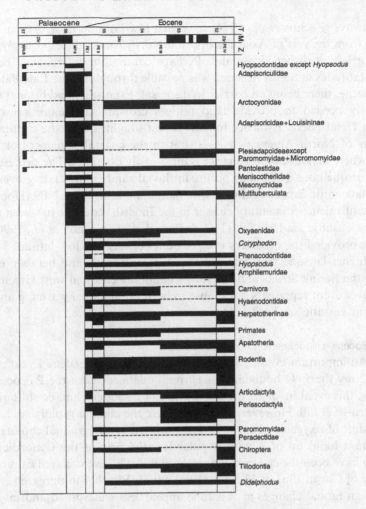

Fig. 22.3. Stratigraphical range diagram as in Fig. 22.1 for well sampled groups in the Late Paleocene and earliest Eocene of NW Europe. Width of Pantolestidae bar = 1 species. The ends of the ranges of Multituberculata and *Hyopsodus* represent their extinction in Europe. Abbreviations: MP6 = Mammalian Reference Level 6 (= Cernaysian ELMA); PE = numbered biozones of the Neustrian ELMA (Hooker in Knox *et al.*, 1996); WALB = Walbeck fauna. Other abbreviations as in Fig. 22.1. Sources: Hooker (1998) and references therein.

has been recorded in mammalian dental enamel and soil carbonates in the oldest Wasatchian fauna (Wa0) in the western USA (Koch *et al.*, 1992). It has also been recognised in soil carbonates in Europe in the Paris Basin (Stott *et al.* in Knox *et al.*, 1996) close in age to the earliest Neustrian mammal faunas (Hooker, 1998).

The sudden appearance of these new mammal faunas has been attributed to higher latitudinal range extension due to the warming event, the newcomers being potentially derived from low latitudes where Late Paleocene faunas are poorly known (Krause & Maas, 1990). It is odd, however, that no major change is detectable in South American mammal faunas at this time. Moreover, the Tethys and other seaways (Zachos *et al.*, 1994) separated low-latitude Asia, Africa, and South America from the rest of Asia, Europe and North America, respectively, thus inhibiting northward dispersal by land. Perhaps continuous interchange between low and mid-latitudes in South America was possible throughout the Late Paleocene and Early Eocene, there being no barrier to dispersal. Even so, an additional mechanism is certainly needed to provide land bridges crossing the major seaways in the Northern Hemisphere. Evidence for this is best sought, not in the continental interior basins of North America or Asia, but in the coastal sequences of north-west Europe, where major sea-level change is most easily observed. The earliest Neustrian fauna of Erquelinnes (Belgium) occurs in fluvial sands immediately overlying fully marine glauconitic deposits calibrated to nannoplankton Zone NP9 (Hooker, 1998). The age-equivalent sedimentary change in the English sequence has been interpreted as a major eustatic sea-level fall (Th4–5: Powell *et al.* in Knox *et al.*, 1996). It could thus have provided the necessary bridges both between the low-latitude land masses and the higher-latitude Northern Hemisphere continents; and between each of the latter via the Bering Straits and the straits bounding east and west Greenland. Thus the turnover event represents a response to both rapid global warming and synchronous major eustatic sea-level fall.

22.3.3   Eocene–Oligocene (E–O) boundary

An important event, referred to as the 'Terminal Eocene Event', post-dates by *ca.* 0.2 my the E–O boundary, as currently defined. Like the Paleocene–Eocene boundary, this event involved the coincidence of a sudden climatic shift and a major eustatic sea-level fall. However, in contrast here the climate rapidly cooled (positive isotopic shift of oxygen and carbon) terminating a 15 my gradual cooling trend and the resultant build up of Antarctic polar ice (the first of the Cenozoic), which is judged to have been the cause of the sea-level fall. The event is well calibrated to the beginning of Chron 13n1 at 33.5 Ma (Brinkhuis & Visscher in Berggren *et al.*, 1995). Mammalian faunal changes at this time appear less widespread and major (except locally) than at the Paleocene–Eocene boundary. This may be because many changes had already taken place gradually in the course of the Mid to Late Eocene in response to the cooling trend. Faunal turnover is best represented in Europe, where it is known as the 'Grande Coupure'. The event is much less marked in North America and Asia, where the Bartonian–Priabonian boundary and a point later in the Early Oligocene, respectively, mark more major turnovers. In South America most of the Late Eocene has apparently yielded no mammals (Flynn & Swisher in Berggren *et al.*, 1995), although it was probably within this time interval that rodents arrived in that continent. Primates probably arrived a little later as their first occurrence as a low diversity group is in the Late Oligocene. Evidence favours dispersal from Africa across the South Atlantic via islands associated with the Walvis and Rio Grande ridges (George & Lavocat, 1993).

In Europe the faunal turnover was considerable. In England (Fig. 22.4), where the faunas on either side of the 'Grande Coupure' can be found in superposed strata, 23 out of 28 (82%) placental species (Hooker *et al.*, 1995) became extinct, although, of these, 8 survived the event unchanged or slightly modified elsewhere in Europe. Of the 22 species that occur immediately after the 'Grande Coupure'

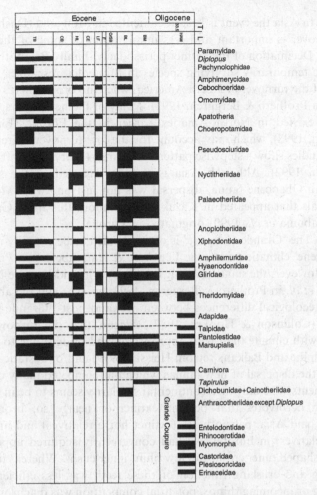

Fig. 22.4. Stratigraphical range diagram as in Fig. 22.1 in the Late Eocene and Early Oligocene of the Hampshire Basin, southern England. Ranges are calibrated to stratal thickness. Dates mark the Grande Coupure (33.5 Ma) and beginning of the Late Eocene (37 Ma). Abbreviations: BL = Bembridge Limestone Formation; BM = Bembridge Marls Member; CB = Colwell Bay Mbr; CE = Cliff End Mbr; F = Fishbourne Mbr; HAM = Hamstead Mbr; HL = Hatherwood Limestone Mbr; L = lithostratigraphical units; LF = Lacey's Farm Limestone Mbr; O/SB = Osborne and Seagrove Bay Mbrs; TB = Totland Bay Mbr. Other abbreviations as in Fig. 22.1. Sources: Hooker in Prothero and Berggren (1992), Hooker *et al.* (1995).

(Hooker in Prothero & Berggren, 1992), 13 have no close relatives in the preceding fauna and only one of those came from elsewhere in Europe (i.e. 55% of the fauna had dispersed from another continent). In southern France a similar pattern is obtained (Remy *et al.*, 1987). Many groups had evolved independently in island Europe during the Eocene after the North Atlantic split from North America at *ca.* 53.5 Ma. These were decimated (Fig. 22.4). The newcomers include hedgehogs (Erinaceidae), hamsters (Cricetidae), beavers (Castoridae), rhinos (Rhinocerotidae) and more advanced types of carnivore.

In Asia the event is best documented in Mongolia (Dashzeveg, 1993), where the turnover is important and sees immigration of some of the same genera as in Europe. Decimation of the extinct perissodactyl family Brontotheriidae, resulting in only one temporarily surviving species in Mongolia, resembles the most important aspect of the turnover in North America, where no brontotheres survived Chron 13n (Emry in Prothero & Berggren, 1992). Many of the newcomers to Asia and Europe already existed in North America earlier in the Eocene (Emry in Prothero & Berggren, 1992), which may account for the low turnover there. Detailed stratigraphical studies show a stepwise pattern of change (Emry, Evanoff *et al.* in Prothero & Berggren, 1992). Although Asia is usually regarded as the source for the new European Oligocene fauna, dispersal was not just one way. Moles (Talpidae) and marsupials that appeared in Kazakhstan at about this time (Gabunia & Gabunia, 1987; Gabunia *et al.*, 1990) originated in Europe.

The 'Grande Coupure' is often claimed to be causally related to the 'terminal Eocene' climatic event (e.g. Legendre & Hartenberger in Prothero & Berggren, 1992). However, the small scale of the turnover in mid-latitude North America (e.g. Evanoff *et al.* in Prothero & Berggren, 1992, fig. 6.5) and the absence of any major obvious ecological differences between pre- and post- 'Grande Coupure' faunas in Europe (Collinson & Hooker, 1987), suggest that the turnovers had less to do directly with climate than with appropriate land bridges due to local tectonics (e.g. in the Alpine and Balkans region: Heissig, 1979) and/or eustatic sea-level fall, which allowed the dispersal of new faunas which out-competed many of the existing endemic elements. The only intercontinental similarity seems to be in the dominant, large browsing herbivores that became extinct or nearly so: brontotheres in North America and Asia, palaeotheres (extinct horse relatives) and anoplotheres (extinct camel relatives) in Europe. All had convergently acquired upper cheek teeth with high W-shaped outer crests and low blunt inner cusps. Whereas it is possible that the chopping and crushing function of these teeth was less efficient than that of the surviving or incoming rhinos, potential competition was diachronous in the different continents, and it is more likely in this case that the rapid cooling destroyed or reduced a keystone food resource.

22.3.4 Comparing the three events

The K–T event in Montana differs from the other two in its small number of immigrant newcomers. The numbers are more balanced in the other two events. Moreover, the K–T event is characterized by rapid diversification of the immigrants and most of the survivors (e.g. Multituberculata) within half a million years following the event (Fig. 22.1). In contrast, at the P–E and E–O events, survivors once diminished remained so until extinction a few million years later (e.g. Multituberculata and Arctocyonidae in Europe) (Fig. 22.3). Although in England no palaeotheres or pseudosciurid rodents survived the E–O event (Fig. 22.4), a few lasted for several million years in other parts of Europe (Remy *et al.*, 1987). Whereas there was little turnover at the P–E and E–O events in certain parts of the world, faunas of Early Paleocene age (where known) are always very different from those of the Late Cretaceous, whether or not a boundary sequence is preserved. Unlike the P–E and E–O events, the K–T event is not unequivocally calibrated with a major eustatic sea-level fall. The P–E event coincides with a rapid warming, the other two events with rapid cooling, according to the marine oxygen and carbon isotope record (e.g. Shackleton, 1986). Diachronism between North America and Europe in the timing of the synchronous decimation of multituberculates and appearance of rodents near the P–E boundary (Figs. 22.2, 22.3), and the similar ecology of the two groups (Krause, 1986), suggest displacement by competitive exclusion. The mainly earlier appearance in North America of groups that, in Europe, coincide with major E–O extinction demonstrates similar diachronism, although in North America the appearances do not coincide with extinction (Stucky in Prothero & Berggren, 1992).

Thus, the K–T event provides evidence of a high level of extinction and disruption of communities in western North America, with the relatively few survivors rapidly speciating and diversifying either in place or after dispersing to and from new areas, where they largely replaced the former inhabitants. It apparently coincides with a major meteorite impact in Mexico (Swisher *et al.*, 1992, Gale and Pickering, this volume), a major phase of extrusive volcanism in India (Officer & Drake, 1985) and a brief low temperature excursion (which could have resulted from either of the first two phenomena). The last of these may have been responsible for establishment of Cenozoic vegetation structure, which differed considerably from that of the Cretaceous (Upchurch & Wolfe, 1987). The mammalian turnover pattern is not inconsistent with a global catastrophic event such as a large meteorite impact. However, without stratigraphically detailed data across the boundary in other areas, and in the absence of clear causal relationships, attributing the turnover to any of these events is highly speculative (Archibald, 1996b).

The diachronism of dispersal and of its correlation with extinction in some groups, overall turnover pattern, coincidence with eustatic sea-level falls and coincidence with a rapid warming suggest that the important P–E event owes its existence to both climate change and low sea level acting together. It is probable that climate-induced northward dispersal would have been inhibited by high sea level; moreover, the high latitude land bridges linking North America to Europe (Greenland) and North America to Asia (Bering Straits) during low sea level would have been less favourable to dispersal without the climatic amelioration. There is evidence of displacement of pre-event faunas by the newcomers, which is greatest in Europe apparently because of the largest number of newcomers, but it is not clear how much influence habitat change had. Land floras based on macrofossils in southern England show little change in composition and in North America show no change in species richness at the time of the event (Collinson & Hooker, 1987; Wing *et al.*, 1995), suggesting that the main climatic factor allowing northward dispersal of the mammals might have been lengthening of the period of annual food availability. However, palynological assemblages show a short-term increase in the proportions of tropical elements both in the North Sea in palynological zones PT19.2 and 19.3 (Schröder, 1992) and in the western London Basin near the base of the Reading Formation (Allen, 1982), both of which are approximately synchronous with the isotope excursion (Powell *et al.*, Aubry *et al.* in Knox *et al.*, 1996; Hooker, 1998).

Turnover patterns similar to the P–E ones, diachronism of dispersal, together with little extinction in North America, and eustatic sea-level fall synchronous with the E–O event suggest that the main cause of the event was the creation of land bridges in parts of Europe and Asia, allowing dispersal into new areas. This affected especially Europe, where endemism had been at a high level, and where displacement best fits the majority of extinctions. The extinctions of brontotheres and in part of palaeotheres and anoplotheres could be due to climate-induced habitat changes as they coincide globally with the sudden cooling.

22.4   CLIMATE-RELATED ECOLOGICAL TRENDS

The role of climate in controlling the geographical distribution of taxa was considered above in relation to extinction and dispersal events. Certain longer duration and thus more gradual changes have been recognized to reflect habitat differences, especially in terms of vegetation type, which in turn are inferred to have responded to climate change. Fleming (1973) demonstrated latitudinal changes in community structure across present-day North and Central America. He created the term ecological diversity for the 'distribution of species in the various classes of body sizes, feeding adaptations and food habits'. Thus mean body size, terrestriality, scansoriality, herbivory and carnivory, as well as overall species richness, increase while aerial and arboreal foraging locomotor adaptations and frugivory (fruit and seed eating) decrease with increasing latitude. These reflect the greater productivity and more closed nature of habitats at low latitudes today. In contrast, the relationship between size and trophic structure may be quite comparable at equivalent latitude (cf. Panama and Malaya – Eisenberg, 1990). Open and closed habitats can, however, also coexist at similar latitudes; and here water availability through precipitation and local edaphic conditions are thought to be the main natural controls (Richards, 1952). Using information on body size, foraging locomotor adaptation and diet, derived from teeth and limb elements, ecological diversity can be assessed for fossil assemblages (Andrews et al., 1979). The reconstruction of palaeocommunities using such ecological parameters is regarded as more reliable than relying on either indicator species (Andrews et al., 1979) or the waxing and waning fortunes of particular taxonomic groups (e.g. Damuth et al., 1992) to infer ancient environments.

Body size has been much used in mammalian studies because of its low intraspecific variability, ecological significance and ease of estimation from isolated elements (e.g. Creighton, 1980; Eisenberg, 1990; Gingerich, 1990). At the level of modern communities, mean body size in closed forest is much smaller and the plot more skewed than on an open plain (Andrews et al., 1979). There is also some evidence that body size in selected species lineages mirrors palaeotemperature deduced from leaf-margin analysis on a scale of less than a million years (Bown et al., 1994). The range and distribution pattern of body size alone (cenogram method) has been used to infer habitat in the European Eocene (Legendre, 1989) and to compare ecomorphological distributions in the North American Paleocene (Maas & Krause, 1994). However, taphonomic and collecting biases are liable to influence the results and the internal testing of bias provided by the three parameter ecological diversity approach is here considered preferable.

Ecological diversity analysis through much of the west European Paleogene suggests a change from closed evergreen forest in the Late Paleocene and Early Eocene to open woodland and floodplain habitats in the Late Eocene and Early Oligocene, mirroring vegetational changes that have been documented in southern England (Collinson & Hooker, 1987; Hooker, 1998; Hooker *et al.*, 1995). It also shows some latitudinal differences in the Late Eocene (Hooker in Prothero & Berggren, 1992) with distinct north–south provincialism. In contrast, no provincialism across a greater span of latitudes (25°) has been observed in the North American Paleocene (Maas & Krause, 1994). This suggests that the Late Eocene cooling was accompanied by a steepening of latitudinal temperature gradients.

The habitat changes inferred from ecological diversity in southern England from the Early Eocene to the Early Oligocene (Collinson & Hooker, 1987) match global isotopically derived climate curves (e.g. Shackleton, 1986). They are also mirrored in North America during a similar timespan (Stucky in Prothero & Berggren, 1992). Such changes are likely to be more evident at mid- than low latitudes and it is in the former that faunal sequences are best known. A selection of these ecological diversity parameters are shown here as percentages of the total species composition of a given fauna (Fig. 22.5(A)–(E)), together with a species richness plot (Fig. 22.5(F)) for a succession of north-west European localities. These are compared with a $\delta^{18}O$ plot from benthic forams at a similar latitude in the south Atlantic as a proxy for temperature (Fig. 22.5(G)). The mammal localities have been chosen for adequate sampling and to combine sufficient temporal coverage with minimal latitudinal variation (all within *ca.* 3°). Species richness correlates positively and quite closely with temperature. The ecological parameters have comparable trends and it is noteworthy that all show strong fluctuations near the E–O boundary. Frugivory, arboreality and scansoriality correlate positively, while herbivory and terrestriality correlate inversely with temperature, suggesting a reduction in availability of fruit as a food source, with concomitant increased reliance on leaves, and a reduction in tree density favouring terrestrial locomotion. The lateness of the frugivory peak with respect to the thermal maximum may be due to a time lag in the evolution of this dietary adaptation following terminal Paleocene extinction of most multituberculates and condylarths (the dominant Paleocene frugivore groups). Most European Eocene frugivores were primates and artiodactyls, a dietary adaptation they evolved during the first half of the epoch.

The ecological parameters correlate less well with the climate curve prior to the thermal maximum. This interval spanning the Paleocene and earliest Eocene is characterized by temperature variation in an ice-free world. It is therefore possible that the parameters documented above are telling us more about seasonality and mean annual range of temperature with the onset of polar glaciation than about overall temperature. Nevertheless, Maas and Krause (1994) noted an increase in herbivores and a reduction in small mammals in the North American Late Paleocene (around 57 Ma) which coincides with a global temperature trough (e.g. Shackleton, 1986). Thus, it is clear that by careful interpretation of data on the ecological composition of ancient mammal faunas, we can gain important insights into the nature and tempo of global climate change.

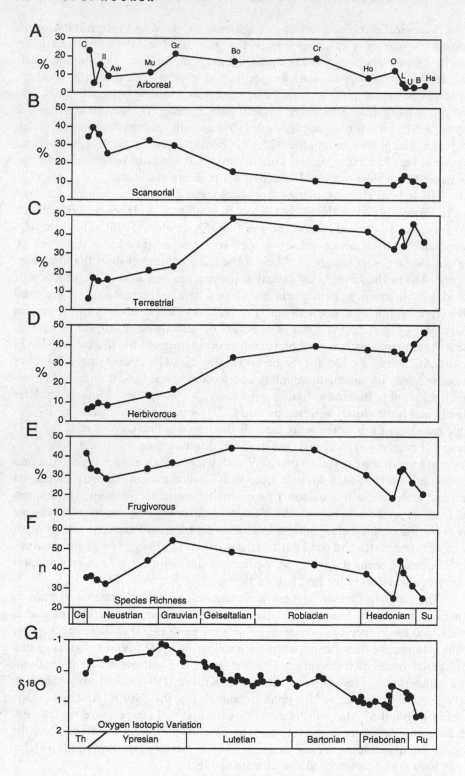

22.5   CONCLUSIONS

The examples given here show that it is possible to unravel the various patterns of mammalian turnover and distribution and distinguish the effects of sea-level change and climatic fluctuation. However, this depends on accurate calibration of the timing of the different parameters. The most significant short-term faunal turnovers are dispersal events facilitated by low sea level. They coincide with extinction of a previously existing fauna either by displacing it or by replacing one decimated by a prior catastrophe. Long term climatic change is an important driving force in community evolution. The change from greenhouse world to icehouse world in the Paleogene saw a great increase and diversification of terrestrial herbivores and a reduction in frugivores, tree dwellers and overall species richness in the progressively less forested mid latitudes. In the oceans, whales and seals radiated to exploit the upwelling zones created by the cooling of ocean bottom waters.

Fig. 22.5. Percentage ecological parameters (A–E) and species numbers (F) for mammal faunas from latest Paleocene to earliest Oligocene of NW Europe; plotted against an oxygen isotope curve (G), based on the benthic foram *Cibicidoides* from ODP Holes 702 and 703 in the South Atlantic (Miller in Prothero and Berggren, 1992). Data plotted on the timescale of Berggren *et al.* (1995) in Ma. Main sources of taxa: Crochet (1980), Franzen and Haubold (1986), Godinot (1988), Hooker in Prothero and Berggren (1992), Hooker (1998), Hooker *et al.* (1995), Jaeger (1971), Louis *et al.* (1983), Sudre *et al.* (1983). Abbreviations: AW = Abbey Wood; B = Bembridge Marls Member; Bo = Bouxwiller; C = Cernay and Berru; Ce = Cernaysian; Cr = Creechbarrow; Gr = Grauves and related sites; Ha = Hamstead Member; Ho = Hordle mammal bed; I = Zone PE I sites; II = Zone PE II sites; L = Bembridge Limestone Formation lower fauna; Mu = Mutigny; O = Osborne and Seagrove Bay Members; Ru = Rupelian; Su = Suevian; Th = Thanetian; U = Bembridge Limestone Fm upper fauna.

23

Response of Old World terrestrial vertebrate biotas to Neogene climate change

PETER J. WHYBROW AND PETER ANDREWS

23.1 INTRODUCTION

. . . Climate appears to limit the range of many animals, though there is some reason to believe that in many cases it is not the climate itself so much as the change of vegetation consequent on climate which produces the effect... . Where barriers have existed from a remote epoch, they will at first have kept back certain animals from coming in contact with each other; but when the assemblage of organisms on the two sides of the barrier have, after many ages, come to form a balanced organic whole, the destruction of the barrier may lead to a very partial intermingling of the peculiar forms of the two regions . . .

Wallace (1876)

At the time that Wallace published his thoughts on *The Geographical Distribution of Animals with a study of the Relations of Living and Extinct Faunas as elucidating the past changes of the Earth's surface,* current belief was in the fixed position of the continents. Since the acceptance of plate tectonic theory during the late 1960s large numbers of publications have appeared concerning the biogeography of Old World Tertiary terrestrial vertebrates. These attempt to take account of the effects of past continental movement, the effects on climate of different land/sea configuration, and the effects of changes in ocean currents on marine and terrestrial ecosystems. This chapter considers how the Neogene vertebrates responded to climate change.

23.2 THE QUALITY OF DATA: DATING AND FAUNAS

Refinement of radioisotopic dating techniques in conjunction with magnetostratigraphy of some new and historic Old World terrestrial vertebrate sites has, in general, provided a greater constraint on the correlation of their chronologies. However, for some critical sites such as those in North Africa (Libya), eastern Saudi Arabia, and the Baluchistan region of Pakistan, such techniques have yet to be utilized. For others, such as in China (see Tedford *et al.*, 1991), there is a lack of suitable material for radioisotopic analyses. In themselves, these techniques do not provide the dating panacea that was once hoped for. The quality of the sample analysed as well as the techniques themselves all produce inherent errors (see Woodburne, 1996, p. 536). For instance, magnetostratigraphical analysis of

Arabian Miocene clastic sediments can produce a dating error of $\pm$ 3 my (Hailwood & Whybrow, 1999), a significant error margin when mammalian dispersal events are being discussed using MPTS (Magnetic Polarity Time Scale; Cande & Kent, 1995).

Vertebrate material collected from sites during the 19th century has a strong collecting bias. Exhibition quality specimens, invariably of large vertebrates, make up a large proportion of geographically scattered collections that, in some cases, require a thorough up-to-date taxonomic reappraisal. Likewise, the small mammals that comprise an important part of any fossil fauna might be lacking in such collections. Consequently, any attempt to detect and interpret faunal change as a response to any scale of climate change should not ignore the quality of data available. Further, there is the common perception that data about past climates from certain sites can be applied to a whole continent. This perception, coupled with the all too easily forgotten fact that the fossil record is incomplete, can lead to the assumption that the biostratigraphical succession is reliable. In Africa, for example, the total area of Neogene fossiliferous localities is approximately 1% of the area of Africa itself and discussion about Africa's Neogene fauna, flora, climate change and notions about mammalian speciation in response to climate change that is based on such minimal data has historically embraced the remaining 99% of the continent. This is especially so for the Kenyan sites which constitute only 0.1% of the African land mass (Hill, 1987) but which are used as the basis for inferring biotic and climatic change for the whole continent.

A further data tool that has been used to explain faunal change and the dispersal of mammals as a response to possible climatic change is the first and last appearances of taxa in certain regions, particularly regions separated by marine, orographic or climate-controlled barriers such as deserts. The basis for this depends on the taxonomic quality of the species used, and in order to extend analyses to a global scale it assumes that species appear or disappear synchronously in different regions. A good example is the '*Hipparion* Datum' for the first appearance of this extinct primitive equid in the Old World between 10.5 and 11 Ma (Berggren & Van Couvering, 1974; see Woodburne, 1996 for discussion on *Hippotherium* Datum). In a recent attempt to overcome taphonomic bias in the terrestrial vertebrate record and to utilize first and last appearances of species (FAD, LAD), Vrba (1995a) has devised complex mathematical models in order to substantiate her notion that speciation, migration and faunal turnover should be principally motivated by climate driven change. Again, the use of first and last appearances assumes that the fossil record is complete, and since it is not it continually produces surprises, for the available data represents incomplete proportions on inadequate samples.

In this chapter we broadly outline major physical events that have reportedly affected Old World Neogene terrestrial faunas: (i) the early Miocene closure of the Tethys epicontinental seaway in the Middle East to form the first landbridge between AfroArabia and south-western Asia; (ii) the so-called Messinian 'salinity crisis' in the Late Miocene that invokes the aridification of part of the Mediterranean basin to produce not only sabkha-type environment but also a terrestrial pathway between northern Africa and south-western Europe for intermigrations of certain continental faunas; and (iii) the uplift of the Tibetan plateau during the Late Miocene at about 8 Ma to produce a possible shift in the pattern of the Indian monsoon. We link to these events an analysis of faunal change for Africa and Eurasia based mainly on ecological changes through the Tertiary. The reader might perceive that we lean towards the idea that tectonic change, orographic evolution and continental disconnections of marine seaways/oceans, had a greater effect on regional climates during the Neogene of the Old World than did advances of Antarctic ice or global sea level lowstands. These processes are not mutually exclusive. However, we feel that the evidence for tectonic change might be more easily interpreted in the terrestrial vertebrate record.

23.3   CONTINENTAL BIOZONATIONS

This chapter uses Mammal Neogene (MN) 'zones' purely as a tool to enable the reader to refer to faunas that make up these zonations. These MN zones have neither biostratigraphical nor chronostratigraphical validity – they are a series of fossil mammalian assemblages from single localities (predominately European) placed in chronological sequence on the basis of stage-of-evolution, entries by migration and exits by extinction of specific taxa (de Bruijn *et al.*, 1992, p. 66).

In the early 1970s it was realized that a scheme to link the chronologies of European mammal faunas with those devised for the marine sequence was needed and international symposia were held in attempt to devise such a scheme (see Steininger *et al.*, 1990). Since the 1970s, the criteria defining MN zonation have been reconsidered, and they are used now (Fig. 23.1) just within their European context (MN 'zones' are only applicable to the Neogene of Iberia, France, Switzerland, Germany, the Netherlands, Hungary, Czechoslovakia, Austria, Italy, Poland, Yugoslavia, Roumania, Bulgaria, Greece, Turkey, south-western Russia (pre-CIS) and Asiatic Turkey). Importantly, MN zones have not been devised for Africa, Arabia, the Middle East and the remaining countries of Asia. Of equal importance is the overall fact that there is still no satisfactory correlation between continental and marine biozones. Most mammalian genera and species have limited geographical ranges, so that first and last occurrences of mammalian taxa are only of local importance (de Bruijn *et al.*, 1992). Although it is recognized that most migrations of land mammals are time transgressive, this need not interfere with the correct usage of MN zonation provided allocation of faunas is based in stage-in-evolution only, although this does not always exclude circular reasoning because it is not always possible to distinguish stage-in-evolution of an assemblage from the 'entries' and 'exits' of particular genera (de Bruijn *et al.*, 1992). A review of mammalian chronostratigraphical principles is given by Woodburne (1996), who recommends the abandoning of last appearance datum (LAD) in mammalian chronostratigraphy.

23.4   INTERPRETING FAUNAL CHANGE

Faunal change is analysed here for Africa and Eurasia, based mainly on ecological changes through the Tertiary. Palaeoenvironmental change is investigated by analysing large and small mammal size, trophic and spatial structure. The method we have used is the one developed by Andrews *et al.* (1979), and the comparative database for this analysis has been extended to cover present-day faunas from habitats across Europe and Asia (Andrews, 1996). Our intention is to compare community patterns of faunas across the Old World through time, both to investigate time successive change and, where overlap in the fossil record occurs, to see if changes are synchronous in different regions.

The database used here is the NOW database (Fortelius *et al.*, 1996). The analysis of the Eurasian faunas includes extensive taxonomic revisions (Bernor *et al.*, 1996). African faunal analysis is based on Pickford (1981) and Andrews (1996) and eastern Asian faunal analysis on Barry *et al.* (1990, 1991). The reconstructions of body size are taken from the NOW database and from the ones previously published (Andrews *et al.*, 1979; Andrews, 1996). The size categories range from small mammals (A = 0–100 g; B = 101–1000 g, etc.) to large mammals (H > 360 kg) (see Fig. 23.1 legend). Spatial analysis is based on the concept of Harrison (1962), whereby the locomotor adaptations of the fauna are inferred from their post-cranial adaptations, and these are interpreted in terms of the space occupied, for example terrestriality or arboreality. The divisions are as follows: terrestrial, locomotion totally restricted to the ground and including both cursorial and springing species; semiarboreal, originally referred to as SGM and including mammals that can live both on the ground and in bushes and low trees; scansorial, large branch arboreality by clawed mammals; arboreal, mainly tree-living by mammals that can grip branches; aerial, mammals with adaptations for flying; fossorial, mammals with adaptations for digging; and aquatic, mammals with adaptations for swimming. Trophic guilds are very general in nature and are based on the major adaptations of the teeth: insectivores with high pointed cusps (or lacking enamel in anteaters); frugivores with low, rounded cusps; browsing herbivores with low crowned (brachydont) teeth; grazing herbivores with high crowned (hypsodont) teeth; carnivores with cutting teeth; and omnivores with large post-canine teeth with low cusp relief (bunodont). Species with mixed diets have been assigned to more than one guild except where such variety is indicated that they should be assigned to the omnivore class.

23.5   MESOZOIC AND PALEOGENE BACKGROUND

Mammals first appeared in the early Mesozoic. During most of this time and into the Late Cretaceous (75–65 Ma), small and medium sized mammals with adaptations to insectivorous, carnivorous, and omnivorous diets were present. Large herbivores were missing from the mammalian faunas (Savage & Russell, 1983), as were aquatic and aerial-adapted species, and mammals were adapted mainly for terrestrial and arboreal/scansorial habits. The community structure of these early mammalian faunas is hard to interpret because they formed only a subset of the contemporaneous terrestrial ecosystems, which at that time were still dominated by the dinosaurs. An analogy can be drawn from extant African faunas between the positions that a comparable spectrum of small mammal species occupy in African ecosystems today relative to the large mammal faunas. For example, assemblages of insectivores and rodents, despite their small size, are a significant a part of terrestrial ecosystems today. Rodents are significant consumers of plants and distributors of plant seeds and insectivores are important consumers of insects; and their main mammalian predators, the viverrids (mongooses and genets), are also important predators of small reptiles and insects. In many habitats these smaller mammals have greater species richness than larger primary consumers, and extending the

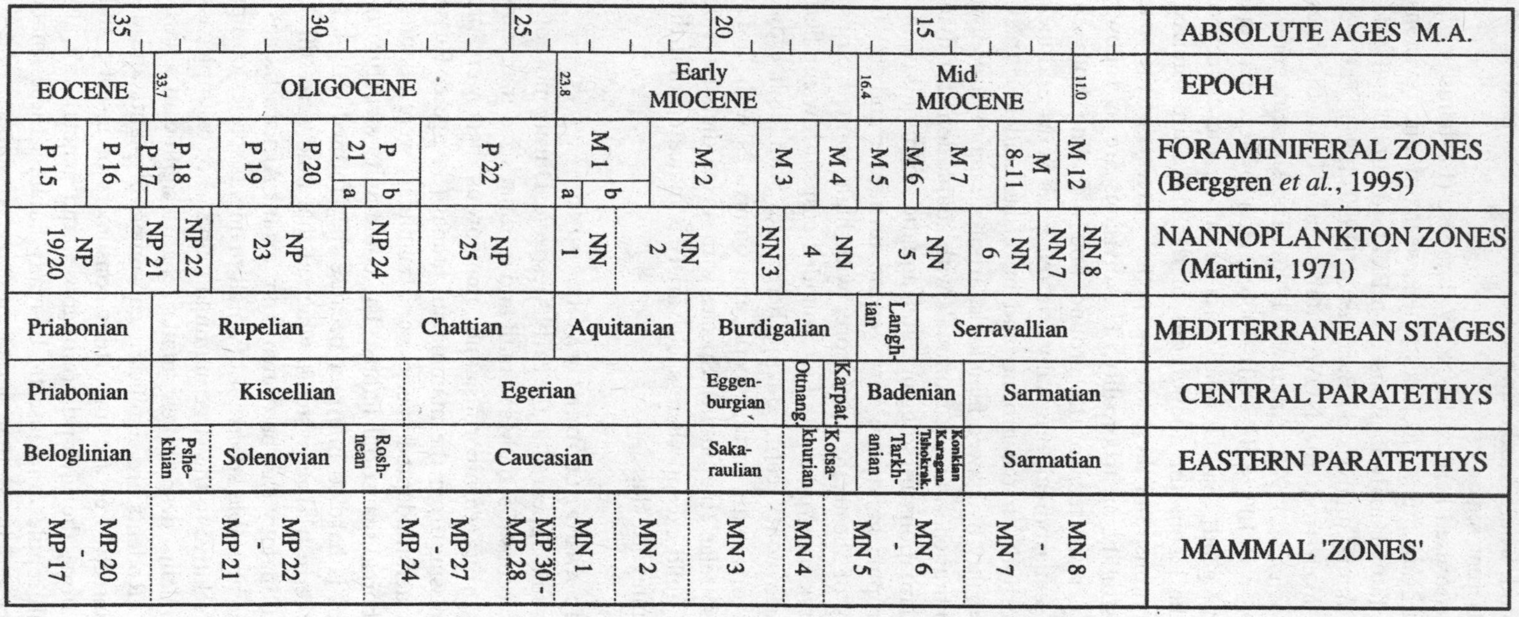

ABSOLUTE AGES M.A.	EPOCH	FORAMINIFERAL ZONES (Berggren et al., 1995)	NANNOPLANKTON ZONES (Martini, 1971)	MEDITERRANEAN STAGES	CENTRAL PARATETHYS	EASTERN PARATETHYS	MAMMAL 'ZONES'
11.0	Mid MIOCENE	M 12	NN 8	Serravallian	Sarmatian	Sarmatian	MN 8 - MN 7
		M 8-11	NN 7				
			NN 6				
15		M 7	NN 5		Badenian	Konkian Karagan. Tshokrak.	MN 6 - MN 5
		M 6		Langhian		Tarkhanian	
16.4		M 5			Karpat.	Kotsakhurian	
	Early MIOCENE	M 4	NN 4	Burdigalian	Ottnang.		MN 4
		M 3	NN 3		Eggenburgian	Sakaraulian	MN 3
20		M 2	NN 2	Aquitanian	Caucasian		MN 2
		M 1 b	NN 1		Egerian		MN 1
23.8		M 1 a					
25	OLIGOCENE	P 22	NP 25	Chattian		Caucasian	MP 30 - MP 28
							MP 27 - MP 24
		P 21 b	NP 24		Kiscellian	Roshnean	
30		P 21 a		Rupelian		Solenovian	MP 22 - MP 21
		P 20	NP 23				
		P 19					
		P 18	NP 22			Pshekhian	
33.7		P 17	NP 21	Priabonian	Priabonian		
35	EOCENE	P 16	NP 19/20			Beloglinian	MP 20 - MP 17
		P 15					

(a)

Fig. 23.1. Relationship of MN 'zones' to chronometric, chromostratigraphical and planktonic zonation. (From Rögl, 1999)

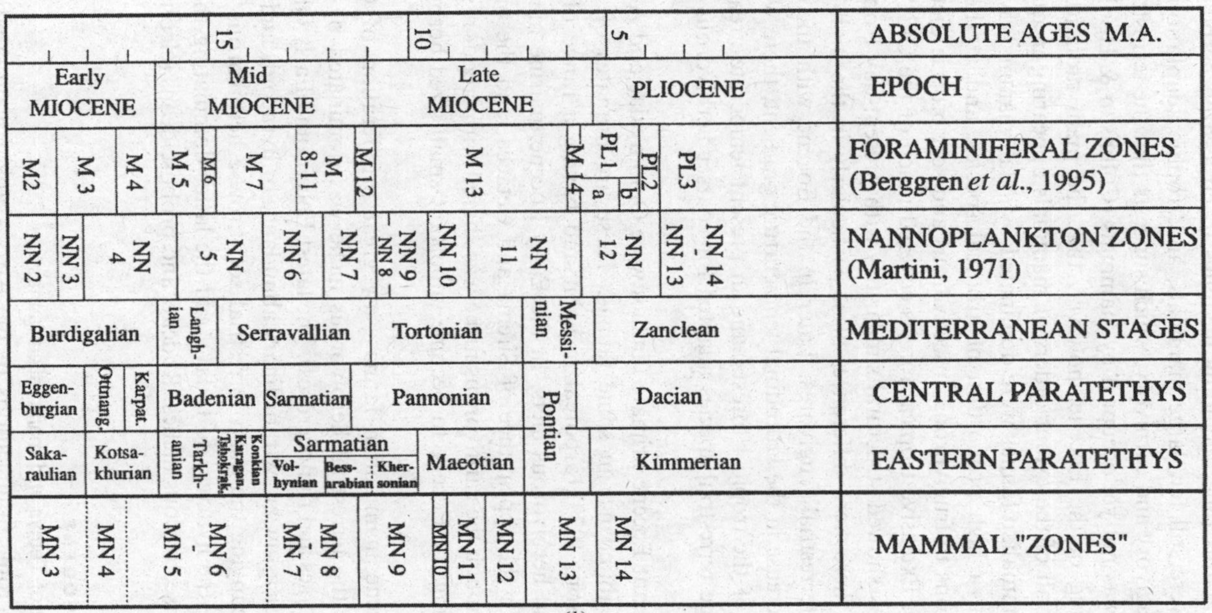

ABSOLUTE AGES M.A.	EPOCH	FORAMINIFERAL ZONES (Berggren et al., 1995)	NANNOPLANKTON ZONES (Martini, 1971)	MEDITERRANEAN STAGES	CENTRAL PARATETHYS	EASTERN PARATETHYS		MAMMAL "ZONES"
	PLIOCENE	PL3	NN 14 - NN 13	Zanclean	Dacian	Kimmerian		MN 14
5		PL2	NN 12					
		PL1 b a						MN 13
		M 14		Messinian	Pontian			
	Late MIOCENE	M 13	NN 11			Maeotian		MN 12
				Tortonian	Pannonian			MN 11
10			NN 10					MN 10
			NN 9			Sarmatian	Kher- sonian	MN 9
			NN 8				Bess- arabian	
		M 12	NN 7	Serravallian	Sarmatian		Vol- hynian	MN 8 - MN 7
	Mid MIOCENE	M 8-11	NN 6			Konkian Karagan. Tshokrak.		
15		M 7	NN 5		Badenian	Tarkh- anian		MN 6
		M 6		Langh- ian				
		M 5				Kotsa- khurian		MN 5
	Early MIOCENE	M 4	NN 4	Burdigalian	Karpat.			MN 4
		M 3	NN 3		Ottnang.	Saka- raulian		MN 3
		M2	NN 2		Eggen- burgian			

(b)

Fig. 23.1 (cont.)

analogy to the Mesozoic, it is likely that early mammals played a greater ecological role in a world where dinosaurs appeared to be dominant.

The early Paleocene faunas maintained an archaic look, but some larger forms appear at this time, including medium- to large-sized herbivores and carnivores. There were still few large mammals, and trophic adaptations were still for insectivory, frugivory and carnivory, which suggests that the terrestrial herbivorous trophic niche was not yet occupied by mammals (Collinson & Hooker, 1987). It is at the beginning of the Eocene, however, that the main radiation of mammals occurred, with all of the extant orders of mammals appearing in quick succession. Again, the mammalian faunas were dominated by small mammals with many small terrestrial taxa, a high proportion of arboreal species, and at the trophic level, many adaptations to frugivorous, insectivorous and soft browse diets (Collinson & Hooker, 1987). Extensive tropical forests covered much of the Old World, and the mammal faunas showed structural similarities with present-day tropical forest faunas, although there were differences also, particularly in the high proportions of arboreal and terrestrial frugivores. Later in the Eocene with the cooling of the climate and increase in the latitudinal temperature gradient, there was great reduction in extent of the tropical forests, and in present temperate areas there was an increase in large terrestrial species adapted for coarser herbivorous diets (Hooker, 1992).

Dominant Eocene primate families were Omomyidae and Adapidae, which reached great abundance in some faunas. These are primates of modern aspect (Szalay & Delson, 1979), arboreal in habit and having a range of insectivorous, frugivorous and herbivorous diets. In the Early Eocene, marine mammals became established with the appearance of Sirenia and Cetacea, and the order Carnivora diversified, although the carnivorous niche was shared with creodonts. Rodents also diversified during the Eocene to become important small sized herbivores and frugivores.

Oligocene mammalian faunas saw great proliferation of Carnivora and Rodentia, and the terrestrial herbivorous niche was dominated by Perissodactyla (present-day horses and rhinoceroses). The latest Eocene and Early Oligocene faunas of northern Africa saw an extraordinary abundance of hyracoids and primates, with the former occupying the main terrestrial soft-browse herbivorous niche (as their living relatives, the hyraxes, do today), and the latter dominating the arboreal frugivorous and insectivorous niches. Rodents and proboscideans (elephants) were also abundant.

23.6   THE NEOGENE
23.6.1   Closure of Tethys and mammalian faunas

The tectonic ramifications of the Afro-Arabian plate creating a subaerial connection with southwestern Asia in the Miocene was the most important single event for the intermingling of Old World Neogene terrestrial biotas. It also brought about vegetation changes to their habitats resulting from closure of the Tethys seaway in Asia and reduction of oceanic influence on terrestrial climates.

The timing of the first marine disconnection of the Tethys seaway in the Middle East, with the inference that a land bridge had formed between Arabia and south-western Asia, is crucial for establishing the time when African continental mammals may have entered Eurasia and vice versa. Adams *et al.* (1983, p. 294), believed '... by mid-Burdigalian times at the latest ... a definite barrier to the dispersal of marine organisms existed between the Indian Ocean and the Mediterranean, and this could have been the land bridge needed for the ... dispersal of mammals.' (see also Adams *et al.*, 1999). The continuity of marine sedimentation in the Mesopotamian–Arabian Gulf region had also been interrupted during the Aquitanian (Adams, 1967) and there was therefore the possibility that an ephemeral landbridge may have existed before mid Burdigalian times (see also Whybrow, 1984, 1987, 1992). Rögl and Steininger (1983, 1984) and Rögl (1999) believe there is good evidence for a reconnection of the Indian Ocean with the Mediterranean during the Langhian at about 16 Ma. Unfortunately, pertinent marine biostratigraphical data that may help resolve this problem lies in the now politically sensitive area of Kurdistan, in the border areas of Iraq, Iran and Turkey. The Langhian reconnection would obviously have formed a barrier to further mammalian dispersal into and out of Afroarabia–Eurasia, and Rögl and Steininger's hypothesis prompted Thomas (1985) to propose two 'Neogene Dispersal Phases' for the dispersal of hominoids out of Africa, the first phase occurring at about 19 Ma and the second, following the Langhian reconnection, at about 15–14 Ma.

Evidence from the continental vertebrates collected from three sites in eastern Saudi Arabia, *ca.* 16–19 Ma (Hamilton *et al.*, 1978; Thomas *et al.*, 1982; Whybrow, 1987) and from Baluchistan, Pakistan, *ca.* 22 Ma (Raza & Meyer, 1981; Downing *et al.*, 1993) strongly suggests that faunal links between southwestern Asia and AfroArabia became established at 20 Ma (Whybrow *et al.*, 1982; Barry *et al.*, 1985) or earlier. The archaic Paleogene mammalian faunas of Africa then changed dramatically by dispersal into Africa of some Eurasian mammals; conversely, Eurasian faunas changed by the immigration of African mammals, notably proboscideans, bovids and hominoids.

From the Late Paleogene, doming prior to rifting was taking place in Kenya and Ethiopia and an anticlockwise shift north-eastwards of the Arabian plate, initiating the formation of the Red Sea and Gulf of Aden rifts, had been in progress since Paleocene times. Topographic variability along the rift margins would have affected, and still does affect, local climates in East Africa, the Horn of Africa and in southern Arabia. In Asia, there was almost continuous non-marine deposition in the Himalayan foredeep throughout the Tertiary; the sediments reach a maximum thickness of some 7000 metres. Vertebrate fossils from these rocks – the Siwalik formations in Pakistan and adjacent areas – provide a remarkable record of Neogene terrestrial life in Asia from about 22 to less than 1 Ma. Here, changes to mammalian community structure are correlated to climatic changes, increasing seasonality and aridity (Barry *et al.*, 1985; Barry, 1995).

During the Eocene, parts of north-western Europe were initially connected to north America and at the Eocene–Oligocene boundary the Tethys Ocean ceased to exist and the Mediterranean and Paratethys formed, the latter an enclosed basin. Tropical floras flourished from the London basin through North Africa to southern Arabia (As-Sarruri *et al.*, 1999). The Oligocene, Early and Mid Miocene of the Mediterranean are characterized by the presence of islands and basins and by the huge Dinarian–Anatolian land mass, sometimes an island and sometimes connected to the southern European mainland in the Alpine region and to Asia in northern Iran (Rögl & Steininger, 1984; Rögl 1999).

23.6.2   Early Miocene mammal faunas – 23 Ma to 16 Ma
Africa

Early Miocene faunas from East Africa were very different from the latest Oligocene north African faunas. They have a distinctly modern look, with community structures essentially like those of the present day (Andrews, 1992a, 1996; Van Couvering, 1980). During the Early Miocene of East Africa, two faunal assemblages have been recognized (Pickford, 1981), dating from 19–20 Ma and about 17 Ma, but structurally they are similar to each other and both are similar to present-day rain forest mammal communities. Where detailed morphological comparisons have been made between elements of the constituent faunas, as has been done for the hominoid primates (Begun et al., 1994), there are minor differences in post-cranial anatomy that indicate reduced reliance on arboreal substrates and greater degree of terrestriality in the younger faunas from Rusinga Island, Kenya, compared with the earlier ones (represented by sites such as Koru and Songhor, Kenya). In both faunas, however, hominoids were one of the more common families of mammal, and they were one of the main parts of the arboreal frugivore niche, although there is evidence that at least one species was more folivorous than the others (Ungar, 1996). Large terrestrial herbivores were also more abundant in the later faunas from the Early Miocene, and together these features indicate a less rich type of forest probably associated with greater seasonality of climate. This change in climate could be related to global cooling during the Early Miocene, but it could equally be due to topographic differences between the two regions of East Africa from which the two faunas come.

A detailed analysis of the community structure of the combined Early Miocene faunas from East Africa is shown in Figs. 23.2 to 23.4. The size distribution of the faunas (Fig. 23.2) shows an even spread in size categories for these faunas, with relative proportions of large vs. small species similar to present-day proportions. The spatial distribution has high numbers of arboreal and semiarboreal species (Fig. 23.3), which is typical of present-day tropical forest faunas, and the trophic analysis similarly has high proportions of frugivorous and browsing herbivorous species (Fig. 23.4), again typical of present-day forest faunas. It is very apparent from this that there is little difference in environment between these Early Miocene forest habitats and present day tropical forest habitats, so that within the limits of the equatorial forest belt there has been little ecological change despite the major climatic change between then and now.

Europe

The Early Miocene faunas of Europe shared many species with the African faunas, but in lacking key elements of the tropical forest faunas their environmental association was very different. No primates, macroscelidids or anomalurid rodents are known for the European Early Miocene so that the community structure lacked the abundance of arboreal frugivores seen to be characteristic of Africa at this time.

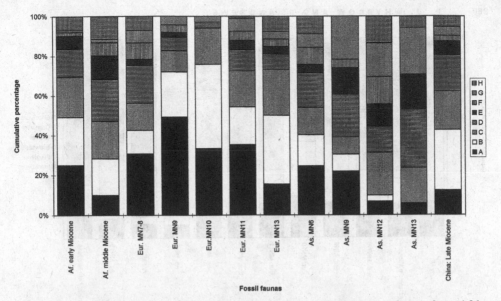

Fig. 23.2. Cumulative distributions of body mass for Miocene mammalian faunas from Africa and Eurasia. Af. Early Miocene: ecological distribution based on Early Miocene faunas from Rusinga Island Hiwegi Formation, Songhor and Koru Chamtwara Formation, Kenya; Af. Mid Miocene: ecological distribution based on Mid Miocene faunas from Maboko Island, Ngorora and Fort Ternan, Kenya; Eur. MN zones: Mid Miocene faunas (MN7–8) and Late Miocene faunas (MN9–13) from west and central Europe, with ecological distributions averaged for each unit (Fortelius *et al.*, 1996); As. MN zones: Mid Miocene faunas (MN6) and Late Miocene faunas (MN9–13) from western Asia, with ecological distributions averaged for each unit (Fortelius *et al.*, 1996); the Chinese Late Miocene distribution is based on the fauna from Lufeng. Size classes are as follows: A, 0–100 g; B, 100–1000 g; C, 1–10 kg; D, 10–45 kg; E, 45–90 kg; F, 90–180 kg; G, 180–360 kg; H, > 360 kg.

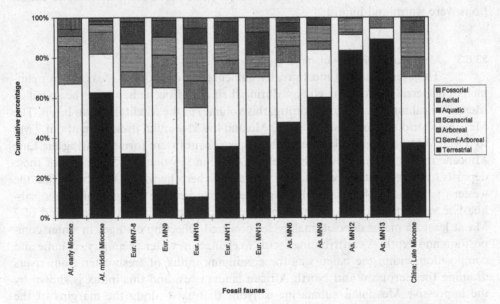

Fig. 23.3. Cumulative distributions of locomotor spatial adaptations for African and Eurasian Miocene mammalian faunas. Categories explained in the text and in Fig. 23.2 caption.

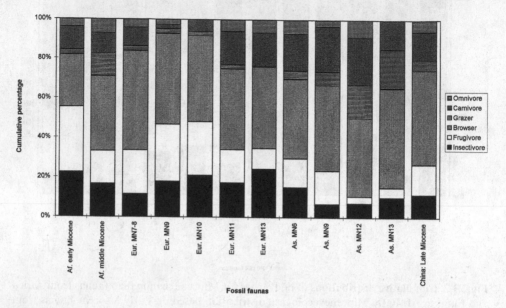

Fig. 23.4. Cumulative distributions of dietary guilds for African and Eurasian Miocene mammalian faunas. Ecological categories are explained in the text and in Fig. 23.2 caption.

Most of the faunas were dominated by large terrestrial mammals, with perissodactyls and artiodactyls both well represented, and carnivores were also abundant. It is not possible to detect faunal change related to global climatic change on present levels of resolution in the European Early Miocene, but indications are that climatic conditions were warm and humid.

23.6.3   Mid to Late Miocene mammal faunas – 16 Ma to 5 Ma

The event that has most greatly influenced and coloured ideas about mammalian dispersal and climate change during the Late Miocene has been the so-called Messinian salinity crisis (see Pickering, this volume) in the Mediterranean basin. The Messinian 'aridification' lasted about 2 Ma and the Messinian itself from about 7 Ma to 5 Ma. We say so-called because although evaporites were formed during the Late Miocene in the Arabian Gulf, the Gulf of Aden and in the Red Sea, none of these deposits has received the accolade of a 'crisis'. Further, the widely held belief that the western part of the Messinian Mediterranean basin became a region of barren sabkha-like salt flats has been replaced by studies indicating that over the period of 2 Ma at least 25 major depositional cycles occurred reflecting changes in water composition and depth. A contributing factor to changes in water chemistry, volume and composition within the basin was the continuing influx of freshwater from rivers draining the European and North African landmasses, and this influx is shown by the impressive Messinian submarine canyons that exist along the margins of the basin (Clauzon *et al.*, 1995; Schreiber & Ryan, 1995). The eastern part (roughly east of Italy) of the Messinian Mediterranean basin appears to have had a different depositional regime, probably because of its connection with the Black Sea (Agusti, 1989; de Bruijn, 1988).

Aridification/desertification associated with the so-called Mediterranean salinity crisis do not appear to have affected faunas and floras of the surrounding landmass. Woodland and forest mammal species existed in northern Africa, Greece, Italy and Iberia. Faunas and floras are associated with some of the sediments deposited by rivers draining towards the Mediterranean basin from southern Europe and northern Africa. However, although the environment outside of the fluvial system is unknown, the perceived continental climate trend is for increasing seasonality not hyperaridity. Similarly, a Late Miocene Mediterranean salinity-crisis hiatus is not recorded from oceanic evidence of aeolian dust (Partridge et al., 1995a, p. 13).

It is likely that the uplift of the Tibetan plateau during the Neogene has influenced the regional climate (and perhaps global climate) and thus the habitats of Late Miocene terrestrial animals in Africa, Arabia and Asia to varying degrees. Certainly an increase in seasonality occurred. Although there has been discussion about the degree of uplift, its timing and therefore its effect on regional climate, uplift of the Tibetan plateau during the Neogene of several thousand metres appears to have been more pronounced at about 8 Ma (Molnar et al., 1993).

In AfroArabia and south-western Asia there are few Late Miocene sites of this age and in the region as a whole relevant comparative sites might be Sahabi, Libya (Boaz et al., 1987); Lothagam, Kenya (Leakey et al., 1996); Tugen Hills, Kenya (Hill, 1995); Baynunah region, United Arab Emirates (Whybrow & Hill, 1999); and the Siwaliks of Pakistan (Barry, 1995).

Today, plants utilize two different photosynthetic pathways and it is assumed that the same pathways were used in the past. These pathways can be differentiated by the relative amounts of two naturally occurring stable carbon isotopes, ^{12}C and ^{13}C. They are referred to as C_3 which is generally a photosynthetic pathway that reflects wooded-forested habitats, and C_4, which generally reflects tropical grassland habitats. At four of the AfroArabian and south-western Asian sites carbon isotope studies of palaeosol carbonates and enamel apatite have been carried out; Lothagam (Leakey et al., 1996, Tugen Hills (Morgan et al., 1994), Baynunah (Kingston, 1999) and the Siwaliks (Quade & Cerling, 1995). Interpretation of the C_3 and C_4 evidence of the dietary preference of herbivores found at these sites appears equivocal, but evidence does suggest an increase in grasslands and/or open woodlands at those sites between 9 Ma and 6 Ma. On the other hand, in Kenya, C_4 grasses are stated to have been present by 15 Ma, although they were not the primary food for herbivores until 7 Ma (Morgan et al., 1994). This might be interpreted as an artefact of the site being within the African rift – atypical for Africa as a whole – or that tectonically induced regional climate change (related to the development of grasslands) occurred earlier, both in Africa and in Asia.

Africa
The Mid Miocene faunas available for study range from 15 to 12 Ma. Late Miocene faunas are poorly known and cannot be included in this survey. There was a substantial change in mammalian faunal composition from Early to Mid Miocene in East Africa. Most important was the spread of true bovids and giraffids, although

tragulids continued as minor elements of the terrestrial browsing fauna. Large terrestrial herbivores generally became more abundant in the Mid Miocene, and their relative abundance increases up through the stages of the Mid Miocene (Shipman *et al.*, 1981). Hypsodont bovids appear at Fort Ternan, Kenya, and monkeys become abundant for the first time in the fossil record, e.g. in the Maboko Island fauna from Kenya (Benefit, 1993).

These changes are reflected in the analysis of community structure (Figs. 23.2–23.4). Small mammals appear to be less abundant in the Mid Miocene faunas of Africa (Fig. 23.2), although this is probably in part the result of taphonomic bias (Shipman *et al.*, 1981). In the spatial analysis (Fig. 23.3) the high proportion of terrestrial mammals can be seen, with many fewer arboreal species, and this probably reflects an opening up of the vegetation structure compared with the Early Miocene forests. Similarly, the trophic analysis (Fig. 23.4) shows reduction in proportions of frugivores and an increase in herbivores, which make up 50% of the fauna compared with less than 30% in the Early Miocene forests. Grazing herbivores appear in relatively high proportions, identified in this case by the presence of increased crown heights of the teeth (Harris, 1993). This analysis indicates that there had been a substantial change in the palaeoenvironment between the Early and Mid Miocene in East Africa, which led to a decrease in forest cover. Since the tectonic activity that later affected this region had yet to produce any major topographic effect, the palaeoenvironmental changes probably reflect an increase in seasonality associated with climatic cooling. During the later Miocene, tectonic movements and uplift in eastern Africa associated with rift valley formation produced fragmentation of the environment leading to the present-day diversity of habitats.

Europe

The Mid Miocene of Europe shows similarities both with the African Mid Miocene and the European Early Miocene. MN stage faunas are shown in the analysis in Figs. 23.2–23.4 based on a compilation of faunas from western Europe (France, Spain) and central Europe (Germany, Switzerland; Poland, Austria, Hungary, Czech Republic, Slovakia). The size distributions show an even distribution, although the MN9 fauna has higher proportions of small mammals which may reflect the greater emphasis on small mammal screening in these sites. The greater proportions of small mammals generally, compared with the African faunas, may again reflect the longer tradition of screening for small mammals in Europe. On the other hand, it is possible that the abundance of small mammals indicates a real ecological event, although the differences observed do not have any recognizable ecological meaning in the context of present-day mammalian faunas. It is also a feature of present-day faunas that the variability within and between different habitats is so great that size distribution has little significance. There is no apparent decrease in small mammal species diversity between MN9 and MN10 (Fig. 23.2).

The spatial analysis of the European Mid Miocene faunas (Fig. 23.3) has also been affected by taphonomic and sampling bias, as shown in the body size analyses, but despite this some interesting patterns emerge. There is a striking similarity between the MN7–8 fauna from Europe and the Early Miocene African faunas, both in terms of the low proportions of terrestrial species and the abundance of arboreal and semiarboreal species. Moreover, the European faunas also have higher proportions of scansorial species to further increase the indications of dense tree cover. These proportions are maintained into the Late Miocene, even increasing in the MN10 faunas, and they are strikingly different from the African Mid Miocene faunas, which have been shown above to be influenced by climatic seasonality leading to more open forest environments. The European faunas have been compared with a variety of present-day mammalian faunas (Fortelius *et al.*, 1996), and they show greatest similarities with present-day African and Asian tropical and subtropical forest faunas. The pattern of reduced terrestriality and increased arboreality compared with Mid Miocene African faunas may indicate less seasonality in the European climates and presence of evergreen tree cover. After MN10, proportions of terrestrial mammals rise, complemented by a decline in arboreal species, indicating a later decrease in tree cover.

The distribution of dietary guilds in the European Mid Miocene faunas gives evidence of taphonomic bias against carnivorous species, which are underrepresented in all faunas (Fig. 23.4). The other faunas are dominated by browsing herbivores, with few grazing species, and the distribution of dietary guilds is at the lower limits of the modern subtropical faunas. There is little change in browsing/grazing proportions through time in the sequence of faunas, for the faunas with the lowest proportions of browsers also have minimal grazers. There is a trend towards increasing insectivory and frugivory. This corresponds with the decreasing terrestriality and increasing arboreality seen in MN9 and MN10 described above. Proportions of frugivores, insectivores and browsing herbivores are similar to those in the Early Miocene African faunas, indicating more closed habitats in Europe at this time, but later in the Miocene (MN11–13) browsers and carnivores increase and frugivores decrease again, indicating opening of the habitat.

There is thus a trend from around 13 to 6 Ma, represented by the faunas from MN7 to MN13, from closed forest habitats in Europe to more open forests. This trend follows on from the indications of warm humid climates in the Early Miocene. The faunal change represents climate change from equable conditions in the lower part of the sequence, similar to that seen in the Early Miocene, to more seasonal conditions at the top. This change corresponds to evidence of climatic cooling during the Early to Late Miocene.

West Asia

Analyses of community structure of Mid Miocene faunas from western Asia (Turkey, Iran) and south-eastern Europe (Greece, Romania, Bulgaria, the former Yugoslavia, Georgia) are also shown on Figs. 23.2–23.4. The faunas range from the Mid to near the end of the Miocene: MN6, about 15Ma to MN13, about 6Ma. They show evidence of taphonomic bias in the size analysis, with varying proportions of large and small mammals. Carnivores and omnivores were almost absent in the MN6 faunas, and small mammals were underrepresented in the MN9 faunas. The MN6 fauna appears least biased, mainly because it includes one well-documented site in Turkey (Paşalar: Andrews, 1990; Bernor & Tobien, 1990). The faunas were dominated by increasing proportions of large terrestrial mammals through time (Fig.

23.3), which indicates a pattern analogous to seasonal woodland and forest faunas living today. The low proportions of arboreal and scansorial species indicate that environments were more open than they were at corresponding periods in Europe. This has interesting consequences for provinciality of faunal change in western Eurasia by which the ecological composition of these eastern faunas in the Mid Miocene (MN6–8) was not seen in western Europe until the Late Miocene (MN11–13).

The distribution of dietary guilds also shows evidence of open country conditions in western Asia throughout the sequence, and this compares with more closed conditions at equivalent periods in Europe. Species with frugivorous adaptations were less abundant at all levels in Asia than in Europe, and the faunas were dominated by browsing herbivores. Proportions of grazing herbivores were higher throughout the Miocene sequence than is seen in the European sequence, and they are higher also than in extant subtropical Asian faunas except for those from the most open deciduous forests. After MN9 the proportion of grazers increased, so that the pattern of the fossil faunas is closest to that found in the South African summer-rainfall woodland habitats. Overall, the habitats indicated vary from open deciduous seasonal forest in MN6 to 9 to increased seasonality and decreased tree cover towards the top of the sequence. This change corresponds in its timing to the change in western Europe, although the actual change involved was different in the two regions. Both changes may be due to global climatic change occurring during this period (15 to 6 Ma).

East Asia

In the long section of the Siwalik deposits of Indo-Pakistan, faunal trends similar to those seen in Europe and western Asia have been described (Barry, 1995; Barry *et al.*, 1985, 1990). These are difficult to correlate with the European and west Asian trends, because the MN zonation does not apply to these faunas. In a study on the numbers of species of rodents and artiodactyls in terrestrial Miocene deposits in Pakistan and India, the major changes in numbers of species occurred between 12.5 to 10.5 Ma, approximately equivalent to MN7–9 of the European biochronology, and after 9.5 Ma, roughly equivalent to MN10 (Barry *et al.*, 1990). High species diversity is indicated in the Mid Miocene before 13.5 Ma, lower diversity from 12.5–9.5 Ma, and lowest diversity from 9.5–7.0 Ma (Barry *et al.*, 1990). Rodents are initially more species-rich than artiodactyls, a relationship indicating environments with closed wooded vegetation (Andrews *et al.*, 1979), and the decline in diversity is a product mainly of loss of rodent species. This change both in species composition and in species richness indicates a degree of opening of the vegetation structure, a conclusion consistent with the isotope record for the same sequence, which showed transition a from C_3 closed vegetation to open country C_4 grasses during the period 9.5–7 Ma (Quade *et al.*, 1989). These ecological changes have been correlated with the development of the Asian monsoon resulting from the uplift of the Himalayan mountain chain and the expansion of the Tibetan Plateau (Quade *et al.*, 1989), and they approximate to the time of marked diversity decline in western Asia in the MN12–13 zones, with biostratigraphical dates of around 7 Ma (de Bruijn *et al.*, 1992; Fortelius *et al.*, 1996).

Arabia

The Arabian plate occupies a pivotal position for ideas about mammalian dispersal out of Africa during the Neogene. It lies at the junction of three continents, Africa, Europe and southern Asia, during the Miocene and it is likely to have been a significant corridor for the movement of faunas between those continents. However, although Arabian sites with terrestrial palaeofaunas and floras are unfortunately rare, the Late Miocene sites of the Baynunah region in the Emirate of Abu Dhabi, United Arab Emirates, are important for several reasons. They are the only Late Miocene terrestrial vertebrate sites in the whole of Arabia, and the most proximal sites with good biotic data to Asian and some northern African sites of a similar age – between 6 and 8 Ma (Hill *et al.*, 1990; Whybrow *et al.*, 1990; Whybrow & Hill, 1999). Vertebrates, invertebrates and plants are found in sediments from a fluvial system that may have been part of a Miocene extension of the Tigris/Euphrates river system. The Miocene river cut into aeolion rocks dated at about 14 Ma. It is tempting to speculate that the change in the climate of Arabia at this time, 6–8 Ma, from hyperarid conditions to grassy woodlands within a river basin might be linked to the regional effects on climate prompted by increasing uplift of the Tibetan plateau.

As might be expected, given the long-established physical link of Arabia with the African plate, its Late Miocene mammalian fauna is predominantly African in nature. Interestingly, the fauna does include some Asian elements such as rodents, pigs and a bovid suggesting that a belt of faunal similarity stretched from parts of Asia, through Arabia, into northern Africa and possibly into eastern Iberia; a belt barred to the north by seas and mountains and, to the south, by ancient deserts and seas. Thomas (1979) first put forward the idea of a southern AfroArabian latitudinal climatic barrier – his 'Saharo-Arabian' belt – and studies of Arabian Late Miocene terrestrial biotopes imply that this idea might be further investigated.

23.7 CONCLUSIONS

The tectonic effects of the AfroArabian and Indian plates impinging on the Eurasian plates prior to and during the Neogene dislocated well-established pre-Tethyan ocean current circulation patterns and produced orographic features that similarly dislocated established atmospheric circulation in the Old World. Within an imprint of global latitudinal climate variation on terrestrial environments, it is likely that tectonically induced orographic events, especially uplift of the Tibetan–Himalayan region in the Late Miocene, influenced regional climate regimes to an extent comparable with changes to polar ice-sheet size or to sea levels. This orographic–climatic influence probably caused the contraction of Paleogene–Early to Mid Miocene habitats and initiated the change to more seasonal forests over much of the Old World. Mammalian community patterns thus evolved in response to expanding–contracting, regional or local vegetation communities.

In Africa there was a shift in the Miocene from faunas adapted for tropical forest environments in the Early Miocene to more seasonal forests in the Mid Miocene. The European Early Miocene faunas appear similar to the later African faunas, but in the Mid Miocene there are greater similarities with tropical and subtropical forest environments increasing in richness into the Late Miocene

(MN9). This trend was reversed after MN9 (Fortelius et al., 1996), indicating climate deterioration and increasing seasonality. In western Asia, environments were more open than at equivalent times in Europe, with seasonal forests giving way to more open conditions after MN10, but the same trend is observed as in Europe. In eastern Asia (Pakistan) there was a gradual decline in mammalian species diversity from 12.5 Ma (temporally equivalent to MN7–8 in the European sequence) reaching its maximum at around 7 Ma (temporally equivalent to MN13), again indicating loss of tree cover as a result of climate deterioration.

Within the limits of present chronostratigraphical resolution, there appears to be no synchroneity of non-catastrophic abiotic global change. Instead, the ramifications of global tectonic events seem to produce a mosaic of unstable marine, atmospheric and orographic conditions that, through time, have had a domino-like effect on the regional and local extinctions, distributions and habitats of terrestrial mammalian biotas. These biotas have a flowing, vicariant, opportunistic relationship within the habitats provided by their physical world. Consequently, linking the imperfect chronostratigraphy of terrestrial abiotic and biotic events on a global scale to infer synchronous mammalian dispersal or species turnover is at present perceived by us as unproductive until prime Old World vertebrate sites can be dated by physical methods to a resolution of less than 1 my. Speculation can and should, however, continue on the response of Neogene terrestrial fossil faunas to climate change in order to achieve a productive debate.

In the long term, more data are needed about past terrestrial environments. For example, better sampling of terrestrial biota to include lower vertebrates, invertebrates and plants; better dating of a biota in order to constrain biostratigraphical sequences; and in-depth systematic revision of certain taxa, perhaps using cladistic rather than phenetic methods. In the short term, we feel that use of the stable carbon isotope technique to analyse the C_3 and C_4 components of palaeosol carbonates and enamel apatite in herbivores (see Cerling et al., 1989; Kingston et al., 1992, 1994; Kingston, 1999; Morgan, 1994: Quade et al., 1992, 1995) should provide more meaningful data about the effects of climate change on Neogene terrestrial ecosystems.

Mammalian response to global change in the later Quaternary of the British Isles

ANDREW CURRANT

24.1 INTRODUCTION

In the area now known as the British Isles, the response of mammalian species to the major environmental changes which took place during the Quaternary Period include some of the most dramatic changes in faunal composition known from any part of the geological succession, anywhere in the world. To illustrate how the distribution of mammals was affected, this chapter focuses on the later Quaternary faunal history of Britain during a major temperate interglacial period and throughout the succeeding cold phase (Devensian).

To an observer standing at the top of London's Northumberland Avenue and looking across Trafalgar Square towards the National Gallery, the only non-human fauna likely to be visible today is a large flock of feral pigeons, Landseer's four bronze lions and the equestrian statue of King Charles I. Yet just beneath the surface of this scene the fossil remains of hippopotamus, straight-tusked elephant and narrow-nosed rhinoceros are abundant (Franks, 1960). An eastwards glance down the Strand past Charing Cross Station would probably show nothing but a mottled red and black scene of buses and taxis, yet beneath the buildings all around lie the bones and teeth of mammoth, reindeer and woolly rhinoceros (Sutcliffe, 1985). All of these fossils date from the later part of the Quaternary Period, the geologically recent past, and the all too obvious message is that things have not always been as they appear today. Here, one is dealing with the interface between geology and the modern world, and the surviving record of comparatively recent events contains very tangible evidence of an ongoing progression of major environmental changes.

24.2 GLOBAL INFLUENCES ON MAMMALIAN DISTRIBUTION

Changes in the British mammalian faunas clearly reflect the complex global climatic and associated environmental changes that occurred during the Quaternary. The Quaternary Period can be defined as that part of the Late Neogene in which orbitally driven cyclic climatic change (see Pickering, this volume) gave rise to the periodic build-up of very large-scale polar and non-polar ice masses. Polar ice masses are known to have existed and undergone major fluctuations in extent over a much longer period, but it is only during the Quaternary that there was a significant mammalian faunal response to these changes. Although we are currently something like 10 000 years into a major interglacial warm interval, for most of the Quaternary period global environments have been significantly colder than they are at present, particularly in the higher latitudes (Ballantyne & Harris, 1994). We are currently at something of a thermal maximum in the known range of Quaternary climatic regimes. If we are right about the underlying pattern of past events, then at some time in the predictably distant future there will be a general world-wide return to colder conditions. While present concerns about the effects of global warming are fully justified, they can create the false impression that we can somehow maintain the environmental *status quo* indefinitely simply by modifying our behaviour. The fossils beneath our feet serve to remind us that this is not likely to be the case.

The cyclic changes in climate during the Quaternary led to major changes in global ice volume, which in turn had a direct effect on sea level. When total ice volume is high, enormous quantities of water are locked up on land and there is a consequent reduction in sea levels. Estimates vary, but during some of the major cold stages global sea levels may well have been over 100 metres lower than they are at present (Zeuner, 1945). Sea-level changes are very important to terrestrial vertebrate populations because they dictate the total land surface available for colonization, and they open or close lowland migration routes between land masses. During the Quaternary, at times of lowered sea level, there was an extensive high latitude land connection between Asia and North America, known in the literature as Beringia, which permitted large scale faunal interchange and provided the orginal route for human expansion into the Americas (Hopkins *et al.*, 1982). Similarly, the area which is now the British Isles was a peninsula of north-west Europe with a greatly extended coastal lowland belt. Remains of mammoth and other cold-stage mammals are commonly dredged from what is now the floor of the North Sea (Leith Adams, 1877–81). Conversely, the occurrence of former phases of high sea level is confirmed by the existence of fossil beach deposits at, or even well above, modern sea level, some of which contain the remains of extinct Quaternary mammals (Lyell, 1865).

As well as global ice volume and related sea levels, another very important factor influencing the distribution of mammals throughout the Quaternary has been the periodic expansion, movement and contraction of areas of moderate to extreme aridity (cf. Lowe & Walker, 1984). It would appear that such desert zones can occur at almost any latitude and are directly related to the concurrent distribution of continental landmasses, global wind patterns and prevailing ocean currents. Direct evidence for former arid environments includes characteristic landforms, large-scale aeolian deposits such as dune sands and loesses, and relict desert soil profiles. Desertification is one of the most important elements of climatic change currently influencing human geography, although predicted rises in sea level resulting from anthropogenically induced global warming could soon challenge this position quite dramatically (Glantz, 1977). Naturally enough, presence or absence of available water has a massive impact on the distribution of plants and animals. It is interesting to observe that evidence for former arid environments can even be found in Quaternary sequences in Britain, although the biological significance of such environments has been largely overlooked (e.g. Matthews, 1970).

At a broad scale, there is no well-developed pattern of correlation between global ice volume and periods of desert extention as was once believed. Some arid regions like the Namib Desert of Namibia are apparently very old (in excess of 6 million years) and support a wide range of animals and plants which are adapted to extreme aridity. These deserts are usually associated with upwelling cold oceanic currents adjacent to major continental landmasses. Other deserts, like the Sahara, are more recent in origin and tend to lack specifically adapted biota. The Sahara has grown phenomenally even during the comparatively short period of recorded human history, engulfing large tracts of once fertile land along the north African coast and encroaching on the woodlands of Chad, Niger and Mali. In this case the development of a major arid region has probably been influenced by changes in global airflow and consequent patterns of rainfall (e.g. Grove & Warren, 1968). The growth and survival of ice masses also shows a degree of non-synchroneity, as is well illustrated by the present development of the Greenland ice cap. The important observation to be made here is that, although the overall pattern of global climate would appear to be driven by some major underlying mechanism, regional effects can have strongly modifying influences within this general pattern (e.g. Occhietti, 1983).

Here at a relatively simple level we have the major factors influencing the distribution of mammals and indeed many other terrestrial life forms across the face of the globe. Heat and cold, wet and dry in their various combinations account for much of the observed environmental change which took place during the Quaternary Period, and it is relatively easy in an environmentally responsive area like Britain to pick out the characteristics of the mammalian faunal groupings which go with each of these combinations. However, the accurate dating and correlation of some of the faunas can be difficult.

The oxygen isotope record (Shackleton & Opdyke, 1973; Pickering, this volume) provides us with a potentially very detailed framework for looking at Quaternary sediments, biotas and events. However, the essentially fragmentary nature of most terrestrial sequences and the absence of reliable, independent dating methods which are applicable to most Quaternary samples mean that correlations are, in practice, rather difficult. For most practical purposes we are reduced to a sophisticated version of 'counting down from the top' from the present day to work out which stage or substage of the oxygen isotope sequence a particular Quaternary sediment or fossil represents. This methodology can only work where there are lots of sites, a good fossil record, and a range of depositional environments to help understand the potential taphonomic biases in fossil assemblages.

24.3   QUATERNARY MAMMALS IN BRITAIN

24.3.1   Faunal assemblages

Historically, much of what has been written about Quaternary mammals and their responses to environmental change has been based on interpretations of the richly fossiliferous cave and river terrace deposits of North West Europe and, in particular, the British Isles. Much of London is built on fossiliferous sands and gravels, which mark earlier stages in the development of the river Thames. During what is known as the Anglian cold stage, currently believed to correspond to oxygen isotope stage 12, the most extensive of the Quaternary ice advances in this part of the world caused glacial ice to reach what is now north London. In doing so, the ice diverted a major drainage route flowing across the centre of southern England slightly southwards to form the present course of the lower Thames Valley (Bridgland, 1994). Since then, the successive changes in mammal faunas brought about by environmental change are well recorded within the deposits of the lower Thames.

Extended a little by earlier evidence from former estuarine and coastal sites, and augmented by finds from other English river systems and a number of important cave sequences, we have been able to reconstruct a fairly good overall picture of the history of the British mammal fauna for much of the last 600 000 years (Currant, 1989). While the record of Quaternary mammal faunas in Britain extends back still further, there are some big gaps in the earlier parts of the sequence. The mammal faunas recovered from successive glacial and interglacial sequences form what may reasonably be claimed as one of the most dramatic palaeoenvironmental contrasts ever observed (Stuart, 1982). Many important members of these faunal assemblages are now totally extinct, but their environmental preferences have been guessed at by drawing comparison with those of extant species which are found in direct association.

During the more temperate periods of the later part of the Quaternary when sea levels were high and mixed oak forest was the dominant vegetation, the British mammal fauna included most of the currently native species. If we take the Last Interglacial, oxygen isotope stage 5, as our model then this includes red deer *Cervus elaphus*, badger *Meles meles*, hedgehog *Erinaceus europaeus*, mole *Talpa europaea*, water vole *Arvicola terrestris*, wood mouse *Apodemus sylvaticus* and bank vole *Clethrionomys glareolus*, etc.; a number of what we would now call exotic species including fallow deer *Dama dama*, wild boar *Sus scrofa*, and during at least one important phase hippopotamus *Hippopotamus amphibius*; and a range of extinct herbivorous megafauna including straight-tusked elephant *Palaeoloxodon antiquus*, narrow-nosed rhinoceros *Stephanorhinus hemitoechus* and aurochs *Bos primigenius* (Sutcliffe, 1960). This list is not complete but it gives a good impression of the distinctive elements of one of these later temperate faunas. It is a thought-provoking observation that, had it not been for local or total extinctions, many happening during the last 10 000 years and most of which can be directly or indirectly linked to human activity, this might well have been the mammal fauna of Britain today.

A contrasting group of fossil remains representing phases of the Last Cold Stage (known in Britain as the Devensian Cold Stage and equivalent to oxygen isotope stages 4, 3 and 2), are commonly encountered in cave and river deposits around the country (Currant & Jacobi, 1997). Species still native to Britain are comparatively few, and some of these are morphologically distinct, for example, a particularly large form of the red deer *Cervus elaphus*, while others are also elements of the warm-stage fauna such as red fox *Vulpes vulpes* and mountain hare *Lepus timidus*. In this case the exotic element consisted of living species like glutton *Gulo gulo*, arctic fox *Vulpes lagopus*, musk ox *Ovibos moschatus*, reindeer *Rangifer tarandus*, Norwegian lemming *Lemmus lemmus*, collared lemming *Dicrostonyx torquatus*, narrow-skulled vole *Microtus gregalis* and northern vole *Microtus oeconomus*, while the extinct component was dominated by woolly mammoth *Mammuthus primigenius* and woolly rhinoceros *Coelodonta antiquitatis*. There are also quite a wide range of species common to both faunas. Bison *Bison priscus* is found in both groups, the form from the temperate assemblages having larger horns. Wild horse *Equus ferus*, which is an important component of other later Quaternary assemblages, appears to be completely absent from Britain during the equivalent of isotope stages 5 and 4. It may be significant that there is no evidence for human activity in Britain during the same period. Humans and horses returned to Britain after a long absence during isotope stage 3, the middle part of the last cold stage.

The other mammals encountered are mainly carnivores, most notably lion *Panthera leo* and spotted hyaena *Crocuta crocuta*, which we could regard as exotics in a modern context, and brown bear *Ursus arctos* and wolf *Canis lupus* which were only exterminated from the modern fauna in recent times. These species are not really climate specific and are commonly associated with both the temperate and the colder faunas.

Although the general picture outlined above is put together with specific reference to Britain, it holds good for most of north-west Europe and the Atlantic seaboard. The Last Interglacial mammal fauna of southern Iberia is substantially the same as that of the British Isles but for minor differences among the smaller species and the presence of an important and climatically ubiquitous montane element in the form of the ibex *Capra ibex*, still found in the region today. During the last cold stage, reindeer, mammoth and woolly rhinoceros are found as far south as the Pyrenees, range extensions equal to or greater than that of the hippopotamus in the preceding warm stage.

24.3.2 General biogeographical considerations

Our interpretations of Quaternary fossil assemblages are, to a very large extent, coloured by the modern distribution of surviving components. For instance, the occurrence of hippopotamus in a British context might give the impression that it was an indicator of tropical or subtropical conditions, and the co-occurrence of such species as lion and spotted hyaena would seem to lend support to this assumption. It is only by reference to other palaeontological data and surviving historical records that we find that this interpretation is wrong. The modern distribution of all three species has been considerably modified by human activity. During the Quaternary, lions and spotted hyaenas had much more extensive ranges including much of Europe and Asia; the lion may even have reached North America, although there is some dispute as to the specific identity of the American fossils. The hippopotamus is essentially a circum-Mediterranean species, which is now limited to the extreme southern part of its former range. While the occurrence of the hippopotamus in Britain is still a remarkable extension of its normal Quaternary range, it is nowhere near as dramatic as if its source region had been in sub-Saharan Africa.

Similarly, the palaeoclimatic significance of the presence of extinct species of elephant and rhinoceros in the British Quaternary fauna is easily misinterpreted. Living members of both these groups again tend to have a tropical distribution, but their recently extinct relatives were once widespread throughout temperate and boreal Europe and Asia, and the elephants and their close relatives were wide-

Table 24.1. *Approximate date of the commencement of isotope stages 5 to 1 with the stage names as used in the text*

Isotope Stage 5	began	*ca.* 128 000 y BP		Last Interglacial
Isotope Stage 4	"	75 000 y BP	,	Early Devensian
Isotope Stage 3	"	64 000 y BP		Mid-Devensian
Isotope Stage 2	"	32 000 y BP		Late Devensian
Isotope Stage 1	"	10 000 y BP		Flandrian

spread in the Americas up to the time of the great Late Pleistocene megafaunal extinctions.

All mammals can maintain their body temperature against an environmental gradient, but there are a wide variety of important physiological adaptations in some species which permit extension of their ranges into environments which would prove inhospitable to others. It would be reasonable to state that the factors which limit the distribution of nearly all living mammals are very poorly understood, and at a species level there is usually phenomenal variation in the spatial distribution of individual members of a population through time. The world over, most mammals exhibit some form of cyclic movement through their environment whether it be diurnal or seasonal, and many species undergo very long migrations from one quite distinct feeding ground to another.

The greatest conservatism in spatial distribution appears to be in equatorial and tropical woodland environments, although movement is comparatively difficult to monitor in such habitats. The larger-scale seasonal movements of mammalian species are usually found in more open environments, the most remarkable of all being the seasonal migrations of the great whales from the their winter breeding grounds near the tropics to their summer feeding grounds towards the polar ice margins. On land too, many herbivorous species travel long distances in their annual round, often to take advantage of the brief but nutritionally valuable flush of vegetation that appears during the short Arctic summer. The North American caribou makes its seasonal migration from its winter feeding grounds in the northern edge of the boreal forests hundreds of kilometres north to the summer breeding grounds on the open tundra of the Arctic Slope. The longitudinal trend of most of the Earth's major vegetation zones dictate that most migrations will have a significant latitudinal trend, the object being to exploit specific resources and/or avoid specific hazards by making use of a range of different environments. In mountain regions and in forests, there is usually a strong vertical component in migratory movement, either a seasonal or diurnal movement up and down slopes, or a primarily diurnal movement between the forest canopy and the forest floor.

24.3.3   Last Interglacial mammals

The known interglacial faunas are broadly similar in overall composition, with a progressive species by species replacement of individual elements through time. Faunal diversity is relatively high. If we look back at the mammals present in Britain during the Last Interglacial, the important elements of which are listed in

Section 24.3.1, we find quite a high proportion of species that are also present in the modern fauna, plus a largely extinct or exotic megafauna (Fig. 24.1). There is every reason to believe that Britain was isolated by high sea levels during the major part of this stage, but it is a big area and we do not find any significant island endemism developing. Admittedly, it is difficult to judge purely from the fossil record, but there does not appear to be a significant migratory component in this interglacial assemblage, even though the latitudinal range of the British Isles would allow the possibility for seasonal movement of this kind. Instead, it seems likely that the necessary environmental diversity came from forest-edge environments which would be developed and maintained by the presence of the larger mammals themselves – just as living populations of hippos maintain wide corridors of open vegetation adjacent to their river homes and modern elephants tend to open up wooded environments during their own feeding activities, which then become available for exploitation by other species.

This faunal grouping contains a high number of elements of the characteristic Mediterranean zone fauna of the later part of the Quaternary.

24.3.4   Last Cold Stage mammals

British interglacial mammal faunas are relatively well known, contrasting quite sharply with cold-stage faunas which are generally very poorly represented in the fossil record, other than that of the Last Cold Stage. What little we know about cold stage faunas further back in the Quaternary suggests that they encompass a wide range of very different kinds of environment. If we take a closer look at the mammals of the Last Cold Stage, we can see something of this environmental diversity.

During the first part of the Last Cold Stage, the equivalent of isotope stage 4, the prevailing environment appears to have been very cold with sparse, open Arctic vegetation. The known fossil assemblages are totally dominated by remains of bison and reindeer. Wolf, red fox, glutton, a particularly large form of brown bear closely resembling the modern polar bear, and Arctic hare are nearly always present in these assemblages, and northern vole is the sole rodent species recovered so far. But for the absence of bison, which has had its range greatly reduced in historical times, this assemblage resembles the modern fauna of Alaska and parts of Arctic Canada today (Sage, 1973). The two dominant species are long-distance migrators and the two dominant carnivores, wolf and bear, almost certainly followed them around quite closely. This is very evidently a cold fauna displaying the low species diversity one might predict under these circumstances. This is the characteristic later Quaternary fauna of the relatively maritime high latitudes.

During the middle part of the Last Cold Stage the picture changes quite dramatically. Isotope stage 3 is known to represent a slight climatic amelioration at the global level, though exactly how this translates in terms of local conditions is difficult to interpret at present. Although the mammal assemblages from this period still contain some bison and reindeer, the position of faunal dominance is taken over by mammoth, woolly rhinoceros and horse. Overall, faunal diversity more than doubles during this period and although associated environmental evidence suggests that conditions remain very cold and essentially open, the environ-

Fig. 24.1. Artist's reconstruction of a summer scene on the South Wales coast during the warmest part of the last (Ipswichian) Interglacial (approx. 125 000 years ago). Global sea levels were high and the climate was temperate with a strong maritime influence. The region supported a high species diversity mammal fauna including spotted hyaena, straight-tusked elephant, narrow-nosed rhinoceros, red deer and fallow deer. (Reproduced with permission from the artist, Peter Snowball, and the Natural History Museum.)

ment is altogether richer. Giant deer and a large form of the red deer (similar in size and morphology to the North American wapiti) are present, and lion and spotted hyaena are important carnivores during this phase. Other species of vole appear, most notably *Microtus gregalis* and the two lemmings, and at least locally the rodent fauna is augmented by populations of ground squirrels. Here we have a mix of migrators, like the mammoths, which are known to have ventured into the High Arctic during the spring, and animals which probably did not have any such response. Here, one must consider the woolly rhinoceros which belonged to a group which has no other leaning towards large scale migratory activity and which did not cross through Beringia into North America unlike most of its contemporaries. The interfaces between environments appear to have been far more complex during this period and allowed for a mix of behavioural responses to the prevailing conditions. Again, it seems likely that the animals themselves contributed towards the nature of their habitat. The taiga forest belt of northern central Asia has only developed to its present extent since the demise of the Quaternary megafauna. The mammals must have maintained relatively open conditions by grazing off most of the young trees.

This is the classical 'cold fauna' of Quaternary literature (e.g. Stuart, 1982) yet in terms of species diversity it has much greater affinities with fully interglacial faunas. It is known that conditions were relatively unstable during stage 3 (Bond *et al.*, 1993), and some of the observed diversity might be due to rapid alternation between bison/reindeer-dominated faunas and mammoth/woolly rhino-dominated faunas, which can no longer be discriminated at the specimen recovery level. But, even if one removed bison and reindeer from the picture entirely, this would still be a relatively high diversity assemblage. Our interpretations of Quaternary faunas are overburdened with assumptions which date back to an early period in the study of the material when we had a very simplistic idea of environmental change. We almost need to start again, and there is no better illustration of that fact than these stage 3 faunas. A small but important element of this stage 3 fauna is man, initially a population of *Homo neanderthalensis* and later replaced by *Homo sapiens*. The fauna appears to spread out from central Asia, specifically from the area north of the Himalayas. It is the characteristic Quaternary mammal fauna of the continental interior, extending during this phase right across to the Atlantic seaboard. In its central Asian homeland, this faunal grouping has a long and relatively uninterrupted history, the huge landmass of Eurasia acting as an environmental buffer against external influences. The continental interior always tends to be relatively dry, and in at least part of the year very cold.

During isotope stage 2 Britain experienced a major phase of ice build up and glacial advance. All of the major upland areas were ice covered, and mobile ice-sheets extended on to adjacent lowland areas. We do not know a great deal about the mammal fauna during this period, but it appears to have been sparse (Fig. 24.2). Reindeer and musk oxen were certainly present.

After the ice-sheets had retreated, north west Europe experienced a brief period of localized warming due in large part to a movement in the position of the flow of the North Atlantic Drift, a warm sea current coming up from the Gulf of Mexico. Reference to a model of the Earth will confirm that Britain experiences a

Fig. 24.2. Artist's reconstruction of an early summer scene on the South Wales coast during the last (Devensian) glacial episode (approx. 18 000 years ago). Global sea levels were low and the climate was cold with tundra-like conditions. The region supported a very low species diversity mammal fauna with reindeer and lemmings. The snowy owl depicted here is still one of the major predators of lemming populations. (Reproduced with permission from the artist, Peter Snowball, and the Natural History Museum.)

rather milder climate today that one might expect when compared with other places at the same latitude. This is almost entirely due to this same phenomenon. By about 12 000 radiocarbon years ago the climate here was similar to that at the start of a major interglacial phase, and the corresponding mammal fauna reflected this amelioration. Horse was the dominant herbivore, with red deer and aurochs as the other larger herbivores. Bear and wolf were the main carnivores during this phase, with a small woodland element appearing in the form of the lynx. There was sporadic human activity in Britain, possibly by small hunting parties, but population densities were still low. A very interesting occurrence during this phase was a brief incursion of the saiga antelope *Saiga tatarica* which is recorded by finds from London and the Mendip Hills in Somerset (Currant, 1987). At about the same time, this species turns up as far away as Alaska, yet its homeland is, and probably always has been, the dry steppes of central southern Asia. The saiga is specifically adapted to the desert-like conditions of its home range, but at times of extreme drought the herds are forced to migrate to the east and west in search of food and water. This one-off pulse of saigas in the fossil record of three continents may represent a specific response to a brief period of quite exceptional aridity.

The last 1000 years of the Last Cold Stage were marked by a short return to fully glacial conditions with readvance of local ice caps and a marked faunal response. The North Atlantic Drift was diverted southwards towards the entrance to the Mediterranean, and Britain was once again surrounded by Arctic waters. Horse remained as the largest herbivore, and in this role appears to have replaced the bison during the post-glacial phase of this stage. Red deer was replaced by reindeer. This short period falls within the range of relatively reliable radiocarbon age determinations and has provided good information on the faunal response to environmental changes taking place at the boundary between a major cold phase and a major interglacial, about 10 000 years ago. At some critical level, response rates appear to be very fast indeed, and it is not too surprising to observe that this appears to be closely related to the change in vegetation from relatively open, to more or less covered conditions.

The fact that cold-stage faunas exhibit greater variation in composition than interglacial faunas is not too surprising when one takes into account the relative lengths of time represented by temperate and cold conditions, particularly when our measure of what is warm and what is cold is related to modern temperatures. The novelty of this observation is that, within cold stages, it is possible to have environments which can support faunas which have high species diversity. As yet, we have no clear understanding of why these faunas are so rich in species, particularly so many large herbivores. One of the reasons behind this is that the intermediate higher latitudes have had a long history of intensive human habitation, and human activity – specifically hunting and farming – has destroyed most of whatever the natural environment once was in these regions. We have virtually nothing existing today which we can use for comparison. It is also very nearly impossible for us to imagine the sheer numbers of individuals that were present in these populations, but one can get a glimpse of the picture from the historical records of herds of six million bison on the plains of North America.

24.4   CONCLUSIONS

In Britain and adjacent parts of Europe there is good evidence to suggest that later Quaternary mammal faunas responded to environmental changes which correlate very closely with the known pattern of the Earth's summed orbital variation. This region of the Earth's surface is particularly susceptible and sensitive to environmental change being at the western extremity of a huge continental landmass, close to one of the major areas of continental ice accumulation and in the path of a major but variable warm ocean current.

In the later part of the Quaternary we can identify faunal groupings which had quite distinctive ecological characteristics; a temperate fauna with strongly Mediterranean affinities; a cold fauna with high-latitude maritime affinities, presumably capable of dealing with relatively high seasonal snow cover; and a continental fauna pre-adapted to surviving wide extremes of temperature variation in a generally arid environment. Temperature must have played a part in this system, but patterns of precipitation may have been far more important in determining which faunal grouping was best suited to a particular region at any one time, particularly whether the precipitation was predominantly in the form of rain or of snow. Ultimately, this has to be related to the dominant vegetation patterns, in this case the changing distribution of woodlands, tundra and steppe-like environments. At a very simple level we can recognize some kind of coherent pattern of faunal responses to major environmental changes.

The briefest look at patterns of distribution in modern mammals suggests we know very little at all about what limits the range of any particular species. It could be the case that present-day mammalian distribution is so badly affected by comparatively recent human interference that we have effectively obscured the boundaries of what were once true animal communities – distinctive species groupings which existed and interacted as functional entities. There is some suggestion of the existence of such communities in the strong affinities seen between successive temperate stage mammal faunas occurring in Britain during the Quaternary. Strong community structure would certainly help explain the apparent total and relatively sudden replacement of one mammalian assemblage by another at critical points in the pattern of ever-changing Quaternary environments. At this stage this is pure surmise.

What we see in the British Quaternary mammalian fossil record is the history of a range of responses to different kinds of change. The interplay of orbital variables and the influence of essentially local factors come together to form a board on which a very complex game is played. Each combination is in some way different. At the present stage of knowledge, all we can do with any confidence is record the moves in the game as accurately as we possibly can and attempt to get them in the right order, because as yet we don't really know the rules. While our continued survival may not hang solely on our understanding of the past, an appreciation of the underlying pattern of longer-term trends in environmental change could prevent us from making serious misjudgements about the effects of some of our present policies and actions, and it may materially assist us in making more informed forecasts and decisions about the future.

25

Human evolution: how an African primate became global

CHRIS STRINGER

25.1 INTRODUCTION

In this chapter, I will review the course of human evolution, and examine possible interactions between hominids and their environments, as these are affected by climatic change, at five important stages:

(i)   The origin of hominines;
(ii)   The origins of the genera *Homo* and *Paranthropus*;
(iii)   The dispersal of early humans to Eurasia;
(iv)   The evolution of the Neanderthals;
(v)   The origin and dispersal of modern humans.

The basic course of human evolution as understood today is represented in Fig. 25.1. An Early Pliocene radiation of bipedal, but ape-like, australopithecines was followed by the Late Pliocene evolution of two more derived clades, perhaps under the influence of drier environments produced by increasing glaciation, or uplift in East Africa. The 'robust australopithecines' (genus *Paranthropus*) underwent gnathic and dental specialization, probably for small hard object feeding (e.g. seeds, nuts), while the other clade became more omnivorous and encephalized, leading to the origin of the genus *Homo*. An early *Homo* species dispersed from Africa, probably during the terminal Pliocene, and its descendants are represented by the *Homo erectus* fossils of eastern Asia and Indonesia. However, the colonization of Europe may have been a Pleistocene event accompanied by amplification of the glacial–interglacial cycles, and while *erectus* persisted in the East, a new species, *Homo heidelbergensis*, evolved in Europe and/or Africa during the Mid Pleistocene. During the later Mid Pleistocene, this species, in turn, gave rise to the temperate–cold-adapted species *Homo neanderthalensis* in western Eurasia, and to *Homo sapiens* in Africa. During the early Late Pleistocene there was biogeographical overlap of these two species in the Levant, but during the later Pleistocene, *Homo sapiens* dispersed widely from Africa. By 25 ka, this species had replaced the last examples of Neanderthals in Europe and of *Homo erectus* in Indonesia to represent the sole surviving species of hominine.

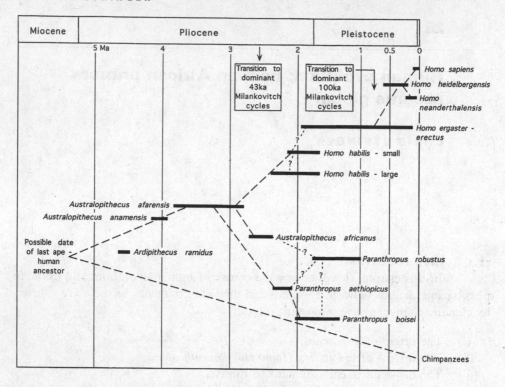

Fig. 25.1. Possible phylogenies of Plio-Pleistocene hominine species.

25.2   THE ORIGIN OF HOMININES

Our knowledge of human evolution has been greatly enhanced by recent discoveries, and by the availability of more detailed contextual information, including palaeoenvironmental and palaeoclimatic data, but the relevant fossil record is still very patchy in both time and space. While finding the 'first hominid' (in the sense of recognizing the earliest members of the human clade) has become a common stated aim of grant applications for field work in Africa and in popular presentations of palaeoanthropology, it is the background data which this search has generated that has enriched this field of research. And, in fact, it would be more accurate to talk about the search for the first hominine, because there is little zoological justification for the retention of the separate family designation Hominidae for the human clade. The close morphological and genetic relationship between living humans and African apes (the chimpanzee species of the genus *Pan*, and *Gorilla*) demand that they should also be classified within Hominidae, with a hominine subfamily for members of the human clade. The evolutionary divergence of apes and humans can be calibrated from genetic data at between 5 and 7 million years ago, but there is little significant fossil evidence from this time period. The nature of the last common ancestor of African apes and humans is thus the source of much speculation – while genetic data slightly favour the separate divergence of the *Gorilla* clade, there is also some morphological evidence pointing to a *Pan–Gorilla* clade (Andrews, 1992*b*). Some workers have suggested that the divergence be treated as a trichotomy for practical purposes (Marks, 1995), and it is therefore not just early evidence of the human clade that is required, but evidence of the early history of the *Pan* and *Gorilla* lineages as well.

Because of the present distribution of our closest living relatives, the African apes, it is also commonly assumed that the equatorial rain forests of Africa were the original home of the hominids, and it has therefore also been generally assumed that hominines moved out of these at an early stage of their evolution. The 'savannah hypothesis' is based on the idea that many hominine characteristics (e.g. bipedalism, human body proportions, loss of body hair, tool-making, omnivory, encephalization) developed through an early adaptation to more open country conditions, but this idea has been increasingly challenged by recent discoveries and interpretations (Shreeve, 1996). The oldest putative hominine, *Ardipithecus ramidus*, is known from Aramis in Ethiopia, based on discoveries made in 1993–4 (White *et al.*, 1994, 1995). The fossil material, which is close to 4.5 Ma, consists of dental, jaw and post-cranial specimens, including a partial skeleton of a single individual which has yet to be described. Dental morphology appears rather chimpanzee-like, although the canine teeth are apparently reduced in size. Basicranial morphology is said to indicate a bipedal adaptation, but body proportions are apparently ape-like (long forearms, short legs?). It is unclear whether *Ardipithecus* is a genuine early hominine which nevertheless preserves morphology from before the *Pan–Homo* lineages diverged, or equally plausibly, could represent an early member of the *Pan* clade. And, even if *Ardipithecus* was bipedal, this could feasibly have been part of the original behavioural repertoire of the *Pan–Homo* last common ancestor. Associated flora and fauna indicate that the environment of *ramidus* was forested rather than open.

A slightly later species, more clearly a hominine, has been discovered in northern Kenya. At Allia Bay and Kanapoi, dental, jaw and postcranial specimens dating from about 4 Ma have been assigned to a new species of *Australopithecus*, *Australopithecus anamensis* (Leakey *et al.*, 1995; Andrews, 1995). In contrast to *Ardipithecus* and *Pan,* this species had larger molars with thicker enamel, and thus more closely resembled Miocene hominoids and later hominines. Tibial and distal humeral specimens, which are assumed to represent the same creature as the separate gnathic material, have been described as fundamentally human in form (but cf. Lague & Jungers, 1996, regarding the humerus). The associated faunas reflect a mosaic environment of woodland and more open country, and while it certainly appears that *anamensis* was a proficient biped, it might still have retained an arboreal climbing capacity, since a recent study suggests that the Kanapoi humerus shows non-human affinities (Lague & Jungers 1996), and a number of later, and possibly descendant, hominine species still show hand bones with signs of such an adaptation.

There has been much recent research on the thermoregulatory benefits of bipedalism, in both avoidance of heat stress and in reduction of water requirements (Wheeler, 1993). There were clear advantages for a human-sized mammal which used evaporative sweating for cooling, in walking upright in open sunlit tropical or subtropical conditions. These include a reduced surface area exposed directly to the sun and to heat radiating from the ground, and an increased air flow around the body for evaporation of sweat. However, if many early hominines were relatively small-bodied and still living in at least partly forested environments, as a growing body of data suggests, the thermoregulatory advantages of bipedalism would have been largely or wholly negated (Ruff, 1993). Until we have more data on the functional anatomy of the postcrania of the earliest hominines, and their usual habitat, the factors behind the adoption of bipedalism will remain uncertain.

25.3   THE ORIGINS OF THE GENERA *HOMO* AND *PARANTHROPUS*

The evolutionary radiation of the australopithecines in the later Pliocene of Africa led to quite extensive speciation and the origin of two new clades, that of the so-called robust australopithecines – genus *Paranthropus* – and that of *Homo*. As evidence has mounted concerning the ape-like anatomy of the australopithecines, and the fact that they had not completed the move to open country environments, attention has switched to the origin of the genus *Homo* as perhaps the real watershed in the adaptive transition to humanity. Regular tool-making, increased brain size, omnivory, posterior dental reduction and 'human' body proportions have all been used to try to identify the first members of the genus *Homo*, generally attributed to the species *Homo habilis*. However, recent analyses of east African fossils attributed to this species (Stringer, 1986; Rightmire, 1993; Wood, 1992) have suggested that at least two species may be represented – a large-brained but also large-faced and toothed species (*Homo rudolfensis* for Wood, 1992) and a smaller-brained, but also smaller-toothed species (*Homo habilis sensu stricto* for Wood, 1992). Moreover, it seems from recent reassessments of postcranial evidence that the appearance of the linear, long-legged physique associated with *Homo* may not have occurred with *Homo rudolfensis* or *Homo habilis,* but with the Late Pliocene human species known as *Homo erectus* (or for Wood, 1992, *Homo ergaster*).

Nevertheless, despite the doubts surrounding the human status of these species, it is clear that there were significant evolutionary changes in hominines, not only with the appearance of early *Homo*, but with *Paranthropus*. These dentally highly derived hominines appear to have been adapted to small hard object feeding in presumed xeric environments in southern and eastern Africa (Grine, 1988). Various hypotheses have been advanced linking the appearance of *Homo* and *Paranthropus* with climatic change, but perhaps the most explicit is Vrba's turnover pulse hypothesis. A conference in 1994 examined the relationship between hominid evolution and climatic change, and Vrba's work was discussed by several participants in the ensuing publication (Vrba *et al.*, 1995). She has argued that the significant shift towards the Earth's recent glacial regime, which occurred about 2.6 Ma, forced major evolutionary changes and extinctions in a number of African mammalian lineages, including antelopes and hominines (Vrba, 1995a). According to Vrba, drier climates led to the break-up of woodlands and the spread of grasslands, and these environmental changes produced speciation, including the appearance of the genus *Homo*. In the case of antelopes, Vrba examined records of nearly 150 African species from the Late Miocene onwards. From her data, there was a major (90%) faunal turnover between 3 and 2 Ma, with a concentration of extinction and first appearances between 2.7 and 2.5 Ma. This period has also been claimed to mark the first significant record of ice-rafted debris in the north Atlantic (Kennett, 1995), and a transition from the domination of 20 000 year Milankovitch cycles to more marked 40 000 year ones, as recorded in ocean cores off the east African coast (deMenocal & Bloemendal, 1995). However, not all African regions nor all mammalian groups appear to show the same turnover patterns, or they may show a less severe turnover over a longer period of time (perhaps between 3 and 2 Ma) (e.g. White, 1995; Behrensmeyer, quoted in Kerr, 1996). Other workers question whether mammals are even especially sensitive to the effects of climatic change (Prothero, quoted in Kerr, 1996). Clearly, the hominine record is itself not rich enough to properly test the turnover pulse hypothesis, and more mammalian records will have to be analysed in order to establish its relevance to human evolution. But, as Partridge *et al.* (1995*b*) noted, the apparent coincidence of the first appearance of both *Homo* and *Paranthropus* in east Africa around 2.5 Ma is still suggestive of some kind of relationship with climatic change, or biogeographical change linked with tectonic uplift, or both.

25.4   THE DISPERSAL OF EARLY HUMANS TO EURASIA

There has been much recent discussion, based on new discoveries and datings, about the antiquity of the first dispersal of humans from Africa to Eurasia. Up to ten years ago, it was generally believed that the species *Homo erectus* first appeared in China and Indonesia about a million years ago, and perhaps never arrived in Europe at all. Thus *Homo erectus*, having evolved in Africa (from *Homo habilis*?) by 1.7 Ma, marked time for over half a million years before emerging from the ancestral hominine homeland. It was assumed that non-African environments, perhaps particularly those of the Middle East corridor, posed too great a challenge to early human adaptive capabilities. Now these views have been compromised by new data. The assumed evolutionary link between *Homo habilis* and *Homo erectus* has been challenged by studies which have demonstrated the large morphological gap between fossils attributed to *habilis/rudolfensis*, and those attributed to *ergaster/*early *erectus*, as well as possible overlap in time (Wood, 1992). Thus it can no longer be assumed that *erectus* could only have evolved in African areas previously inhabited by *habilis/rudolfensis*. Could the real precursor species have even been non-African, implying a yet undiscovered Late Pliocene dispersal of hominines from Africa?

The skeleton from Nariokotome, Kenya, attributed to *Homo ergaster* or *erectus*, displays a tall, linear physique which was apparently well adapted to hot, dry climates, and long distance walking or running (Ruff, 1993). The very human form of this 1.5 million year old skeleton below the neck appears to have been well suited to tropical or subtropical conditions, and during warm intervals this species would seem to have had the capability of spreading from north-east Africa through subtropical environments to the east. New dating work in China and Indonesia suggests that this could have happened, or that the dispersal could even have been in the reverse direction (Larick & Ciochon, 1996). Sediments putatively associated with Indonesian *erectus* fossils have been radiometrically dated at 1.8–1.6 Ma (Swisher *et al.*, 1994), while dated fauna, artefacts and palaeomagnetism associated with, or stratified above, an apparent early hominine from Longgupo Cave, China, has been used to suggest a similar antiquity (Culotta, 1995). Both these claims remain controversial, and more supporting data are required, but they have by no means been shown to be implausible.

In the case of Europe, humans assigned to the species *H. heidelbergensis* had certainly arrived there by the middle part of the Mid Pleistocene (oxygen isotope stage 13 – about 500 ka), based on well-attested evidence of artefacts, butchered fauna and rare fossil evidence (Roebroeks & Van Kolfschoten, 1994). However, how much earlier they might have arrived than this is an unresolved question. Higher-latitude environments were challenging to tropically adapted primates, because of lower mean temperatures and also the greater seasonality and longer winters. Europe had also been colonized successfully by newly evolved large carnivore guilds in the Plio–Pleistocene, which would have been formidable competitors to early hominines with only nascent hunting abilities (Turner, 1992), although they might alternatively have been provided with excellent scavenging opportunities. A fierce debate has developed between workers who favour a short chronology (Middle Pleistocene onwards) for European human colonization (e.g. Roebroeks & Van Kolfschoten, 1994) and those favouring a long chronology (Early Pleistocene or even Pliocene onwards, e.g. Rolland, 1992; Bonifay & Vandermeersch, 1991; Carbonell *et al.*, 1995). The argument of the former workers has been that archaeological sites claimed to date prior to about 500 ka have isolated and poorly characterized 'arte-facts' in disturbed contexts, without human fossils, while from the middle part of the Mid Pleistocene, sites have rich and unquestionable archaeological assemblages in primary contexts, in association with human fossils. In their view the archaeological and chronological evidence from sites such as Orce (Spain) and Soleilhac and Le Vallonet (France) has been assessed too uncritically by proponents of their great antiquity.

However, in the last few years, new evidence has strengthened the claim for at least an intermittent earlier human occupation of southern Europe (Bower, 1997). The site of Dmanisi in Georgia has produced a human mandible and artefacts in association with Villafranchian (Plio–Pleistocene) fauna, claimed to date from 1.8–1.6 Ma (Gabunia & Vekua, 1995). Recent investigation of the site shows evidence of a more complex stratigraphy, including the fact that the human and faunal remains are apparently contained within animal burrows (Ferring *et al.*, 1996). The Dmanisi mandible itself has been claimed to show resemblances to African Plio–Pleistocene specimens assigned to *Homo habilis* and *Homo ergaster* (Gabunia & Vekua, 1995), but a recent comparative study (Bräuer & Schultz, 1996) demonstrated its primary affinity to later *Homo erectus* mandibles, perhaps favouring an age closer to 1 Ma.

More extensive hominine remains have recently been excavated from the Gran Dolina site in the Sierra de Atapuerca, Spain (Carbonell *et al.*, 1995). Cranial, dental and postcranial parts of several individuals, as yet not formally assigned to a species, are claimed to date from about 800 ka. The associated mammalian fauna is largely early Mid Pleistocene in character, but with some archaic elements which have been used to support the palaeomagnetic age assignment. Some have argued that these archaic fauna are endemics which persisted in the Iberian peninsula long after their extinction elsewhere in Europe, and hence do not support an Early, rather than Mid Pleistocene age assignment. However, the hominine fossils themselves are said to show archaic traits compared with Mid Pleistocene homologues, and the question of their exact antiquity remains unresolved at the moment. Nevertheless, at minimum, they do appear to provide evidence of a pre-stage 13 human occupation of Europe.

As we move on to undisputed Mid Pleistocene evidence of humans in Europe, probably correlated with oxygen isotope stage 13, about 500 ka, we also move into the period when the glacial–interglacial cycles had become dominated by the longer 100 000 year Milankovitch signal, producing greater climatic extremes, especially in higher latitudes, closer to the major polar ice caps. Humans occupied Britain and north western Europe during the oxygen isotope stage 13 interglacial, as evidenced at sites like Boxgrove (England) where ancient land surfaces preserve episodes of stone tool manufacture and butchering of megafauna (Roberts *et al.*, 1994). These humans probably represented the species *Homo heidelbergensis*, based on the penecontemporaneous mandible from the Mauer sandpit near Heidelberg, Germany, and were evidently capable enough hunter–gatherers to sustain long-term occupation of the region. The Boxgrove tibia (Roberts *et al.*, 1994) is massively built and seems to indicate body proportions which were physically adapted to temperate–boreal climates (Trinkaus *et al.*, 1996). There is some recent evidence that these early European populations were able to maintain at least sporadic occupation of Britain during the succeeding (Anglian?) cold stage, as stone tool knapping events are recorded in subsequent deposits of this age at Boxgrove, and butchered faunal remains have been excavated from correlated cave deposits at Westbury-sub-Mendip, England (Andrews *et al.*, 1999).

25.5 THE EVOLUTION OF THE NEANDERTHALS

As the Mid Pleistocene progressed in Europe, *Homo heidelbergensis* gave way to *Homo neanderthalensis*, probably through *in situ* evolution. A marvellous intermediate skeletal sample is known from the Sima de los Huesos site at Atapuerca, dating from around 300 ka (Arsuaga *et al.*, 1993), but it was probably only after oxygen isotope stage 7 that the Neanderthals developed their full physical and (presumed) cultural adaptations to European periglacial conditions. Neanderthal fossils or indicative Mousterian stone tools younger than stage 7 are often found associated with cold-adapted or steppe–tundra faunas across western Eurasia, and there is strong evidence that Neanderthals were physically adapted to low temperatures through their body proportions, which parallel those of recent cold-adapted peoples (Ruff, 1993). Their bodies were characterized by relatively wide and long trunks, with shortened distal extremities (ulna/radius and tibia/fibula), minimizing body surface area compared with volume. This physique contrasts strongly with that present in the Early Pleistocene *Homo ergaster*/early *Homo erectus* skeleton from Nariokotome in Kenya discussed earlier. There is some evidence that there was a cline in cold-adapted body shape when comparing the Neanderthals of western Europe with those of the Middle East (Iraq/Syria/Israel). Neanderthal populations which lived in southern Europe, especially during inter-glacials, were presumably also less extreme in this respect. In this regard, it will be interesting to know more about the body proportions of the 'Sima' Atapuerca skeletons, in order to learn whether the apparent boreal–temperate adaptation esti-mated for the Boxgrove individual may have been maintained right through to the later Neanderthals, or whether *heidelbergensis* and *neanderthalensis* developed this in parallel.

The climate of Europe was predominantly cooler and drier during the Neanderthal occupation there, but evidence from Greenland ice cores and north Atlantic ocean floor cores shows rapid and severe fluctuations of climate through oxygen isotope stages 4-2 associated with Heinrich events – massive periodic iceberg discharges from the Laurentide, and possibly other, ice-sheets (Lehman, 1996, and Pickering, this volume). These brief events occurred about every 6000 years and produced a sudden drop in ocean temperatures associated with the release of large quantities of cold fresh meltwater into the surface of the Atlantic, followed by a rapid warming. The impacts of these events are beginning to be picked up in the terrestrial record and may well have been global to judge by the widespread nature of changes associated with the last of these fluctuations – the Younger Dryas. Their rapidity and severity means that they must surely have had a significant impact on human populations, but these are difficult to detect from the existing sparse and poorly dated data.

25.6   THE ORIGIN AND DISPERSAL OF MODERN HUMANS

Our species, *Homo sapiens*, appears in the fossil record by at least 100 ka at sites in Africa and Israel (Grün & Stringer, 1991). Its precise origins are still unclear, although there are potential ancestral populations sampled in various parts of Africa in the later Mid Pleistocene (300–130 ka). It may well be that the severity of oxygen isotope stage 6 (190–130 ka) in Africa led to desiccation and desertification, isolating sub-Saharan populations both from those further north and from each other, thus forcing the morphological transition to *Homo sapiens* from the preceding late archaic species (? late *Homo heidelbergensis* or a distinct species based on the South African Florisbad fossil *Homo helmei*). However, much more fossil and palaeoclimatic evidence will be required from this period in order to test these and related ideas.

Early *Homo sapiens* of Africa are known from sites such as the Klasies River Mouth Caves in South Africa and Omo Kibish in Ethiopia. They show the presence of a robust modern human cranial morphology with great size variation and variable development of 'modern' features such as a chin. There is little post-cranial evidence, but what there is suggests that these people had not completely shed their archaic heritage of a robust skeleton. The penecontemporaneous samples from the Levant retained some archaic cranial features in their large and broad faces, wide noses, large teeth and, in some cases, well developed supraorbital brow ridges. Their post-cranial architecture showed some reduction in robusticity compared with archaic humans, and their body proportions contrasted with those of the Neanderthals in apparently being warm adapted, with a physique like that of the more ancient Nariokotome skeleton. Whether they coexisted with Neanderthals in the Levant or alternated in occupation cannot be determined from available chronological data, but it is clear that at least some early modern humans preceded the later Neanderthals in the area. The Levant can thus be considered as an area of biogeographical overlap between two sister clades, the northern origin Neanderthals and southern origin early *Homo sapiens*, perhaps driven by climatic changes to the north (western Eurasia) and south (north Africa), but the Levant had also long been a

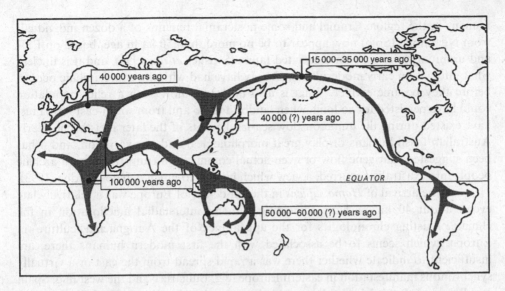

Fig. 25.2. A map showing the reconstructed pattern of the dispersal of *Homo sapiens* over the last 100 ka, based on fossil, archaeological and genetic data. The routes shown are conjectural.

corridor for south–north population dispersal. There is no evidence for Neanderthal penetration of Africa, but by 40 ka, *Homo sapiens* had dispersed widely in Eurasia, and had apparently already reached Australia (Fig. 25.2).

How many routes of dispersal from Africa were used is currently unclear, but Lahr (1996) has argued that a separate tropical–subtropical coastal route via a southern Arabian landbridge could have been used at a time of lower sea level by the first Australian colonizers. Archaeological evidence suggests that northern Australia was colonized by 50 ka, but even with the lowest sea levels of glacial maxima, there were always a number of 50 km ocean gaps between the New Guinea–Australian continent and the islands of south-east Asia. It is not known whether the eastern route via Timor, or the western route via Java, was primarily used, but seagoing craft must have been constructed long before there is any direct archaeological evidence of the existence of boats. Since the Australasian continent would never have been physically visible to potential colonizers, it is surmised that accidental 'sweepstake' transport of people attempting to travel to visible islands was involved where they were swept away by winds or currents. Many potential colonizers must have perished against the few that survived.

The Australian continent would have been challenging to these first human colonizers, with dense tropical–subtropical forests in the north (although reduced during drier phases), and an arid heartland, but archaeology shows that people had established at least intermittent occupation of the interior by the time of the last glacial maximum (*ca.* 20 ka), as well as of the temperate–boreal landscapes of Tasmania, which was only an island during interglacial stages.

Recently, startling evidence has emerged that a late *Homo erectus* population may have persisted in Java (Indonesia) at least until the arrival of modern humans in the region. Cranial and some postcranial remains of a dozen individuals from Ngandong (Solo) now appear to be no more than 50 ka in age, based on ESR and uranium series ages on associated fauna (Swisher *et al.*, 1996), and it is unclear what interaction early modern humans may have had with these late archaic people, before they became extinct. What is also unclear is how such a relict population could have persisted for so long, when landbridges to and from south-east Asia must have existed during the numerous low sea-level events of the later Pleistocene. Early Australian fossil humans display great morphological and size variation, and it has been suggested that gene flow or even actual colonization from Indonesian archaic people into Australia occurred, a view which has led to much debate (Pardoe, 1993).

The arrival of *Homo sapiens* in the cul-de-sac of Europe was a relatively late event, about 40 ka, perhaps coinciding with an interstadial amelioration in the climate. Existing chronologies for the appearance of the Aurignacian 'culture' in Europe, which seems to be associated with the first modern humans there, are insufficient to indicate whether there was a rapid spread from the east or a virtually synchronous manifestation in eastern Europe (e.g. Bulgaria) and the west (e.g. Spain and Belgium). But it is generally assumed that modern humans entered Europe from the east, because there is no archaeological evidence of dispersal across the Mediterranean via regions like Gibraltar and Sicily at this time. The manufacturers of the Aurignacian, and the somewhat later Gravettian of eastern Europe, were strongly built modern humans, who shared some plesiomorphies (retained primitive features) with the early African/Levantine samples, including a warm-adapted physique. Over the next ten millennia, *Homo neanderthalensis* and *Homo sapiens* coexisted in Europe during a period of repeated rapid climatic fluctuations linked with the Heinrich events mentioned earlier. But despite being cold adapted, and having survived such conditions previously, the Neanderthals had apparently become extinct over their whole Eurasian range soon after 30 ka. Direct evidence supporting the distinctiveness of the Neanderthals from all living humans, including Europeans, has recently emerged through the extraction of mitochondrial DNA from the type specimen of *Homo neanderthalensis* – the Neander Valley skeleton (Krings *et al.*, 1997).

How the two populations interacted is unclear, but during this period of coexistence, some Neanderthals began to produce new stone tool assemblages which seemed to follow the model of the Aurignacian industry. Either they produced these independently, perhaps under the challenge of competition from the newcomers, or these changes represent the result of Neanderthal acculturation under their influence. However, the geographical range of the final manifestations of Neanderthal behaviour was limited, and it appears that they were gradually marginalised in Europe, perhaps only persisting in upland areas such as the southern Iberian peninsula and the Alps after 30 ka. Following that, *Homo sapiens* was the sole occupant of Europe, although it is unclear how much continuity there was between successive modern human populations. It is only in the terminal Pleistocene, after the last glacial maximum (about 20 ka), that skeletal samples start to resemble recent Europeans in morphology and body proportions, and it is likely that north-western Europe was completely depopulated immediately south of the glacial margins, and recolonized during the so-called 'late glacial interstadial', after about 14 ka.

The tempo and mode of human colonization of the Americas are still very much in dispute between those who favour a late arrival across Beringia, after the last glacial maximum (about 20 ka), and those who argue that it occurred earlier, perhaps as far back as 40 ka. Whichever date turns out to be the more accurate, it appears that several separate migrations into the continent took place, based on linguistic, dental and genetic evidence (Gibbons, 1996). Possibly related populations are sampled from the Chinese Late Pleistocene at Upper Cave Zhoukoudian and Liujiang, although none of these appears 'Mongoloid' in affiliation. The same may well be true for the earliest colonizers of the Americas according to some researchers. Lahr (1996) has argued that the Ona–Patagonian aborigines who lived near the southern end of South America were relics of a robust non-Mongoloid first wave of colonization that was marginalized or replaced through most of the continent by later, more typical, Native American colonizers. By 11 ka, humans were well established in areas as widespread as the United States (e.g. the Clovis industry) and Brazil (Hecht, 1996). The extent to which the major megafaunal extinctions were due to the impact of human predation vs. Late Pleistocene–early Holocene climatic changes is also still the subject of continuing fierce debate.

25.7 CONCLUSIONS

The emerging detailed record of Late Pleistocene climatic changes, combined with the growing body of fossil material, and genetic data from modern human populations, means that we have the potential to study comprehensively the effect of climatic changes on later human evolution. In Europe, work on mitochondrial DNA (mtDNA) has allowed reconstruction of ancestral Pleistocene patterns (e.g. Richards et al., 1996), although recovery of significant ancient DNA from human fossils is still a rare event. The dispersal pattern of modern humans was probably very complex, with later colonisations superimposed on earlier ones, and varying degrees of replacement–gene flow in different regions (Lahr, 1996). Populations represented today by geographically restricted peoples such as the Khoisan, Andamanese Islanders, Veddahs and Ainu may formerly have been much more widespread, and they may have originated in quite different regions from those indicated by their present more limited distributions. And many early dispersals may have been eradicated by the severity or rapidity of climatic or biogeographical changes. One particularly significant volcanic event was the massive eruption of Mount Toba in south east Asia about 74 ka, the largest so far known in the Cenozoic (Pickering, this volume). This explosive eruption may have played a role in forcing the oxygen isotope stage 5–4 transition, and could have led to a brief period of very severe climatic deterioration through the production of large quantities of atmospheric dust. There is genetic evidence of a Late Pleistocene bottleneck in human population numbers which might be related to this event (S. Ambrose, personal communication; Rogers & Jorde, 1995; Takahata, 1995).

Our abilities to reconstruct earlier human population history in comparable detail are limited, not just by a sparser fossil record, but also by factors such as poorer chronological control and lack of the genetic data, which we now have available for Homo sapiens. Nevertheless it is clear that earlier human populations would have been victims of the vagaries of global change to an even greater extent than their modern human successors, and it is apparent from the decline in hominoid species numbers since the Miocene, and in hominine species numbers since the Pliocene, that our species perhaps deserves the name 'Homo fortunatus' as much as Homo sapiens.

26

The biotic response to global change: a summary

STEPHEN J. CULVER AND PETER F. RAWSON

26.1   INTRODUCTION

What comes over so clearly in the contributions to this volume is that the nature and rates of biotic response to global change have been very variable. While many biotic groups show a marked response to the physical variables of global change considered in Chapters 2 and 3, in others the main changes over the last 145 Ma have been driven primarily by intrinsic or extrinsic biological factors. Furthermore, any organisms that contributed substantially to carbon burial (e.g. trees, corals, nannofossils) helped to drive global change. In this chapter we bring together some of the common strands that have emerged from the preceding chapters.

26.2   ENVIRONMENTAL CONSIDERATIONS AND BIOTIC RESPONSE

Climate change, itself a product of other components of global change, is a key forcing mechanism for global biotic change. The time interval embraced by this volume witnessed a major change in Earth's climatic history, from a very warm 'greenhouse' state in the Cretaceous through a long, but far from monotonic, decline through much of the Cenozoic to the present-day icehouse world. During the 'greenhouse' phase, climate varied but apparently changed only slowly over many thousand to several million years. But as global climate cooled a clearly marked cyclicity developed, driven by Milankovitch and sub-Milankovitch scale cycles. The rate of climate change increased, and, in the Pleistocene to Recent icehouse interval, rapid change became the norm, with evidence, for example, that sea-surface temperature could rise by as much as 9 °C in only 50 years (Parker, Chapter 18).

Although Cretaceous climate was not as equable across latitudes as often portrayed, polar areas were far less cold than at the present day (Spicer, Chapter 17). The subsequent deterioration of global climate was accompanied by an increasing temperature gradient between polar and tropical areas.

The world's biota had to cope not only with this evolution of our climate but also with dramatic changes in sea level from a Late Cretaceous peak when so much continental crust was flooded to create huge areas of shelf sea, to the much lower sea levels with rapid oscillations characteristic of the Pleistocene and

Holocene. All the evidence indicates that current climates and sea levels are atypical of those of much of our Phanerozoic past.

How do organisms react to environmental change of a rate or magnitude that precludes *in situ* adaptation? There are two possibilities; they can migrate or, if that is not possible or sufficient, they become extinct. In looking at these two responses more closely, however, it is evident that there are many possible variations on these themes. Entire communities could migrate *en masse* as they track suitable environments. Alternatively, a community might undergo gradual break-up as individual species migrate independently when their particular threshold conditions are crossed. The complexity of migration patterns means that should the same environmental conditions return to a particular location then it is not guaranteed that the same community of organisms would return. Migration may not be possible for all species in a community, and local or even total extinction of species may take place as environmental thresholds are passed. In this context, the rate of environmental change is of considerable importance in determining whether migration or extinction occurs.

The complexities of environmental change are such that it is extremely difficult to identify proximal causes. For example, we may state that organisms react in some way to a change in the climate. But that climatic change may well be a result of change in sea level which, in turn, may be driven by plate tectonic processes which are dependent on mantle processes. Add to these internal, typically long-term factors, external, short-term factors such as Milankovich cyclicity or bolide collision and the complexity is further increased.

Of course, an organism's environment includes biotic as well as abiotic variables and so another level of complexity is added. Climatic changes can lead to changes in the composition, structure and geographical distribution of plant communities which can represent a fundamental and sometimes fatal change in the biotic environment of terrestrial animals.

Given the above, it is for good reason that the authors of this book have exercised great caution when trying to determine which environmental variables were the most influential in any particular case. For convenience, in the following summary abiotic and biotic variables, and terrestrial and marine organisms, are dealt with under different headings.

26.3   BIOTIC RESPONSE TO ABIOTIC VARIABLES - TERRESTRIAL ORGANISMS

All the chapters dealing with terrestrial plants and animals emphasize the importance of climatic changes. Currant (Chapter 24) points out that precipitation patterns (related to vegetation patterns) may have been more important than temperature in determining Quaternary mammalian community composition in England. Currant considered the successive communities to be distinct functional entities maintained by extensive interactions. Whybrow and Andrews (Chapter 23) similarly consider Neogene mammalian communities in Europe and Asia to be responsive to orographically forced climatic variations (again resulting in changing vegetation patterns). These authors, however, recognized a less structured 'flowing, vicariant, opportunistic relationship' to the physical world via local extinctions, migration and range contractions and extensions. Hooker (Chapter 22), concurs and notes that climatic change (related to vegetation changes) was an important long-term driving force in Paleogene terrestrial mammalian community evolution. On a shorter term, Hooker recognizes mammalian faunal turnovers as dispersal events facilitated by lowered sea level. The history of our own species and its hominid predecessors has also been affected by the vissicitudes of climate change. Stringer (Chapter 25) describes the strong climatic (including sea level) imprint on human evolution and dispersal and notes how many early dispersals of *Homo sapiens* out of Africa may have been wiped out by rapid or severe climate change. Our species survived a severe population bottleneck about 74 ka when Mount Toba in south-east Asia erupted on a massive scale. The injection of particulate matter into the atmosphere led to a very brief period of severe climatic deterioration, which may have come close to exterminating our ancestors and, indeed, our species.

Farther back in time, it is harder to relate global climatic change directly to particular biotic responses in terrestrial vertebrates. But the importance of other factors can be recognized. Milner, Milner and Evans (Chapter 21) conclude that endemism, dispersals and extinctions of amphibian, reptile and bird faunas were the result of plate tectonically and eustatically driven isolations and reconnections of land areas. Forey (Chapter 8) also ascribes biogeographical responses of freshwater fish faunas to plate tectonically driven environmental change but points out that the patchy nature of the fish fossil record means that 'equations of cause and effect can usually only be offered in the most general terms'.

Insect faunas, like vertebrate faunas, exhibit migration and/or extinction in response to climatic changes throughout the time interval treated in this book (Ross, Jarzembowski and Brooks, Chapter 19; Coope, Chapter 20). Of particular interest is Coope's conclusion that Quaternary insect species have remained morphologically and physiologically stable as a result of the frequent splitting and joining of populations caused by sudden and intensive climatic changes and disruption of habitats.

We have already noted the relationship between climatic change and changes in vegetation patterns in the Cenozoic. Lupia, Crane and Lidgard (Chapter 15) recognize the late Early Cretaceous parallel diversification of distinct groups of plants as a response to climatic factors rather than to evolutionary innovations. Collinson (Chapter 16) relates extinctions and range restrictions of Cenozoic plants to climate deterioration, whilst plate tectonically driven geographical changes and sea-level falls have enabled floral interchanges during the Cenozoic. The very detailed pollen data from the south-east of Britain (Parker, Chapter 18) illustrates well the dynamic nature of floral response to climate-related environmental change. Early postglacial colonizer trees were replaced by thermophilous trees as the Holocene climate warmed and provided soil stability and freer drainage. Later decline of elm was associated in part with a deterioration in climate. Our knowledge of modern floras and their relationship to current, climatic variables means that pollen analysis (Parker, Chapter 18) is a powerful tool in reconstructing past Quaternary environments. But detailed palaeoclimatic reconstructions are also possible in the Cretaceous back to 100 million years ago using the technique described by Spicer (Chapter 17) and based on the foliar physiognomy of woody dicotyledonous angiosperms.

26.4   BIOTIC RESPONSE TO ABIOTIC VARIABLES - MARINE ORGANISMS

The related variables of plate tectonics, sea level, palaeoceanography and climate figure strongly in chapters on marine organisms. Chapman (Chapter 6) shows that around 2.5 Ma enhanced climatic variability, related to Milankovitch – driven ice-volume variations, affected both surface and deep water circulation patterns and coincided with a major change in planktonic foraminiferal communities. About 40% of species found in the North Atlantic became extinct, not in a single pulse, but over some 800 ky. All major biogeographical realms and depth habitats were affected and resulted in an impoverished fauna for several hundred thousand years. After 2.3 Ma, periodic arrivals of pre-existing taxa from the Indo-Pacific augmented diversity. Chapman considers that the variable time delay (150 ky to 1.4 my) of these migrations may have been partly due to the tectonic re-emergence of the isthmus of Panama, a barrier to species dispersal.

The appearance of the Panama Isthmus also led to a divergence of Atlantic and Pacific marine gastropod faunas (Morris and Taylor, Chapter 11). These authors further note the coincidence of gastropod radiations with major sea-level changes. Major sea-level falls across the Jurassic–Cretaceous and Cretaceous–Paleocene boundaries led to the development of extensive low-lying areas of non-marine sedimentation and are coincident with the origination of several groups of freshwater snails. In contrast, high sea levels in the intervening mid Cretaceous correlate with increased diversity of many marine gastropod groups.

The interaction of general Cenozoic climatic deterioration and the extensive north–south oriented continental margins, resulting from Mesozoic continental break-up, played an important role in the process of taxonomic diversification of bivalves (Crame, Chapter 10). As temperatures declined, temperature-related biogeographical provinces became progressively established on continental shelves, leading to increased taxic diversity which, through time, became concentrated in the tropics.

The diversity and biogeography of pelagic molluscs, the cephalopods, was also intimately related to sea-level change and climate during the Cretaceous. Rawson (Chapter 7) notes the relationship between diversity and area of available shelf sea but also points out the importance of geographical changes related to the break-up of Gondwana and the opening of the Atlantic in greatly influencing the palaeobiogeography of cephalopods. In contrast, the palaeobiogeographical distribution of Cretaceous nannofossils is considered to be dominated by latitudinally distributed floral provinces that originated in direct response to climatic change, whilst the group show little evidence of vicariance (i.e. endemism related to barriers to dispersal). As a result, even rather imperfect datasets can be used to infer first-order global change trends as demonstrated by analysis of Cretaceous Indian Ocean coccoliths (Burnett, Young and Bown, Chapter 4).

Interestingly, Rawson is the only author to mention the end-Cretaceous bolide impact in his conclusions. This reflects the opinions of many of the authors of this volume (including Rawson) that the end Cretaceous mass extinction was, in fact, a latest Cretaceous mass extinction with many floral and faunal groups in decline throughout the latest Maastrichtian. The subject has been covered in detail by MacLeod *et al.* (1997), who showed that many groups passed across the K–T boundary with only minor diversity changes, and only a few microfossil groups experienced major turnovers.

Culver and Buzas (Chapter 9) discuss palaeocommunities of shallow water benthic foraminifera and note that communities changed in response particularly to sea-level rises and falls. Species responded individually to environmental change so that successive communities in an embayment on the North American mid-Atlantic shelf differed in composition as a result of climatically mediated differential recruitment of taxa from a species pool. Just as for bivalves (Crame, Chapter 10), species diversity generally increased throughout the Cenozoic, perhaps as a result of the differentiation of several temperature-related biogeographical provinces.

26.5   BIOTIC RESPONSE TO BIOTIC VARIABLES

An organism's external environment is characterized not only by abiotic variables but also by biotic variables – these can be termed *extrinsic biotic factors*. But biota also respond to *intrinsic biotic factors*, such as key evolutionary innovations.

26.5.1   Response to extrinsic biotic factors

We have already noted in this summary the obvious close relationship between changes in vegetation and the history of terrestrial vertebrates (Hooker, Chapter 22; Whybrow and Andrews, Chapter 23; Currant, Chapter 24). Towards the end of the Paleogene, climate-related reduction of forests in the mid latitudes led to a great increase and diversification of terrestrial mammalian herbivores at the expense of tree-dwellers and frugivores (Hooker, Chapter 22). In the Neogene, with the exception of a reversal in trend in the Mid to Late Miocene, mammalian diversity dropped in eastern Asia as a result of loss of tree cover, whereas African mammal communities changed from faunas adapted to tropical forests to those adapted to more seasonal forests (Whybrow and Andrews, Chapter 23). In the Quaternary, the climate-related changing distribution of woodlands, tundra and steppe-like environments was of great importance in determining which mammalian faunal grouping was best suited to inhabit Britain at any particular time (Currant, Chapter 24).

All authors dealing with the Quaternary note the effect of our own species on the biota. Currant (Chapter 24) suggests that recent human interference in Europe has obscured the boundaries of what were once true terrestrial mammal communities. The response of insect species (Coope, Chapter 20) to artificial or natural global change in the near future may well be unpredictable, given the increased isolation of populations as a result of human activities such as vegetation clearing and large-scale farming. Climatic tracking by species will become increasingly difficult until glacial conditions return and anthropogenic modifications to Northern Hemisphere vegetation patterns are obliterated. Human effects on vegetation in Britain can be traced back over 6000 years with early phases of clearance associated with agricultural activity. Extensive regional woodland clearance commenced about 3500 years BP in the late Bronze age and continued into Roman times (Parker, Chapter 18).

26.5.2   Response to intrinsic biotic factors

In his chapter on bryozoans, Taylor (Chapter 14) notes that extinctions are usually explained in the context of physical and, to a lesser extent, biotic environmental variables, whereas radiations are more likely to be explained in the context of evolutionary innovations. Taylor sees little evidence for a strong signature of environmental change on the global bryozoan fossil record during the past 145 million years. Biogeographical ranges of several warm-water groups contracted in concert with late Cenozoic cooling but the latest Cretaceous extinction had little effect on Bryozoa. The greatest change in the Mesozoic–Cenozoic bryozoan fossil record is the steep rise in family diversity beginning in the mid Cretaceous, perhaps related to high sea levels, but more likely due to an intrinsic biological trigger in the Cheilostomata.

Key evolutionary innovations greatly affected the history of Mesozoic–Cenozoic plant communities (Collinson, Chapter 16) although not in isolation from the physical environment (Lupia, Crane and Lidgard, Chapter 15). The arrival of the flowering plants, with their distinctive functional biology, and their subsequent diversification and coevolution with insects and mammals resulted in true global change of terrestrial ecology. Evolutionary innovation in gastropods during the mid Cretaceous also resulted in great diversification. The development of predation by several groups of gastropods, which together prey on most marine phyla, caused profound changes in marine benthic community structure (Morris and Taylor, Chapter 11). The results of evolutionary innovation may not be immediately apparent, however. Crame (Chapter 10) suggests that the Mesozoic–Cenozoic diversification of bivalves can be attributed to macroevolutionary lag following the development of various adaptive breakthroughs in hard and soft part morphology.

Intrinsic biotic factors were clearly of great importance in determining the way planktonic and benthic foraminifera responded to severe environmental change in the Paleogene. MacLeod (Chapter 5) shows that morphological change during three episodes of rapid environmental perturbations most likely occurred as a byproduct of selection for ecological aspects of the foraminiferal holophenotype. Both benthic and planktonic foraminifera altered their developmental and life history patterns in response to environmental selection pressures during the Cretaceous–Tertiary, Paleocene–Eocene and Late Eocene episodes of global environmental change.

26.6   CONSTRAINTS ON OUR KNOWLEDGE OF BIOTIC RESPONSE TO PAST GLOBAL CHANGE

Intrinsic to any palaeontologist's research should be an understanding of the limitations of the fossil record. In this book several authors specifically note various aspects of these limitations. Taylor (Chapter 14) points out the need for further alpha taxonomy so that the geographical and temporal distributions of bryozoan taxa can be better understood. Almost all of the authors in this book could easily have reiterated that need. Taylor continues with a call for further phylogenetic analysis to enable ghost lineages to be inferred. This is echoed in the chapters by Smith and Jeffery (Chapter 13), MacLeod (Chapter 5) and Whybrow and Andrews (Chapter 23). An explicitly phylogenetic approach to the field of biotic response to global change is exemplified by Forey's chapter on fossil fish, where he correlates theories of phylogenetic history with ecological and geological history (Chapter 8).

The patchy nature of the fish fossil record is described by Forey (most species of fishes are found at a single locality only) who also notes that occasional lagerstätten may conceal true patterns of origination and/or extinction. The sparse vertebrate fossil record and the resulting limited ability to correlate with global environmental changes is also lamented by Stringer (Chapter 25) and Milner, Milner and Evans (Chapter 21). The marine invertebrate fossil record also has its limitations. Rosen (Chapter 12) shows that corals characterized by algal symbiosis must have survived the environmental perturbations across the Cretaceous–Tertiary boundary despite the lack of a fossil record, and absence of reefs, during this interval. At least 11 Lazarus genera survived, perhaps in refugia in the western Pacific. Preservational and sampling biases are also noted by Smith and Jeffery (Chapter 13) as complicating factors in the study of echinoid diversity and species richness over the past 145 million years. These authors consider that major sea level cycles, strongly influencing preservation potential, largely controlled perceived echinoid taxic diversity during the Mesozoic and Cenozoic.

Changes in biotic patterns can only begin to be related to regional or global environmental events when detailed chronostratigraphical resolution is available. Lupia, Crane and Lidgard (Chapter 15), Hooker (Chapter 22), Whybrow and Andrews (Chapter 23), and Stringer (Chapter 25) all explicitly mention the need for improved stratigraphical resolution so that reliable correlations between biotic patterns and environmental and/or intrinsic biological changes can be made. Coincident change does not necessarily imply common causal relationships, but this is a useful starting point for the development of testable hypotheses.

26.7   CONCLUDING COMMENTS

The authors of this book were given a broad remit to consider to what extent the organisms that they specialize on can be used to address questions of biotic response to global change over the past 145 million years. The varying scientific backgrounds, research philosophies and methodologies of the authors, together with the extremely variable nature of the fossil record (from oceanic microplankton where cores provide continuous sequences and effectively infinite numbers of specimens to hominids where the record consists of a few hundred scattered specimens) and the variable amount of data available, have led to a wide variety of approaches to the subject of this book. None of these approaches is necessarily preferable to others. Indeed, the multiplicity of approach is a strength of this research field.

Simple generalities about plant, animal and protistan response to global change are hard to find in the preceding chapters. Some groups of organisms become extinct during severe environmental perturbations whilst others continue, seemingly independent of their surroundings. Some groups exhibit strong correlations between variations in their diversity and patterns of abiotic environmental changes. Others show a seemingly greater response to biotic environmental variables. Some fossil groups exhibit fundamental changes in their biotic patterns that are more likely explained by intrinsic evolutionary factors than by external environmental factors.

The nature of biotic response to global change is, therefore, extremely variable. If one can imagine a response (or lack of one) then there will be a group of organisms that at some time, in some place, will behave in that way (but not always). While what we learn about the past cannot serve as a blueprint for the future, it does give us an appreciation of how biotic response to natural or anthropogenic global change has occurred and may again take place. A true appreciation of the complexity of possibilities, together with a similar appreciation of our limited understanding, is paramount as we make policies and take actions that will affect the future of all organisms on this Earth.

References

Aaby, B. (1976). Cyclic climatic variation in climate over the last 5,500 years. *Nature*, **263**, 241–84.

Aagaard, K. & Carmack, E.C. (1994). The Arctic Ocean and climate: a perspective. In *The Polar Oceans and Their Role in Shaping the Global Environment*, ed. O.M. Johannessen, R.D. Muench & J.E. Overland. American Geophysical Union, Geophysical Monograph, **85**, 5–20.

Aalto, M.M., Coope, G.R. & Gibbard, P.L. (1984). Late-Devensian river deposits beneath the Floodplain Terrace of the River Thames at Abingdon, Berkshire, England. *Proceedings of the Geologists' Association*, **95**, 65–79.

Adams, C.G. (1967). Tertiary foraminifera in the Tethyan, American and Indo-Pacific provinces. In *Aspects of Tethyan Biogeography*, ed. C.G. Adams & D.V. Ager. Systematics Association Publication, **7**, 195–217.

Adams, C.G. (1970). A reconsideration of the East Indian Letter Classification of the Tertiary. *Bulletin of the British Museum (Natural History) Geology*, **19**, 87–137.

Adams, C.G. (1989). Foraminifera as indicators of geological events. *Proceedings of the Geologists' Association*, **100**, 297–311.

Adams, C.G., Bayliss, D.D. & Whittaker, J.E. (1999). The terminal Tethyan event: a critical review of conflicting age determinations for the disconnection of the Mediterranean from the Indian Ocean. In *Fossil Vertebrates of Arabia*, ed. P.J. Whybrow & A. Hill, pp. 477–84. New Haven: Yale University Press.

Adams, C.G., Gentry, A.W. & Whybrow, P.J. (1983). Dating the terminal Tethyan events. In *Reconstruction of Marine Environments*, ed. J. Meulenkamp. Utrecht Micropaleontological Bulletins, **30**, 273–98.

Adams, C.G., Lee, D.E. & Rosen, B.R. (1990). Conflicting isotopic and biotic evidence for tropical sea-surface temperatures during the Tertiary. *Palaeogeography, Palaeoclimatology, Palaeoecology*, **77**, 289–313.

Ager, D.V. (1971). Space and time in brachiopod history. In *Faunal Provinces in Space and Time*, ed. F.A. Middlemiss, P.F. Rawson & G. Newall. Geological Journal Special Issue, **4**, 95–110.

Aguirre Urreta, M.B. & Rawson, P.F. (1997). The ammonite sequence in the Agrio Formation (Lower Cretaceous), Neuquén Basin, Argentina. *Geological Magazine*, **134**, 449–58.

Agusti, J. (1989). On the peculiar distribution of some muroid taxa in the Western Mediterranean. *Bollettino della Societá Paleontologica Italiana*, **28**, 147–54.

Aharon, P., Goldstein, S.L., Wheeler, C.W. & Jacobson, G. (1993). Sea-level events in the South Pacific linked with the Messinian salinity crisis. *Nature*, **21**, 771–75.

Alberch, P., Gould, S.J., Oster, G.F. *et al.* (1979). Size and shape in ontogeny and phylogeny. *Paleobiology*, **5**, 296–317.

Ali, M.S.M. (1985). On some Pliocene echinoids from the area north of Mersa Alam, Red Sea coast, Egypt. *Paläontologische Zeitschrift*, **59**, 277–300.

Alifanov, V. (1993). Some peculiarities of the Cretaceous and Palaeogene lizard faunas of the Mongolian People's Republic. *Kaupia*, **3**, 9–13.

Allen, L.O. (1982). Palynology of the Palaeocene and early Eocene of the London Basin. Unpublished PhD thesis, University of London.

Allen, P. (1981). Pursuit of Wealden models. *Journal of the Geological Society, London*, **138**, 375–405.

Allen, T.G. (1995). *Lithics and Landscape: Archaeological Discoveries on the Thames Water Pipeline at Gatehampton Farm, Goring, Oxfordshire 1985–92*, Thames Valley Landscape Monograph No. 7, Oxford Archaeological Unit/Oxford University Committee for Archaeology, Oxford.

Allen, T.G. & Robinson, M.A. (1993). *The Prehistoric and Iron Age Enclosed Settlement at Mingies Ditch, Hardwick-with-Yelford*, Thames Valley Landscapes: the Windrush Valley Vol 2, Oxford Archaeological Unit/Oxford University Committee for Archaeology.

Alley, R.D., Meese, D.A., Schumm, C.A., Gow, A.J., Taylor, K.C., Grootes, P.M., White, J.W.C., Ram., Waddington, E.D., Mayeski, P.A. & Zielinski, G.H. (1993). Abrupt increases in Greenland snow accumulation at the end of the Younger Dryas event. *Nature*, **362**, 527–9.

Allmon, W.D. & Ross, R.M. (1990). Specifying causal factors in evolution: the paleontological contribution. In *Causes of Evolution*, ed. R.M. Ross & W.D. Allmon, pp. 1–17. Chicago: University of Chicago Press.

Alvarez, L.W., Alvarez, W., Asaro, F. & Michel, H.V. (1980). Extraterrestrial cause for the Cretaceous–Tertiary extinction. *Science*, **208**, 1095–108.

Andors, V.A. (1992). Reappraisal of the Eocene groundbird *Diatryma* (Aves: Anserimorphae). *Los Angeles County Museum of Natural History, Science Series*, **36**, 109–26.

Andrewartha, H.G. & Birch, L.C. (1954). *The Distribution and Abundance of Animals*. Chicago: University of Chicago Press.

Andrews, J.T. & Tedesco, K. (1992). Detrital carbonate-rich sediments, north western Labrador Sea, implications for ice-sheet dynamics and iceberg rafting (Heinrich) events in the North Sea. *Geology*, **20**, 1087–90.

Andrews, P. (1990). Palaeoecology of the Miocene fauna from Pasalar, Turkey. *Journal of Human Evolution*, **19**, 569–82.

Andrews, P. (1992*a*). Community evolution in forest habitats. *Journal of Human Evolution*, **22**, 423–38.

Andrews, P. (1992*b*). Evolution and environment in the Hominoidea. *Nature*, **360**, 641–6.

Andrews, P. (1995). Ecological apes and ancestors. *Nature*, **376**, 555–6.

Andrews, P. (1996). Palaeoecology and hominid palaeoenvironments. *Biological Review*, **71**, 257–300.

Andrews, P., Cook, J., Currant, A. & Stringer, C. (eds.) (1999). *The Pleistocene Cave at Westbury-sub-Mendip, England*. Bristol: Western Academic and Specialist Press.

Andrews, P, Lord, J.M. & Nesbit Evans, E.M. (1979). Patterns of ecological diversity in fossil and modern mammalian faunas. *Biological Journal of the Linnean Society*, **11**, 177–205.

Anning, T., Nimer, N., Merrett, M.J. & Brownlee, C. (1996). Costs and benefits of calcification in coccolithophorids. *Journal of Marine Systems*, **9**, 45–56.

Archibald, J.D. (1987). Stepwise and non-catastrophic late Cretaceous terrestrial extinctions in the Western Interior of North America: testing observations in the context of an historical science. *Mémoires de la Société Géologique de France*, (NS) **150**, 45–52.

Archibald, J.D. (1993). The importance of phylogenetic analysis for the assessment of species turnover: a case history of Palaeocene mammals in North America. *Paleobiology*, **19**, 1–27.

Archibald, J.D. (1994). Metataxon concepts and assessing possible ancestry using phylogenetic systematics. *Systematic Biology*, **43**, 27–40.

Archibald, J .D. (1996a). *Dinosaur Extinction and the End of an Era: What the Fossils Say.* New York: Columbia University Press.

Archibald, J.D. (1996b). Fossil evidence for a Late Cretaceous origin of 'hoofed' mammals. *Science*, **272**, 1150–3.

Archibald, J.D. (1996c). Testing extinction theories at the Cretaceous–Tertiary boundary using the vertebrate fossil record. In *Cretaceous–Tertiary Mass Extinctions: Biotic and Environmental Changes*, ed. N. MacLeod & G. Keller, pp. 373–97. New York: W.W. Norton & Co.

Archibald, J.D. & Bryant, L.J. (1990). Differential Cretaceous–Tertiary extinctions of non-marine vertebrates: evidence from northeastern Montana. In *Global Catastrophes in Earth History: An Interdisciplinary Conference in Impacts, Volcanism, and Mass Mortality*, ed. V.L. Sharpton & P.D. Ward. Geological Society of America Special Paper, **247**, 549–62.

Archibald, J.D. & Clemens, W.A. (1984). Mammal evolution near the Cretaceous–Tertiary boundary. In *Catastrophes and Earth History. The New Uniformitarianism*, ed. W.A. Berggren & J.A. Van Couvering, pp. 339–71. Princeton: Princeton University Press.

Arnaud-Vanneau, A. & Arnaud, H. (1990). Hauterivian to Lower Aptian carbonate shelf sedimentation and sequence stratigraphy in the Jura and northern Subalpine chains (southeastern France and Swiss Jura). *Special Publications of the International Association of Sedimentologists*, **9**, 203–33.

Arsuaga, J.L., Martinez, I. Gracia, A. *et al.* (1993). Three new human skulls from the Sima de los Huesos Middle Pleistocene site in Sierra de Atapuerca, Spain. *Nature*, **362**, 534–7.

Arthur, M.A., Dean, W.E. & Pratt, L.M. (1988) Geochemical and climatic effects of increased marine organic carbon burial at the Cenomanian/Turonian boundary. *Nature*, **335**, 714–17.

Arthur, M.A., Dean, W.E., & Schlanger, S.O. (1985). Variations in the global carbon cycle during the Cretaceous related to climate, volcanism, and changes in atmospheric CO_2. In *The Carbon Cycle and Atmospheric CO_2: Natural Variations Archean to Present*. American Geophysical Union, Geophysical Monograph, **32**, 504–29.

Arthur, M.A., Jenkyns, H.C., Brumsack, H.-J. & Schlanger, S.O. (1990). Stratigraphy, geochemistry and Paleoceanography of organic carbon-rich Cretaceous sequences. In *Cretaceous Resources, Events and Rhythms*, ed. R.N. Ginsburg & B. Beaudoin, pp. 75–119. Dordrecht: Kluwer Academic Publishers.

Asgaard, U. (1979). The irregular echinoids and the boundary in Denmark. In *Cretaceous/Tertiary Boundary Events Symposium. I. The Maastrichtian and Danian of Denmark*, ed. T. Birkelund & R.G. Bromley, pp. 74–7. Copenhagen: University of Copenhagen.

Askevold, I.S. (1990). Classification of Tertiary fossil Donaciinae of North America and their implications about evolution of Donaciinae (Coleoptera : Chrysomelidae). *Canadian Journal of Zoology*, **68**, 2135–45.

As-Sarruri, M.L., Whybrow, P.J. & Collinson, M.E. (1999). Vertebrates (?Sirenia), fruits and seeds from the Kaninah Formation (Middle Eocene), Republic of Yemen. In *Fossil Vertebrates of Arabia*, ed. P.J. Whybrow & A. Hill, pp. 443–53. New Haven: Yale University Press.

Atkinson, M.J., Carlson, B. & Crow, G.L. (1995). Coral growth in high-nutreint, low-pH seawater: a case study of corals cultured at the Waikiki Aquarium, Honolulu, Hawaii. *Coral Reefs*, **14**, 215–23.

Atkinson, T.C., Briffa, K.R. & Coope, G.R. (1987). Seasonal temperatures in Britain during the past 22,000 years, reconstructed using beetle remains. *Nature*, **325**, 587–92.

Aubry, M-P. (1992). Late Paleogene calcareous nannoplankton evolution: a tale of climatic deterioration. In *Eocene–Oligocene Climatic and Biotic Evolution*, ed. P.R. Prothero & W.A. Berggren, pp. 272–309. Princeton: Princeton University Press.

Aubry, M-P., Hailwood, A.E. & Townsend, H.A. (1986). Magnetic and calcareous nanno-fossil stratigraphy of Lower Paleogene formations of Hampshire and London basins. *Journal of the Geological Society, London*, **143**, 729–35.

Bailey, I.W. & Sinnot, E.W. (1915). A botanical index of Cretaceous and Tertiary climates. *Science*, **41**, 831–34.

Ballantyne, C.K. & Harris, C. (1994). *The Periglaciation of Great Britain*. Cambridge: Cambridge University Press.

Ballmann, P. (1969). Les oiseaux miocènes de La Grive-Saint-Alban (Isère). *Geobios*, **2**, 157–204.

Bambach, R.K. (1977). Species richness in marine benthic habitats through the Phanerozoic. *Paleobiology*, **3**, 152–67.

Bandel, K. (1991). Gastropods from brackish and fresh water of the Jurassic–Cretaceous transition (a systematic evaluation). *Berliner Geowissenschaftliche Abhandlugen*, **134**, 9–55.

Bandel, K. (1993). Caenogastropoda during Mesozoic times. *Scripta Geologia* Special Issue, **2**, 7–56.

Bandel, K. (1996). Some heterostrophic gastropods from Triassic St. Cassian Formation with a discussion on the classification of the Allogastropoda. *Paläontologische Zeitschrift*, **70**, 325–65.

Banner, F.T. (1982). A classification and introduction to the Globigerinacea. In *Aspects of Micropalaeontology*, ed. F.T. Banner & A.R. Lord, pp. 145–239. London: George Allen and Unwin.

Banner, F.T. & Lowry, F.M.D. (1985). The stratigraphical record of planktonic foraminifera and its evolutionary implications. *Special Papers in Palaeontology*, **33**, 117–30.

Bannikov, A.G. (1961). *Biology of the Saiga*. Moscow. [English translation]: 1967, Jerusalem: Israel Program for Scientific Translations.

Barber K.E. (1981). *Peat Stratigraphy and Climatic Change – A Palaeoecological Test of the Theory of Cyclic Peat Bog Regeneration*. Rotterdam: A.A. Balkema.

Barber K.E. (1982). Peat-bog stratigraphy as a proxy climate record. In *Climatic Change in Later Pre-History*, ed. A.F. Harding, pp. 103–33. Edinburgh: Edinburgh University Press.

Barber, K.E., Chambers, F.M., Maddy, D., Stoneman, R. & Brew, J.S. (1994). A sensitive high-resolution record of late Holocene climatic change from a raised bog in northern England. *The Holocene*, **4**, 198–205.

Barber K.E. & Clarke, M.J. (1987). Cranes Moor, New Forest: palynology and macrofossil stratigraphy. In *Wessex and the Isle of Wight: Field Guide*, ed. K.E. Barber, pp. 33–44. Cambridge: Quaternary Research Association.

Barber K.E. & Coope, G.R. (1987). Climatic history of the Severn valley during the last 18 000 years. In *Palaeohydrology in Practice*, ed. K.J. Gregory, J. Lewin & J.B. Thornes, pp. 201–16. Chichester: John Wiley.

Barber, K.E., Dumayne, L. & Stoneman, R. (1993). Climatic change and human impact during the late Holocene in northern England. In *Climate Change and Human Impact on the Landscape*, ed. F.M. Chambers, pp. 225–46. London: Chapman & Hall.

Bard, E., Arnold, M., Maurice, P., Dupont, J., Moyes, J. & Duplessey, J.C. (1987). Retreat velocity of the North Atlantic polar front during the last deglaciation determined by ^{14}C accelerator mass spectrometry. *Nature*, **328**, 791–4.

Barrera, E. (1994). Global environmental changes preceding the Cretaceous–Tertiary boundary: early-late Maastrichtian transition. *Geology*, **22**, 877–80.

Barrera, E. & Huber, B.T. (1991). Paleogene and early Neogene oceanography of the southern Indian Ocean: Leg 119 foraminiferal stable isotope results. *Proceedings of the Ocean Drilling Program, Scientific Results*, **119**, 693–717.

Barrera, E. & Keller, G. (1990). Foraminiferal stable isotope evidence for gradual decrease of marine productivity and Cretaceous species survivorship in the earliest Danian. *Paleoceanography*, **5**, 867–70.

Barrera, E. & Keller, G. (1994). Productivity across the Cretaceous–Tertiary boundary in high latitudes. *Geological Society of America Bulletin*, **106**, 1254–66.

Barrett, P.J., Adams, C.J., McIntosh, W.C., Swisher, C.C. & Wilson, G.S. (1992). Geochronological evidence supporting Antarctic deglaciation three million years ago. *Nature*, **359**, 816–8.

Barrier, P., Casale, V., Costa, B., Di Geronimo, I., Olivieri, O. & Rosso, A. (1987). La sezione plio-pleistocenica di Pavigliana (Reggio Calabria). *Bollettino della Società Paleontologica Italiana*, **25** [for 1986], 107–44.

Barron, E.J. (1983). A Warm Equable Cretaceous: the Nature of the Problem. *Earth-Science Reviews*, **19**, 305–38.

Barron, E.J., Fawcett, P.J., Peterson, W.H. *et al.* (1995). A 'simulation' of mid-Cretaceous climate. *Paleoceanography*, **10**, 953–62.

Barron, E.J., Fawcett, P.J., Pollard, D. & Thompson, S. (1993). Model simulations of Cretaceous climates: the role of geography and carbon dioxide. *Philosophical Transactions of the Royal Society of London*, B, **341**, 307–16.

Barron, E.J., Hay, W.W. & Thompson, S. (1989). The hydrological cycle: a major variable during earth history. *Palaeoceanography, Palaeoclimatology, Palaeoecology*, **75**, 157–74.

Barron, E.J. & Peterson, W.H. (1990). Mid-Cretaceous ocean circulation: results from model sensitivity studies. *Paleoceanography*, **5**, 319–37.

Barron, E.J. & Washington, W.M. (1984). The role of geographic variables in explaining paleoclimates: results from Cretaceous climate model sensitivity studies. *Journal of Geophysical Research*, **89**, 1267–79.

Barry, J.C. (1995). Faunal turnover and diversity in the terrestrial Neogene of Pakistan. In *Paleoclimate and Evolution with Emphasis on Human Origins*, ed. E.S. Vrba, G.H. Denton, T.C. Partridge & L.H. Burckle, pp. 115–34. New Haven: Yale University Press.

Barry, J.C., Flynn, L.J. & Pilbeam, D.R. (1990). Faunal diversity and turnover in a Miocene terrestrial sequence. In *Causes of Evolution: A Paleontological Perspective*, ed. R. Ross & W. Allmon, pp. 381–421. Chicago: University of Chicago Press.

Barry, J.C., Johnson, N.M., Raza, S.M. & Jacobs, L.L. (1985). Neogene mammalian faunal change in southern Asia: correlations with climate, tectonic and eustatic events. *Geology*, **13**, 637–40.

Barry, J.C., Morgan, M.E., Winkler, A.J., Flynn, L.J., Lindsay, E.H., Jacobs, L.L. & Pilbeam, D. (1991). Faunal interchange and Miocene terrestrial vertebrates of southern Asia. *Paleobiology*, **17**, 231–45.

Bartek, L.B., Sloan, L.C., Anderson, J.B. *et al.* (1992). Evidence from the Antarctic continental margin of Late Paleogene ice sheets: a manifestation of plate reorganization and synchronous changes in atmospheric circulation over the emerging southern ocean? In *Eocene–Oligocene Climatic and Biotic Evolution*, ed. D.R. Prothero & W.A. Berggren. Princeton: Princeton University Press.

Baumann, K-H. & Meggers, H. (1996). Palaeoceanographical changes in the Labrador Sea during the last 3.1 My: evidence from calcareous plankton records. In *Microfossils and Oceanic Environments*, ed. A. Moguilevsky & R.C. Whatley, pp. 131–53. Aberystwyth: University of Wales Press.

Bé, A.W.H. (1977). An ecological, zoogeographic and taxonomic review of Recent planktonic foraminifera. In *Oceanic Micropalaeontology*, ed. A.T.S. Ramsay, pp. 1–100. London: Academic Press.

Bé, A.W.H. & Tolderlund, D.S. (1971). Distribution and ecology of living planktonic foraminifera in surface waters of the Atlantic and Indian Oceans. In *Micropalaeontology of the Oceans*, ed. B.M. Funnell & W.R. Riedel, pp. 105–49. Cambridge: Cambridge University Press.

Beauvais, L. & Beauvais, M. (1974). Studies on the world distribution of the Upper Cretaceous corals. In *Proceedings of the Second International Symposium on Coral Reefs, October 1974*, ed. A.M. Cameron *et al.*, Great Barrier Reef Committee, Brisbane, **1**, 475–94.

Becker, B., Kromer, B. & Trimborn, P. (1991). A stable-isotope tree-ring timescale of the Late Glacial/Holocene boundary. *Nature*, **353**, 647–9.

de Beer, G. (1958). *Embryos and Ancestors*. Oxford: Oxford University Press (Clarendon).

Begun, D.R., Teaford, M.F. & Walker, A. (1994). Comparative and functional anatomy of *Proconsul* phalanges from the Kaswamga primate site, Rusinga Island, Kenya. *Journal of Human Evolution*, **26**, 89–165.

Behre, K.E. (ed.) (1986). *Anthropogenic Indicators in Pollen Diagrams*. Rotterdam: A.A. Balkema.

Behrensmeyer, A.K., Damuth, J.D., DiMichele, W.A. *et al.* (ed.) (1992). *Terrestrial Ecosystems Through Time*. Chicago: University of Chicago Press.

Bell, M. & Walker, M.J.C. (1992). *Late Quaternary Environmental Change: Physical and Human Perspectives*. Harlow: Longman.

Bender, M., Sowers, T., Dickson, M-L., Orchardo, J., Grootes, P., Mayewski, P.A. & Meese, D.A. (1994). Climate correlations between Greenland and Antarctica during the past 100,000 years. *Nature*, **372**, 663–6.

Benefit, B. (1993). The permanent dentition and phylogenetic position of *Victoriapithecus* from Maboko Island, Kenya. *Journal of Human Evolution*, **25**, 83–172.

Bennett, M.R. & Doyle, P. (1996). Global Cooling inferred from dropstones in the Cretaceous: fact or wishful thinking? *Terra Nova*, **8**, 182–5.

Benson, R.H. (1975). The origin of the psychrosphere as recorded in changes of deep-sea ostracodes. *Lethaia*, **8**, 69–83.

Benson, R.N. (1984). Structure contour map of pre-Mesozoic basement, landward margin of Baltimore Canyon Trough. *Delaware Geological Survey*, Miscellaneous Map Series No. 2.

Benton, M.J. (1990). The causes of the diversification of life. In *Major Evolutionary Radiations*, ed. P.D. Taylor & G.P. Larwood, pp. 409–40. Oxford: Clarendon Press.

Benton, M.J. (ed.) (1993). *The Fossil Record 2*. London: Chapman & Hall.

Berger, W.H. (1982). Increase of carbon dioxide in the atmosphere during deglaciation: the coral reef hypothesis. *Naturwissenschaften*, **69**, 87–8.

Berger, W.H. & Jansen, E. (1994). Mid-Pleistocene climate shift – the Nansen connection. In *The Polar Oceans and Their Role in Shaping the Global Environment*, ed. O.M. Johannessen, R.D. Muench & J.E. Overland. American Geophysical Union, Geophysical Monograph, **85**, 295–311.

Berger, W.H. & Kier, R.S. (1984). Glacial–Holocene changes in atmospheric CO_2 and the deep-sea record. In *Climate Processes and Climate Sensitivity*, ed. J.E. Hansen & T. Takahashi. American Geophysical Union, Geophysical Monograph, **29**, 337–51.

Berger, W.H., Vincent, E., & Thierstein, H.R. (1981). The deep-sea record: major steps in Cenozoic ocean evolution. In *The Deep Sea Drilling Project: A Decade of Progress*, ed. J.E. Warme, R.G. Douglas & E.L. Winterer. Society of Economic Paleontologists and Mineralogists Special Publication No. **32**, 489–505.

Berger, W.H. & Wefer, G. (1992). Klimageschichte aus Tiefseesedimenten: neues vom Ontong Java Platea (Westpazifik). *Naturwissenschaften*, **79**, 541–50.

Berggren, W.A. (1969). Rates of evolution in some Cenozoic planktonic foraminifera. *Micropaleontology*, **15**, 351–6.

Berggren, W.A. & Van Couvering, J.A. (1974). *The Late Neogene: Biostratigraphy, Geochronology and Paleoclimatology of the Last 15 Million Years*. Amsterdam: Elsevier.

Berggren, W.A. & Hollister, C.D. (1977). Plate tectonics and paleocirculation: commotion in the ocean. *Tectonophysics*, **38**, 11–48.

Berggren, W.A., Kent, D.V., Flynn, J.J. & van Couvering, J.A. (1985). Cenozoic geochronology. *Geological Society of America Bulletin*, **96**, 1407–18.

Berggren, W.A., Kent, D.V., Aubry, M-P. & Hardenbol, J. (eds.) (1995). *Geochronology, Time Scales and Global Stratigraphic Correlation*. SEPM (Society for Sedimentary Geology) Special Publication, **54**.

Berggren, W.A., Kent, D.V., Swisher, C.C., III and Aubry, M.-P. (1995). A revised Cenozoic geochronology and chronostratigraphy, in *Geochronology, Time Scales and Global Stratigraphic Correlation*, ed. W.A. Berggren, D.V. Kent, M-P. Aubry & J. Hardenbol. SEPM (Society for Sedimentary Geology) Special Publication, **54**, 129–212.

Berglund, B.E. (1983). Palaeohydrological studies in lakes and mires a palaeohydrological research strategy. In *Background to Palaeohydrology*, ed. K.J. Gregory, pp. 237–54. Chichester, UK: John Wiley.

Bernard, F.R., Cai, Ying-Ya & Morton, B. (1993). *Catalogue of the Living Marine Bivalve Molluscs of China*. Hong Kong: Hong Kong University Press.

Bernecker, M. & Weidlich, O. (1990). The Danian (Paleocene) coral limestone of Fakse, Denmark: a model for ancient aphotic, azooxanthellate coral mounds. *Facies*, **22**, 103–38.

Berner, R.A. (1990). Atmospheric carbon dioxide levels over Phanerozoic time. *Science*, **249**, 1382–6.

Bernor, R.L., Fahlbusch, V. & Mittmann, H. (1996). *Evolution of Western Eurasian Neogene Mammal Faunas*. Columbia: Columbia University Press.

Bernor, R.L., Kovar-Eder, J., Lipscomb, D. *et al.* (1988). Systematic, stratigraphic, and palaeoenvironmental contexts of first-appearing Hipparion in the Vienna Basin, Austria. *Journal of Vertebrate Paleontology*, **8**, 427–52.

Bernor, R.L. & Tobien, H. (1990). The mammalian geochronology and biogeography of Pasalar (Middle Miocene, Turkey). *Journal of Human Evolution*, **19**, 551–68.

Bieler, R. (1992). Gastropod phylogeny and systematics. *Annual Review of Ecology and Systematics*, **2**, 311–38.

Bilotte, M. (1984). Les grandes foraminiferes benthiques du Cretace superieur Pyrenéen. Biostratigraphie. Reflexion sur les correlations Mesogeenes. In *Benthos '83; 2nd International Symposium on Benthic Foraminifera*, ed. H.J. Oertli, pp. 41–3. Pau: EPF Aquitaine, REP, and Total CFP.

Birks, H.J.B. (1986). Late-Quaternary biotic changes in terrestrial and lacustrine environments, with particular reference to north-west Europe. In *Handbook of Holocene Palaeoecology and Palaeohydrology*, ed. B.E. Berglund, pp. 3–65. Chichester, UK: Wiley.

Birks, H.J.B. (1989). Holocene isochron maps and patterns of tree-spreading in the British Isles. *Journal of Biogeography*, **16**, 503–40.

Blair, F.W. (ed.) (1972). *Evolution in the Genus Bufo*. Austin, London: University of Texas Press.

Blake, D.B. & Zinsmeister, W.J. (1991). A new marsupiate cidaroid echinoid from the Maastrichtian of Antarctica. *Palaeontology*, **34**, 629–36.

Bland, P., Gilmor, I. & Kelley, S. (1996). Mass extinctions and asteroids. *Geoscientist*, **6**, 5–6.

Boaz, N.T., El-Arnauti, A., Gaziry, A., de Heinzelin, J. & Boaz, D.D. (1987). *Neogene Paleontology and Geology of Sahabi*. New York: Alan Liss.

Böcher, J. (1995). Palaeoentomology of the Kap København Formation, a Plio-Pleistocene sequence in Peary Land, North Greenland. *Meddelelser om Grønland, Geoscience*, **33**, 1–82.

Bolli, H.M., Beckmann, J-P. & Saunders, J.B. (1994). *Benthic Foraminiferal Biostratigraphy of the South Caribbean Region*. Cambridge: Cambridge University Press.

Bolli, H.M. & Saunders, J.B. (1985). Oligocene to Holocene low latitude planktic foraminifera. In *Plankton Stratigraphy*, ed. H.M. Bolli, J.B. Saunders & K. Perch-Nielsen, pp. 155–262. Cambridge: Cambridge University Press.

Boltovskoy, E. & Wright, R. (1976). *Recent Foraminifera*. The Hague: Dr W. Junk.

Bonaparte, J.F. (1986). History of the terrestrial Cretaceous vertebrates of Gondwana. *IV Congresso Argentino de Paleontología y Bioestratigraphía, Mendoza, Argentina*. **2**, 63–95.

Bonaparte, J.F. & Novas, F.E. (1985). *Abelisaurus comahuensis*, n.g., n.sp., Carnosauria del Cretacico tardio de Patagonia. *Ameghiniana*, **21**, 259–65.

Bonaparte, J.F., Franchi, M.R., Powell, J.E. & Sepulveda, E.G. (1984). La Formación Los Alamitos (Campaniano–Maastrichtiano) del sudeste de Río Negro, con descripción de *Kritosaurus australis* n. sp. (Hadrosauridae). Significado paleogeográphico de los Vertebrados. *Revista de la Sociedad Geologica Argentina*, **39**, 284–99.

Bond, G. & Lotti, R. (1995). Iceberg discharges into the North Atlantic on millennial time scales during the last glaciation. *Science*, **267**, 1005–10.

Bond, G., Broecker, W., Johnsen, S., McManus, J., Labeyrie, L., Jouzel, J. & Bonani, G. (1993). Correlations between climate records from North Atlantic sediments and Greenland ice. *Nature*, **365**, 143–7.

Bond, G., Heinrich, H., Broecker, W., Labeyrie, L., McManus, J., Andrews, J., Huon, S., Jantschik, R., Clasen, S., Simet, C., Tedesco, K., Klas, M., Bonani, G. & Ivy, S. (1992). Evidence for massive discharges of icebergs into the North Atlantic ocean during the last glacial period. *Nature*, **360**, 245–9.

Bond, W.J. (1989). The tortoise and the hare: ecology of angiosperm dominance and gymnosperm persistence. *Biological Journal of the Linnean Society*, **36**, 227–49.

Bonifay, E. & Vandermeersch, B. (ed.) (1991). *Les Premiers Europeens*. Paris: Editions du C.T.H.S.

Bookstein, F.L. (1986). Size and shape spaces for landmark data in two dimensions. *Statistical Science*, **1**, 181–242.

Borsuk-Biarynicka, M. (1991). Cretaceous lizard occurrences in Mongolia. *Cretaceous Research*, **12**, 607–8.

Böttcher, R. (1987). Neue funde von *Andrias scheuchzeri* (Cryptobranchidae, Amphibia) aus der süddeutschen Molasse (Miozän). *Stuttgarter Beiträge zur Naturkunde, Series B Geologie und Paläontologie*, **131**, 1–38.

Bottjer, D.J. (1985). Bivalve paleoecology. In *Mollusks, Notes for a Short Course*, ed. T.W. Broadhead, pp. 122–37. University of Tennessee Department of Geological Sciences: Studies in Geology 13.

Bouchet, P. (1997). Inventorying the molluscan diversity of the world: what is our rate of progress? *The Veliger*, **40**, 1–11.

Bouchet, P. & Perrine, D. (1996). More gastropods feeding at night on fishes. *Bulletin of Marine Science*, **59**, 224–8.

Boucot, A.J. (1983). Does evolution take place in an ecological vacuum? II. *Journal of Paleontology*, **57**, 1–30.

Boulter, M.C. & Fisher, H.C. (1994). *Cenozoic plants and climates of the Arctic*. Berlin: Springer Verlag.

Boulter, M.C. & Hubbard, R.N.L.B. (1982). Objective paleoecological and biostratigraphic interpretation of Tertiary palynological data by multivariate statistical analysis. *Palynology*, **6**, 55–68.

Boulton, G.S. (1993). The spectrum of Quaternary global change – understanding the past and anticipating the future. *Geoscientist*, **2**, 10–12.

Bower, B. (1997). Ancient roads to Europe. *Science News*, **151**, 12–13.

Bown, P.R. (1987). Taxonomy, evolution and biostratigraphy of late Triassic and Lower Jurassic calcareous nannofossils. *Special Papers in Palaeontology*, **38**, 1–118.

Bown, P.R. (1992). Late Triassic–Early Jurassic Calcareous Nannofossils of the Queen Charlotte Islands, British Columbia. *Journal of Micropalaeontology*, **11**, 177–88.

Bown, P.R., Burnett, J.A. & Gallagher, L.T. (1991). Critical events in the evolutionary history of calcareous nannoplankton. *Historical Biology*, **5**, 279–90.

Bown, T.M., Holroyd, P.A. & Rose, K.D. (1994). Mammal extinctions, body size, and paleo-temperature. *Proceedings of the National Academy of Science, USA*, **91**, 10403–6.

Box, J.S. (1981). *Macroclimate and Plant Forms: An Introduction to Predictive Modelling in Phytogeography*. The Hague: Dr W. Junk.

Bradley, R.S. (1985). *Quaternary Palaeoclimatology*. London: Unwin Hyman.

Bradley, R.S. & Jones, P.D. (1993). 'Little Ice Age' summer temperature variations: their nature and relevance to recent global warming. *The Holocene*, **3**, 367–76.

Bralower, T.J., Arthur, M.A., Leckie, R.M. *et al.* (1994). Timing and paleoceanography of oceanic Dysoxia/Anoxia in the late Barremian to early Aptian (Early Cretaceous). *Palaios*, **9**, 335–69.

Bralower, T.J., Zachos, J.C., Thomas, E. *et al.* (1995). Late Paleocene to Eocene paleoceano-graphy of the equatorial Pacific Ocean: stable isotopes recorded at ODP Site 865, Allison Guyot. *Paleoceanography*, **10**, 841–65.

Brand, L.E. (1994). Physiological ecology of coccolithophores, in *Coccolithophores*, ed. A. Winter & W.G. Siesser, pp. 39–50. Cambridge: Cambridge University Press.

Brasier, M.D. (1980). *Microfossils*. London: George Allen & Unwin.

Brasier, M.D. (1988). Foraminifera extinction and ecological collapse during global biological events. In *Extinction and Survival in the Fossil Record*, ed. G.P. Larwood. Systematics Association Special Volume No. **34**, 37–64.

Brass, G.W., Southam, J.R. & Peterson, W.H. (1982). Warm saline bottom water in the ancient ocean. *Nature*, **296**, 620–2.

Bräuer, G. & Schultz, M. (1996). The morphological affinities of the Plio-Pleistocene mandible from Dmanisi, Georgia. *Journal of Human Evolution*, **30**, 445–81.

Breheret, J.G. & Delamette, M. (1989). Correlations between Mid-Cretaceous Vocontian Black Shales and Helvetic Phosphorites in the Western External Alps. In *Cretaceous of the Western Tethys* ed. J. Wiedmann, pp. 637–55. Stuttgart: E. Schweitzerbart'sche Verlagsbuchhandlung.

Brenner, G.J. (1963). The spores and pollen of the Potomac Group of Maryland. *Department of Geology, Mines and Water Resources, State of Maryland, Bulletin*, **27**, 1–215.

Brenner, G.J. (1976). Middle Cretaceous floral provinces and early migrations of angiosperms. In *Origin and Early Evolution of Angiosperms*, ed. C.B. Beck, pp. 23–47. New York: Columbia University Press.

Brenner, G.J. (1996). Evidence for the earliest stages of angiosperm pollen evolution; a paleoe-quatorial section from Israel, in *Flowering Plant Origin, Evolution and Phylogeny*, ed. D.W. Taylor & L.J. Hickey, pp. 91–115. New York: Chapman and Hall.

Bridge, M.C., Haggart, B.A. & Lowe, J.J. (1990). The history and palaeoclimatic significance of sub-fossil remains of *Pinus sylvestris* in blanket peat from Scotland. *Journal of Ecology*, **78**, 77–91.

Bridgland, D.R. (1994). *Quaternary of the Thames*. London: Chapman & Hall.

Briggs, D.J., Coope, G.R. & Gilbertson, D.D. (1985). The Chronology and Environmental Framework of Early Man in the Upper Thames Valley: a New Model. *British Archaeological Reports, British Series*, **137**.

Brodin, Y-W. (1982). Palaeoecological studies of the recent development of Lake Vaxjosjon IV. Interpretation of the eutrophication process through the analysis of subfossil chironomids. *Archives für Hydrobiologia*, **93**, 313–26.

Brodin, Y-W. & Gransberg, M. (1993). Responses of insects, especially Chironomidae (Diptera), and mites to 130 years of acidification in a Scottish lake. *Hydrobiologia*, **250**, 201–12.

Broecker, W.S. (1994). Massive iceberg discharges as triggers for global climate change. *Nature*, **372**, 421–4.

Broecker, W.S. & Denton, G.H. (1990). What drives glacial cycles? *Scientific American*, **262**, 42–50.

Broecker, W.S., Kennett, J.P., Flower, B.P., Teller, J.T., Trumbore, S., Bonami, G. & Wolfli, W. (1989). Routing of meltwater from the Laurentide ice sheet during the Younger Dryas. *Nature*, **341**, 318–21.

Broecker, W.S., Peteet, D.M. & Rind, U. (1985). Does the ocean-atmosphere system have more than one stable mode of operation? *Nature*, **315**, 21–5.

Brooks, S.J. (1996). Three thousand years of environmental history in a Cairngorms Lochan revealed by analysis of non-biting midges. *Botanical Journal of Scotland*, **48**, 89–90.

Brooks, S.J., Mayle, F.E. & Lowe, J.J. (1997). Chironomid-based lateglacial climatic reconstruction for southeast Scotland. *Journal of Quaternary Science*, **12**, 161–7.

Brown, D.A. (1952). *The Tertiary cheilostomatous Polyzoa of New Zealand*. London: British Museum (Natural History).

Brown, D.A. (1958). Fossil Cheilostomatous Polyzoa from south-west Victoria. *Memoirs of the Geological Survey of Victoria*, **20**, 1–90.

Brown, P.M., Miller, J.A. & Swain, F.M. (1972). Structural and stratigraphic framework, and spatial distribution of permeability of the Atlantic Coastal Plain, North Carolina to New York. *U.S. Geological Survey Professional Paper*, **796**, 1–79.

Browning, J.V., Miller, K.G. & Pak, D.K. (1996). Global implications of lower to middle Eocene sequence boundaries on the New Jersey coastal plain: the icehouse cometh. *Geology*, **24**, 639–42.

Bruijn, H. de (1988). Smaller mammals from the upper Miocene and lower Pliocene of the Strimon basin. *Abstracts, International Workshop 'Continental faunas at the Mio/Pliocene boundary'*, **10**, Faenza.

Bruijn, H. de, Daams, R., Daxner-Höck, G., Fahlbusch, V., Ginsburg, L., Mein, P., Morales, J., Heinzmann, E., Mayhew, D.F., Meulen, A.J.van der, Schmidt-Kittler, N. & Telles Antunes, M. (1992). Report of the RCMNS working group on fossil mammals, Reisenburg 1990. *Newsletters on Stratigraphy*, **26**, 65–118.

Brummer, G.J.A. & van Eijden, A.J.M. (1992). 'Blue-ocean' paleoproductivity estimates from pelagic carbonate mass accumulation rates. *Marine Micropaleontology*, **19**, 99–117.

Bryan, J.R. (1991). A Paleocene coral–algal–sponge reef from southwestern Alabama and the ecology of Early Tertiary reefs. *Lethaia*, **24**, 423–38.

Budd, A.F. & Coates, A.G. (1992). Nonprogressive evolution in a clade of Cretaceous *Montastraea*-like corals. *Paleobiology*, **18**, 425–46.

Budd, A.F., Johnson, K.G. & Stemann, T.A. (1996). Plio-Pleistocene turnover and extinctions in the Caribbean reef-coral fauna. In *Evolution and Environment in Tropical America*, ed. J.B.C. Jackson, A.F. Budd & A.G. Coates, pp. 168–204. Chicago: University of Chicago Press.

Budd, A.F., Stemann, T.A. & Johnson, K.G. (1994). Stratigraphic distributions of genera and species of Neogene to Recent Caribbean reef corals. *Journal of Paleontology*, **68**, 951–77.

Buddemeier, R.W. & Fautin, D.G. (1993). Coral bleaching as an adaptive mechanism. *BioScience*, **43**, 320–6.

Buffetaut, E. (1982*a*). Radiation évolutive, paléoécologie et biogéographie des crocodiliens Mésosuchiens. *Mémoires de la Société Géologique de France*, No. **142**, 1–88.

Buffetaut, E. (1982*b*). A ziphodont mesosuchian crocodile from the Eocene of Algeria and its implications for vertebrate dispersal. *Nature, London*, **300**, 176–8.

Buffetaut, E. (1990). Vertebrate extinctions and survival across the Cretaceous–Tertiary Boundary. *Tectonophysics*, **171**, 337–45.

Buffetaut, E. & Rage, J-C. (1993). Chapter 7. Fossil amphibians and reptiles and the Africa–South America connection, In *The Africa–South America Connection*, ed. W. George & R. Lavocat, pp. 87–99. Oxford: Clarendon Press.

Buffetaut, E. & Taquet, P. (1977). The giant crocodilian *Sarcosuchus* in the Early Cretaceous of Brazil and Niger. *Palaeontology*, **20**, 203–8.

Buge, E. (1982). Influence des facteurs chronologiques et climatologiques sur l'apparition de la faune actuelle de bryozoaires d'Europe occidentale. *Bulletin de la Société Zoologique de France*, **107**, 289–95.

Bulot, L. & Thieuloy, J-P. (1994). Les biohorizons du Valanginien du Sud-Est de la France: un outil fondamental pour les corrélations au sein de la Téthys occidentale. *Géologie Alpine*, Mem. H.S. **20**, 15–41.

Bulot, L.G., Thieuloy, J-P., Blanc, E. & Klein, J. (1993). Le cadre stratigraphique du Valanginien supérieur et de l'Hauterivien du Sud-Est de la France: définition des biochronozones et caractérisation de nouveaux biohorizons. *Géologie Alpine*, **68**, 13–56.

Burger, W.C. (1981). Why are there so many kinds of flowering plants? *Bioscience*, **31**, 577–81.

Burnett, J.A. (1991). A new nannofossil zonation scheme for the Boreal Campanian. *INA Newsletter*, **12**, 67–70.

Burnett, J.A. (1996). Nannofossils and Upper Cretaceous (sub-) stage boundaries – state of the art. *Journal of Nannoplankton Research*, **18**, 23–32.

Busack, S.D. (1986). Biogeographic analysis of the herpetofauna separated by the formation of the Strait of Gibraltar. *National Geographic Research*, **2**, 17–36.

Bush, M.B. (1993). An 11 400 year palaeoecological history of a British Chalk grassland. *Journal of Vegetation Science*, **4**, 47–66.

Buzas, M.A. (1965). The distribution and abundance of Foraminifera in Long Island Sound. *Smithsonian Miscellaneous Collections*, **149**, 1–89.

Buzas, M.A. & Culver, S.J. (1989). Biogeographic and evolutionary patterns of continental margin benthic foraminifera. *Paleobiology*, **15**, 11–19.

Buzas, M.A. & Culver, S.J. (1994). Species pool and dynamics of marine paleocommunities. *Science*, **264**, 1439–41.

Buzas, M.A. & Culver, S.J. (1998). Assembly, disassembly and balance in marine paleocommunities. *Palaios*, **13**, 263–75.

Buzas, M.A. & Gibson, T.G. (1969). Species diversity: benthonic foraminifera in western north Atlantic. *Science*, **163**, 72–5.

Cande, S.C. & Kent, D.V. (1995). Revised calibration of the geomagnetic polarity time scale for the late Cretaceous and Cenozoic. *Journal of Geophysical Research*, **100**, 6093–5.

Canu, F. & Bassler, R.S. (1920) North American Early Tertiary Bryozoa. *Bulletin of the United States National Museum*, **106**, 1–879.

Canu, F. & Bassler, R.S. (1923). North American Later Tertiary and Quaternary Bryozoa. *Bulletin of the United States National Museum*, **125**, 1–244.

Canudo, J.I. (1997). El Kef blind test I results. *Marine Micropaleontology*, **29**, 73–6.

Cappetta, H. (1987). Chondrichthyes II. Mesozoic and Cenozoic Elasmobranchii. In *Handbook of Paleoichthyology, Volume 3B*, ed. H-P. Schultze. Stuttgart: Springer-Verlag.

Cappetta, H., Duffin, C. & Zidek, J. (1993). Chondrichthyes. In *The Fossil Record 2*, ed. M. Benton, pp. 593–609. London: Chapman & Hall.

Carbone, F., Matteucci, R., Pignatti, J.S. & Russo, A. (1993). Facies analysis and biostratigraphy of the Auradu Limestone Formation in the Berbera–Sheikh area, northwestern Somalia. *Geologica Romana*, **29**, 213–35.

Carbonell, E., Bermudez de Castro, J., Arsuaga, J. *et al.* (1995). Lower Pleistocene hominids and artefacts from Atapuerca-TD6 (Spain). *Science*, **269**, 826–9.

Carlisle, D.B. & Braman, D.R. (1991). Nanometre-size diamonds in the Cretaceous/Tertiary boundary clay of Alberta. *Nature*, **352**, 708–9.

Caron, M. (1985). Cretaceous planktic foraminifera. In *Plankton Stratigraphy*, ed. H.M. Bolli, J.B. Saunders & K. Perch-Nielsen, pp. 17–86. Cambridge: Cambridge University Press.

Carpenter, F.M. (1992). Superclass Hexapoda. *Treatise on Invertebrate Palaeontology, Part R, Arthropoda 4*, **3 & 4**. Boulder, CO and Lawrence, KA: Geological Society of America and University of Kansas Press.

Carpenter, J.M. & Rasnitsyn, A.P. (1990). Mesozoic Vespidae. *Psyche*, **97**, 1–20.

Carter, J.G. (1980). Environmental and biological controls of bivalve shell mineralogy and microstructure. In *Skeletal Growth of Aquatic Organisms*, ed. D.C. Rhoads & R.A. Lutz, pp. 69–113. New York: Plenum.

Carvahlo, M.R. De (1996). Higher-level elasmobranch phylogeny, basal squaleans, and paraphyly. In *Interrelationships of Fishes*, ed. M. Stiassny, G.D. Johnson & L. Parenti, pp. 35–62. New York: Academic Press.

Cathles, L.M. & Hallam, A. (1991). Stress-induced changes in plate density, Vail sequences, epeirogeny, and short-lived global sea-level fluctuations. *Tectonics*, **10**, 659–71.

Cavender, T. (1986). Review of the fossil history of North American freshwater fishes. In *The Zoogeography of North American Freshwater Fishes*, ed. C.H. Hocutt & E.O. Wiley, pp. 699–724. New York: John Wiley.

Cerling, T.E., Quade, J., Wang, Y. & Bowman, J.R. (1989). Carbon isotopes in soil and paleosols as ecologic and paleoecologic indicators. *Nature*, **341**, 138–9.

Cevallos-Ferriz, S.R.S., Stockey, R.A., & Pigg, K.B. (1991). The Princeton chert: evidence for in situ aquatic plants. *Review of Palaeobotany and Palynology*, **70**, 173–85.

Chamberlin, T.C. (1898). The influence of great epochs of limestone formation upon the constitution of the atmosphere. *Journal of Geology*, **6**, 609–21.

Chambers, F.M. & Price, S.M. (1985). Palaeoecology of *Alnus* (alder): early post-glacial rise in a valley mire, north-west Wales. *New Phytologist*, **101**, 333–44.

Chandler, R.M. (1994). The wings of *Titanis walleri* (Aves: Phorusrhacidae) from the late Blancan of Florida. *Bulletin of the Florida State Museum of Natural History, Biology Series*, **36**, 175–80.

Chang, M-M. & Chow, C. (1986). Stratigraphic and geographic distributions of the late Mesozoic and Cenozoic fishes of China. In *Indo-Pacific Fish Biology: Proceedings of the Second International Conference on Indo-Pacific Fishes*, ed. T. Uyeno, R. Arai, T. Taniuchi & K. Matsuura, pp. 529–39. Tokyo: Ichthyological Society of Japan.

Chapman, M.R., Funnell, B.M. & Weaver, P.P.E. (1996a). High resolution Pliocene planktonic foraminiferal biozonation of the tropical North Atlantic. In *Microfossils and Oceanic Environments*, ed. A. Moguilevsky & R.C. Whatley, pp. 307–16. Aberystwyth: University of Wales Press.

Chapman, M.R., Funnell, B.M. and Weaver P.P.E. (1998). Isolation, extinction and migration within Late Pliocene populations of the planktonic foraminiferal lineage *Globorotalia* (*Globoconella*) in the North Atlantic. *Marine Micropaleontology*, **33**, 203–22.

Chapman, M.R., Shackleton, N.J., Zhao, M. & Eglinton, G. (1996*b*). Faunal and alkenone reconstructions of subtropical North Atlantic surface hydrography and paleotemperature over the last 28 kyr. *Paleoceanography*, **11**, 343–57.

Charlson, R.J., Lovelock, J.E., Andreae, M.O. & Warren, S.G. (1987). Oceanic phytoplankton, atmospheric sulphur, cloud albedo and climate. *Nature*, **326**, 655–61.

Chase, M.W., Soltis, D.E., Olmstead, R.G. *et al.* (1993). Phylogenetics of seed plants: an analysis of nucleotide sequences from the plastid gene rbcL. *Annals of the Missouri Botanical Garden*, **80**, 528–80.

Cheetham, A.H. (1967). Paleoclimatic significance of the bryozoan *Metrarabdotos*. *Transactions of the Gulf Coast Association of Geological Societies*, **17**, 400–7.

Chepstow-Lusty, A. (1996). The last million years of the discoasters: a global synthesis for the Upper Pliocene. In *Microfossils and Oceanic Environments*, ed. A. Moguilevsky & R. Whatley, pp. 165–75. Aberystwyth: University of Wales Press.

Chesner, C.A., Rose, W.I., Deino, A., Drake, R. & Westgate, J. (1991). Eruptive history of Earth's largest Quaternary caldera (Toba, Indonesia) clarified. *Geology*, **19**, 200–3.

Chiappe, L. (1995). The first 85 million years of avian evolution. *Nature*, **378**, 349–55.

Churkin, M. (1972). Western boundary of the North American continental plate in Asia. *Bulletin of the Geological Society of America*, **83**, 1027–36.

Cifelli, R.S. (1969). Radiation of Cenozoic planktonic foraminifera. *Systematic Zoology*, **18**, 154–68.

Cifelli, R.S. (1976). Evolution of ocean climate and the record of planktonic foraminifera. *Nature*, **264**, 431–2.

Cifelli, R.S. & Bénier, C.S. (1976). Planktonic foraminifera from near the West African coast and a consideration of faunal parcelling in the North Atlantic. *Journal of Foraminiferal Research*, **6**, 258–73.

Cifelli, R.S. & Scott, G.H. (1986). Stratigraphic record of the Neogene globorotalid radiation (Planktonic Foraminiferida). *Smithsonian Contributions to Paleobiology*, **58**, 1–101.

Clarke, A. (1988). Seasonality in the Antarctic marine environment. *Comparative Biochemistry and Physiology*, **90B**, 461–73.

Clarke, A. (1990). Temperature and evolution: Southern Ocean cooling and the Antarctic marine fauna. In *Antarctic Ecosystems. Ecological Change and Conservation*, ed. K.R. Kerry & G. Hempel, pp. 9–22. Berlin: Springer-Verlag.

Clauzon, G., Gautier, F., Berger, A. & Loutre, M-F. (1995). An alternative model for a new interpretation of the Mediterranean Messinian Salinity Crisis. *Abstracts, International Conference on the Biotic and Climatic Effects of the Mediterranean Event on the Circum Mediterranean, Benghazi, 1995*, p. 27.

Cleevely, R.J. & Morris, N.J. (1988). Taxonomy and ecology of Cretaceous Cassiopidae (Mesogastropoda). *Bulletin of the British Museum (Natural History), Geology Series*, **44**, 233–91.

Clements, F.E. (1916). *Plant Succession, an Analysis of Development of Vegetation*. Publication No. **242**. Washington, DC: Carnegie Institution.

Clemens, W.A. & Nelms, L.G. (1993). Paleoecological implications of Alaskan terrestrial vertebrate fauna in latest Cretaceous time at high paleolatitudes. *Geology*, **21**, 503–6.

Clifford, E., Coram, R., Jarzembowski, E.A. & Ross, A.J. (1994). A supplement to the insect fauna from the Purbeck Group of Dorset. *Proceedings of the Dorset Natural History and Archaeological Society*, **115**, 143–6.

CLIMAP (1976). The surface of the ice-age earth. *Science*, **191**, 1131–7.

CLIMAP (1984). The last interglacial ocean. *Quaternary Research*, **21**, 123–224.

Clyde, W. (1990). Benthic Foraminiferal Turnover across the K/T Boundary at the Brazos River Locality in East-Central Texas. Senior Thesis, Princeton University.

Clyde, W.C., Stamatakos, J. & Gingerich, P.D. (1994). Chronology of the Wasatchian Land-Mammal Age (Early Eocene): magnetostratigraphic results from the McCullough Peaks section, northern Bighorn Basin, Wyoming. *Journal of Geology*, **102**, 367–77.

Coates, A.G. (1973). Cretaceous Tethyan coral-rudist biogeography related to the evolution of the Atlantic Ocean, in *Organisms and Continents Through Time. A Joint Symposium of the Geological Society, Palaeontological Association and Systematics Association*, ed. N.F. Hughes. *Special Papers in Palaeontology*, **12**, 169–74.

Coates, A.G. (1977). Jamaican Cretaceous coral assemblages and their relationships to rudist frameworks. In *Second Symposium International sur Coraux et Récifs Coralliens Fossiles. Mémoires du Bureau de Recherches Géologiques et Miniéres*, **89**, 336–41.

Coates, A.G. & Jackson, J.B.C. (1985). Morphological themes in the evolution of clonal and aclonal marine invertebrates. In *Population Biology and Evolution of Clonal Organisms*, ed. J.B.C. Jackson, L.W. Buss & R.E. Cook, pp. 67–105. New Haven: Yale University Press.

Coates, A.G. & Jackson, J.B.C. (1987). Clonal growth, algal symbiosis, and reef formation by corals. *Paleobiology*, **13**, 363–78.

Coates, A.G. & Oliver, W.A., Jr. (1973). Coloniality in zoantharian corals. In *Animal Colonies; Development and Function through Time*, ed. R.S. Boardman, A.H. Cheetham & W.A. Oliver, Jr., pp. 3–27. Stroudsburg, Pennsylvania: Dowden, Hutchinson and Ross, Inc.

Coccioni, R. & Galeotti, S. (1994). K–T boundary extinction: geological instantaneous or gradual event? Evidence from deep-sea benthic foraminifera. *Geology*, **22**, 779–82.

Coffin, M.F. & Eldholm, O. (1993). Large igneous provinces. *Scientific American*, **269**, 26–33.

Cogger, H.G. & Heatwole, H. (1981). Chapter 49. The Australian reptiles: origins, biogeography, distribution patterns and island evolution. In *Ecological Biogeography of Australia*, ed. A. Keast, pp. 1333–73. The Hague; Dr W. Junk.

Cogger, H.G. & Zweifel, R.G. (1998). *Encylopedia of Reptiles and Amphibians*. 2nd ed. Sydney and San Diego: Weldon Owen and Academic Press.

Coleman, R.G. (1993). *Geologic Evolution of the Red Sea*, Oxford Monographs on Geology and Geophysics No. 24. Oxford: Oxford University Press.

Collins, T.M., Frazer, K., Palmer, A.R., Vermeij, G.J. & Brown, W.M. (1996). Evolutionary history of northern hemisphere *Nucella* (Gastropoda, Muricidae): molecular, morphological, ecological and paleoecological evidence. *Evolution*, **50**, 2287–304.

Collinson, M.E. (1983). *Fossil Plants of the London Clay*. Palaeontological Association Field Guides to Fossils Number 1. London: Palaeontological Association.

Collinson, M.E. (1990). Plant evolution and ecology during the early Cainozoic diversification. *Advances in Botanical Research*, **17**, 1–98.

Collinson, M.E. (1993). Taphonomy and fruiting biology of Recent and fossil Nypa. *Special papers in Palaeontology*, **49**, 165–80.

Collinson, M.E. (1996a). 'What use are fossil ferns?' – 20 years on: with a review of the fossil history of extant pteridophyte families and genera. In *Pteridology in Perspective*, ed. J.M. Camus, M. Gibby & R.J. Johns, pp. 349–94. Kew: Royal Botanic Gardens.

Collinson, M.E. (1996b). Plant macrofossils from the Bracklesham group (Early and Middle Eocene), Bracklesham Bay, West Sussex, England: review and significance in the context of coeval British Tertiary Floras. *Tertiary Research*, **16**, 175–202.

Collinson, M.E. (1999). Evolution of angiosperm fruit and seed architecture and associated functional biology : status in the Late Cretaceous and Palaeogene. In *Evolution of Plant Architecture*, ed. M. Kurman & A.R. Hemsley. Kew: Royal Botanic Gardens.

Collinson, M.E., Boulter, M.C. & Holmes, P.L. (1993). Magnoliophyta ('Angiospermae'). In *The Fossil Record 2*, ed. M.J. Benton, pp. 809–41. London: Chapman and Hall.

Collinson, M.E. & Hooker, J.J. (1987). Vegetational and mammalian faunal changes in the Early Tertiary of southern England. In *The Origins of Angiosperms and Their*

Biological Consequences, ed. E.M. Friis, W.G. Chaloner & P.R. Crane, pp. 259–304. Cambridge: Cambridge University Press.

Collinson, M.E. & Hooker, J.J. (1991). Fossil evidence of interactions between plants and plant-eating mammals. *Philosophical Transactions of the Royal Society of London*, B, **333**, 197–208.

Cook, P.L. & Chimonides, P.J. (1983). A short history of the lunulite Bryozoa. *Bulletin of Marine Science*, **33**, 566–81.

Coope, G.R. (1959). A Late Pleistocene insect fauna from Chelford, Cheshire. *Proceedings of the Royal Society of London*, B, **151**, 70–86.

Coope, G.R. (1961). On the study of glacial and interglacial insect faunas. *Proceedings of the Linnaean Society of London*, **172**, 62–71.

Coope, G.R. (1970a). Climatic interpretations of Late Weichselian Coleoptera from the British Isles. *Revue de Geographie Physique et de Géologie Dynamique*, **12**, 149–55.

Coope, G.R. (1970b). Interpretations of Quaternary insect fossils. *Annual Review of Entomology*, **15**, 97–120.

Coope, G.R. (1973). Tibetan Species of Dung Beetle from Late Pleistocene Deposits in England. *Nature*, **245**, 335–6.

Coope, G.R. (1976). Assemblages of fossil Coleoptera from the terraces of the Upper Thames, near Oxford. In *Field Guide to the Oxford Region*, ed. D. Roe, pp. 20–22. Oxford: Quaternary Research Association.

Coope, G.R. (1978). Constancy of insect species versus inconstancy of Quaternary environments In *Diversity of Insect Faunas*, ed. L.A. Mound & N. Waloff, pp. 176–87. Oxford: Blackwell (Symposia of the Royal Entomological Society of London, 9).

Coope, G.R. (1979). Late Cenozoic fossil Coleoptera: evolution, biogeography and ecology. *Annual Review of Ecology and Systematics*, **10**, 247–67.

Coope, G.R. (1995). Insect faunas in ice age environments: why so little extinction?. In *Extinction Rates*, ed. J.H. Lawton & R.M. May, pp. 55–74. Oxford: Oxford University Press.

Coope, G.R. & Angus R.B. (1975). An ecological study of a temperate interlude in the middle of the last glaciation, Based on fossil Coleoptera from Isleworth, Middlesex. *Journal of Animal Ecology*, **44**, 365–91.

Coope, G.R. & Brophy, J.A. (1972). Late-glacial environmental changes indicated by a Coleopteran succession from North Wales. *Boreas*, **1**, 97–142.

Coope, G.R., Gibbard, P.L., Hall, A.R., Preece, R.C.,Robinson, J.E. & Suttcliffe, A.J. (1997). Climatic and Environmental Reconstructions based on fossil assemblages from Middle Devensian (Weichselian) deposits of the River Thames at South Kensington, Central London. *Quaternary Science Reviews*, **16** (10), 1163–96.

Coope, G.R., Shotton F.W. & Strachan, I. (1961). A Late Pleistocene fauna and flora from Upton Warren, Worcestershire. *Philosophical Transactions of the Royal Society of London*, B, **244**, 379–421.

Copper, P. (1989). Enigmas in Phanerozoic reef development. In *Fossil Cnidaria 5. Proceedings of the Fifth International Symposium on Fossil Cnidaria including Archaeocyatha and Spongiomorphs held in Brisbane, Queensland, Australia, 25–29 July 1988*, ed. P.A. Jell & J.W. Pickett. *Memoirs of the Association of Australasian Palaeontologists*, **8**, 371–85.

Coram, R., Jarzembowski, E.A. & Ross, A.J. (1995). New records of Purbeck fossil insects. *Proceedings of the Dorset Natural History and Archaeological Society*, **116**, 145–50.

Corliss, B.H. (1985). Microhabitats of benthic foraminifera within deep-sea sediments. *Nature*, **314**, 435–38.

Corliss, B.H. & Chen, C. (1988). Morphotype patterns of Norwegian Sea deep-sea benthic foraminifera and ecological implications. *Geology*, **16**, 716–19.

Couper, R.A. (1958). British Mesozoic microspores and pollen grains, a systematic and stratigraphic study. *Palaeontographica*, **103**(B), 75–179.

Courtillot, V.E. (1990). A volcanic eruption. *Scientific American*, **263**, 85–92.

Courtillot, V., Jaeger, J-J., Yang, Z., Féraud, G. & Hofmann, C. (1996). The influence of continental flood basalts on mass extinctions: where do we stand? In *The Cretaceous–Tertiary Event and Other Catastrophes in Earth History*, ed. G. Ryder, D. Fastovsky, & S. Gartner. The Geological Society of America, Special Paper, **307**, 513–25.

Cowen, R. (1988). The role of algal symbiosis in reefs through time. *Palaios*, **3**, 221–7.

Crabtree, D.R. (1987). Angiosperms of the northern Rocky Mountains: Albian to Campanian (Cretaceous) megafossil floras. *Annals of the Missouri Botanical Garden*, **74**, 707–47.

Cracraft, J. (1974). Phylogeny and evolution of ratite birds. *Ibis*, **115**, 494–521.

Crame, J.A. (1996a). Evolution of high-latitude molluscan faunas. In *Origin and Evolutionary Radiation of the Mollusca*, ed. J.D. Taylor, pp. 119–31. Oxford: Oxford University Press.

Crame, J.A. (1996b). Antarctica and the evolution of taxonomic diversity gradients in the marine realm. *Terra Antarctica*, **3**, 121–34.

Crame, J.A. (1997). An evolutionary framework for the polar regions. *Journal of Biogeography*, **24**, 1–9.

Crame, J.A. & Clarke, A. (1997). The historical component of marine taxonomic diversity gradients. In *Marine Biodiversity: Patterns and Processes*, ed. R.F.G. Ormond & J.D. Gage, pp. 258–73. Cambridge: Cambridge University Press.

Crane, P.R. (1985). Phylogenetic analysis of seed plants and the origin of angiosperms. *Annals of the Missouri Botanical Garden*, **72**, 716–93.

Crane, P.R. (1987). Vegetational consequences of the angiosperm diversification. In *The Origins of Angiosperms and their Biological Consequences*, ed. E.M. Friis, W.G. Chaloner & P.R. Crane, pp. 107–40. Cambridge: Cambridge University Press.

Crane, P.R. (1988). Major clades and relationships in the 'higher' gymnosperms. In *Origin and Evolution of Gymnosperms*, ed. C.B. Beck, pp. 218–72. New York: Columbia University Press.

Crane, P.R. (1996). The fossil history of the Gnetales. *International Journal of Plant Sciences*, **157** Suppl., 50–7.

Crane, P.R. & Blackmore, S. (eds.) (1989). *Evolution, Systematics, and Fossil History of the Hamamelidae* Volumes 1 and 2. Systematics Association Special Volumes, Nos. 40A, 40B.

Crane, P.R., Friis, E.M. & Pedersen, K.R. (1995). The origin and early diversification of angiosperms. *Nature*, **374**, 27–33.

Crane, P.R. & Lidgard, S. (1989). Angiosperm diversification and palaeolatitudinal gradients in Cretaceous floristic diversity. *Science*, **246**, 675–8.

Crane, P.R. & Lidgard, S. (1990). Angiosperm radiation and patterns of Cretaceous palynological diversity. In *Major Evolutionary Radiations*, ed. P.D. Taylor & G.P. Larwood, pp. 377–407. Oxford: Clarendon Press.

Crane, P.R., Pedersen, K.R., Friis, E.M. *et al.* (1993). Early Cretaceous (Early to Middle Albian) platanoid inflorescences associated with Sapindopsis leaves from the Potomac group of eastern North America. *Systematic Botany*, **18**, 328–44.

Crane, P.R. & Upchurch, G.R. (1987). *Drewria potomacensis* gen. et sp. nov., an early Cretaceous member of the Gnetales from the Potomac Group of Virginia. *American Journal of Botany*, **74**, 1722–36.

Creighton, G.K. (1980). Static allometry of mammalian teeth and the correlation of tooth size and body size in contemporary mammals. *Journal of Zoology, London*, **191**, 435–43.

Crepet, W.L. & Nixon, K.C. (1998). Fossil Clusiaceae from the late Cretaceous (Turonian) of New Jersey and implications regarding the history of bee pollination. *American Journal of Botany*, **85**, 1122–33.

Crochet, J-Y. (1980). *Les Marsupiaux du Tertiaire d'Europe*. Paris: Singer Polignac. 279 pp.

Croizat, L. (1958). *Panbiogeography*. Caracus: published by the author.

Cronquist, A. (1968). *The Evolution and Classification of Flowering Plants*. Boston: Houghton Mifflin.

Cronquist, A. (1981). *An Integrated System of Classification of Flowering Plants*. New York: Columbia University Press.

Crosskey, R.W. & Taylor, B.J. (1986). Fossil blackflies from Pleistocene interglacial deposits in Norfolk, England. *Systematic Entomology*, **11**, 401–12.

Crowley, T.J. (1981). Temperature and circulation changes in the Eastern North Atlantic during the last 150,000 years: evidence from the planktonic foraminiferal record. *Marine Micropaleontology*, **6**, 97–129.

Culotta, E. (1995). Asian hominids grow older. *Science*, **270**, 1116–7.

Culver, S.J. (1990). Benthic foraminifera of Puerto Rican mangrove-lagoon systems: potential for palaeoenvironmental interpretations. *Palaios*, **5**, 34–51.

Culver, S.J. & Goshorn, J. (1996). Foraminifera and paleoenvironments of the Eastover Formation (Upper Miocene, Virginia, USA). *Journal of Foraminiferal Research*, **26**, 300–23.

Currant, A.P. (1987). Late Pleistocene Saiga Antelope *Saiga tatarica* on Mendip. *Proceedings of the University of Bristol Spelaeological Society*, **18**, 74–80.

Currant, A.P. (1989). The Quaternary Origins of the modern British mammal fauna. *Biological Journal of the Linnean Society*, **38**, 23–30.

Currant, A.P. & Jacobi, R.M. (1997). Vertebrate faunas of the British Late Pleistocene and the chronology of human settlement. *Quaternary Newsletter*, **82**, 1–8.

Curry, D., King, A.D., King, C. & Stinton, F.C. (1977). The Bracklesham Beds (Eocene) of Bracklesham Bay and Selsey, Sussex. *Proceedings of the Geologists' Association*, **88**, 243–54.

Curtis, C.D. (1990). Aspects of climatic influence on the clay mineralogy and geochemistry of soils, palaeosols and clastic sedimentary rocks. *Journal of the Geological Society, London*, **147**, 351–7.

Damuth, J.D., Jablonski, D., Harris, J.A., *et al.* (1992). Taxon-free characterization of animal communities. In *Terrestrial Ecosystems through Time*, ed. A.K. Behrensmeyer, J.D. Damuth, W.A. DiMichele *et al.*, pp. 183–203. Chicago: University of Chicago Press.

Dansgaard, W., Clausen, H.B., Gundestrup, N., Hammer, C.U., Johnsen, S.J., Kristinsdottir, P.M. & Reeh, N. (1992). A new Greenland deep ice core. *Science*, **218**, 1273–77.

Dansgaard, W., Johnsen, S.J., Claussen, H.B., Dahl-Hensen, D., Gundestrup, N.S. & Hammer, C.V. (1984). North Atlantic climate oscillations revealed in deep Greenland ice cores. In *Climate Processes and Climate Sensitivity*. American Geophysical Union, Geophysical Monograph, **29**, 288–98.

Dansgaard W., Johnsen, S.J., Clausen, H.B., Dahl-Jensen, D., Gundestrup, N.S., Hammer, C.U., Hvidberg, C.S., Steffensen, J.P., Sveinbjörnsdottier, A.E., Jouzel, J. & Bond, G. (1993). Evidence for general instability of past climate from a 250-kyr ice core record. *Nature*, **364**, 218–20.

Dansgaard, W., White, J.W.C. & Johnsen, S.J. (1989). The abrupt termination of the Younger Dryas climate event. *Nature*, **339**, 532–4.

Darrell, J.G., Gili, E., Rosen, B.R., Skelton, P.W. & Valdeperas, F.X. (1995). 4th Day (10/09/1995) Stop 4. Les Collades de Basturs, Tremp area: rudist and coral formations, in *Field Trip C. Bioconstructions of the Eocene South Pyrenean Foreland Basin (Vic and Igualada areas) and of the Upper Cretaceous South Central Pyrenees (Tremp area)*, ed. A. Perejón & P. Busquets, VII International Symposium on Fossil Cnidaria and Porifera, [Madrid].

Dashzeveg, D. (1988). Holarctic correlation of non-marine Palaeocene-Eocene boundary strata using mammals. *Journal of the Geological Society, London*, **145**, 473–78.

Dashzeveg, D. (1993). Asynchronism of the main mammalian faunal events near the Eocene–Oligocene boundary. *Tertiary Research*, **14**, 141–9.

Daugherty, C.H. & Maxson, L.R. (1982). A biochemical assessment of the evolution of myobatrachine frogs. *Herpetologica*, **38**, 341–8.

David, B. & Mooi, R. (1990). An echinoid that 'gives birth': morphology and systematics of a new Antarctic species, *Urechinus mortenseni* (Echinodermata, Holasteroida). *Zoomorphology*, **110**, 75–89.

Davis, J.C. (1986). *Statistics and Data Analysis in Geology*. New York: John Wiley.

Day, S.P. (1991). Post-glacial vegetational history of the Oxford region. *New Phytologist*, **119**, 445–70.

DeLaca, T.E., Lipps, J.H. & Hessler, R.R. (1980). The morphology and ecology of new large agglutinated Antarctic foraminifera (Textulariina: Notodendrodidae *nov.*). *Zoological Journal of the Linaean Society*, **69**, 205–24.

deMenocal, P. & Bloemendal, J. (1995). Plio-Pleistocene climatic variability in subtropical Africa and the paleoenvironment of hominid evolution: a combined data-model approach. In *Paleoclimate and Evolution, with Emphasis on Human Origins*, ed. E. Vrba, G. Denton, T. Partridge & L. Burckle, pp. 262–88. New Haven: Yale University Press.

Devoy, R.J. (1979). Flandrian sea-level changes and vegetational history of the lower Thames estuary. *Philosophical Transactions of the Royal Society of London*, B, **285**, 355–410.

Dilcher, D.L. (1979). Early angiosperm reproduction: an introductory report. *Review of Palaeobotany and Palynology*, **27**, 291–328.

Dimbleby, G.W. & Evans, J.G. (1974). Pollen and land-snail analysis of calcareous soils. *Journal of Archaeological Science*, **1**, 117–33.

Disney, R.H.L. & Ross, A.J. (1997). *Abaristophora* and *Puliciphora* (Diptera: Phoridae) from Dominican amber (Oligocene) and revisionary notes on modern species. *European Journal of Entomology*, **93**, 127–35.

Ditchfield, P. & Marshall, J.D. (1989). Isotopic variation in rhythmically bedded chalks: paleotemperature variation in the Upper Cretaceous. *Geology*, **17**, 842–5.

Ditchfield, P.W., Marshall, J.D. & Pirrie, D. (1994). High latitude paleotemperature variation: new data from the Tithonian to Eocene of James Ross Island, Antarctica. *Palaeogeography, Palaeoclimatology, Palaeoecology*, **107**, 79–101.

Dodd, J.R. & Stanton, R.J., Jr. (1990). *Paleoecology; Concepts and Applications*, 2nd edn. New York: John Wiley.

Domning, D.P. (1976). An ecological model for late Tertiary sirenian evolution in the north Pacific Ocean. *Systematic Zoology*, **25**, 352–62.

Done, T.J., Ogden, J.C., Wiebe, W.J. & Rosen, B.R. (with contributions from the BIOCORE Working Group) (1996). In *Biodiversity and Ecosystem Function of Coral Reefs*, ed. H.A. Mooney, J.H. Cushman, E. Medina, O.E. Sala & E-D. Schulze, pp. 393–429. Chichester, UK: John Wiley.

Donovan, S.K. (1993). Jamaican Cenozoic Echinoidea. *Geological Society of America Memoir*, **182**, 371–411.

Douglas, R.G. & Savin, S.M. (1975). Oxygen and Carbon isotope analysis of Tertiary and Cretaceous microfossils from Shatsky Rise and other sites in the North Pacific Ocean. *Initial Reports of the Deep Sea Drilling Project*, **32**, 509–20.

Douglas, R.G. & Woodruff, F. (1981). Deep sea benthic foraminifera. In *The Oceanic Lithosphere, The Sea* Vol. 7, ed. C. Emiliani, pp. 1233–1327. New York: Wiley Interscience.

Douvillé, H. (1904). Mollusques fossiles. In *Mission Scientifique en Perse*, ed. J. de Morgan, *4*, *Etudes Geologiques Paleontologie*, pp. 191–380.

Downing, K.F., Lindsay, E.H., Downs, W.R. & Speyer, S.E. (1993). Lithostratigraphy and vertebrate biostratigraphy of the early Miocene Himalayan Foreland, Zinda Pir Dome, Pakistan. *Sedimentary Geology*, **87**, 25–37.

Dowsett, H.J. (1989). Application of the graphic correlation method to Pliocene marine sequences. *Marine Micropaleontology*, **14**, 3–32.

Dowsett, H.J. & Loubere, P. (1992). High resolution Late Pliocene sea-surface temperature record from the Northeast Atlantic Ocean. *Marine Micropaleontology*, **20**, 91–105.

Dowsett, H.J., Thompson, R.S., Barron, J.A., Cronin, T.M., Fleming, R.F., Ishman, S.E., Poore, R.Z., Willard, D.A. & Holtz, T.R. (1994). Joint investigation of the middle Pliocene climate I: PRISM paleoenvironmental reconstructions. *Global and Planetary Climate Change*, **9**, 169–95.

Doyle, J.A. (1969). Cretaceous angiosperm pollen of the Atlantic Coastal Plain and its evolutionary significance. *Journal of the Arnold Arboretum*, **50**, 1–35.

Doyle, J.A. (1978). Origin of angiosperms. *Annual Review of Ecology and Systematics*, **9**, 365–92.

Doyle, J.A. & Donoghue, M.J. (1986). Seed plant phylogeny and the origin of angiosperms: an experimental cladistic approach. *Botanical Review*, **52**, 321–431.

Doyle, J.A. & Donoghue, M.J. (1993). Phylogenies and angiosperm diversification. *Paleobiology*, **19**, 141–67.

Doyle, J.A., Donoghue, M.J. & Zimmer, E.A. (1994). Integration of morphological and ribosomal RNA data on the origin of angiosperms. *Annals of the Missouri Botanical Garden*, **81**, 419–50.

Doyle, J.A. & Hickey, L.J. (1976). Pollen and leaves from the mid-Cretaceous Potomac Group and their bearing on early angiosperm evolution. In *Origin and Early Evolution of Angiosperms*, ed. C.B. Beck, pp. 139–206. New York: Columbia University Press.

Doyle, J.A., Jardiné, S. & Doerenkamp, A. (1982). *Afropollis*, a new genus of early angiosperm pollen, with notes on the Cretaceous palynostratigraphy and paleoenvironments of northern Gondwana. *Bulletin des Centres de Recherches Exploration-Production Elf-Aquitaine*, **6**, 39–117.

Doyle, J.A. & Robbins, E.I. (1977). Angiosperm pollen zonation of the continental Cretaceous of the Atlantic Coastal Plain and its application to deep wells in the Salisbury Embayment. *Palynology*, **1**, 43–78.

Doyle, P. (1992). A review of the biogeography of Cretaceous belemnites. *Palaeogeography, Palaeoclimatology, Palaeoecology*, **92**, 207–16.

Doyle, P., Donovan D.T. & Nixon, M. (1994). Phylogeny and systematics of the Coleoidea. *The University of Kansas Paleontological contributions*, NS **5**, 1–15.

Doyle, P. & Kelly, S.R.A. (1988). The Jurassic and Cretaceous belemnites of Kong Karls Land, Svalbard. *Norsk Polarinstitutt Skrifter*, **189**, 77 pp.

Drobne, K. Ogorelec, B., Pleniar, M., Zucchi-Stolfa, M.L. & Turnek, D. (1988). Maastrichtian, Danian and Thanetian beds in Dolnja Vas (NW Dinarides, Yugoslavia): microfacies, foraminifers, rudists and corals. *Razprave IV. Razreda Sazu*, **29**, 147–224.

Dubois, A.D. & Ferguson, D.K. (1985). The climatic history of pine in the Cairngorms based on radiocarbon dates and stable isotope analysis, with an account of the events leading up to its colonisation. *Review of Palaeobotany and Palynology*, **46**, 55–80.

Duellman, W.E. & Trueb, L. (1986). *Biology of Amphibians*. New York: McGraw-Hill.

Duncan, P.M. (1880). Sind fossil corals and Alcyonaria. *Memoirs of the Geological Survey of India. Palaeontologica Indica*. Series 7 and 14. **1**(1), 1–110.

Duncan, P.M. & Sladen, W.P. (1882). The fossil Echinoidea from the Ranikot Series of nummulitic strata in western Sind. *Memoirs of the Geological Survey of India. Palaeontologia Indica*. Series 14. **1**(3), 21–100.

Duncan, P.M. & Sladen, W.P. (1884). The fossil Echinoidea from the Khirthar Series of nummulitic strata in western Sind. *Memoirs of the Geological Survey of India. Palaeontologia Indica.* Series 14. **1**(3), 101–246.

Durham, J.W. & McNeil, F.S. (1967). Cenozoic migrations of marine invertebrates through the Bering Strait region. In *The Bering Land Bridge*, ed. D.M. Hopkins, pp. 326–49. Stanford: Stanford University Press.

Dury, G.H. (1965). Theoretical implications for underfit streams. *U.S. Geological Survey Professional Paper*, **452-A**.

Dyke, A.S. & Prest, V.K. (1986). Late Wisconsin and Holocene retreat of the Laurentide ice sheet. *Geological Survey of Canada Map, Scale 1: 5,000,000*, Map 1702A.

Eaton, J.G. (1993). Therian mammals from the Cenomanian (upper Cretaceous) Dakota Formation, southwestern Utah. *Journal of Vertebrate Paleontology*, **13**, 105–24.

Edwards, K.J. (1982). Man, space and the woodland edge – speculations on the detection and interpretation of human impact in pollen profiles. In *Archaeological Aspects of Woodland Ecology*, ed. M. Bell & S. Limbrey. *British Archaeological Reports, International Series*, **146**, 5–22.

Edwards, K.J. (1988). The hunter–gatherer/agricultural transition and the pollen record in the British Isles. In *The Cultural Landscape: Past, Present and Future*, ed. H.H. Birks, H.J.B. Birks, P.E. Kaland & D. Moe, pp. 255–66. Cambridge: Cambridge University Press.

Edwards, K.J. (1989). The cereal pollen record and early agriculture. In *The Beginnings of Agriculture*, ed. A. Miles, D. Williams & N. Gardner. *British Archaeological Reports, International Series*, **496**, 113–35.

Edwards, K.J. & Hirons, K.R. (1984). Cereal pollen grains in pre-elm decline deposits: implications for the earliest agriculture in Britain and Ireland. *Journal of Archaeological Science*, **11**, 71–80.

Edwards, K.J. & Ralston, I. (1984). Postglacial hunter–gatherers and vegetational history in Scotland. *Proceedings of the Society of Antiquaries of Scotland*, **114**, 15–34.

Egge, J.K. & Aksnes, D.L. (1992). Silicate as regulating nutrient in phytoplankton competition. *Marine Ecology Process Series*, **83**, 281–9.

Eicher, U. & Siegenthaler, U. (1976). Palynological and oxygen isotope investigations on Late-Glacial sediment cores from Swiss lakes. *Boreas*, **5**, 109–17.

Eisenberg, J.F. (1990). The behavioral/ecological significance of body size in the Mammalia. In *Body Size in Mammalian Paleobiology: Estimation and Biological Implications*, ed. J. Damuth & B.J. MacFadden, pp. 25–37. Cambridge: Cambridge University Press.

El Hajjaji, K. (1992). Les bryozoaires du Miocène Supérieur du Maroc Nord-oriental. *Documents des Laboratoires de Géologie de la Faculté des Sciences de Lyon*, **123**, 1–355.

Elias, S.A. (1994). *Quaternary Insects and their Environments*. Washington: Smithsonian Institution Press.

Elias, S.A., Anderson, K.H. & Andrews, J.T. (1996). Late Wisconsin climate in the northeastern USA and southeastern Canada, reconstructed from fossil beetle assemblages. *Journal of Quaternary Science*, **11**, 417–21.

Elmslie, S.D. (1995). A catastrophic death assemblage of a new species of cormorant and other seabirds from the Pliocene of Florida. *Journal of Vertebrate Paleontology*, **15**, 313–30.

Emlet, R.B. (1989). Apical skeletons of sea urchins (Echinodermata: Echinoidea): two methods for inferring mode of larval development. *Paleobiology*, **15**, 223–54.

Emlet, R.B. (1990). World patterns of developmental mode in echinoid echinoderms. In *Advances in invertebrate reproduction*, 5, ed. M. Hosht & O. Yamashita, pp. 329–35. Amsterdam: Elsevier.

Emlet, R.B., McEdward, L.R. & Strathmann, R.R. (1987). Echinoderm larval ecology viewed from the egg. *Echinoderm Studies*, **2**, 55–136.

Endress, P.K. (1987). The Chloranthaceae: reproductive structures and phylogenetic position. *Botanische Jahrbücher für Systematik, Pflansengeschichte und Pflanzengeographie*, **109**, 153–75.

Endress, P.K. (1989). Evolution of reproductive structures and functions in primitive angiosperms. *Memoirs of the New York Botanical Garden*, **55**, 5–34.

Endress, P.K. & Friis, E.M. (eds.) (1994). *Early Evolution of Flowers*. Vienna: Springer Verlag.

Epshteyn, O.G. (1978). Mesozoic–Cenozoic climates of northern Asia and glacial–marine deposits. *International Geology Revue*, **20**, 49–58.

Erba, E. (1992). Middle Cretaceous calcareous nannofossils from the western Pacific (Leg 129): evidence for paleoequatorial crossings. *Proceedings of the Ocean Drilling Program, Scientific Results*, **129**, 189–201.

Erba, E., Castradori, D., Guasti, G. & Ripepe, M. (1992). Calcareous nannofossils and Milankovitch cycles: the example of the Albian Gault Clay Formation (southern England). *Palaeogeography, Palaeoclimatology, Palaeoecology*, **93**, 47–69.

Erbacher, J. (1994). Entwicklung und Paläoozeanographie mittelkretazischer Radiolarien der westlichen Tethys (Italien) und des Nordatlantiks. PhD dissertation, Institut und Museum für Geologie und Paläontologie der Universität Tübingen.

Ericson, D.B. & Wollin, G. (1970). Pleistocene climates in the Atlantic and Pacific oceans: a comparison based on deep-sea sediments. *Science*, **167**, 1524–5.

Eskov, K. (1987). A new archaeid spider (Chelicerata: Araneae) from the Jurassic of Kazakhstan, with notes on the so-called 'Gondwanan' ranges of recent taxa. *Neues Jahrbuch für Geologie und Palaontologie Abhandlungen*, **175**, 81–106.

Eskov, K. (1992). Archaeid spiders from Eocene Baltic amber (Chelicerata: Araneida: Archaeidae) with remarks on the so-called 'Gondwanan' range of Recent taxa. *Neues Jahrbuch für Geologie und Paläontologie Abhandlungen*, **185**, 311–28.

Espinosa-Arrubarrena, L. & Applegate, S.P. (1996). A paleoecological model of the vertebrate bearing beds in the Tlayúa Quarries, near Tepexi de Rodrígues, Puebla, México. In *Mesozoic Fishes – Systematics and Paleoecology*, ed. G. Arratia & G. Viohl, pp. 539–50. München: Verlag Dr. Freidrich Pfeil.

Estes, R. & Baéz, A. (1985). Chapter 6. Herpetofaunas of North and South America during the Late Cretaceous and Cenozoic: evidence for interchange. In *The Great American Biotic Interchange*, ed. F.G. Stehli & S.D. Webb, pp. 139–97. New York and London: Plenum Press.

Estes, R. & Hutchison, J.H. (1980). Eocene lower vertebrates from Ellesmere Island, Canadian Arctic Archipelago. *Palaeogeography, Palaeoclimatology, Palaeoecology*, **30**, 325–47.

Evans, S.E. (1990). The skull of *Cteniogenys*, a choristodere (Reptilia: Archosauromorpha) from the Middle Jurassic of Oxfordshire. *Zoological Journal of the Linnean Society*, **99**, 209–37.

Evans, S.E. (1993). Jurassic lizard assemblages. *Revue de Paléobiologie (Suisse) (2nd Georges Cuvier Symposium), Volume spéciale*, **7**, 55–65.

Evans, S.E. & Sigogneau-Russell, D. (1997). New sphenodontians (Diapsida: Lepidosauromorpha: Rhynchocephalia) from the Early Cretaceous of North Africa. *Journal of Vertebrate Paleontology*, **17**, 45–51.

Evenhuis, N. (1994). *Catalogue of the Fossil Flies of the World*. Leiden: Backhuys.

Faegri, K. & Iverson, J. (1989). *Textbook of Pollen Analysis*, 4th ed. Chichester, UK: John Wiley.

Fagerstrom, J.A. (1987). *The Evolution of Reef Communities*. New York: John Wiley.

Fastovsky, D.E. (1987). Paleoenvironments of vertebrate-bearing strata during the Cretaceous–Paleogene transition, eastern Montana and western North Dakota. *Palaios*, **2**, 282–95.

Feduccia, A. (1996). *The Origin and Evolution of Birds*. New Haven: Yale University Press.

Ferguson, D.K. (1985). The origin of leaf assemblages, new light on an old problem. *Review of Palaeobotany and Palynology*, **46**, 117–88.

Ferring, C., Swisher, C., Bosinski, G. *et al.* (1996). Progress report on the geology of the Plio-Pleistocene Dmanisi site and the Diliska Gorge, Republic of Georgia. *Abstracts of the Paleoanthropology Society Meeting, New Orleans, 1996*, 5–6.

Field, C.B., Chapin, F.S. III, Matson, P.A., & Mooney, H.A. (1992). Response of terrestrial ecosytems to the changing atmosphere: a resource-based approach. *Annual Review of Ecology and Systematics*, **23**, 201–35.

Fink, W.L. (1988). Phylogenetic analysis and the detection of ontogenetic patterns. In *Heterochrony in Evolution: A Multidisciplinary Approach*, ed. M.L. McKinney, pp. 71–92. New York: Plenum Press.

Fischer, A.G. (1981). Climatic oscillations in the biosphere. In *Biotic Crises in ecological and evolutionary time*, ed. M. Nitecki, pp. 103–31. London: Academic Press.

Fischer, A.G. (1984). The two Phanerozoic supercycles. In *Catastrophes and Earth history*, ed. W.A. Berggren & J.A. Van Couvering, pp. 129–50. Princeton: Princeton University Press.

Fisher, N.S. & Honjo, S. (1991). Intraspecific differences in temperature and salinity responses in the coccolithophore *Emiliania huxleyi*. *Biological Oceanography*, **6**, 355–61.

Fisher, R.V. & Schmincke, H-U. (1984). *Pyroclastic Rocks*. Berlin: Springer-Verlag.

Flannery, T.F., Archer, M., Rich, T.H. & Jones, R. (1995). A new family of monotremes from the Cretaceous of Australia. *Nature*, **377**, 418–20.

Fleming, T.H. (1973). Numbers of mammal species in North and Central American forest communities. *Ecology*, **54**, 555–63.

Flessa, K.W. & Jablonski, D. (1995). Biogeography of Recent marine, bivalve molluscs and its implications for palaeobiogeography and the geography of extinction: a progress report. *Historical Biology*, **10**, 25–47.

Flessa, K.W. & Jablonski, D. (1996). The geography of evolutionary turnover: a global analysis of extant bivalves. In *Evolutionary Paleobiology. In Honor of James W. Valentine*, ed. D. Jablonski, D.H. Erwin & J.H. Lipps, pp. 376–97. Chicago: University of Chicago Press.

Flexer, A. & Reyment, R.A. (1989). Note on Cretaceous Transgressive Peaks and their relation to Geodynamic Events for the Arabo–Nubian and the Northern African Shields. *Journal of African Earth Sciences*, **8**, 65–73.

Flower, R.H. (1988). Progress and changing concepts in cephalopod and particularly nautiloid phylogeny and distribution. In *Cephalopods Present and Past*, ed. J. Wiedmann & J. Kullmann, pp. 17–24. Stuttgart: E. Schweizerbart'sche Verlagsbuchhandlung.

Foote, M. (1993). Discordance and concordance between morphological and taxonomic diversity. *Paleobiology*, **19**, 185–204.

Ford, L.S. & Cannatella, D.C. (1993). The major clades of frogs. *Herpetological Monographs*, **7**, 94–117.

Forey, P.L. & Young, V.T. (1999). Late Miocene fishes of the Emirate of Abu Dhabi, United Arab Emirates. In *Fossil Vertebrates of Arabia*, ed. P.J. Whybrow & A. Hill, pp. 120–35. New Haven: Yale University Press.

Fortelius, M., Werdelin, L., Andrews, P., Bernor, R., Gentry, A.W., Humphrey, L., Mittman, W. & Viranta, S. (1996). Provinciality, diversity, turnover and paleoecology in land mammal faunas of the later Miocene of western Eurasia. In *Evolution of Neogene Continental Biotopes in Europe and the Eastern Mediterranean*, ed. R. Bernor, V. Fahlbusch & H. Mittman, pp. 414–48. New York: Columbia University Press.

Fox, R.C. & Youzwyshyn, G.P. (1994). New primitive carnivorans (Mammalia) from the Paleocene of western Canada, and their bearing on relationships of the order. *Journal of Vertebrate Paleontology*, **14**, 382–404.

Frakes, L.A. (1979). *Climates throughout Geologic Time*. New York: Elsevier.

Frakes, L.A. & Francis, J.E. (1988). A guide to Phanerozoic cold polar climates from high-latitude ice-rafting in the Cretaceous. *Nature*, **333**, 547–9.

Frakes, L.A. & Francis, J.E. (1990). Cretaceous palaeoclimates. In *Cretaceous Resources, Events and Rhythms*, ed. R.N. Ginsburg & B. Beaudoin, pp. 273–87. Dordrecht: Kluwer Academic Publishers.

Frakes, L.A., Francis, J.E. & Syktus, J.I. (1992). *Climate Modes of the Phanerozoic: The History of the Earth's Climate Over the Past 600 Million Years*. Cambridge: Cambridge University Press.

Francis, J.E. (1986). Growth rings in Cretaceous and Tertiary wood from Antarctica and their palaeoclimatic implications. *Palaeontology*, **29**, 665–84.

Francis, J.E. (1991). Palaeoclimatic significance of Cretaceous–Early Tertiary fossil forests of the Antarctic Peninsula. In *Geological Evolution of Antarctica*, ed. M.R.A. Thomson, J.A. Crame & J.W. Thomson, pp. 623–8. Cambridge: Cambridge University Press.

Francis, J.E. & Frakes, L.A. (1993). Cretaceous climates. In *Sedimentology Review 1*, ed. V.P. Wright, pp. 17–30. Oxford: Blackwell Scientific Publications.

Franks, J.W. (1960). Interglacial deposits at Trafalgar Square, London. *New Phytologist*, **59**, 145–52.

Franzen, J-L. & Haubold, H. (1986). The middle Eocene of European mammalian stratigraphy. Definition of the Geiseltalian. *Modern Geology*, **10**, 159–70.

Fraser, R.H. & Currie, D.J. (1996). The species richness-energy hypothesis in a system where historical factors are thought to prevail: coral reefs. *The American Naturalist*, **148**, 138–59.

Frederiksen, N. (1991). Nineteenth Professor Birbal Sahni memorial lecture: Rates of floral turnover and diversity change in the fossil record. *The Palaeobotanist*, **39**, 127–39.

Frederiksen, N. (1994). Paleocene flora diversities and turnover events in eastern North America and their relation to diversity models. *Review of Palaeobotany and Palynology*, **82**, 225–38.

Frenzel, B. (1966). Climatic change in the Atlantic/Sub-Boreal transition on the northern hemisphere: botanical evidence. In *World Climate from 8000 to 0 BC*, ed. J.S. Sawyer, Proceedings of the International Symposium, Imperial College London 1966, Royal Meteorological Society of London, pp. 99–123.

Friis, E.M. (1984). Preliminary report of Upper Cretaceous angiosperm reproductive organs and their level of organization. *Review of Palaeobotany and Palynology*, **39**, 161–88.

Friis, E.M., Chaloner, W.G. & Crane, P.R. (eds.) (1987). *The Origins of Angiosperms and their Biological Consequences*. Cambridge: Cambridge University Press.

Friis, E.M., Crane, P.R. & Pedersen, K.R. (1997). Fossil history of magnoliid angiosperms. In *Evolution and Diversification of Land Plants*, ed. K. Iwatsuki & P.H. Raven, pp. 121–56. Tokyo: Springer-Verlag.

Fursich, F.L. & Jablonski, D. (1984). Late Triassic naticid drillholes: carnivorous gastropods gain a major adaptation but fail to radiate. *Science*, **224**, 78–80.

Gabunia, L.K. & Gabunia, V.J. (1987). [On the origin of moles (Talpinae)]. *Bulletin of the Academy of Sciences of the Georgian SSR*, **125**, 649–51. [In Russian].

Gabunia, L.K., Shevyreva, N.S. & Gabunia, V.J. (1990). A new opossum (Didelphidae, Marsupialia, Metatheria, Mammalia) from the lowermost Oligocene in the Zaysan Basin (eastern Kazakhstan). *Paleontological Journal*, **24**, 61–8.

Gabunia, L.K. & Vekua, A. (1995). A Plio-Pleistocene hominid from Dmanisi, East Georgia, Caucasus. *Nature*, **373**, 509–12.

Gale, A.S. (1995). Cyclostratigraphy and correlation of the Cenomanian Stage in Western Europe. In *Orbital Forcing Timescales and Cyclostratigraphy*, ed. M.R. House & A.S. Gale. Geological Society Special Publication No. **85**, 177–197.

Gale, A.S. & Christensen, W.K. (1996). Occurrence of the belemnite *Actinocamax plenus* in the Cenomanian of SE France and its significance. *Bulletin of the Geological Society of Denmark*, **43**, 68–77.

Gale, A.S., Jenkyns, H.C., Kennedy, W.J. & Corfield, R.M. (1993). Chemostratigraphy versus biostratigraphy: data from around the Cenomanian–Turonian boundary. *Journal of the Geological Society, London*, **150**, 29–32.

Gale, A.S., Kennedy, W.J., Burnett, J.A., Caron, M. & Kidd, B.E. (1996). The Late Albian to Early Cenomanian succession at Mont Risou, near Rosans (Drôme, SE France): an integrated study (ammonites, inoceramids, planktonic foraminifera, nannofossils, oxygen and carbon isotopes). *Cretaceous Research*, **17**, 515–606.

Gao, K. & Fox, R.C. (1991). New teiid lizards from the Upper Cretaceous Oldman Formation (Judithian) of Southeastern Alberta, Canada, with a review of the Cretaceous record of teiids. *Annals of the Carnegie Museum*, **60**, 145–62.

Gao, K. & Hou, L. (1995). Iguanians from the Upper Cretaceous Djadochta Formation, Gobi Desert, China. *Journal of Vertebrate Paleontology*, **15**, 57–78.

Garzanti, E., Critelli, S. & Ingersoll, R.V. (1996). Paleogeographic and paleotectonic evolution of the Himalayan Range as reflected by detrital modes of Tertiary sandstones and modern sands (Indus transect, India and Pakistan). *Geological Society of America Bulletin*, **108**, 631–42.

Gaudant, J. (1988). L'Ichthyfaune éocène de Messel et du Geiseltal (Allemagne): Essai d'approche paléobiogéographique. *Courier Forschungsinstitüt Senckenberg*, **107**, 355–67.

Gaudant, J. (1993). The Eocene freshwater fish–fauna of Europe: from palaeobiogeography to palaeoclimatology. *Kaupia*, **3**, 231–44.

Gayet, M., Rage, J-C., Sempere, T. & Gagnier, P-Y. (1992). Modalités des échanges de vertébrés continentaux entre l'Amérique du Nord et l'Amérique du Sud au Crétacé supérieur et au Paléocene. *Bulletin de la Societé géologique de France*, **163**, 781–91.

Genin, A., Lazar, B. & Brenner, S. (1995). Vertical mixing and coral death in the Red Sea following the eruption of Mount Pinatubo. *Nature*, **377**, 507–10.

George, W. & Lavocat, R. (eds.) (1993). *The Africa–South America Connection*. Oxford Monographs on Biogeography 7. Oxford: Clarendon Press.

Gheerbrant, E., Cappetta, H., Feist, M. *et al.* (1993). La succession des faunes de vertébrés d'âge paléocène supérieur et éocène inférieur dans le Bassin d'Ouarzazate, Maroc. Contexte géologique, portée biostratigraphique et paléogéographique. *Newsletters on Stratigraphy*, **28**, 33–58.

Gheerbrant, E., Sudre, J. & Cappetta, H. (1996). A Palaeocene proboscidean from Morocco. *Nature*, **383**, 68–70.

Gibbard, P.L. (1977). Pleistocene history of the Vale of St Albans. *Philosophical Transactions of the Royal Society of London*, B, **280**, 445–483.

Gibbard, P.L. & Hall, A.R. (1982). Late Devensian river gravel deposits in the Lower Colne valley, West London, England. *Proceedings of the Geologists' Association*, **93**, 291–9.

Gibbons, A. (1996). The peopling of the Americas. *Science*, **274**, 31–3.

Gibbs, R.H. Jr (1985). The stomioid genus *Eustomias* and the oceanic species concept. In *Pelagic Biogeography*, ed. A.C. Pierrot-Bults, S. van der Spoel, B.J. Zahuranec & R.K. Johnson, *Unesco Technical Papers in Marine Science*, 49, pp. 98–106.

Gibson, T.G. (1967). Stratigraphy and paleoenvironment of the phosphatic Miocene strata of North Carolina. *Geological Society of America Bulletin*, **78**, 631–50.

Gibson, T.G. (1983). Stratigraphy of Miocene through lower Pleistocene strata of the United States Central Atlantic Coastal Plain. In *Geology and Paleontology of the Lee Creek Mine North Carolina, I*, ed. C.E. Ray. *Smithsonian Contributions to Paleobiology, No.* **53**, 35–80.

Gibson, T.G. & Buzas, M.A. (1973). Species diversity: patterns in modern and Miocene foraminifera of the eastern margin of North America. *Geological Society of America Bulletin*, **84**, 217–38.

Gibson, T.G., Bybell, L.M. & Govoni, D.L. (1991). Paleocene and Eocene strata of the central Atlantic Coastal Plain. In *Paleocene–Eocene Boundary Sedimentation in the Potomac River Valley, Virginia and Maryland*, ed. T.G. Gibson & L.M. Bybell, I.G.C.P. Project 308, Field Trip Guidebook, pp. 1–13.

Gili, E., Masse, J-P. & Skelton, P.W. (1995a). Rudists as gregarious sediment-dwellers, not reef-builders, on Cretaceous carbonate platforms. *Palaeogeography, Palaeoclimatology, Palaeoecology*, **118**, 245–67.

Gili, E., Skelton, P.W., Vicens, E. & Obrador, A. (1995b). Corals to rudists – an environmentally induced assemblage succession. *Palaeogeography, Palaeoclimatology, Palaeoecology*, **119**, 127–36.

Gingerich, P.D. (1990). Prediction of body mass in mammalian species from long bone lengths and diameters. *Contributions from the Museum of Paleontology, University of Michigan*, **28**, 79–92.

Gingerich, P.D. & Gunnell, G.F. (1995). Rates of evolution in Paleocene–Eocene mammals of the Clarks Fork Basin, Wyoming, and a comparison with Neogene Siwalik lineages of Pakistan. *Palaeogeography, Palaeoclimatology, Palaeoecology*, **115**, 227–47.

Givnish, T.J. (1979). On the adaptive significance of leaf form. In *Topics on Plant Population Biology*, ed. O.T. Solbrig, S. Jain, G.B. Johnson & P.H. Raven, pp. 375–407. New York: Columbia University Press.

Givnish, T.J. (1986). *On the Economy of Plant Form and Function*. Cambridge: Cambridge University Press.

Givnish, T.J. & Vermeij, G.J. (1976). Sizes and shapes of liana leaves. *American Naturalist*, **975**, 743–78.

Glantz, M.H., (ed.) (1977). *Desertification: Environmental Degradation in and around Arid Lands*. Boulder: Westview Press.

Gleason, H.A. (1926). The individualistic concept of the plant association. *Bulletin of the Torres Botanical Club*, **53**, 7–26.

Glenn, C.R. and Kelts, K. (1991). Sedimentary rhythems in lake deposits. In *Cycles and Events in Stratigraphy*, ed. G. Einsele, W. Ricken & A. Seilacher, pp. 188–221. Berlin: Springer Verlag.

Godinot, M. (1988). Les primates adapidés de Bouxwiller (Eocène Moyen, Alsace) et leur apport à la compréhension de la faune de Messel et à l'évolution des Anchomomyini. *Courier ForschungsInstitut Senckenberg*, **107**, 383–407.

Godwin, H. (1964). Late Weichselian conditions in south-eastern Britain: organic deposits from Colney Heath, Hertfordshire. *Proceedings of the Royal Society of London*, B, **160**, 258–75.

Godwin, H. (1975). *History of the British Vegetation: a Factual Basis for Phytogeography*. Cambridge: Cambridge University Press.

Goody, R. (1980). Polar processes and world climate (a brief overview). *Monthly Weather Review*, **108**, 1935–42.

Goudie, A.S. (1983). *Environmental Change*, 2nd edn. Oxford: Clarendon Press.

Goudie, A.S. & Parker, A.G. (1996). *The Geomorphology of the Cotswolds*. Oxford: Cotswold Naturalists' Field Club.

Goudie, A.S., Viles, H.A. & Pentecost, A. (1993). The late-Holocene tufa decline in Europe. *The Holocene*, **3**, 181–6.

Gould, S.J. (1977). *Ontogeny and Phylogeny*. Cambridge: Harvard University Press.

de Graciansky, P.C., Brosse, E., Deroo, G., Herbin, J-P., Montadert, C., Müller, C., Sigal, J. & Schaaf, A. (1987). Organic-rich sediments and palaeoenvironmental reconstructions of the Cretaceous North Atlantic. In *Marine Petroleum Source*

Rocks, ed. J. Brooks & A.J. Fleet. Geological Society Special Publication No. **26**, 317–44.

Gradstein, F.M., Agterberg, F.P., Ogg, J.G., Hardenbol, J., Van Veen, P., Thierry, J. & Huang, Z. (1995). A Triassic, Jurassic and Cretaceous Time Scale. In *Geochronology, Time Scales and Global Stratigraphic Correlation*, ed. W.A. Berggren, D.V. Kent, M-P. Aubry & J. Hardenbol. SEPM (Society for Sedimentary Geology), Special Publication, **54**, 95–126.

Grande, L. (1984). Paleontology of the Green River Formation, with a review of the fish fauna. 2nd edn, *Bulletin of the Geological Survey of Wyoming*, **63**, 1–333.

Grande, L. (1985). The use of paleontology in systematics and biogeography, and a time control refinement for historical biogeography. *Paleobiology*, **11**, 234–43.

Grande, L. (1989). The Eocene Green River Lake system, Fossil Lake, and the history of the North American fish fauna. In *Mesozoic/Cenozoic Vertebrate Paleontology: Classic Localities, Contemporary Approaches*, ed. J. Flynn, International Geological Congress fieldtrip guidebook T322, pp. 18–28. Washington DC: American Geophysical Union.

Grande, L. (1994). Repeating patterns in nature, predictability, and 'impact' in science. In *Interpreting the Hierarchy of Nature*, ed. L. Grande & O. Rieppel, pp. 61–84. San Diego: Academic Press.

Grande, L. & Bemis, W.E. (1998). A comprehensive phylogenetic study of Amiid fishes (Amiidae) based on comparative skeletal anatomy. An empirical search for interconnected patterns of natural history. *Memoirs of the Society of Vertebrate Paleontology*, **4**.

Grande, L. & Micklich, N. (1993). Paleobiogeography of the Eocene Messel and Geiseltal fish faunas. *Kaupia*, **3**, 245–55.

Gravesen, P. (1979). Remarks on the regular echinoids in the Upper Maastrichtian and Lower Danian of Denmark. In *Cretaceous/Tertiary Boundary Events Symposium. I. The Maastrichtian and Danian of Denmark*, ed. T. Birkelund & R.G. Bromley, pp. 72–3. Copenhagen: University of Copenhagen.

Greenwood, D.R. & Basinger, J.F. (1994). The paleoecology of high-latitude Eocene swamp forests from Axel Heiberg Island, Canadian High Arctic. *Review of Palaeobotany and Palynology*, **81**, 83–97.

Greenwood, P.H. (1974). The cichlid fishes of Lake Victoria, East Africa: the biology and evolution of a species flock. *Bulletin of the British Museum (Natural History), Zoology*, **Suppl. 6**, 1–134.

Gregory, J.W. (1930). The fossil fauna of the Samana Range and some neighbouring areas: Part VII. The Lower Eocene corals. *Memoirs of the Geological Survey of India. Palaeontologica Indica*, (NS) **15**, 81–128.

Gregory, R.T., Douthitt, C.B., Duddy, I.R. *et al.* (1989). Oxygen isotope composition of carbonate concretions from the lower Cretaceous of Victoria, Australia: implications for the evolution of meteoric waters on the Australian continent in a paleopolar environment. *Earth and Planetary Science Letters*, **92**, 27–42.

Greig, J. (1982). Past and present lime woods across Europe. In, *Archaeological Aspects of Woodland Ecology*, ed. M. Bell & S. Limbrey. *British Archaeological Reports, International Series*, **146**, 23–55.

Greig, J. (1992). The deforestation of London. *Review of Palaeobotany and Palynology*, **73**, 71–86.

Grieve, R.A.F. (1991). Terrestrial impact; the record in the rocks. *Meteorics*, **26**, 175–94.

Grieve, R., Rupert, J., Smith, J. & Therriault, A. (1996). The record of terrestrial impact cratering. *GSA Today*, **5**, 193–5.

Grimaldi, D.A., Shedrinsky, A., Ross, A. & Baer, N.S. (1995) Forgeries of fossils in 'amber': history, identification and case studies. *Curator*, **37**, 251–74.

Grine, F. (ed.) (1988). *Evolutionary History of the 'Robust' Australopithecines*. New York: Aldine de Gruyter.

Grove, A.T. & Warren, A. (1968). Quaternary landforms and climate on the south side of the Sahara. *Geographical Journal*, **134**, 194–208.

Grove, J.M. (1979). The glacial history of the Holocene. *Progress in Physical Geography*, **3**, 1–54.

Grove, J.M. (1988). *The Little Ice Age*. London: Methuen.

Grün, R. & Stringer, C. (1991). Electron spin resonance dating and the evolution of modern humans. *Archaeometry*, **33**, 153–99.

Gullan, P.J. & Cranston, P.S. (1994). *The Insects: An Outline of Entomology*. London: Chapman & Hall.

Haedrich, R.L. & Merrett, N.R. (1988). Summary atlas of deep-living demersal fishes in the North Atlantic Basin. *Journal of Natural History*, **22**, 1325–62.

Haffer, J. (1974). Avian specialisation in tropical South America with a systematic survey of the toucans (Ramphastidae) and jacamars (Gabulidae). *Publications of the Nuttall Ornithological Club*, **14**. Cambridge, MA.

Hailwood, E.A. & Whybrow, P.J. (1999). Palaeomagnetic correlation and dating of the Baynunah and Shuwaihat Formations, Emirate of Abu Dhabi, United Arab Emirates. In *Fossil Vertebrates of Arabia*, ed. P.J. Whybrow & A. Hill, pp. 75–87. New Haven: Yale University Press.

Håkansson, E., Bromlet, R.G. & Perch-Nielsen, K. (1974). Maastrichtian Chalk of north-west Europe – a pelagic shelf sediment. In *Pelagic Sediments: On Land and Under the Sea*, ed. K.J. Hsü & H.C. Jenkyns. Special Publication of the International Association of Sedimentologists, **1**, 211–33.

Håkansson, E. & Thomsen, E. (1979). Distribution and types of bryozoan communities at the boundary in Denmark. In *Cretaceous–Tertiary Boundary Events. I. The Maastrichtian and Danian of Denmark*, ed. T. Birkelund & R.G. Bromley, pp. 78–91. Copenhagen: University of Copenhagen.

Hallam, A. (1984). Continental humid and arid zones during the Jurassic and Cretaceous. *Palaeogeography, Palaeoclimatology, Palaeoecology*, **47**, 195–223.

Hallam, A. (1985). A review of Mesozoic climates. *Journal of the Geological Society, London*, **142**, 433–45.

Hallam, A. (1992). *Phanerozoic Sea-level Changes*. New York: Columbia University Press.

Hallam, A. & Miller, A.I. (1988). Extinction and survival in the Bivalvia. In *Extinction and Survival in the Fossil Record*, ed. G.P. Larwood. Systematics Association Special Volume No. **34**, 121–38.

Hallock, P. (1987). Fluctuations in the trophic resource continuum: a factor in global diversity cycles. *Paleoceanography*, **2**, 457–71.

Hallock, P., Röttger, R. & Wetmore, K. (1991). Hypotheses on form and function in foraminifera. In *Biology of the Foraminifera*, ed. J.J. Lee & O.R. Anderson, pp. 41–72. London: Academic Press.

Hallock, P. & Schlager, W. (1986). Nutrient excess and the demise of reefs and carbonate platforms. *Palaios*, **1**, 389–98.

Ham, R. van der (1988). Echinoids from the early Palaeocene (Danian) of the Maastricht area (NE Belgium, SE Netherlands): preliminary results. *Mededelingen van de Werkgroep voor Tertiaire en Kwartaire Geologie*, **25**, 127–61.

Ham, R. van der, Wit, W. de, Zuidema, G. & Birgelen, M. van (1987). *Zeeëgels uit het Krijt en Tertiair van Maastricht Luik en Aken*. Maastricht: Natuurhistorisch Genootschap in Limburg.

Hamilton, W.R., Whybrow, P.J. & McClure, H.A. (1978). Fauna of fossil mammals from the Miocene of Saudi Arabia. *Nature*, **274**, 248–9.

Hancock, J.M. (1976). The petrology of the Chalk. *Proceedings of the Geologists' Association*, **86**, 499–535.

Hancock, J.M. (1989). Sea-level changes in the British region during the Late Cretaceous. *Proceedings of the Geologists' Association*, **100**, 565–94.

Hancock, J.M. & Kauffman, E.G. (1979). The great transgressions of the Late Cretaceous. *Journal of the Geological Society, London*, **136**, 175–86.

Handlirsch, A. (1906–08). *Die fossilen Insekten und die Phylogenie der rezenten Formen.* Leipzig: Wilhelm Engelmann.

Hann, B.J., Warner, B.G. & Warwick, W.F. (1992). Aquatic invertebrates and climatic change: a comment on Walker *et al.* (1991). *Canadian Journal of Fisheries and Aquatic Science*, **49**, 1274–6.

Haq, B.U. (1980). Biogeographic history of Miocene calcareous nannoplankton and paleoceanography of the Atlantic Ocean. *Micropaleontology*, **26**, 414–43.

Haq, B.U. (1984). Paleoceanography: A synoptic Overview of 200 Million Years of Ocean History. In *Marine Geology and Oceanography of Arabian Sea and Coastal Pakistan*, ed. B.U. Haq & J.D. Milliman, pp. 201–31. New York: Van Nostrand Reinhold.

Haq, B.U., Hardenbol, J. & Vail, P.R. (1987). Chronology and fluctuating sea levels since the Triassic. *Science*, **235**, 1156–66.

Haq, B.U., Hardenbol, J. & Vail, P.R. (1988). Mesozoic and Cenozoic chronostratigraphy and cycles of sea-level change. In *Sea-Level Changes: An Integrated Approach*, ed. C.K. Wilgus, B.S. Hastings, C.A. Ross, H.W. Posamentier, J. Van Wagoner & G.St.G. Kendall. Society of Economic Paleontologists and Mineralogists, Special Publication, **42**, 39–45.

Haq, B.U. & Lohmann, G.P. (1976). Early Cenozoic calcareous nannoplankton biogeography of the Atlantic Ocean. *Marine Micropaleontology*, **1**, 119–94.

Haq, B.U., Lohmann, G.P. & Wise, S.W. (1977). Calcareous nannoplankton biogeography and its paleoclimatic implications: Cenozoic of the Falkland Plateau (DSDP Leg 36) and Miocene of the Atlantic Ocean. *Initial Reports of the Deep Sea Drilling Project*, **36**, 745–59.

Haq, B.U. & Malmgren, B.A. (1982). Potential of calcareous nannoplankton in paleoenvironmental interpretations – a case study of the Miocene of the Atlantic Ocean. *Stockholm Contributions in Geology*, **37**, 79–98.

Harmelin, J-G. (1992). Facteurs historiques et environnementaux de la biodiversité de la Méditerranée: l'exemple des bryozoaires. *Revue de Paléobologie*, **11**, 503–11.

Harmelin, J-G., Boronat, J., Moissette, P. & Rosso, A. (1989). *Distansescharella seguenzai* Cipolla, 1921 (Bryozoa, Cheilostomata), nouvelles données morphologiques et écologiques tirées de spécimens fossiles (Miocène, Pliocène) et actuelles de Méditerranée. *Geobios*, **22**, 485–501.

Harmelin, J-G. & d'Hondt, J-L. (1993). Transfers of bryozoan species between the Atlantic Ocean and the Mediterranean Sea via the Strait of Gibraltar. *Oceanologica Acta*, **16**, 63–72.

Harries, P.J., Kauffman, E.G. & Hansen, T.A. (1996). Model for biotic survival following mass extinction. In *Biotic Recovery from Mass Extinction Events*, ed. M.B. Hart. Geological Society Special Publication No. **102**, 41–60.

Harris, J. (1993). Ecosystem structure and growth of the African savanna. *Global and Planetary Change*, **8**, 231–48.

Harrison, J.L. (1962). The distribution of feeding habits among animals in a tropical rain forest. *Journal of Animal Ecology*, **31**, 53–63.

Hart, M.B. (1980). A water depth model for the evolution of the planktonic Foraminiferida. *Nature*, **286**, 252–4.

Hart, M.B. (1985). Oceanic anoxic event 2, on-shore and off-shore SW England. *Proceedings of the Ussher Society*, **6** (2), 183–90.

Hart, M.B. (1996). Recovery of the food chain after the Late Cenomanian extinction event. In *Biotic Recovery from Mass Extinction Events*, ed. M B. Hart, Geological Society Special Publication No. **102**, 265–77.

Hart, M.B. & Bigg, P.J. (1981). Anoxic events in the late Cretaceous chalk seas of North-West Europe. In *Microfossils from Recent and Fossil Shelf Seas*, ed. J.W. Neale & M.D. Brasier, pp. 177–85. Chichester, UK: Ellis Horwood.

Harvey, L.D.D. (1989). Modelling the Younger Dryas. *Quaternary Science Reviews*, **8**, 137–49.

Haskins, L.E. (1978). The Vegetational History of South-East Dorset. Unpublished PhD thesis, University of Southampton.

Haszprunar, G. (1988). On the origin and evolution of major gastropod groups, with special reference to the Streptoneura (Mollusca). *Journal of Molluscan Studies*, **54**, 367–441.

Haubold, H. (1990). Dinosaurs and fluctuating sea levels during the Mesozoic. *Historical Biology*, **4**, 75–106.

Hay, W.W. (1995). Cretaceous Paleoceanography. *Geologica Carpathica*, **46**, 257–66.

Hays, J.D., Imbrie, J. & Shackleton, N.J. (1976). Variations in the Earth's orbit: pacemaker of the Ice Ages. *Science*, **194**, 1121–32.

Hays, J.D. & Pitman III, W.C. (1973). Lithospheric plate motion, sea level changes and climatic and ecological consequences. *Nature*, **246**, 18–22.

Hayward, P.J. (1975). Observations on the bryozoan epiphytes of *Posidonia oceanica* from the island of Chios (Aegean Sea). *Documents des Laboratoires de Géologie de la Faculté des Sciences de Lyon, Hors Série*, **3**, 347–56.

Heard, S.B. & Hauser, D.L. (1995). Key evolutionary innovations and their ecological mechanisms. *Historical Biology*, **10**, 151–73.

Hecht, J. (1996). You take the coast road. *New Scientist*, 27 April, 21.

Hecht, M.K. (1992). A new choristodere (Reptilia: Diapsida) from the Oligocene of France: an example of the Lazarus effect. *Geobios*, **25**, 115–31.

Heckel, P.H. (1974). Carbonate buildups in the geologic record: a review. In *Reefs in Time and Space. Selected Examples from the Recent and Ancient*, ed. L.F. Laporte. Society of Economic Paleontologists and Mineralogists, Special Publication, **18**, 90–154.

Heinrich, H. (1988). Origin and consequences of cyclic ice rafting in the northeast Atlantic Ocean during the past 130,000 years. *Quaternary Research*, **29**, 143–52.

Heissig, K. (1979). Die hypothetische Rolle Sudosteuropas bei den Säugetierwanderungen im Eozän und Oligozän. *Neues Jahrbuch für Geologie und Paläontologie Monatshefte*, **2**, 89–96.

Hemleben, C., Spindler, M. & Anderson, O.R. (1988). *Modern Planktonic Foraminifera*. New York: Springer Verlag.

Hendrickson, D.A. (1986). Congruence of bolitoglossine biogeography and phylogeny with geologic history: paleotransport on displaced suspect terranes? *Cladistics*, **2**, 113–29.

Hennig, W. (1966). *Phylogenetic Systematics*. Chicago: Chicago University Press.

Hennig, W. (1966). Spinnenparasiten der Familie Acroceridae im Baltischen Bernstein. *Stuttgarter Beitrage zur Naturkunde*, No. **165**, 21 pp.

Hennig, W. (1967). Die sogenannten 'niederen Brachycera' im Baltischen Bernstein (Diptera: Fam. Xylophagidae, Xylomyidae, Rhagionidae, Tabanidae). *Stuttgarter Beitrage zur Naturkunde*, No. **174**, 51 pp.

Herbert, T.D. & Fischer, A.G. (1986). Milankovitch climatic origin of mid-Cretaceous black shale rhythms in central Italy. *Nature*, **321**, 739–43.

Herbert, T.D. & Sarmiento, J.L. (1991). Ocean nutrient distribution and oxygenation: limits on the formation of warm saline bottom water over the past 91 m.y. *Geology*, **19**, 702–5.

Herendeen, P.S. & Crane, P.R. (1995). The fossil history of the monocotyledons. In *Monocotyledons: systematics and evolution*, ed. P.J. Rudall, P.J. Cribb, D.F. Cutler & C.J. Humphries, pp. 1–21. Kew: Royal Botanic Gardens.

Herendeen, P.S., Crane, P.R. & Drinnan, A.N. (1995). Fagaceous flowers, fruits and cupules from the Campanian (late Cretaceous) of central Georgia, USA. *International Journal of Plant Sciences*, **156**, 93–116.

Herendeen, P.S. & Dilcher, D.L. (eds.) (1992). *Advances in Legume Systematics Part 4 The Fossil Record*. Kew: Royal Botanic Gardens.

Herman, A.B. (1994). Diversity of the Cretaceous platanoid plants of the Anadyr'-Koryak subregion in relation to climatic changes. *Stratigraphy and Geological Correlation*, **2**, 365–78.

Herman, A.B. & Shczepetov, S.V. (1992). The Mid-Cretaceous flora of the Anadyr River basin (Tchukotka, NE Siberia). In *Palaeovegetational Development in Europe and Regions Relevant to its Palaeofloristic Evolution*, ed. J. Kovar-Eder, Proceedings of the Pan-European Palaeobotanical Conference, Vienna, 1991. Museum of Natural History, Vienna, pp. 273–9.

Herman, A.B. & Spicer, R.A. (1996). Palaeobotanical evidence for a warm Cretaceous Arctic Ocean. *Nature*, **380**, 330–3.

Herman, A.B. & Spicer, R.A. (1997). New quantitative palaeoclimate data for the Late Cretaceous Arctic: evidence for a warm polar ocean. *Palaeogeography, Palaeoclimatology, Palaeoecology*, **128**, 227–51.

Herngreen, G.F. W. & Chlonova, A.F. (1981). Cretaceous microfloral provinces. *Pollen et Spores*, **23**, 442–555.

Hickey, L.J. & Doyle, J.A. (1977). Early Cretaceous fossil evidence for angiosperm evolution. *Botanical Review*, **43**, 3–104.

Hildebrand, A.R., Penfield, G.T., Kring, D.A., Pilkingotn, M. & Jacobsen, S.B. (1991). Chicxulub Crater: a possible Cretaceous/Tertiary boundary impact crater on the Yucatán Peninsula, Mexico. *Geology*, **19**, 867–71.

Hill, A. (1987). Causes of perceived faunal change in the later Neogene of East Africa. *Journal of Human Evolution*, **16**, 583–96.

Hill, A. (1995). Faunal and environmental change in the Neogene of East Africa: Evidence from the Tugen Hills Sequence, Baringo District, Kenya. In *Paleoclimate and Evolution, with Emphasis on Human Origins*, ed. E.S. Vrba, G.H. Denton, T.C. Partridge & L.H. Burkle, pp. 178–93. New Haven: Yale University Press.

Hill, A., Whybrow, P.J., & Yasin al-Tikriti, W. (1990). Late Miocene primate fauna from the Arabian Peninsula: Abu Dhabi, United Arab Emirates. *American Journal of Physical Anthropology*, **81**, 240–1.

Hill, C.R. (1996). New plant with flower-like organs from the Wealden of the Weald (Lower Cretaceous). *Cretaceous Research*, **17**, 27–38.

Hill, M.O. (1973). Reciprocal averaging: an eigenvector method of ordination. *Journal of Ecology*, **61**, 237–49.

Hill, M.O. (1974). Correspondence analysis: a neglected multivariate method. *Applied Statistics*, **23**, 340–54.

Hill, R.S. (ed.) (1994). *History of the Australian Vegetation: Cretaceous to Recent*. Cambridge: Cambridge University Press.

Hodell, D.A. & Kennett, J.P. (1986). Late Miocene–Early Pliocene stratigraphy and paleoceanography of the South Atlantic and Southwest Pacific Oceans: a synthesis. *Paleoceanography*, **1**, 285–311.

Hoedemaeker, Ph. J. (1995). Ammonite evidence for long-term sea-level fluctuations between the 2nd and 3rd order in the lowest Cretaceous. *Cretaceous Research*, **16**, 231–41.

Hofmann, W. (1986). Chironomid analysis. In *Handbook of Holocene Palaeoecology and Palaeohydrology*, ed. B. Berglund, PP. 715–27. New York: Wiley.

Holdridge, L.R. (1947). Determination of world plant formations from simple climatic data. *Science*, **105**, 367–8.

Holligan, P.M. (1992). Do marine phytoplankton influence global climate?. In *Primary Productivity and Biogeochemical Cycles in the Sea*, ed. P.G. Falkowski & A.D. Woodhead, pp. 487–501. New York: Plenum Press.

Holligan, P.M., Fernandez, E., Aiken, J. *et al.* (1993). A biogeochemical study of the coccolithophore *Emiliania huxleyi* in the North Atlantic. *Global Biogeochemical Cycles*, **7**, 879–900.

Holligan, P.M., Viollier, M., Harbour, D.S. *et al.* (1983). Satellite and ship studies of coccolithophore production along a continental shelf edge. *Nature*, **304**, 339–42.

Holman, J.A. (1995). Pleistocene amphibians and reptiles in North America. *Oxford Monographs on Geology and Geophysics*, **32**. Oxford: Clarendon Press.

Hooker, J.J. (1992). British mammalian paleocommunities across the Eocene–Oligocene transition and their environmental implications. In *Eocene–Oligocene Climatic and Biotic Evolution*, ed. D.R. Prothero & W.A. Berggren. Princeton: Princeton University Press.

Hooker, J.J. (1998). Mammalian faunal change across the Paleocene–Eocene transition in Europe. in *Late Paleocene–Early Eocene Climatic and Biotic Events in the Marine and Terrestrial Records*, ed. M-P. Aubry, S.G. Lucas & W.A. Berggren, pp. 428–50. New York: Columbia University Press.

Hooker, J.J., Collinson, M.E., Van Bergen, P.F. *et al.* (1995). Reconstruction of land and freshwater palaeoenvironments near the Eocene–Oligocene boundary, southern England. *Journal of the Geological Society, London*, **152**, 449–68.

Hooker, J.J., Milner, A.C. & Sequeira, S.E.K. (1991). An ornithopod dinosaur from the Late Cretaceous of West Antarctica. *Antarctic Science*, **3**, 331–2.

Hopkins, D.M., Matthews, J.V., Schweger, C.E. & Young, (ed.) (1982). *Paleoecology of Beringia*. New York: Academic Press.

Horowitz, A.S. & Pachut, J.F. (1994). Lyellian bryozoan percentages and the fossil record of the Recent bryozoan fauna. *Palaios*, **9**, 500–5.

Horowitz, A.S. & Pachut, J.F. (1996). Diversity of Cenozoic Bryozoa: a preliminary report. In *Bryozoans in Space and Time*, ed. D.P. Gordon, A.M. Smith & J.A. Grant-Mackie, pp. 133–7. Wellington: New Zealand Institute of Water and Atmospheric Research.

Hottinger, L. (1987). Conditions for generating carbonate platforms. *Memorie della Societê Geologica d'Italia*, **40**, 265–71.

House, M.R. (1993). Fluctuations in ammonoid evolution and possible environmental controls. In *The Ammonoidea: Environment, Ecology and Evolutionary Change*, ed. M. R.House. Systematics Association Special Volume, **47**, 13–34.

Hovan, S.A., Rea, D.K., Pisias, N.G. & Shackleton, N.J. (1989). A direct link between the China loess and marine records: aeolian flux to the north Pacific. *Nature*, **340**, 296–8.

Hsü, K.J., Montadert, L., Bernouilli, D., Cita, M.B., Erickson, A., Garrison, R.E., Kidd, R.B., Melieres, F., Muller, C. & Wright, R. (1977). History of the Mediterranean salinity crisis. *Nature*, **267**, 399–403.

Huang, C. (1981). Observations on the interior of some late Neogene planktonic foraminifera. *Journal of Foraminiferal Research*, **11**, 173–90.

Huber, B.T. (1994). Ontogenetic morphometrics of some late Cretaceous trochospiral planktonic foraminifera from the austral realm. *Smithsonian Contributions to Paleobiology*, **77**, 1–85.

Huber, B.T., Hodell, D.A. & Hamilton, C.P. (1995). Middle-late Cretaceous climates of the southern high latitudes: stable isotopic evidence for minimal equator-to-pole thermal gradients. *Bulletin of the Geological Society of America*, **107**, 1164–91.

Huber, B.T. & Watkins, D.K. (1992). Biogeography of Campanian–Maastrichtian calcareous plankton in the region of the Southern Ocean: paleogeographic and paleoclimatic implications. *Antarctic Research Series*, **56**, 31–60.

Hughes, N.F. (1976). *Palaeobiology of Angiosperm Origins*. Cambridge: Cambridge University Press.

Hughes, N.F. (1994). *The Enigma of Angiosperm Origins*. Cambridge: Cambridge University Press.

Hunt, A.P., Lockley, M.G., Lucas, S.G. & Meyer, C.A. (1994). The global sauropod fossil record. *Gaia*, No.**10**, 261–79.

Huntley, B. (1994). The use of climate response surfaces to reconstruct palaeoclimate from Quaternary pollen and plant macrofossil data. In *Palaeoclimates and their Modelling*, ed. J.R.L. Allen, B.J. Hoskins, B.W. Sellwood, R.A. Spicer & P.J. Valdes, pp. 7–16. London: The Royal Society and Chapman & Hall.

Huntley, B. & Birks, H.J.B. (1983). *An Atlas of Past and Present Pollen Maps of Europe: 0–13 000 Years Ago*. Cambridge: Cambridge University Press.

Hutchison, J.H. (1982). Turtle, crocodilian and champsosaur diversity changes in the Cenozoic of the North-Central region of Western United States. *Palaeogeography, Palaeoclimatology, Palaeoecology*, **37**, 149–64.

Hüssner, H. (1994). Reefs, an elementary principle with many complex realizations. *Beringeria. Würzburger Geowissenschaftliche Mitteilungen*, **11**, 3–99.

Imbrie, J., Boyle, E.A., Clemens, S.C. *et al.* (1992). On the structure and origin of major glaciation cycles 1. linear responses to Milankovitch forcing. *Paleoceanography*, **7**, 701–38.

Imbrie, J., Hays, J.D., Martinson, D.G., McIntyre, A., Mix, A.C., Morley, J.J., Pisias, N.G., Prell, W.L., & Shackleton, N.J. (1984). The orbital theory of Pleistocene climate: support from a revised chronology of the marine $\delta^{18}O$ record. In *Milankovitch and Climate*, ed. A.L. Berger *et al.*, pp. 269–305. Dordrecht: Reidel Publishing Company.

Imbrie, J. & Kipp, N.G. (1971). A new micropalaeontological method for quantitative paleo-climatology: application to a late Pleistocene Caribbean core. In *The Late Cenozoic Glacial Ages*, ed. K.K. Turekian, pp. 71–181. New Haven: Yale University Press.

Insalaco, E. (1996). Upper Jurassic microsolenid biostromes of northern and central Europe: facies and depositional environment. *Palaeogeography, Palaeoclimatology, Palaeoecology*, **121**, 169–94.

Insalaco, E. (1998). The descriptive nomenclature and classification of growth fabrics in scleractinian reefs. *Sedimentary Geology*, **118**, 159–86.

Ivany, L.C., Portell, R.W. & Jones, D.S. (1990). Animal–plant relationships and paleobiogeo-graphy of an Eocene seagrass community from Florida. *Palaios*, **5**, 244–58.

Iversen, J. (1941). Landnam I Danmarks stenalder (land occupation in Denmark's Stone Age). *Danmarks Geologische Undersøgelse*, series II, **66**, 1–68.

Iversen, J. (1949). The influence of prehistoric man on vegetation. *Danmarks Geologische Undersøgelse*, series IV, **3**, 1–23.

Izett, G.A., Cobban, W.A., Obradovich, J.D. & Kunk, M.J. (1993). The Manson impact structure $^{40}Ar/^{39}Ar$ age and its distal impact ejecta in the Pierre Shale in southeastern South Dakota. *Science*, **262**, 729–32.

Jablonski, D. (1986*a*). Background and mass extinctions: the alternation of macroevolution-ary regimes. *Science*, **231**, 129–33.

Jablonski, D. (1986*b*). Causes and consequences of mass extinctions; a comparative approach. In *Phanerozoic Diversity Patterns. Profiles in Macroevolution*, ed. J.W. Valentine, pp. 335–54. Princeton: Princeton University Press.

Jablonski, D. (1995). Extinctions in the fossil record. In *Extinction Rates*, ed. J.H. Lawton & R.M. May, pp. 25–44. Oxford: Oxford University Press.

Jablonski, D. & Bottjer, D.J. (1990). Onshore–offshore trends in marine invertebrate evolu-tion. In *Causes of Evolution*, ed. R.M. Ross & W.D. Allmon, pp. 21–75. Chicago: University of Chicago Press.

Jablonski, D., Lidgard, S. & Taylor, P.D. (1997). Comparative ecology of bryozoan radiations: origin of novelties in cyclostomes and cheilostomes. *Palaios*, **12**, 505–23.

Jablonski, D. & Lutz, R.A. (1983). Larval ecology of marine benthic invertebrates: paleobiological implications. *Biological Review*, **58**, 21–89.

Jablonski, D. & Raup, D. (1995). Selectivity of end-Cretaceous marine bivalve extinctions. *Science*, **268**, 389–91.

Jackson, J.B.C. (1991). Adaptation and diversity of reef corals. *BioScience*, **41**, 475–82.

Jackson, J.B.C. (1992). Pleistocene perspectives on coral reef community structure. *American Zoologist*, **32**, 719–31.

Jackson, J.B.C. (1994). Community unity? *Science*, **264**, 1412–13.

Jackson, J.B.C., Jung, P. Coates, A.G. & Collins, L.S. (1993). Diversity and extinction of tropical American mollusks and emergence of the Isthmus of Panama. *Science*, **260**, 1624–26.

Jaeger, J-J. (1971). La faune de mammifères du Lutétien de Bouxwiller (Bas-Rhin) et sa contribution à l'élaboration de l'échelle des zones biochronologique de l'Eocène européen. *Bulletin de la Service de la Carte géologique d'Alsace et Lorraine*, **24**, 93–105.

Jaeger, J-J., Cortillot, V. & Tapponier, P. (1989). Paleontological view of the ages of the Deccan Traps, the Cretaceous/Tertiary boundary, and the India–Asia collision. *Geology*, **17**, 316–9.

Jaeger, J-J. & Rage, J-C. (1990). Reply to comments on 'Paleontological view of the ages of the Deccan Traps, the Cretaceous/Tertiary boundary, and the India–Asia collision'. *Geology*, **18**, 186–8.

Jagt, J.W.M. & Ham, R. van der (1994). Early Palaeocene marsupiate regular echinoids from NE Belgium. In *Echinoderms through Time*, ed. B. David, A. Guille, J-P. Feral & M. Roux, pp. 725–9. Rotterdam: A.A. Balkema.

Jagt, J.W.M. & Michels, G.P.H. (1990). Additional note on the echinoid genus *Cyclaster* from the Late Maastrichtian of northeastern Belgium. *Geologie en Mijnbouw*, **69**, 179–85.

James, N.P. (1983). Reef environment. In *Carbonate Depositional Environments*, ed. P.A. Scholle, D.G. Bebout & C.H. Moore. Memoirs of the American Association of Petroleum Geologists, **33**, 345–462.

Janis, C.M. (1989). A climatic explanation for the patterns of evolutionary diversity in ungulate mammals. *Palaeontology*, **32**, 463–81.

Jansen, E., & Sjoholm, J. (1991). Reconstruction of glaciation over the past 6 Myr from ice-borne deposits in the Norwegian Sea. *Nature*, **349**, 600–3.

Jarvis, I. Carson, G.A., Cooper, M.K.E., Hart, M.B., Leary, P.N., Tocher, B.A., Horne, D. & Rosenfeld, A. (1988). Microfossil assemblages and the Cenomanian–Turonian (late Cretaceous) oceanic anoxic event. *Cretaceous Research*, **9**, 2–103.

Jarzembowski, E.A. (1984). Early Cretaceous insects from Southern England. *Modern Geology*, **9**, 71–93.

Jarzembowski, E.A. (1995). Early Cretaceous insect faunas and palaeoenvironment. *Cretaceous Research*, **16**, 681–93.

Jarzembowski, E.A. & Ross, A.J. (1993). Time flies: the geological record of insects. *Geology Today*, **9**, 218–23.

Jarzembowski, E.A. & Ross, A.J. (1996). Insect origination and extinction in the Phanerozoic. In *Biotic Recovery from Mass Extinction Events*, ed. M.B. Hart. Geological Society Special Publication No. **102**, 65–78.

Jefferies, R.P.S. (1962). The palaeoecology of the *Actinocamax plenus* subzone (lowest Turonian) in the Anglo-Paris Basin. *Palaeontology*, **4**, 609–47.

Jeffery, C.H. (1997). All change at the K-T boundary? Echinoids from the Maastrichtian and Danian of the Mangyshlak Peninsula, Kazakhstan. *Palaeontology*, **40**, 659–712.

Jeffery, C.H. (1998). Carrying on regardless: the echinoid genus *Cyclaster* at the Cretaceous Tertiary boundary. *Lethaia*, **31**, 149–57.

Jeffery, C.H. & Smith, A.B. (1998). Estimating extinction levels for echinoids across the Cretaceous–Tertiary boundary. In *Echinoderms*, ed. R. Mooi & M. Telford, pp. 695–701. Rotterdam: A.A. Balkema.

Jell, P.A. & Duncan, P.M. (1986). Invertebrates, mainly insects, from the freshwater, Lower Cretaceous, Koonwarra Fossil Bed (Korumburra Group), South Gippsland, Victoria. *Memoir of the Association of Australasian Palaeontologists*, No. **3**, 111–205.

Jenkins, D.G. & Shackleton, N.J. (1979). Parallel changes in species diversity and palaeotemperature in the Lower Miocene. *Nature*, **278**, 50–1.

Jenkyns, H.C., Gale, A.S. & Corfield, R.M. (1994). Carbon- and oxygen-isotope stratigraphy of the English Chalk and Italian Scaglia and its palaeoclimatic significance. *Geological Magazine*, **131**, 1–34.

Jensen, K.R. (1993). Evolution of buccal apparatus and diet radiation in the Sacoglossa. Opisthobranchia). *Bolletino Malacologico*, **2**, 147–72.

Jerzykiewicz, T. & Russell, D.A. (1991). Late Mesozoic stratigraphy and vertebrates of the Gobi Basin. *Cretaceous Research*, **12**, 345–77.

Johnsen, S.J., Clausen, H.B., Dansgaard, W., Fuhrer, K., Gunerstrup, N., Hammer, C.U., Iversen, P., Jouzel, J., Stauffer, B. & Steffensen, J.P. (1992). Irregular glacial interstadials recorded in a new Greenland ice core. *Nature*, **359**, 311–13.

Johnson, C.C., Baron, E.J., Kauffman, E.G., Arthur, M.A., Fawcett, P.J. & Yasuda, M.K. (1996). Middle Cretaceous reef collapse linked to ocean heat transport. *Geology*, **24**, 376–80.

Johnson, C.C. & Kauffman, E.G. (1990). Originations, radiations and extinctions of Cretaceous rudistid bivalve species in the Caribbean Province. In *Extinction Events in Earth History*, ed. E.G. Kauffman & O.H. Walliser, Lecture Notes in Earth History, **30**, Springer, Berlin, pp. 305–24.

Johnson, C.C. & Kauffman, E.G. (1996). Maastrichtian extinction patterns of Carribean province rudistids. In *The Cretaceous–Tertiary Mass Extinction: Biotic and Environmental Changes*, ed. N. MacLeod & G. Keller, pp. 231–74. New York: W.W. Norton & Co.

Johnson, G.D. & Patterson, C. (1993). Percomorph phylogeny: a survey of acanthomorphs and a new proposal. *Bulletin of Marine Science*, **52**, 554–626.

Johnson, K. (1992). Leaf-fossil evidence for extensive floral extinction at the Cretaceous–Tertiary boundary, North Dakota, USA. *Cretaceous Research*, **13**, 91–117.

Johnson, K.G. (1998). A phylogenetic test of accelerated turnover in Neogene Caribbean brain corals (Scleractinia: Faviidae). *Palaeontology*, **41**, 1247–68.

Johnson, K.G., Budd, A.F. & Stemann, T.A. (1995). Extinction selectivity and ecology of Neogene Caribbean reef corals. *Paleobiology*, **21**, 52–73.

Johnson, M.R.W. (1994). Volume balance of erosional loss and sediment deposition related to Himalayan uplifts. *Journal of the Geological Society, London*, **151**, 217–20.

Johnson, R.G. & Maclure, B.T. (1976). A model for northern hemisphere continental ice sheet variation. *Quaternary Research*, **6**, 325–53.

Jones, D.S. (1988). Sclerochronology and the size versus age problem. In *Heterochrony in Evolution*, ed. M.L. McKinney, pp. 93–108. New York: Plenum Press.

Jones, M.H. (1990). *Middle Eocene Foraminifera from the Piney Point Formation of the Virginia and Maryland Coastal Plain*. Unpublished thesis, Old Dominion University, Norfolk, VA.

Jouzel, J., Lorius, C., Petit, J.R., Genthon, C., Barkov, N.I., Kotlyakov, V.M. & Petrov, V.M. (1987). Vostok ice core: a continuous isotope temperature record over the last climatic cycle (160,000 years). *Nature*, **329**, 403–7.

Jung, P. (1987). Giant gastropoods of the genus *Campanile* from the Caribbean Eocene. *Eclogae Geologicae Helvetiae*, **80**, 889–96.

Kabat, A.R. (1990). Predatory ecology of naticid gastropods with a review of shell boring predation. *Malacologia*, **32**, 155–93.

Kauffman, E.G. (1984). The fabric of Cretaceous extinctions. In *Catastrophes and Earth History: The New Uniformitarianism*, ed. W.A. Berggren & J.A. Van Couvering, pp. 151–246. Princeton: Princeton University Press.

Kauffman, E.G. & Fagerstrom, J.A. (1993). The Phanerozoic evolution of reef diversity. In *Species Diversity in Ecological Communities: Historical and Geographical Perspectives*, ed. R.E. Ricklefs & D. Schluter, pp. 315–29. Chicago: The University of Chicago Press.

Kauffman, E.G. & Harries, P.J. (1996). The importance of crisis progenitors in recovery from mass extinction. In *Biotic Recovery from Mass Extinction Events*, ed. M.B. Hart. Geological Society Special Publication No. **102**, 15–39.

Kauffman, E.G. & Hart, M.B. (1996). Cretaceous bio-events. In *Global Events and Event Stratigraphy in the Phanerozoic: results of the International Interdisciplinary co-operation in the IGCP Project 216 'Global Events in Earth History'*, ed. O.H. Walliser, pp. 285–304. New York: Springer-Verlag.

Kauffman, E.G. & Sohl, N.F. (1974). Structure and evolution of Antillean rudist frameworks. *Verhandlungen der Naturforschenden Gesellschaft in Basel*, **84**, 399–467.

Kay, E.A. (1996). Evolutionary radiations in the Cypraeidae. In *Origin and Evolutionary Radiation of the Mollusca*, ed. J.D. Taylor, pp. 211–20. Oxford: Oxford University Press.

Keen, A.M. (1971). *Sea Shells of Tropical West America*. Stanford: Stanford University Press.

Keigwin, L.D. (1978). Pliocene closing of the Isthmus of Panama, based on biostratigraphic evidence from nearby Pacific Ocean and Caribbean Sea cores. *Geology*, **6**, 630–4.

Keigwin, L.D. (1987). Stable isotope record of Deep Sea Drilling Project Site 606: sequential events of ^{18}0 enrichment beginning at 3.1 Ma. *Initial Reports of the Deep Sea Drilling Project*, **94**, 911–17.

Keller, G. (1988a). Extinction, survivorship and evolution of planktic foraminifera across the Cretaceous/Tertiary boundary at El Kef Tunisia. *Marine Micropaleontology*, **13**, 239–63.

Keller, G. (1988b). Biotic turnover in benthic foraminifera across the Cretaceous/Tertiary boundary at El Kef, Tunisia. *Palaeogeography, Palaeoclimatology, Palaeoecology*, **66**, 153–71.

Keller, G. (1989). Extended Cretaceous/Tertiary boundary extinctions and delayed population change in planktonic foraminiferal faunas from Brazos River, Texas. *Paleoceanography*, **4**, 287–332.

Keller, G. (1993). The Cretaceous–Tertiary boundary transition in the Antarctic Ocean and its global implications. *Marine Micropaleontology*, **21**, 1–46.

Keller, G., Barrera, E., Schmitz, B. & Mattson, E. (1993). Gradual mass extinction, species survivorship, and long-term environmental changes across the Cretaceous–Tertiary boundary in high latitudes. *Geological Society of America Bulletin*, **105**, 979–97.

Keller, G. & Lindinger, M. (1989). Stable isotope, TOC and $CaCO_3$ record across the Cretaceous/Tertiary boundary at El Kef, Tunisia. *Palaeogeography, Palaeoclimatology, Palaeoecology*, **73**, 243–65.

Kemp, J.J. (1991). Late Oligocene Pacific-wide tectonic event. *Terra Nova*, **3**, 65–9.

Kemper, E. (1987). Das Klima der Kreide-Zeit. *Geologisches Jahrbuch*, **A96**, 5–185.

Kemper, E., Rawson, P.F. & Thieuloy, J-P. (1981). Ammonites of Tethyan ancestry in the early Lower Cretaceous of north-west Europe. *Palaeontology*, **24**, 251–311.

Kemper, E. & Wiedenroth, K. (1987). Klima und Tier-Migrationen am Beispiel der frühekre-tazischen Ammoniten. *Geologisches Jahrbuch*, **A96**, 315–63.

Kendall, M.A. (1996). Are Arctic soft-sediment macrobenthic communities impoverished? *Polar Biology* **16**, 393–9.

Kennedy, E.M. (1996). A palaeoclimate perspective on two Cretaceous fossil floras from New Zealand. In *Memorial Conference Dedicated to Vsevolod Andreevich Vakhrameev (Abstracts and Proceedings) November 13–14th, 1996* M.: GEOS, pp. 34–37.

Kennedy, W.J. & Cooper, M.A. (1975). Cretaceous ammonite distributions and the opening of the South Atlantic. *Journal of the Geological Society, London*, **131**, 283–8.

Kennett, J.P. (ed.) (1985). *The Miocene Ocean: Paleoceanography and Biogeography*. The Geological Society of America, Memoir **163**.

Kennett, J.P. (1995). A review of polar climatic evolution during the Neogene, based on the marine sediment record. In *Paleoclimate and Evolution, with Emphasis on Human Origins*, ed. E. Vrba, G. Denton, T. Partridge & L. Burckle, pp. 49–64. New Haven: Yale University Press.

Kennett, J.P. & Srinivasan, M.S. (1983). *Neogene Planktonic Foraminifera*. Stroudsburgh: Hutchinson Ross Publishing Company.

Kennett, J.P. & Stott, L.D. (1990). Proteus and Proto-Oceanus: Ancestral Paleogene oceans as revealed from Antarctic stable isotopic results; ODP Leg 113. *Proceedings of the Ocean Drilling Program, Scientific Results*, **113**, 865–80.

Kennett, J.P. & Stott, L.D. (1991). Abrupt deep-sea warming, paleoceanographic changes and benthic extinctions at the end of the Palaeocene. *Nature*, **353**, 225–9.

Kent, M. & Coker, P. (1992). *Vegetation Description and Analysis*. London: Bellhaven Press.

Kerr, R. (1996). New mammal data challenge evolutionary pulse theory. *Science*, **273**, 431–2.

Kidwell, S.M. & Baumiller, T. (1990). Experimental disintegration of regular echinoids: roles of temperature, oxygen and decay thresholds. *Paleobiology*, **16**, 247–72.

Kielan-Jaworowska, Z., Bown, T.M. & Lillegraven, J.A. (1979). Eutheria. In *Mesozoic Mammalia, the First Two-thirds of Mammalian History*, ed. J.A. Lillegraven, Z. Kielan-Jaworowska & W.A. Clemens, pp. 112–58. Berkeley: University of California Press.

Kielan-Jaworowska, Z. & Dashzeveg, D. (1989). Eutherian mammals from the Early Cretaceous of Mongolia. *Zoologica Scripta*, **18**, 347–55.

Kier, P.M. (1967). Sexual dimorphism in an Eocene echinoid. *Journal of Paleontology*, **41**, 988–93.

Kier, P.M. (1968). Echinoids from the middle Eocene Lake City Formation of Georgia. *Smithsonian Miscellaneous Collections*, **153**, (2), 45 pp., 10 pls.

Kier, P.M. (1969). Sexual dimorphism in fossil echinoids. In *Sexual dimorphism in fossil metazoa and taxonomic implications*, ed. G.E.G. Westermann, pp. 215–22. Stuttgart: E. Schweitzerbart'sche Verlag.

Kier, P.M. (1975a). Evolutionary trends and their functional significance in the post-Paleozoic echinoids. *The Paleontological Society Memoir 5; Supplement to Journal of Palaeontology*, **48(3)**, 1–95.

Kier, P.M. (1975b). The echinoids of Carrie Bow Cay, Belize. *Smithsonian Contributions to Zoology*, **206**, 1–45.

Kier, P.M. (1980). The echinoids of the Middle Eocene Warley Hill Formation, Santee Limestone, and Castle Hayne Limestone of North and South Carolina. *Smithsonian Contributions to Paleobiology*, **39**, 1–102.

Kilham, P. & Kilham, S.S. (1980). The evolutionary ecology of phytoplankton. In *The Physiological Ecology of Phytoplankton*, ed. I. Morris, pp. 571–97. Berkeley: University of California Press.

Kingston, J. (1999). Isotopes and environments of the Baynunah Formation, Emirate of Abu Dhabi, United Arab Emirates. In *Fossil Vertebrates of Arabia*, ed. P.J. Whybrow & A. Hill, pp 389–407. New Haven: Yale University Press.

Kingston, J., Hill, A. & Marino, B. (1992). Isotopic evidence of late Miocene/Pliocene vegetation in the east African Rift Valley. *American Journal of Physical Anthropology. Supplement No.* **14**, 100–1.

Kingston, J., Hill, A. & Marino, B.D. (1994). Isotopic evidence for Neogene hominid paleoenvironments in the Kenya Rift Valley. *Science*, **264**, 955–9.

Kipp, N.G. (1976). New transfer function for estimating past sea-surface conditions from seabed distribution of planktonic foraminifera assemblages in the North Atlantic. In *Investigations of Late Quaternary Paleoceanography and Paleoclimatology*, ed. R.M. Cline & J.D. Hays. Geological Society of America, Memoir, 3–42.

Klink, A. (1989). The Lower Rhine: Palaeoecological analysis. In *Historical Change of Large Alluvial Rivers: Western Europe*, ed. G.E. Petts, pp. 183–201. London: John Wiley.

Knobloch, E., Kvacek, Z., Buzek, C. *et al.* (1993). Evolutionary significance of floristic changes in the northern hemisphere during the late Cretaceous and Palaeogene, with particular reference to Central Europe. *Review of Palaeobotany and Palynology*, **78**, 41–54.

Knoll, A.H. (1984). Patterns of extinction in the fossil record of vascular plants In *Extinctions*, ed. M.H. Nitecki, pp. 21–68. Chicago: University of Chicago Press.

Knoll, A.H. (1986). Patterns of change in plant communities through geologic time. In *Community Ecology*, ed. J. Diamond & T.J. Case, pp. 126–41. New York: Harper and Row.

Knox, R.W.O'B., Corfield, R.M. & Dunay, R.E. (ed.) (1996). *Correlation of the Early Paleogene in Northwest Europe*. Geological Society of London Special Publication No. **101**.

Koç, N. & Jansen, E. (1994). Response of the high-latitude Northern Hemisphere to orbital climatic forcing: evidence from the Nordic Seas. *Geology*, **22**, 523–6.

Koç, N., Jansen, E & Haflidason, H. (1992). Palaeooceanic reconstructions of surface ocean conditions in the Greenland, iceland and Norwegian Seas through the last 14ka based on diatoms. *Quaternary Science Reviews*, **12**, 115–40.

Koch, P.L. (1998). Isotopic reconstruction of past continental environments. *Annual Review of Earth and Planetary Sciences*, **26**, 573–613.

Koch, P.L., Zachos, J.C. & Dettman, D.L. (1995). Stable isotope stratigraphy and paleoclimatology of the Paleogene Bighorn Basin (Wyoming, USA). *Palaeogeography, Palaeoclimatology, Palaeoecology*, **115**, 61–89.

Koch, P.L., Zachos, J.C. & Gingerich, P.D. (1992). Correlation between isotope records in marine and continental carbon reservoirs near the Paleocene/Eocene boundary. *Nature*, **358**, 319–22.

Kohn, A.J. (1990). Tempo and mode of evolution in Conidae. *Malacologia*, **32**, 55–67.

Kolodny, Y. & Raab, M. (1988). Oxygen isotopes in phosphatic fish remains from Israel: paleothermometry of tropical Cretaceous and Tertiary shelf waters. *Palaeogeography, Palaeoclimatology, Palaeoecology*, **64**, 59–67.

Koutsoukos, E.A.M., Mello, M.R., De Azambuja Filho, N.C., Hart, M.B. & Maxwell, J.R. (1991). The Upper Aptian-Albian succession of the Sergipe Basin, Brazil: an integrated paleoenvironmental assessment. *American Association of Petroleum Geologists Bulletin*, **35**, 479–97.

Kovach, W.L. & Spicer, R.A. (1995). Canonical correspondence analysis of leaf physiognomy: a contribution to the development of a new palaeoclimatological tool. *Palaeoclimates*, **1**, 125–38.

Kovar-Eder, J., Kvacek, Z., Zastawniak, E. *et al.* (1996). Floristic trends in the vegetation of Paratethys surrounding areas during Neogene time. In *The Evolution of Western*

Eurasian Neogene Mammal Faunas, ed. R.L. Bernor, V. Fahlbusch & H-W. Mittmann, pp. 395–413. New York: Columbia University Press.

Krantz, D.E. (1991). A chronology of Pliocene sea-level fluctuations: the U.S. Middle Atlantic coastal plain record. *Quaternary Science Reviews*, **10**, 163–74.

Krause, D.W. (1982). Jaw movement, dental function and diet in the Paleocene multituberculate *Ptilodus*. *Paleobiology*, **8**, 265–81.

Krause, D.W. (1986). Competitive exclusion and taxonomic displacement in the fossil record: the case of rodents and multituberculates in North America. *Contributions to Geology, University of Wyoming, Special Papers*, **3**, 95–117.

Krause, D.W. & Maas, M.C. (1990). The biogeographic origins of late Paleocene-early Eocene mammalian immigrants to the Western Interior of North America. Geological Society of America Special Paper, **243**, 71–105.

Krings, M., Stone, A., Schmitz, R., Krainitzki, H., Stoneking, M. and Pääbo, S. (1997). Neanderthal DNA sequences and the origin of modern humans. *Cell*, **90**, 19–30.

Krogh, T.E., Kamo, S.L. & Bohor, B.F. 1993. Fingerprinting the K/T impact site and determining the time of impact by U-Pb dating of single shocked zircons from distal ejecta. *Earth and Planetary Science Letters*, **119**, 425–9.

Kudrass, H.R., Erlekeuser, H., Vollbrecht, R. & Weiss, W. (1992). Global nature of the Younger Dryas cooling event from oxygen isotope data from the Sulu Sea cores. *Nature*, **349**, 406–9.

Kukla, G., An, Z.S., Melice, J.L., Gavin, J. & Xiao, J.L. (1990). Magnetic susceptibility record of Chinese loess. *Transactions of the Royal Society of Edinburgh: Earth Sciences*, **81**, 263–88.

Kürschner, W.M. (1996). *Leaf stomata as biosensors of palaeoatmospheric CO_2 levels*. LPP Contributions Series No. **5**, LPP Foundation, Utrecht, pp. 1–153.

Kurochkin, E.N. (1995). Synopsis of Mesozoic birds and early evolution of the class Aves. *Archaeopteryx*, **13**, 47–66.

Kvacek, Z. & Buzek, C. (1995). Endocarps and foliage of the flowering plant family Icacinaceae from the Tertiary of Europe. *Tertiary Research*, **15**, 121–38.

Kvacek, Z. & Walther, H. (1989). Paleobotanical studies in Fagaceae of the European Tertiary. *Plant Systematics and Evolution*, **162**, 213–29.

Labandeira, C.C. (1995). A compendium of fossil insect families. *Milwaukee Public Museum Contributions in Biology and Geology*, No. **88**, 71 pp.

Labandeira, C.C. & Sepkoski, J.J. (1993). Insect diversity in the fossil record. *Science*, **261**, 310–15.

Lagaaij, R. (1963). *Cupuladria canariensis* (Busk)–portrait of a bryozoan. *Palaeontology*, **6**, 172–217.

Lague, M. & Jungers, W. (1996). Morphometric variation in Plio-Pleistocene hominid distal humeri. *American Journal of Physical Anthropology*, **101**, 401–27.

Lahr, M. (1996). *The Evolution of Modern Human Diversity: a Study of Cranial Variation*. Cambridge: Cambridge University Press.

Lamb, H.H. (1977). The late Quaternary history of the climate of the British Isles. In *British Quaternary Studies: Recent Advances*, ed. F.W. Shotton, pp. 283–298. Oxford: Clarendon Press.

Lamb, H.F., Gasse, F., Benkaddour, A., El Hamouti, N., van der Kaars, S., Perkins, W.T., Pearce, N.J. & Roberts, C.N. (1995). Relation between century-scale Holocene arid intervals in tropical and temperate zones. *Nature*, **373**, 134–7.

Lambert, J. (1933). Échinides de Madagascar communiqués par M.H. Besairie. *Annales géologiques du service des mines*, **3**, 7–49.

Lambrick, G. & Robinson, M.A. (1979). *The Iron Age and Roman Riverside Settlement at Farmoor*. Council for British Archaeology, Research Report, **32**.

Lambrick, G. & Robinson, M.A. (1988). The development of the floodplain grassland in the Upper Thames Valley. In *Archaeology and the Flora of the British Isles*, ed. M. Jones. Oxford University Committee for Archaeology Monograph **14**, pp. 55–75.

Landman, N.H., Tanabe, K. & Shigeta, Y. (1996), Ammonoid embryonic development. In *Ammonoid Paleobiology*, ed. N.H. Landman, K. Tanabe & R.A. Davis, New York and London: Plenum Press, pp. 344–405.

Lang, P.J., Scott, A.C. & Stephenson, J. (1995). Evidence of plant–arthropod interactions from the Eocene Branksome Sand formation, Bournemouth, England: introduction and description of leaf mines. *Tertiary Research*, **15**, 145–74.

Larick, R. & Ciochon, R. (1996). The African emergence and early Asian dispersals of the genus *Homo*. *American Scientist*, **84**, 538–51.

Larson, R.L. (1991). Latest pulse of Earth: evidence for a mid-Cretaceous superplume. *Geology*, **19**, 547–50.

Larson, R.L., Fischer, A.G., Erba, E. & Premoli Silva, I. (eds.) (1993). APTICORE-ALBICORE: *A Workshop Report on Global Events and Rhythms of the Mid-Cretaceous, 4–9 October 1992. Perugia, Italy.* Washington: Joint Oceanographic Institutions Inc.

Lauenberger, M., Siegenthaler, U. & Langway, C.C. (1992). Carbon isotope composition of atmospheric CO_2 during the last ice age from an Antarctic ice core. *Nature*, **357**, 488–90.

Lawyer, L.A., Gahagan, L.M. & Coffin, M.F. (1992). The development of paleoseaways around Antarctica. In *The Antarctic Paleoenvironment: A Perspective on Global Change*, ed. J.P. Kennett & D.A. Warnke. American Geophysical Union, Geophysical Monograph, **56**, 7–30.

Le Loeuff, J. & Buffetaut, E. (1991). *Tarascosaurus salluvicus* nov. gen., nov. sp., dinosaure théropode du Crétacé supérieur du Sud de la France. *Geobios*, **25**, 585–94.

Leakey, M.G., Feibel, C.S., Bernor, R.L., Harris, J.M., Cerling, T.E., Stewart, K.M., Storrs, G.W., Walker, A., Werdelin, L. & Winkler, A.J. (1996). Lothagam: a record of faunal change in the late Miocene of East Africa. *Journal of Vertebrate Paleontology*, **16**, 556–70.

Leakey, M., Feibel, C., McDougall, I. & Walker (1995). A. new four-million-year-old hominid species from Kanapoi and Allia Bay, Kenya. *Nature*, **376**, 565–71.

Leckie, R.M. (1989). A paleoceanographic model for the early evolutionary history of planktonic foraminifera. *Palaeogeography, Palaeoclimatology, Palaeoecology*, **73**, 107–38.

Leemans, R. (1989). *Global Holdridge Life Zone Classifications*. Digital Raster on a 0.5-degree Geographic (lat/long) 360x720 grid. Laxenberg, Austria:IIASA. Floppy Disk, 0.26MB.

Legendre, S. (1989). Les communautés de mammifères du Paléogène (Eocène supérieur et Oligocène) d'Europe occidentale: structures, milieux et évolution. *Münchner Geowissenschaftliche Abhandlungen*, (A) **16**, 1–110.

Lehman, S. (1996). True grit spells double trouble. *Nature*, **382**, 25–7.

Lehman, S.J. & Keigwin, L.D. (1992). Sudden changes in North Atlantic circulation during the last deglaciation. *Nature*, **356**, 757–62.

Leith Adams, A. (1877–81). *Monograph on the British Fossil Elephants*. London: Palaeontographical Society.

Levesque, A.J., Mayle, F.E., Walker, I.R. & Cwynar, L.C. (1993). A previously unrecognised late-glacial cold event in eastern North America. *Nature*, **361**, 623–6.

Li, G.-Q. (1994). Systematic position of the Australian fossil osteoglossoid fish *Phareodus (= Phareoides) queenslandicus* Hills. *Memoirs of the Queensland Museum*, **37**, 287–300.

Li, G-Q & Wilson, M.V.H. (1996). Phylogeny of the Osteoglossomorph. In *Interrelationships of Fishes*, ed. M. Stiassny, G.D. Johnson & L. Parenti, pp. 163–174. New York: Academic Press.

Li, Z.P. (1990). Bryozoaires de Montbrison-Fontbonau (Drome) et comparaison avec les autres faunes Miocènes du Bassin Rhodanien Méridional. *Nouvelles Archives du Muséum d'Histoire Naturelle de Lyon*, **27**, 1–126.

Lichtenhaler, H.K. (1985). Differences in morphology and chemical composition of leaves grown at different light intensities and qualities. In *Control of Leaf Growth*, ed. N.R. Baker, W.J. Davies & C.K. Ong, pp. 201–21. Cambridge: Cambridge University Press.

Lidgard, S. & Crane, P.R. (1988). Quantitative analyses of the early angiosperm radiation. *Nature*, **331**, 344–6.

Lidgard, S. & Crane, P.R. (1990). Angiosperm diversification and Cretaceous floristic trends: a comparison of palynofloras and leaf macrofloras. *Paleobiology*, **16**, 77–93.

Lidgard, S., McKinney, F.K. & Taylor, P.D. (1993). Competition, clade replacement, and a history of cyclostome and cheilostome bryozoan diversity. *Paleobiology*, **19**, 352–71.

Lindegaard, C. (1995). Classification of water-bodies and pollution. In *The Chironomidae: Biology and Ecology of Non-biting Midges*, ed. P.S. Armitage, P.S. Cranston & L.C.V. Pinder. London: Chapman & Hall.

Lini, A., Weissert, H. & Erba, E. (1992). The Valanginian carbon isotope event: a first episode of Greenhouse climate conditions during the Cretaceous. *Terra Nova*, **4**, 374–84.

Lipps, J.H. (1979). Ecology and paleoecology of planktic foraminifera. In *Foraminiferal Ecology and Paleoecology*, ed. J.H. Lipps, W.H. Berger, M.A. Buzas *et al.*, Society of Economic Paleontologists and Mineralogists, Short Course **6**, Houston, pp. 62–104.

Lipps, J.H. & Hickman, C.S. (1982). Origin, age and evolution of Antarctic and deep-sea faunas. In *The Environment of the Deep Sea*, ed. W.G. Ernst & J.G. Morris, pp. 324–56. Englewood Cliffs, NJ: Prentice-Hall Inc.

Lipps, J.H. & Mitchell, E. (1976). Trophic model for the adaptive radiations and extinctions of pelagic marine mammals. *Paleobiology*, **2**, 147–55.

Liu, C. & Olsson, R.K. (1992). Evolutionary radiation of microperforate planktonic foraminifera following the K/T mass extinction. *Journal of Foraminiferal Research*, **22**, 328–46.

Lloyd, C.R. (1982). The mid-Cretaceous earth: paleogeography; ocean circulation and temperature; atmospheric circulation. *Journal of Geology*, **90**, 393–413.

Lockwood, J.G. (1979). Water balance of Britain, 50 000 BP to the present day. *Quaternary Research*, **12**, 297–310.

Loconte, H. & Stevenson, D.W. (1990). Cladistics of the Spermatophyta. *Brittonia*, **42**, 197–211.

Lofgren, D.L. (1995). The Bug Creek problem and the Cretaceous–Tertiary transition at McGuire Creek, Montana. *University of California Publications in Geological Sciences*, **140**, 1–185.

Loubere, P. & Moss, K. (1986). Late Pliocene climatic change and the onset of Northern Hemisphere glaciation as recorded in the northeast Atlantic Ocean. *Geological Society of America Bulletin*, **97**, 818–28.

Louis, P., Laurain, M., Bolin, C. *et al.* (1983). Nouveau gisement de vertébrés dans le Cuisien supérieur de Saint-Agnan (Aisne) ses relations stratigraphiques avec les autres gisements yprésiens du Bassin parisien. *Bulletin d'Information des Géologues du Bassin de Paris*, **20**, 3–20.

Lovelock, J.E. (1991). Geophysiology of the oceans. In *Ocean Margin Processes in Global Change*, ed. R.F.C. Mantoura, J-M. Martin & R. Wollast, pp. 419–31. Chichester, UK: John Wiley.

Lowe, J.J. & Walker, M.J.C. (1984). *Reconstructing Quaternary Environments*. Harlow: Longman.

Lowenstam, H.A. (1964). Paleotemperatures of the Permian and Cretaceous Periods. In *Problems in Paleoclimatology*, ed. A.E.M. Nairn, pp. 227–48. New York: John Wiley.

Lowenstam, H.A. & Epstein, S. (1954). Paleotemperatures of the post-Aptian Cretaceous as determined by the oxygen isotope method. *Journal of Geology*, **62**, 207–48.

Lu, G. & Keller, G. (1993). The Paleocene–Eocene transition in the Antarctic Ocean: Inference from planktic foraminifera. *Marine Micropaleontology*, **21**, 101–42.

Lu, G. & Keller, G. (1995*a*). Planktic foraminiferal faunal turnovers in the subtropical Pacific during the Late Paleocene to Early Eocene. *Journal of Foraminiferal Research*, **25**, 97–116.

Lu, G. & Keller, G. (1995*b*). Ecological stasis and saltation: species richness change in planktic foraminifera during the Late Paleocene to Early Eocene, DSDP Site 577. *Palaeogeography, Palaeoclimatology, Palaeoecology*, **117**, 211–27.

Lu, G. & Keller, G. (1996). Separating ecological assemblages using stable isotope signals: Late Paleocene to Early Eocene planktic foraminifera, DSDP Site 577. *Journal of Foraminiferal Research*, **26**, 103–12.

Lu, G., Keller, G., Adatte, T. *et al.* (1995). Abrupt change in the upwelling system along the southern margin of the Tethys during the Paleocene-Eocene transition event. *Israeli Journal of Earth Sciences*, **44**, 185–95.

Lucas, S.G. & Hunt, A.P. (1989). *Alamosaurus* and the sauropod hiatus in the Cretaceous of the North American western interior. In *Paleobiology of the Dinosaurs*, ed. J. Farlow. Geological Society of America Special Paper, **283**, 75–85.

Lupia, R. (1994). Estimating plant diversity and abundance using macrofossils, mesofossils and microfossils: an example from the Campanian of Georgia. *Geological Society of America Abstracts with Programs*, **26**, A124.

Lupia, R. (1997). Palynological Record of the Cretaceous Angiosperm Radiation: Diversity, Abundance and Morphological Patterns. PhD thesis, University of Chicago.

Luyendyk, B.P., Forsyth, D. & Phillips, J.D. (1972). Experimental approach to the paleocirculation of the oceanic surface waters. *Geological Society of America Bulletin*, **83**, 2649–64.

Lyell, C. (1865). *Elements of Geology*, 6th edn. London: John Murray.

Maas, M.C. & Krause, D.W. (1994). Mammalian turnover and community structure in the Paleocene of North America. *Historical Biology*, **8**, 91–128.

MacArthur, R.H. (1972). *Geographical Ecology*. New York: Harper & Row.

MacDonald, G.M. & Walters, N.M. (1987). An evaluation of automated and mapping algorithms for the analysis of Quaternary pollen data. *Review Palaeobotany and Palynology*, **51**, 289–307.

Mackensen, A. & Ehrmann, W.U. (1992). Middle Eocene through early Oligocene climate history and paleoceanography in the Southern Ocean: stable oxygen and carbon isotopes from ODP sites on Maud Rise and Keguelen Plateau. *Marine Geology*, **108**, 1–27.

Macklin, M.G. & Lewin, J. (1991). Holocene river alluviation in Britain. *Proceedings of the Second International Geomorphology Conference*, Frankfurt.

MacLean, J.A. (1992). Morphometric analysis of dwarfing in foraminifera across the Cretaceous–Tertiary boundary. Junior thesis, Princeton University.

MacLeod, N. (1990). Effects of Last Eocene impacts on planktic foraminifera. In *Global Catastrophes in Earth History: an Interdisciplinary Conference on Impacts, Volcanism, and Mass Mortality*, ed. V.L. Sharpton & P.D. Ward. Geological Society of America Special Paper, **247**, 595–606.

MacLeod, N. (1996). Nature of the Cretaceous–Tertiary (K–T) planktonic foraminiferal record: stratigraphic confidence intervals, Signor-Lipps effect, and patterns of survivorship. In *The Cretaceous–Tertiary Mass Extinction: Biotic and Environmental Changes*, ed. N. MacLeod & G. Keller, pp. 85–138. New York: W.W. Norton & Co.

MacLeod, N. & Keller, G. (1990). Foraminiferal phenotypic response to environmental changes across the Cretaceous–Tertiary boundary. *Geological Society of America Abstracts with Programs*, **22**, A106.

MacLeod, N. & Keller, G. (1994). Comparative biogeographic analysis of planktic foraminiferal survivorship across the Cretaceous/Tertiary (K/T) boundary. *Paleobiology*, **20**, 143–77.

MacLeod, N., Keller, G. & Kitchell, J.A. (1990). Progenesis in Late Eocene populations of *Subbotina linaperta* (Foraminifera) from the western Atlantic. *Marine Micropaleontology*, **16**, 219–40.

MacLeod, N. & Kitchell, J. (1990). Morphometrics and evolutionary inference: a case study involving ontogenetic and developmental aspects of evolution. In *Proceedings of the Michigan Morphometrics Workshop*, ed. F.J. Rohlf & F.L. Bookstein, The University of Michigan Museum of Zoology, Special Publication **2**, Ann Arbor, pp. 283–99.

MacLeod, N. & Ortiz, N. (1995). Comparison of patterns of phenotypic variation in planktonic and benthonic foraminifera across the Cretaceous–Tertiary and Paleocene–Eocene event horizons. *Geological Society of America Abstracts with Programs*, **26**.

Macleod, N., Rawson, P.F., Forey, P.L., Banner, F.T., Boudagher-Fadel, M.K., Bown, P.R., Burnett, J.A., Chambers, P., Culver, S., Evans, S.E., Jeffery, C., Kaminski, M.A., Lord, A.R., Milner, A.C., Milner, A.R., Morris, N., Owen, E., Rosen, B.R., Smith, A.B., Taylor, P.D., Urquhart, E. & Young, J.R. (1997). The Cretaceous–Tertiary biotic transition. *Journal of the Geological Society, London*, **154**, 265–92.

MacLeod, N. & Rose, K.D. (1993). Inferring locomotor behavior in Paleogene mammals via eigenshape analysis. *American Journal of Science*, **293-A**, 300–55.

Magurran, A.E. (1988). *Ecological Diversity and its Measurement*. London: Croom Helm.

Maisey, J.G. (1991). *Santana Fossils: an Illustrated Atlas*. New Jersey: T.F.H. Publications.

Malin, G., Liss, P.S. & Turner, S.M. (1994). Dimethyl sulphide production and atmospheric consequences In *The Haptophyte Algae*, ed. J.C. Green & B.S.C. Leadbeater. Systematics Association Special Volume No. **51**, 303–320.

Malmgren, B.A. Berggren, W.A. (1987). Evolutionary changes in some Late Neogene planktonic foraminiferal lineages and their relationships to paleoceanographic changes. *Paleoceanography*, **2**, 445–56.

Malmgren, B.A., Berggren, W.A. & Lohmann, G.P. (1983). Evidence for punctuated gradualism in the Neogene *Globorotalia tumida* lineage of planktonic foraminifera. *Paleobiology*, **9**, 377–89.

Malmgren, B.A. & Kennett, J.P. (1981). Phyletic gradualism in a Late Cenozoic planktonic foraminiferal lineage; DSDP Site 284, southwest Pacific. *Paleobiology*, **7**, 230–40.

Manchester, S.R. (1987). The fossil history of the Juglandaceae. *Monographs in Systematic Botany from the Missouri Botanical Garden*, **21**, 1–137.

Manchester, S.R. (1989). Early history of the Juglandaceae. *Plant Systematics and Evolution*, **162**, 231–50.

Manchester, S.R. (1994). Fruits and seeds of the Middle Eocene Nut Beds Flora, Clarno Formation, Oregon. *Palaeontographica Americana*, **58**, 1–205.

Manchester, S.R. (1999). Biogeographical relationships of North American Tertiary floras. *Annals of the Missouri Botanical Garden*, **86**, 472–522.

Manchester, S.R. & Kress, W.J. (1993). Fossil bananas (Musaceae): *Ensete oregonense* sp nov, from the Eocene of western North America and its phylogenetic significance. *American Journal of Botany*, **80**, 1264–72.

Mannion, A.M. (1991). *Global Environmental Change*. Harlow: Longman.

Marks, J. (1995). Learning to live with a trichotomy. *American Journal of Physical Anthropology*, **98**, 211–13.

Markwick, P .J. (1994). 'Equability', continentality, and Tertiary 'climate': the crocodilian perspective. *Geology*, **22**, 613–16.

Marshall, J.D. (1994). Climatic and oceanographic isotope signals from the carbonate rock record and their preservation. *Geological Magazine*, **129**, 143–60.

Martill, D.M. (1993). *Fossils of the Santana and Crato formations, Brazil*. London: The Palaeontological Association.

Martin, R.E. (1996). Secular increase in nutrient levels through the Phanerozoic: implications for productivity, biomass, and diversity of the marine biosphere. *Palaios*, **11**, 209–19.

Martinez-Delclos, X. & Martinell, J. (1995). The oldest known record of social insects. *Journal of Paleontology*, **69**, 594–9.

Martini, E. (1971). Standard Tertiary and Quaternary calcareous nannoplankton zonation. In *Proceedings II Planktonic Conference, Roma*, 1970, ed. A. Farinacci, **2**, 739–85.

Masse, J.P. (1977). Les constructions à Madréporaires des calcaires urgoniens (Barrémien-Bédoulien) de Provence (SE de la France). In *Second Symposium International sur Coraux et Récifs Coralliens Fossiles. Mémoires du Bureau de Recherches Géologiques et Miniéres*, **89**, 322–35.

Masters, B.A. (1997). El Kef blind test II results. *Marine Micropaleontology*, **29**, 77–9.

Matthews, B. (1970). Age and origin of aeolian sand in the Vale of York. *Nature*, **227**, 1234–6.

Matthews, J.V.Jr. (1974). Quaternary Environments at Cape Deceit (Seward Peninsula, Akaska): Evolution of a Tundra Ecosystem. *Bulletin of the Geological Society of America*, **85**, 1353–84.

Matthews, J.V.Jr (1976). Insect fossils from the Beaufort Formation: geological and biological significance. *Geological Survey of Canada Papers*, **76–1B**, 217–27.

Maxson, L.R. (1976). The phylogenetic status of phyllomedusine frogs (Hylidae) as evidenced from immunological studies of their serum albumins. *Experientia*, **32**, 1149–50.

McArthur, J.M., Kennedy, W.J., Chen, M., Thirlwall, M.F. & Gale, A.S. (1994). Strontium isotope stratigraphy for Late Cretaceous time: direct numerical calibration of the Sr isotope curve based on the US Western Interior. *Palaeogeography, Palaeoclimatology, Palaeoecology*, **108**, 95–119.

McCall, J., Rosen, B.R. & Darrell, J.G. (1994). Carbonate deposition in accretionary prism settings: Early Miocene coral limestones and corals of the Makran mountain range in southern Iran. *Facies*, **31**, 141–78.

McCormick, M.P., Thomason, L.W. & Trepte, C.R. (1995). Atmospheric effects of the Mt. Pinatubo eruption. *Nature*, **373**, 399–404.

McElwain, J. & Chaloner, W.G. (1995). Stomatal density and index of fossil plants track atmospheric carbon dioxide in the Paleozoic. *Annals of Botany*, **76**, 389–95.

McGowran, B. (1989). Silica burp in the Eocene ocean. *Geology*, **17**, 857–60.

McGowran, B., Moss, G. & Beecroft, A. (1992). Late Eocene and early Oligocene in Southern Australia: local neritic signals of global oceanic changes. In *Eocene–Oligocene Climatic and Biotic Evolution*, ed. D.R. Prothero & W.A. Berggren, pp. 178–201. Princeton: Princeton University Press.

McIntyre, A. & Bé, A.W.H. (1967). Modern Coccolithophoridae of the Atlantic Ocean–I. Placoliths and Cyrtoliths. *Deep-Sea Research*, **14**, 561–97.

McIntyre, A., Bé, A.W.H. & Roche, M.B. (1970). Modern Pacific Coccolithophorida: a Paleontological Thermometer. *Transactions of the New York Academy of Sciences*, **32**, 720–31.

McKinney, F.K. & Jackson, J.B.C. (1989). *Bryozoan Evolution*. London: Allen & Unwin.

McKinney, F.K., Lidgard, S., Sepkoski, J.J.Jr & Taylor, P.D. (1998). Decoupled temporal patterns of evolution and ecology in two post-Paleozoic clades. *Science*, **281**, 807–9.

McKinney, M.L. (1988). Classifying heterochrony: allometry, size, and time. In *Heterochrony in Evolution*, ed. M.L. McKinney, pp. 17–34. New York: Plenum Press.

McKinney, M.L. & McNamara, K.J. (1991). *Heterochrony: The Evolution of Ontogeny*. New York: Plenum Press.

McKinney, M.L., McNamara, K.J., Carter, B.D. & Donovan, S.K. (1992). Evolution of Paleogene echinoids: a global and regional view. In *Eocene–Oligocene Climatic and Biotic Evolution*, ed. D.R. Prothero & W.A. Berggren, pp. 349–67. Princeton: Princeton University Press.

McNamara, K.J. (1994). Diversity of Cenozoic marsupiate echinoids as an environmental indicator. *Lethaia*, **27**, 257–68.

McNamara, K.J. (1988). The abundance of heterochrony in the fossil record. In *Heterochrony in Evolution: A Multidisciplinary Approach*, ed. M.L. McKinney, pp. 287–326. New York: Plenum Press.

McVean, D.N. (1956). Ecology of *Alnus glutinosa* (L.) Gaertn. VI. post-glacial history. *Journal of Ecology*, **44**, 331–3.

Meyerhoff, A.A., Lyons, J.B. & Officer, C.B. (1994). Chicxulub structure: a volcanic sequence of late Cretaceous age. *Geology*, **24**, 3–4.

Miles, D. (1986). *The Archaeology at Barton Court Farm, Abingdon, Oxon*. Council for British Archaeology, Research Report, **50**.

Miller, A.I. & Sepkoski, J.J., Jr (1988). Modeling bivalve diversification: the effect of interaction on a macroevolutionary system. *Paleobiology*, **14**, 364–9.

Miller, K.G. (1982). Cenozoic benthic foraminifera: case histories of paleoceanographic and sea-level changes. In *Foraminifera, Notes for a Short Course*, ed. T.W. Broadhead, University of Tennessee, Studies in Geology, **6**, pp. 107–26.

Miller, K.G. (1992). Middle Eocene to Oligocene stable isotopes, climate, and deep-water history: the Terminal Eocene event?. In *Eocene–Oligocene Climatic and Biotic Evolution*, ed. D.R. Prothero & W.A. Berggren, pp. 160–77. Princeton: Princeton University Press.

Miller, K.G., Fairbanks, R.G. & Mountain, G.S. (1987a). Tertiary oxygen isotope synthesis, sea-level history, and continental margin erosion. *Paleoceanography*, **2**, 1–19.

Miller, K.G., Janecek, T.R., Katz, M.E. *et al.* (1987b). Abyssal circulation and benthic foraminiferal changes near the Paleocene/Eocene boundary. *Paleoceanography*, **2**, 741–61.

Miller, K.G., Mountain, G.S. *et al.* (1996). Drilling and dating New Jersey Oligocene–Miocene sequences: ice volume, global sea-level and Exxon records. *Science*, **271**, 1092–5.

Miller, K.G., Wright, J.D. & Fairbanks, R.G. (1991). Unlocking the Ice House: Oligocene–Miocene oxygen isotopes, eustasy, and margin erosion. *Journal of Geophysical Research*, **96**, 6829–48.

Miller, W., III (1991). Hierarchical concept of reef development. *Neues Jahrbuch für Geologie und Paläontologie, Abhandlungen*, **182**, 21–35.

Milner, A.C. (1998). Cretaceous dinosaurs and Gondwanan biogeography. *Journal of Vertebrate Paleontology*, **18**, (3, suppl.), 64 pp.

Milner, A.R. (1983). The biogeography of salamanders in the Mesozoic and early Caenozoic: a cladistic–vicariance model. In *Evolution, Time and Space, The Emergence of the Biosphere*, ed. R. Sims, J. Price & P. Whalley. Systematics Association Special Volume No. **23**, 431–68.

Minckley, W.L., Hendrickson, D.A. & Bond, C. (1986). Geography of Western North American freshwater fishes: description and relationships to intracontinental tectonism. In *The Zoogeography of North American Freshwater Fishes*, ed. C.H. Hocutt & E.O. Wiley, pp. 519–613. New York: John Wiley.

Młynarski, M. (1977). New notes on the amphibian and reptilian fauna of the Polish Pliocene and Pleistocene. *Acta Zoologia Cracoviensis*, **22**, 13–36.

Moffett, L., Robinson, M.A. & Straker, V. (1989). Cereals, fruit and nuts: charred plant remains from Neolithic sites in England and Wales and the Neolithic economy. In *The Beginnings of Agriculture*, ed. A. Milles, D. Williams & N. Gardner. *British Archaeological Reports, International Series*, **496**, 243–261.

Moissette, P. (1988). Faunes de Bryozoaires du Messinien d'Algérie Occidentale. *Documents des Laboratoires de Géologie de la Faculté des Sciences de Lyon*, **102**, 1–351.

Moissette, P., Delrieu, B. & Tsagaris, S. (1993). Bryozoaires du bassin néogène d'Héraklion (Crète centrale, Grèce). Le Miocène supérieur: premiers résultats. *Neues Jahrbuch für Geologie und Paläontologie, Abhandlungen*, **190**, 75–123.

Moissette, P. & Pouyet, S. (1987). Bryozoan faunas and the Messinian salinity crisis. *Annals of the Hungarian Geological Institute*, **70**, 447–53.

Moissette, P. & Spjeldnaes, N. (1995). Plio-Pleistocene deep-water bryozoans from Rhodes, Greece. *Palaeontology*, **38**, 771–99.

Molfino, B., Kipp, N.G. & Morley, J.J. (1982). Comparison of foraminiferal, coccolithophorid and radiolarian paleotemperature equations: assemblage coherency and estimate concordancy. *Quaternary Research*, **17**, 279–313.

Molnar, P., England, P. & Martinod, J. (1993). Mantle dynamics, uplift of the Tibetan Plateau, and the Indian monsoon. *Reviews of Geophysics*, **31**, 357–96.

Moore, P.D. (1977). Ancient distribution of lime trees. *Nature*, **268**, 13–14.

Moore, P.D. (1986). Hydrological changes in mires. In *Handbook of Holocene Paleaoecology and Palaeohydrology*, ed. B.E. Berglund, pp. 91–107. Chichester, UK: John Wiley.

Moore, P.D., Webb, J.A. & Collinson, M.E. (1991). *Pollen Analysis*. Oxford: Blackwell.

Morgan, A.V. (1987). Late Wisconsin and early Holocene paleoenvironments of east-central North America based on assemblages of fossil Coleoptera. In *The Geology of North America, Vol. k-3 North America and Adjacent Oceans during the Last Deglaciation*, ed. W.F. Ruddiman & H.E. Wright, Geological Society of America, Boulder, Colorado.

Morgan, A.V., Morgan, A., Ashworth, A.C. & Matthews, J.V. Jr. (1984). Late Wisconsin fossil beetles in North America. In *Late Quaternary Environments of the United States, Vol. 1: The Late Pleistocene*, ed. S.C. Porter, Geological Society of America, Boulder, Colorado.

Morgan, M.E., Kingston, J.D. & Marino, B.D. (1994). Carbon isotopic evidence for the emergence of C4 plants in the Neogene from Pakistan and Kenya. *Nature*, **367**, 162–5.

Morley, R.J. (1999). *Origin and Evolution of Tropical Rainforests*. Chichester: John Wiley.

Morris, N.J. & Skelton, P.W. (1995). Late Campanian–Maastrichtian rudists from the United Arab Emirates–Oman Border region. *Bulletin of the British Museum of Natural History (Geology)*, **51**, 277–305.

Morton, B. (1991). Aspects of predation by *Tonna zonatum* (Prosobranchia: Tonnoidea) feeding on holothurians in Hong Kong. *Journal of Molluscan Studies*, **57**, 11–19.

Morton, B. (1996). The evolutionary history of the Bivalvia. In *Origin and Evolutionary Radiation of the Mollusca*, ed. J.D. Taylor, pp. 337–59. Oxford: Oxford University Press.

Mosbrugger, V. (1995). New methods and approaches in Tertiary palaeoenvironmental research. *Abhandlungen des Staatlichen Museums für Mineralogie und Geologie zu Dresden*, **41**, 41–52.

Mourer-Chauviré, C. (1981). Première indication de la presénce de Phorusracidés, famille d'oiseaux géants d'Amérique du Sud, dans la Tertiare européen: *Ameghinornis* nov. gen. (Aves Ralliformes) des Phosphorites de Quercy, France. *Geobios*, **14**, 637–47.

Moussavian, E. & Vecsei, A. (1995). Paleocene reef sediments from the Maiella carbonate platform, Italy. *Facies*, **32**, 213–22.

Muizon, C. de (1994). A new carnivorous marsupial from the Palaeocene of Bolivia and the problem of marsupial monophyly. *Nature*, **370**, 208–11.

Müller, D.R., Royer, J-Y. and Lawver, L.A. (1993). Revised plate motions relative to the hotspots from combined Atlantic and Indian Ocean hotspot tracks. *Geology*, **21**, 275–8.

Muller, J. (1970). Palynological evidence on early differentiation of angiosperms. *Biological Reviews of the Cambridge Philosophical Society*, **45**, 417–50.

Muller, J. (1981). Fossil Pollen records of extant angiosperms. *Botanical Review*, **47**, 1–146.

Muller, J. (1985). Significance of fossil pollen for angiosperm history. *Annals of the Missouri Botanical Garden*, **71**, 419–43.

Mutterlose, J. (1992a). Biostratigraphy and palaeobiogeography of Early Cretaceous calcareous nannofossils. *Cretaceous Research*, **13**, 167–89.

Mutterlose, J. (1992b). Migration and evolution patterns of floras and faunas in marine Early Cretaceous sediments of NW Europe. *Palaeogeography, Palaeoclimatology, Palaeoecology*, **94**, 261–82.

Naidin, D.P. (1960). The stratigraphy of the Upper Cretaceous of the Russian Platform. *Stockholm Contributions in Geology*, **6 (4)**, 39–61.

Naumann, I.D. (ed.) (1995). *Systematic and Applied Entomology*. Melbourne: Melbourne University Press.

Naylor, B.G. (1979). A new species of *Taricha* (Caudata: Salamandridae), from the Oligocene John Day Formation of Oregon. *Canadian Journal of Earth Sciences*, **16**, 970–3.

Nazarov, V.I. (1984). A reconstruction of the Anthropogene landscapes in the northeastern part of Byelorussia according to palaeoentomological data. *Works of the Palaeontological Institute, Nauka Press, Moscow*, 95 pp.

Nebelsick, J.H. (1992). The northern Bay of Safaga (Red Sea, Egypt): an actuopaläontological approach. III. Distribution of echinoids. *Beiträge zur Paläontologie von Österreich*, **17**, 5–79.

Nebelsick, J.H. (1996). Biodiversity of shallow-water Red Sea echinoids: implications for the fossil record. *Journal of the Marine Biological Association, UK*, **76**, 185–94.

Negretti, B. (1984). Echinides Neogenes du littoral de la Nerthe. *Travaux du Laboratoire de Stratigraphie et de Paleoecologie, Université de Provence, Marseille, nouvelle série*, **2**, 1–139, pls 1–10.

Nel, A. & Paicheler, J-C. (1993). Les Isoptera fossiles. In *Cahiers de Paleontologie*, CNRS Editions, Paris, 102–79.

Nelson, C.S., Hyden, F.M., Keane, S.L., Leask, W.L. & Gordon, D.P. (1988). Application of bryozoan zoarial growth-form studies in facies analysis of non-tropical carbonate deposits in New Zealand. *Sedimentary Geology*, **60**, 301–22.

Nelson, G. (1985). A decade of challenge the future of biogeography. In *Plate Tectonics and Biogeography*, ed. A.E. Levington & H.L. Aldrich, *Journal of the Earth Science Natural History Society*, **4**, 187–96.

Nelson, G. & Platnick, N. (1981). *Systematics and Biogeography, Cladistics and Vicariance*. New York: Columbia University Press.

Nelson, J.S. (1994). *Fishes of the World*, 3rd ed. New York: John Wiley.

Néraudeau, D. (1993). Sexual dimorphism in mid-Cretaceous hemiasterid echinoids. *Palaeontology*, **36**, 311–18.

Néraudeau, D. & Moreau, P. (1989). Paléoécologie et paléobiogéographie des faunes d'échinides au Cénomanien nord-aquitain (Charente-Maritime, France). *Geobios*, **22**, 293–324.

Nessov, L.A., Sigogneau-Russell, D. & Russell, D.E. (1994). A survey of Cretaceous tribosphenic mammals from middle Asia (Uzbekistan, Kazakhstan and

Tajikistan), of their geological setting, age and faunal environment. *Palaeovertebrata*, **23**, 51–92.

Neumayr, M. (1883). Uber klimatische Zonen während der Jura- und Kreidezeit. *Denkschriften. Akadademie der Wissenschaften in Wien (Mathematisch-Naturwissenschaften Klasse)*, **47**, 277–310.

Newell, N.D. (1971). An outline history of tropical organic reefs. *American Museum Novitates*, **2465**, 1–37.

Niklas, K.J., Tiffney, B.H. & Knoll, A.H. (1980). Apparent changes in the diversity of fossil plants: a preliminary assessment. In *Evolutionary Biology*, ed. M.K. Hecht, W.C. Steere & B. Wallace, pp. 1–89. New York: Plenum Press.

Niklas, K.J., Tiffney, B.H. & Knoll, A.H. (1983). Patterns in vascular land plant diversification. *Nature*, **303**, 614–6.

Niklas, K.J., Tiffney, B.H. & Knoll, A.H. (1985). Patterns in vascular land plant diversification: a factor analysis at the species level. In *Phanerozoic Diversity Patterns: Profiles in Macroevolution*, ed. J.W. Valentine, pp. 97–128. Princeton: Princeton University Press.

Ninkovich, D., Shackleton, N.J., Abdel-Monem, A., Obradovich, J.D. & Izett, G. (1978). K-Ar age of the late Pleistocene eruption of Toba, north Sumatra. *Nature*, **276**, 574–7.

Nixon, K.C., Crepet, W.L., Stevenson, D. & Friis, E.M. (1994). A reevaluation of seed plant phylogeny. *Annals of the Missouri Botanical Garden*, **81**, 484–533.

Noble, G.K. (1928). Two new fossil amphibians of zoogeographic importance from the Miocene of Europe. *American Museum Novitates*, **303**, 1–13.

Noonan, G.R. (1988). Faunal relationships between eastern North America and Europe as shown by insects. *Memoirs of the Entomological Society of Canada*, No. **144**, 39–53.

Norris, R.D. (1990). Iterative evolution in planktonic foraminifera. PhD thesis, Harvard University.

Norris, R.D. (1991). Biased extinction and evolutionary trends. *Paleobiology*, **17**, 388–99.

Norris, R.D. (1992). Extinction selectivity and ecology in planktonic foraminifera. *Palaeogeography, Palaeoclimatology, Palaeoecology*, **95**, 1–17.

Novacek, M.J. & Marshall, L.G. (1976). Early biogeographic history of the ostariophysan fishes. *Copiea*, **1976**, 1–12.

Nur, A. & Ben-Avraham, Z. (1981). Lost Pacifica: a mobilistic speculation. In *Vicariance Biogeography: A Critique*, ed. G. Nelson & D.E. Rosen, pp. 341–58. New York: Columbia University Press.

Obradovich, J., Snee, L.W. & Izett, G.A. (1989). Is there more than one glassy layer in the late Eocene? *Geological Society of America Abstracts with Programs*, **21**, 134.

Occhietti, S. (1983). Laurentide ice sheet: oceanic and climatic implication. *Palaeoeography, Palaeoclimatology, Palaeoecology*, **44**, 1–22.

Oerlemans, J. (1982). Glacial cycles and ice-sheet modelling. *Climatic Change*, **4**, 353–74.

Officer, C. (1993). Victims of volcanoes. *New Scientist*, February 20, 34–8.

Officer, C.B. & Drake, C.L. (1985). Terminal Cretaceous environmental events. *Science*, **227**, 1161–7.

Okada, H. & Honjo, S. (1973). The distribution of oceanic coccolithophorids in the Pacific. *Deep-Sea Research*, **20**, 355–74.

Olander, H. (1992). Subfossil chironomid stratigraphy of a small acid lake in southern Finland. *Bulletin of the Geological Society of Finland*, **64**, 183–8.

Olander, H., Korhola, A. & Blom, T. (1997). Surface sediment Chironomidae (Insecta: Diptera) distributions along an ecotonal transect in subarctic Fennoscandia: developing a tool for palaeotemperature reconstructions. *Journal of Paleolimnology*, **18**, 45–9.

Olsen, J.S., Watts, J.A. & Allison, L.J. (1983). *Carbon in Live Vegetation of Major World Ecosystems*. ORNL-5862, Oak Ridge National Laboratory, Oak Ridge.

Olson, S.L. (1985). The Fossil Record of Birds. In *Avian Biology*, ed. D.S. Farner, J.R. King & K.C. Parkes. **8**, 79–252.

Olson, R.K. (1997). El Kef blind test III results. *Marine Micropaleontology*, **29**, 80–4.

Olsson, R., Gibson, T.G., Hansen, H.J. & Owens, J.P. (1988). Geology of the northern Atlantic Coastal Plain: Long Island to Virginia. In *The Geology of North America, The Atlantic Continental Margin*, ed. R.E. Sheridan & J.A. Crow, U.S. Geological Survey, Boulder, Colorado, **1–2**, pp. 87–105.

Opdyke, N.D., Glass, B., Hays, J.D. & Foster, J. (1966). Paleomagnetic study of Antarctic deep-sea cores. *Science,* **154**, 349–57.

Ortiz, N. (1995). Differential patterns of benthic foraminiferal extinctions near the Paleocene/Eocene boundary in the North Atlantic and the western Tethys. *Marine Micropaleontology*, **26**, 341–59.

Orue-Extebarria, X. (1997). El Kef blind test IV results. *Marine Micropaleontology*, **29**, 85–8.

Osborne, P.J. (1972). Insect faunas of Late Devensian and Flandrian age from Church Stretton, Shropshire. *Philosophical Transactions of the Royal Society of London*, B, **263**, 327–67.

Osborne, P.J. (1976). Evidence from the insects of climatic variation during the Flandrian period: a preliminary note. *World Archaeology*, **8**, 150–58.

Osborne, P.J .(1982). Some British later prehistoric insect faunas and their climatic implications. In *Climatic Change in Later Prehistory*, ed. A.F. Harding, pp. 68–74. Edinburgh: Edinburgh University Press.

Owen, H.G. (1976). Continental displacement and expansion of the Earth during the Mesozoic and Cenozoic. *Philosophical Transactions of the Royal Society of London*, A, **281**, 223–91.

Paasche, E. (1962). Coccolith formation. *Nature*, **193**, 1094–5.

Page, K. (1996). Mesozoic Ammonoids in Space and Time. In *Ammonoid Paleobiology*, ed. N.H. Landman, K. Tanabe & R.A. Davis, pp. 756–94. New York and London: Plenum Press.

Palmer, M.R. & Elderfield, H. (1986). Rare earth elements and neodynium isotopes in ferromanganese oxide coatings of Cenozoic foraminifera from the Atlantic Ocean. *Geochimica et Cosmochimica Acta*, **50**, 409–17.

Pandolfi, J.M. (1992). Successive isolation rather than evolutionary centres for the origination of Indo-Pacific reef corals. *Journal of Biogeography*, **19**, 593–609.

Pardoe, C. (1993). Competing paradigms and ancient human remains: the state of the discipline. *Archaeology in Oceania*, **26**, 79–85.

Parenti, L.R. (1991). Ocean basins and the biogeography of freshwater fishes. *Australian Systematic Botany*, **4**, 137–49.

Parker, A.G. (1995*a*). Late Quaternary Palaeoenvironments of the Upper Thames Basin, Central Southern England. Unpublished DPhil thesis, University of Oxford.

Parker, A.G. (1995*b*). Pollen analysis. In *Lithics and Landscape: Archaeological Discoveries on the Thames Water Pipeline at Gatehampton Farm, Goring, Oxfordshire 1985–92*, ed. T.G. Allen, Thames Valley Landscape Monograph No. 7, pp. 109–13. Oxford: Oxford Archaeological Reports/Oxford University Committee for Archaeology.

Parker, A.G. (1996). Pollen analysis. In Mudd, A. The excavation of a late Bronze Age/early Iron Age site at Eight Acre Field, Radley, Oxon. *Oxoniensia*, **60**, 51–3.

Parker, A.G. & Anderson, D.E. (1996). A note on the peat deposits at Minchery Farm, Littlemore, Oxford, and the implications for environmental reconstruction. *Proceedings of the Cotteswold Naturalists' Field Club*, **41**, 129–38.

Parkhurst, D. & Loucks, O. (1972). Optimal leaf size in relation to environment. *Journal of Ecology*, **60**, 505–37.

Parrish, J.T. & Curtis, R.L. (1982). Atmospheric circulation, upwelling, and organic-rich rocks in the Mesozoic and Cenozoic eras. *Palaeogeography, Palaeoclimatology, Palaeoecology*, **40**, 31–66.

Parrish, J.T. & Spicer, R.A. (1988*a*). Middle Cretaceous Wood from the Nanushuk Group, Central North Slope, Alaska. *Palaeontology*, **31**, 19–34.

Parrish, J.T. & Spicer, R.A. (1988*b*). Late Cretaceous terrestrial vegetation: a near-polar temperature curve. *Geology*, **16**, 22–5.

Parrish, J.T., Ziegler, A.M. & Scotese, C.R. (1982). Rainfall patterns and the distribution of coals and evaporites in the Mesozoic and Cenozoic. *Palaeogeography, Palaeoclimatology, Palaeoecology*, **40**, 67–101.

Partridge, T.C., Bond, G.C., Hartnady, C.J.H., deMenocal, P.B. & Ruddiman, W.F. (1995*a*). Climatic effects of Late Neogene Tectonism and Volcanism. In *Paleoclimate and Evolution, with Emphasis on Human Origins*, ed. E.S. Vrba, G.H. Denton, T.C. Partridge & L.H. Burckle, pp. 8–23. New Haven: Yale University Press.

Partridge, T.C., Wood, B. & deMenocal, P. (1995*b*). The influence of global climatic change and regional uplift on large-mammalian evolution in East and Southern Africa. In *Paleoclimate and Evolution, with Emphasis on Human Origins*, ed. E.S. Vrba, G.H. Denton, T.C. Partridge & L.H. Burckle, pp. 331–55. New Haven: Yale University Press.

Pascual, R., Vucetich, M.G., Scillato-Yané, G.J. *et al.* (1985). Main pathways of mammalian diversification in South America. In *The Great American Biotic Interchange*, ed. F.G. Stehli & S.D. Webb, pp. 219–47. New York: Plenum Press.

Paterson, K. (1976). Excursion to the Vale of White Horse. In *Field Guide to the Oxford Region*, ed. D. Roe, pp. 50–2. Oxford: Quaternary Research Association.

Patterson, C. (1993*a*). Osteichthyes: Teleostei. In *The Fossil Record 2*, ed. M. Benton, pp. 621–56. London: Chapman & Hall.

Patterson, C. (1993*b*). An overview of the early fossil record of acanthomorphs. *Bulletin of Marine Science*, **52**, 29–59.

Patterson, C. (1994). Bony fishes. In *Major Features of Vertebrate Evolution, Short Course in Paleontology, number 7*, ed. R.S. Spencer, published for the Paleontological Society, Knoxville, The University of Tenessee, pp. 57–84.

Patterson, C. & Smith, A.B. (1989). Periodicity in extinction: the role of systematics. *Ecology*, **70**, 802–11.

Paul, C.R.C. (1992). Milankovitch cycles and microfossils: principles and practice of palaeoecological analysis illustrated by Cenomanian chalk–marl rhythms. *Journal of Micropalaeontology*, **11**, 95–105.

Paul, C.R.C., Marshall, J.D., Leary, P.N., Gale, A.S., Duane, A.M. & Ditchfield, P.N. (1994). Palaeoceanographic events in the Middle Cenomanian of northwest Europe. *Cretaceous Research*, **15**, 707–38.

Paulay, G., (1997). Diversity and distribution of reef organisms. In *Life and Death of Coral Reefs*, ed. C.E. Birkeland, pp. 298–352. London: Chapman and Hall.

Paus, A. (1989). Late Weichselian vegetation, climate, and floral migration at Liastemmen, North Rogaland, south-west Norway. *Journal of Quaternary Science*, **4**, 223–42.

Pedersen, T.F. & Calvert, S.E. (1990). Anoxia vs. productivity: what controls the formation of organic-carbon-rich sediments and sedimentary rocks? *American Association of Petroleum Geologists Bulletin*, **74**, 454–66.

Pennington, W. (1977). The Late Devensian flora and vegetation of Britain. *Philosophical Transactions of the Royal Society of London*, B **280**, 247–71.

Peryt, D. & Lamolda, M. (1996). Benthonic foraminiferal mass extinction and survival assemblages from the Cenomania–Turonian Boundary Event in the Menoyo section, northern Spain. In *Biotic Recovery from Mass Extinction Events*, ed. M.B. Hart. Geological Society Special Publication No. **102**, 245–58.

Petite-Marie, N., Fontugne, M. & Rouland, C. (1991). Atmospheric methane ratio and environmental changes in the Sahara and Sahel during the last 130 k yr. *Palaeogeography, Palaeoclimatology, Palaeoecology*, **86**, 197–204.

Petitot, M-L.(1961). Contribution à l'étude des échinides fossiles du Maroc (Jurassique et Crétacé). *Notes et Memoires du Service geologique*, **146**, 1–183.

Peucker-Ehrenbrink, B., Ravizza, G. & Hofmann, A.W. (1995). The marine $^{187}Os/^{186}Os$ record of the past 80 million years. *Earth and Planetary Science Letters*, **130**, 155–67.

Pickering, K.T. & Owen, L.A. (1997). *An Introduction to Global Environmental Issues*, 2nd edn. London and New York: Routledge.

Pickering, K.T., Souter, C., Oba, T., Taira, A., Schaaf, M. & Platzman, E. (1999). Glacio-eustatic control on deep-marine clastic forearc sedimentation, Pliocene–mid-Pleistocene (*c*. 1180–600 ka), Kazusa Group, S.E. Japan. *Journal of the Geological Society, London*, **156**, 125–36.

Pickering, K.T., Underwood, M.B. & Taira, A. (1993). Stratigraphic synthesis of the DSDP–ODP sites in the Shikoku Basin, Nankai Trough and accretionary prism. *Proceedings of the Ocean Drilling Program, Scientific Results*, **131**, 313–30.

Pickford, M. (1981). Preliminary Miocene mammalian biostratigraphy for western Kenya. *Journal of Human Evolution*, **10**, 73–97.

Pierce, R.L. (1961). Lower Upper Cretaceous plant microfossils from Minnesota. *University of Minnesota and Minnesota Geological Survey, Bulletin*, **42**, 1–86.

Pigg, K.B. & Stockey, R.A. (1991). Platanaceous plants from the Paleocene of Alberta, Canada. *Review of Palaeobotany and Palynology*, **70**, 125–46.

Pigott, C.D. (1969). The status of *Tilia cordata* and *Tilia platyphyllos* on the Derbyshire limestone. *Journal of Ecology*, **57**, 491–504.

Pike, E.M. (1994). Historical changes in insect community structure as indicated by hexapods of Upper Cretaceous Alberta (Grassy Lake) amber. *Canadian Entomologist*, **126**, 695–702.

Pirrie, D. & Marshall, J.D. (1990). High-paleolatitude late Cretaceous paleotemperatures: new data from James Ross Island, Antarctica. *Geology*, **18**, 31–4.

Poag, C.W. (1985). Depositional history and stratigraphic reference section for central Baltimore Canyon Trough. In *Geologic Evolution of the United States Atlantic Margin*, ed. C.W. Poag, pp. 217–64. New York: Van Nostrand Reinhold.

Poag, C.W. & Commeau, J.A. (1995). Paleocene to middle Miocene planktic foraminifera of the southwestern Salisbury embayment, Virginia and Maryland: biostratigraphy, allostratigraphy and sequence stratigraphy. *Journal of Foraminiferal Research*, **25**, 134–55.

Poag, C.W., Powars, D.S., Poppe, L.J. & Mixon, R.B. (1994). Meteoroid mayhem in Ole Virginny: source of the North American tektite strewn field. *Geology*, **22**, 691–4.

Poag, C.W., Powars, D.S., Poppe, L.J., Mixon, R.B., Edwards, l.E., Folger, D.W. & Bruce, S. (1992). Deep Sea Drilling Project Site 612 bolide event: new evidence of a late Eocene impact-wave deposit and a possible impact site, U.S. East Coast. *Geology*, **20**, 771–4.

Poag, C.W. & Ward, L.W. (1987). Cenozoic unconformities and depositional supersequences of North Atlantic continental margins: testing the Vail model. *Geology*, **15**, 159–62.

Podubiuk, R.H. & Rose, E.P.F. (1984). Relationships between mid-Tertiary echinoid faunas from the central Mediterranean and eastern Caribbean and their palaeobiogeographic significance. *Annales Géologiques des Pays Helléniques*, **32**, 115–28.

Poinar, G.O. (1992). *Life in Amber*. Stanford: Stanford University Press.

Pole, M.S. & Macphail, M.K. (1996). Eocene Nypa from Regatta Point, Tasmania. *Review of Palaeobotany and Palynology*, **92**, 55–67.

Ponder, W.F. & Lindberg, D.R. (1996). Gastropod phylogeny – challenges for the 90s. In *Origin and Evolutionary Radiation of the Mollusca*, ed. J.D. Taylor, pp. 135–54. Oxford: Oxford University Press.

Ponomarenko, A.G. (1976). A new insect from the Cretaceous of Transbaikalia, a possible parasite of pterosaurians. *Paleontological Journal*, **10**, 339–43.

Poole, I. (1992). Pyritized twigs from the London Clay, Eocene, of Great Britain. *Tertiary Research*, **13**, 71–85.

Poole, I. (1993). A Dipterocarpaceous twig from the Eocene London Clay Formation of southeast England. *Special Papers in Palaeontology*, **49**, 155–63.

Porter, J.W. (1976). Autotrophy, heterotrophy, and resource partitioning in Caribbean reef-building corals. *The American Naturalist*, **110**, 731–42.

Pouyet, S. (1991). Bryozoaires cheilostomes Neogenes du Bassin du Rhone. Gisement Burdigalien de Caumont Picabrier (Vaucluse, France). *Revue de Paléobiologie*, **10**, 389–421.

Povey, D.A.R. (1995). Palaeobotanical Determinations of Tertiary Palaeoelevation in Western North America. Unpublished DPhil thesis, Oxford University.

Povey, D.A.R., Spicer, R.A. & England, P.C. (1994). Palaeobotanical Investigation of early Tertiary Palaeoelevations in northeastern Nevada: Initial Results. *Review of Palaeobotany and Palynology*, **81**, 1–10.

Prell, W.L. (1984). Covariance patterns of foraminiferal $\delta^{18}O$: an evaluation of Pliocene ice volume changes near 3.2 million years ago. *Science*, **226**, 692–4.

Prentice, I.C. (1990). Bioclimatic distribution of vegetation for general circulation models studies. *Journal of Geophysical Research*, **95**, 811–30.

Prentice, I.C., Cramer, W., Harrison, S.P., Leemans, R., Monserud, R.A. & Solomon, A.M. (1992). A global biome model based on plant physiology and dominance, soil properties and climate. *Journal of Biogeography*, **19**, 117–34.

Prothero, D.R. (1994). *The Eocene–Oligocene Transition: Paradise Lost*. New York: Columbia University Press.

Prothero, D.R. & Berggren, W.A. (eds.) (1992). *Eocene–Oligocene Climatic and Biotic Evolution*. Princeton: Princeton University Press.

Quade, J. & Cerling, T. (1995). Expansion of C_4 grasses in the late Miocene of northern Pakistan: evidence from stable isotopes in paleosols. *Palaeogeography, Palaeoclimatology, Palaeoecology*, **115**, 91–116.

Quade, J., Cerling, T.E., Andrews, P. & Alpagut, B. (1995). Paleodietary reconstruction of Miocene faunas from Pasala, Turkey using carbon and oxygen isotopes of fossil tooth enamel. *Journal of Human Evolution*, **28**, 373–84.

Quade, J., Cerling, T.E., Barry, J.C., Morgan, M.E., Pilbeam, D.R., Chivas, A.R., Lee-Thorp, K.A. & van der Merwe, N.J. (1992) A 16 Ma record of paleodiet using carbon and oxygen isotopes in fossil teeth from Pakistan. *Chemical Geology*, **94**, 183–92.

Quade, J., Cerling, T.E. & Bowman, J. (1989). Development of Asian monsoon revealed by marked ecological shift during the latest Miocene in northern Pakistan. *Nature*, **342**, 163–6.

Quine, M. & Bosence, D. (1991). Stratal geometries, facies and sea-floor erosion in Upper Cretaceous Chalk, Normandy, France. *Sedimentology*, **38**, 1113–52.

Raff, R.A. (1992). *The Evolution of Life Histories*. New York: Chapman & Hall.

Raff, R.A. (1996). *The Shape of Life: Genes, Development, and the Evolution of Animal Form*. Chicago: The University of Chicago Press.

Raff, R.A. & Kauffman, T.C. (1983). *Embryos, Genes, and Evolution*. New York: MacMillan.

Rage, J-C. (1984). Are the Ranidae (Anura, Amphibia) known prior to the Eocene? *Amphibia–Reptilia*, **5**, 281–8.

Rage, J-C. (1988). Gondwana, Tethys and terrestrial vertebrates during the Mesozoic and Cainozoic. In *Gondwana and Tethys*, ed. M.G. Audley-Charles & A. Hallam. Geological Society Special Publication No. **37**, 255–73.

Rage, J-C. & Auge, M. (1993). Squamates from the Cainozoic of the Western Part of Europe. A review. *Revue de Paléobiologie*, Special Volume 7, 199–216.

Rahmstorf, S. (1994). Rapid climate transitions in a coupled ocean–atmosphere model. *Nature*, **372**, 82–5.

Rampino, M.R. & Self, S. (1992). Volcanic winter and accelerated glaciation following the Toba super-eruption. *Nature*, **359**, 50–52.

Rasnitsyn, A.P. (1996). Conceptual issues in phylogeny, taxonomy, and nomenclature. *Contributions to Zoology*, **66**, 3–41.

Rasnitsyn, A.P., Jarzembowski, E.A. & Ross, A.J. (1998). Wasps (Insecta: Vespida = Hymenoptera) from the Purbeck and Wealden (Lower Cretaceous) of southern England and their biostratigraphical and palaeoenvironmental significance. *Cretaceous Research*, **19**, 329–91.

Rasnitsyn, A.P. & Michener, C.D. (1991). Miocene fossil bumble bee from the Soviet Far East with comments on the chronology and distribution of fossil bees (Hymenoptera: Apidae). *Annals of the Entomological Society of America*, **84**, 583–9.

Rat, P. (1989). The Iberian Cretaceous: climatic implications. In *The Cretaceous of Western Tethys*, ed. J. Wiedmann, pp. 17–25. Stuttgart: E. Schweitzerbart'sche Verlagsbuchhandlung.

Raunkiaer, C. (1934). *The Life Forms of Plants and Statistical Plant Geography*. Oxford: Clarendon Press.

Raup, D.M. & Jablonski, D. (1993). Geography of end-Cretaceus marine bivalve extinctions. *Science*, **260**, 971–3.

Raven, P.H. (1977). A suggestion concerning the Cretaceous rise to dominance of the angiosperms. *Evolution*, **31**, 451–2.

Ravizza, G. (1993). Variations of the $^{187}Os/^{186}Os$ ratio of seawater over the past 28 million years as inferred from metalliferous carbonates. *Earth and Planetary Science Letters*, **118**, 335–48.

Ravven, W. (1991). Asteroid impact emptied Gulf of Mexico. *New Scientist*, **129**, 1762, 14.

Rawson, P.F. (1973). Lower Cretaceous (Ryazanian–Barremian) marine connections and cephalopod migrations between the Tethyan and Boreal Realms. In *The Boreal Lower Cretaceous*, ed. R. Casey & P.F. Rawson. Geological Journal Special Issue, **5**, 131–44.

Rawson, P.F. (1981). Early Cretaceous ammonite biostratigraphy and biogeography. In *The Ammonoidea*, ed. M.R. House & J.R. Senior. Systematics Association Special Volume No. **18**, pp. 499–529.

Rawson, P.F. (1993). The influence of sea level changes on the migration and evolution of Lower Cretaceous (pre-Aptian) ammonites. In *The Ammonoidea: Environment, Ecology and Evolutionary Change*, ed. M.R. House. Systematics Association Special Volume No. **47**, 227–42.

Rawson, P.F. (1994). Sea level changes and their influence on ammonite biogeography in the European Early Cretaceous. In *Proceedings of the 3rd Pergola International Symposium*, ed. G. Pallini. Paleopelagos Special Publication 1, 317–26.

Rawson, P.F. (1995). Biogeographical affinities of NW European Barremian ammonite faunas and their palaeogeographical implications. *Memorie Descrittive della Carta Geologica d'Italia*, **51**, 131–36.

Rawson, P.F. & Riley, L.A. (1982). Latest Jurassic–Early Cretaceous events and the 'Late Cimmerian unconformity' in North Sea area. *The American Association of Petroleum Geologists Bulletin*, **66**, 2628–48.

Raymo, M.E. & Ruddiman, W.F. (1992). Tectonic forcing of late Cenozoic climate. *Nature*, **359**, 117–22.

Raymo, M.E., Ruddiman, W.F., Backman, J., Clement, B.M. & Martinson, D.G. (1989). Late Pliocene variation in Northern Hemisphere ice sheets and North Atlantic deep water circulation. *Paleoceanography*, **4**, 413–46.

Rayner, R.J., Waters, S.B., McKay, I.J., Dobbs, P.N. & Shaw, A.L. (1991). The mid-Cretaceous palaeoenvironment of central southern Africa (Orapa, Botswana). *Palaeogeography, Palaeoclimatology, Palaeoecology*, **88**, 147–56.

Raza, M.S. & Meyer, G.E. (1981). Early Miocene geology and paleontology of the Bugti Hills, Pakistan. *Memoirs of Geological Survey of Pakistan*, **11**, 43–63.

Reboulet, S. (1995). L'évolution des ammonites du Valanginien–Hauterivien inférieur du bassin Vocontien et de la Plate-Forme Provencale (Sud-Est de la France): relations avec la stratigraphie séquentielle et implications biostratigraphiques. *Documents dels Laboratoires de Géologie Lyon*, **237**, 371 pp.

Reboulet, S. & Atrops, F. (1995). Rôle du climat sur les migrations et la composition des peuplements d'ammonites du Valanginien Supérieur du Bassin Vocontien (S-E de la France). *Geobios*, N.S. 18, 357–65.

Regal, P.J. (1977). Ecology and evolution of flowering plant dominance. *Science*, **196**, 622–9.

Reid, D.G. (1996). *Systematics and Evolution of* Littorina. London: The Ray Society.

Reid, R.G. & Brand, D.G. (1984). Sulfide-oxidising symbiosis in lucinaceans: implications for bivalve evolution. *Veliger*, **29**, 3–24.

Remy, J.A., Crochet, J-Y., Sigé, B. *et al.* (1987). Biochronologie des Phosphorites du Quercy: mise à jour des listes fauniques et nouveaux gisements de mammifères fossiles. *Münchner Geowissenschaftlicher Abhandlungen*, (A) **10**, 169–88.

Reyment, R.A. & Bengtson, P. (1986). *Events of the mid-Cretaceous*. Oxford: Pergamon Press.

Reyment, R.A. & Dingle, R.V. (1987). Palaeogeography of Africa in the Cretaceous period. *Palaeogeography, Palaeoclimatology, Palaeoecology*, **59**, 93–116.

Reynoso, V.H. (1995). Lepidosaurian reptiles from the Cantera Tlayua (Albian), Tepexi de Rodríguez, Puebla, Mexico. II *International Symposium on Lithographic Limestones, Cuenca, Spain, 1995, Extended Abstracts*. Ediciones de la Universidad Autónoma de Madrid, pp. 131–2.

Reys, J. (1972). Recherches géologiques sur le Crétacé inférieur de l'Estremadura. *Mémoires du Serviços geologicos de Portugal*, **21**, 1–471, pls 1–22.

Rich, T.H.V., Molnar, R.E. & Rich, P.V. (1983). Fossil vertebrates from the Late Jurassic or Early Cretaceous Kirkwood Formation, Algoa Basin, Southern Africa. *Transactions of the Geological Society of South Africa*, **86**, 281–91.

Richards, M., Corte-Real, H., Forster, P. *et al.* (1996). Palaeolithic and Neolithic lineages in the European Mitochondrial Gene Pool. *American Journal of Human Genetics*, **59**, 185–203.

Richards, P.W. (1952). *The Tropical Rain Forest: An Ecological Study*. Cambridge: Cambridge University Press.

Riedel, F. (1996). An outline of cassoidean phylogeny. *Contributions to Tertiary and Quaternary Geology*, **32**, 97–132.

Riezebos, P.A. & Slotboom, R.T. (1984). Three-fold subdivision of the Allerod chronozone. *Boreas*, **13**, 347–53.

Rigby, J.K.jr, Newman, K.R., Smit, J. *et al.* (1987). Dinosaurs from the Paleocene part of the Hell Creek Formation, McCone County, Montana. *Palaios*, **2**, 296–302.

Rightmire, G.P. (1993). Variation among early *Homo* crania from Olduvai Gorge and the Koobi Fora region. *American Journal of Physical Anthropology*, **90**, 1–33.

Roberts, M., Stringer, C. & Parfitt, S. (1994). A hominid tibia from Middle Pleistocene sediments at Boxgrove, UK. *Nature*, **369**, 311–3.

Robertson, J.E., Robinson, C., Turner, D.R. *et al.* (1994). The impact of a coccolithophore bloom on oceanic carbon uptake in the northeast Atlantic during summer 1991. *Deep-Sea Research*, **41**, 297–314.

Robinson, M.A. (1981). Investigations of Palaeoenvironments in the Upper Thames Valley, Oxfordshire. Unpublished PhD thesis, University of London.

Robinson, M.A. (1992). Environmental archaeology of the river gravels: past achievements and future directions. In *Developing Landscapes of Lowland Britain. The Archaeology of the British Gravels: a Review*, ed. M. Fulford & E. Nichols, pp. 47–62. Society of Antiquities Occasional Paper **14**.

Robinson, M.A. & Lambrick, G. (1984). Holocene alluviation and hydrology in the Upper Thames basin. *Nature,* **308**, 809–14.

Robinson, M.A. & Wilson, B. (1987). A survey of the environmental archaeology in the south Midlands. In *Environmental Archaeology: A Regional Review, Vol II*, ed. H.C.M. Keeley, Historic Buildings and Monuments Commission for England, Occasional Paper No **1**, pp. 16–84.

Roček, Z. (1996). The salamander *Brachycormus noachicus* from the Oligocene of Europe, and the role of neoteny in the evolution of salamanders. *Palaeontology*, **39**, 477–95.

Roček, Z. & Lamaud, P. (1995). *Thaumastosaurus bottii* de Stefano 1903, an anuran with Gondwanan affinities from the Eocene of Europe. *Journal of Vertebrate Paleontology,* **15**, 506–15.

Roebroeks, W. & Van Kolfschoten, T. (1994). The earliest occupation of Europe: a short chronology. *Antiquity*, **68**, 489–503.

Rogers, A. & Jorde, L. (1995). Genetic evidence on modern human origins. *Human Biology*, **67**, 1–36.

Rögl, F. (1999). Oligocene and Miocene palaeogeography and stratigraphy of the circum-Mediterranean region. In *Fossil Vertebrates of Arabia*, ed. P.J. Whybrow & A. Hill, pp. 485–500. New Haven: Yale University Press.

Rögl, F. & Steininger, F.F. (1983). Vom Zerfall der Tethys zu Mediterran und Paratethys. Die neogene Paläogeographie und Palinspastik des zirkum-mediterranen Raumes. *Annalen des Naturhistorischen Museums, Wien*, **85A**, 135–63.

Rögl, F. & Steininger, F.F. (1984). Neogene Paratethys, Mediterranean and Indo-Pacific seaways. Implications for the paleobiogeography of marine and terrestrial biotas. In *Fossils and Climate*, ed. P.J. Brenchley. Geological Journal Special Issue, **11**, 171–200.

Rolland, N. (1992). The Palaeolithic colonization of Europe: an archaeological and biogeographic perspective. *Trabajos de Prehistoria*, **49**, 69–111.

Roman, J. (1984). Les échinidés et la crise Crétacé–Tertiaire. *Bulletin de la Section des Sciences, 1984*, **6**, 133–47.

Roman, J. & Strougo, A. (1994). Echinoïdes du Libyen (Eocene inferieur) d'Egypte. *Revue de Paléobiologie*, **13**, 29–57.

Romano, S.L. & Palumbi, S.R. (1996). Evolution of scleractinian corals inferred from molecular systematics. *Science*, **271**, 640–2.

Roniewicz, E. & Morycowa, E. (1993). Evolution of the Scleractinia in the light of microstructural data. In *Proceedings of the VI International Symposium on Fossil Cnidaria and Porifera held in Münster, Germany 9–14 September 1991*, ed P. Oekentorp-Küster, *Courier Forschungsinstitut Senckenberg*, **164**, 233–40.

Rose, J., Turner, C., Coope, G.R. & Bryan, M.D. (1980). Channel changes in a lowland river catchment over the last 13 000 years. In *Timescales in Geomorphology*, ed. R.A. Cullingford, D.A. Davidson & J. Lewin, pp. 156–76. Chichester, UK: John Wiley.

Rosen, B.R. (1977). The depth distribution of Recent hermatypic corals and its palaeontological significance. In *Second Symposium International sur Coraux et Récifs Coralliens Fossiles. Mémoires du Bureau de Recherches Géologiques et Miniéres*, **89**, 507–517.

Rosen, B.R. (1981). The tropical high diversity enigma – the corals' eye view. In *Chance, Change and Challenge: the Evolving Biosphere*, ed. P.H. Greenwood & P.L. Forey, pp. 103–29. Cambridge and London: Cambridge University Press and British Museum (Natural History).

Rosen, B.R. (1984). Reef coral biogeography and climate through the late Cainozoic: just islands in the sun or a critical pattern of islands? In *Fossils and Climate*, ed. P.J. Brenchley. Geological Journal Special Issue, **11**, 201–62.

Rosen, B.R. (1988*a*). Biogeographical patterns: a perceptual overview, in *Analytical Biogeography; an Integrated Approach to the Study of Animal and Plant Distributions*, ed. A.A. Myers & P.S. Giller, pp. 23–55. London: Chapman & Hall.

Rosen, B.R. (1988*b*). Progress, problems and patterns in the biogeography of reef corals and other tropical marine organisms. *Helgoländer Meeresunters*, **42**, 269–301.

Rosen, B.R. (1992). Empiricism and the biogeographical black box: concepts and methods in marine palaeobiogeography. In *Biogeographic Patterns in the Cretaceous Ocean*, ed. B.A. Malmgren & P. Bengtson. *Palaeogeography, Palaeoclimatology, Palaeoecology*, **92**, 171–205.

Rosen B.R. & Turnšek, D. (1989). Extinction patterns and biogeography of scleractinian corals across the Cretaceous/Tertiary boundary. In *Fossil Cnidaria 5. Proceedings of the Fifth International Symposium on Fossil Cnidaria including Archaeocyatha and Spongiomorphs held in Brisbane, Queensland, Australia, 25–29 July 1988*, ed. P.A. Jell & J.W. Pickett, *Memoirs of the Association of Australasian Palaeontologists*, **8**, 355–70.

Rosen, D.E. (1976) A vicariance model of Caribbean biogeography. *Systematic Zoology*, **24**, 431–64.

Rosen, D.E. (1978). Vicariant patterns and historical explanation in biogeography. *Systematic Zoology*, **27**, 159–88.

Rosen, D.E. (1985). Geological hierarchies and biogeographic congruence in the Caribbean. *Annals of the Missouri Botanical Garden*, **72**, 636–59.

Ross, A.J. (1997). Insects in amber. *Geology Today*, **13**, 24–8.

Ross, A.J. & Jarzembowski, E.A. (1993). Arthropoda (Hexapoda; Insecta). In *The Fossil Record 2*, ed. M.J. Benton, pp. 363–426. London: Chapman & Hall.

Ross, A.J. & Jarzembowski, E.A. (1996). A provisional list of fossil insects from the Purbeck Group of Wiltshire. *Wiltshire Archaeological and Natural History Magazine*, **89**, 106–15.

Ross, D.J. & Skelton, P.W. (1993). Cretaceous Climates. In *Sedimentology Review 1*, ed. V.P. Wright, pp. 73–91. Oxford: Blackwell.

Ross, D.J. & Skelton, P.W. (1996). Rudist formations of the Cretaceous; a palaeoecological, sedimentological and stratigraphical review. *Sedimentology Review*, **1**, 73–91.

Ross, J.R.P. & Ross, C.A. (1996). Bryozoan evolution and dispersal and Paleozoic sea-level fluctuations. In *Bryozoans in Space and Time*, ed. D.P. Gordon, A.M. Smith & J.A. Grant-Mackie, pp. 243–58. Wellington: NIWA.

Roth, J.L. & Dilcher, D.L. (1978). Some considerations in leaf size and leaf margin analysis of fossil leaves. *Courier Forschunginstitüt Senckenberg*, **30**, 165–71.

Roth, P.H. & Bowdler, J.L. (1981). Middle Cretaceous calcareous nannoplankton biogeography and oceanography of the Atlantic Ocean. Society of Economic Paleontologists and Mineralogists, Special Publication No. **32**, 517–546.

Roy, K., Jablonski, D. & Valentine, J.W. (1995). Thermally anomalous assemblages revisited: patterns in the extraprovincial latitudinal range shifts of Pleistocene marine mollusks. *Geology*, **23**, 1071–4.

Ruddiman, W.F. & Duplessey, J.C. (1985). Conference on the last deglaciation: timing and mechanism. *Quaternary Research*, **23**, 1–17.

Ruddiman, W.F. & Kutzbach, J.E. (1989). Forcing of late Cenozoic northern hemisphere climate by plateau uplift in southern Asia and the American West. *Journal of Geophysical Research*, **94**, 18,409–18, 427.

Ruddiman, W.F. & Kutzbach, J.E. (1991). Plateau Uplift and Climatic Change. *Scientific American*, **264** (3), 42–50.

Ruddiman, W.F. & McIntyre, A. (1976). Northeast Atlantic Palaeoclimatic changes over the past 600,000 years. *Memoir Geological Society of America*, **145**, 111–46.

Ruddiman, W.F. & McIntyre, A. (1981). The north Atlantic during the last deglaciation. *Palaeogeography, Palaeoclimatology, Palaeoecology*, **35**, 145–214.

Ruddiman, W.F. & Raymo, M.E. (1988). Northern Hemisphere climatic regimes during the past 3 Ma: possible tectonic connections. *Philosophical Transactions of the Royal Society of London*, B, **318**, 411–30.

Ruddiman, W.F., Raymo, M.E., Martinson, D.G., Clement, B. & Backman, J. (1989). Pleistocene evolution: Northern Hemisphere ice sheets and North Atlantic Ocean. *Paleoceanography*, **4**, 353–412.

Ruddiman, W.F., Raymo, M.E. & McIntyre, A. (1986a). Matuyama 41,000-year cycles: North Atlantic Ocean and northern hemisphere ice sheets. *Earth and Planetary Science Letters*, **80**, 117–29.

Ruddiman, W.F., Sancetta, C.D. & McIntyre, A. (1977). Glacial/Interglacial response rate of subpolar North Atlantic waters to climatic change: the record in ocean sediments. *Philosophical Transactions of the Royal Society of London*, B, **280**, 119–242.

Ruddiman, W.F., Shackleton, N.J. & McIntyre, A. (1986b). North Atlantic sea-surface temperatures for the last 1.1 million years. In *North Atlantic Palaeoceanography*, ed. C.P. Summerhayes & N.J. Shackleton, pp. 155–73. Oxford: Blackwell, Oxford.

Ruff, C. (1993). Climatic adaptation and hominid evolution: the thermoregulatory imperative. *Evolutionary Anthropology*, **2**, 53–60.

Ruffell, A.H. & Batten, D.J. (1990). The Barremian–Aptian arid phase in northern Europe. *Palaeogeography, Palaeoclimatology, Palaeoecology*, **80**, 197–212.

Ruffell, A.H. & Rawson, P.F. (1994). Palaeoclimate control on sequence stratigraphic patterns in the late Jurassic to mid-Cretaceous, with a case study from eastern England. *Palaeogeography, Palaeoclimatology, Palaeoecology*, **110**, 43–54.

Ruffell, A.H., Ross, A.J. & Taylor, K. (1996). Early Cretaceous environments of the Weald. *Geologists' Association Guide*, No. **55**, 81 pp.

Russell, D.A. (1993). The role of Central Asia in dinosaurian biogeography. *Canadian Journal of Earth Sciences*, **30**, 2002–2012.

Russell, D.A., Russell, D.E. & Sweet, A.R. (1993). The end of the dinosaurian era in the Nanxiong Basin. *Vertebrata Palasiatica*, **31**, 139–45.

Russell, D.E. & Zhai Ren-jie (1987). The Paleogene of Asia: mammals and stratigraphy. *Mémoires du Muséum National d'Histoire Naturelle*, Paris (C)**52**, 1–488.

Ryan, W.B.F. (1973). Geodynamic implications of the Messinian crisis of salinity. In *Messinian events in the Mediterranean*, ed. C.W. Drooger *et al.*, pp. 26–38. Amsterdam: North Holland.

Rybnicek, K. & Rybnickova, E. (1987). Palaeogeobotanical evidence of middle Holocene stratigraphic hiatuses in Czechoslovakia and their explanation. *Folia Geobotanica et Phytotaxonomica*, **22**, 313–27.

Ryland, J.S. (1970). *Bryozoans*. London: Hutchinson.

Sage, B.L. (1973). *Alaska and its Wildlife*. London and New York: Hamlyn.

Sahni, A., Gheerbrant, E. & Khajuria, C.K. (1994). Eutherian mammals from the Upper Cretaceous (Maastrichtian) Intertrappean Beds of Naskal, Andhra Pradesh, India. *Journal of Vertebrate Paleontology*, **14**, 260–77.

Sahni, A., Kumar, K., Hartenberger, J-L., Jaeger, J-J., Rage, J-C., Sudre, J. & Vianey-Liaud, M. (1982). Microvertébrés nouveaux des Trappes du Deccan (Inde): Mise en évidence d'une voie de communication terrestre probable entre la Laurasie et l'Inde à la limite Crétacé–Tertiare. *Bulletin de la Société geologique de France*, **24**, 1093–9.

Saito, T. (1976). Geologic significance of coiling direction in the planktonic foraminifera *Pulleniatina. Geology*, **4**, 305–9.

Saltzman, E.S. & Barron, E.J. (1982). Deep circulation in the late Cretaceous: oxygen isotope paleotemperatures from *Inoceramus* remainsin D.S.D.P. cores. *Palaeogeography, Palaeoclimatology, Palaeoecology*, **40**, 167–81.

Sampson, S.D., Krause, D.W., Forster, C.A. & Dodson, P. (1996). Non-avian theropod dinosaurs from the Late Cretaceous of Madagascar and their paleobiological implications. *Journal of Vertebrate Paleontology*, **16** (3, suppl.), 62A.

Samylina, V.A. (1963). The Mesozoic flora of the lower course of the Aldan River: Academiya Nauk S.S.S.R., *Paleobotanica*, ser. 8, **4**, 58–139 [in Russian].

Samylina, V.A. (1973). Correlation of Lower Cretaceous Continental deposits of Northeast U.S.S.R. based on paleobotanical data. *Sovietskaya Geologiya*, no. **8**, 42–57 [in Russian].

Samylina, V.A. (1974). Early Cretaceous floras of Northeast U.S.S.R. (Problems of establishing Cenophytic floras). *Komarovskiyc chteniya*, **27**, Izd-vo Nauka, Leningrad [in Russian].

Sanchíz, F.B. (1977). La familia Bufonidae (Amphibia, Anura) en el terciaric Europeo. *Trabajos del Departamento de Paleontologia (Madrid)*, **8**, 75–111.

Sandford, K. (1965). Note on the gravels of the Upper Thames Flood Plain between Lechlade and Dorchester. *Proceedings of the Geologists' Association*, **76**, 61–75.

Sarnthein, M. (1978). Sand deserts during the Glacial Maximum and climatic optimum. *Nature*, **273**, 43–6.

Savage, D.E. & Russell, D.E. (eds.) (1983). *Mammalian Paleofaunas of the World*. Reading: Addison Wesley Publishing Co.

Savage, J.M. (1982). The enigma of the Central American herpetofauna: dispersals or vicariance? *Annals of the Missouri Botanic Garden*, **69**, 464–547.

Savage, R.J.G., Domning, D.P. & Thewissen, J.G.M. (1994). Fossil Sirenia of the west Atlantic and Caribbean region. V. The most primitive known sirenian, *Prorastomus sirenoides* Owen, 1855. *Journal of Vertebrate Paleontology*, **14**, 427–49.

Savin, S.M. (1977). The history of the earth's surface temperature during the past 100 million years. *Annual Review of Earth and Planetary Science*, **5**, 319–55.

Savin, S.M., Abel, L. *et al.* (1985). The evolution of the Miocene surface and near-surface marine temperatures: oxygen isotope evidence. In *The Miocene Ocean: Paleoceanography and Biogeography*, ed. J.P. Kennett. *Geological Society of America, Memoir*, **163**, 49–82.

Scaife, R.G. (1982). Late-Devensian and early Flandrian vegetation changes in southern England. In *Archaeological Aspects of Woodland Ecology*, ed. M. Bell & S. Limbrey, *British Archaeological Reports, International Series*, **146**, 57–74.

Scaife, R.G. (1987). A review of later Quaternary microfossil and macrofossil research in southern England; with special reference to environmental archaeological evidence. In *Environmental Archaeology: a Regional Review Vol II*, ed. H.C.M. Keeley. Historic Buildings and Monuments Commission for England, Occasional Paper No 1, pp. 125–79.

Scanlon, J. (1993). Madtsoiid snakes from the Eocene Tingmarra fauna of eastern Queensland. *Kaupia*, **3**, 3–8.

Schaaf, M. & Thurow, J. (1996). Late Pleistocene–Holocene climatic cycles recorded in Santa Barbara Basin sediments: interpretation of colour density logs from Site 893. *Proceedings of the Ocean Drilling Program, Scientific Results*, **146**, 31–44.

Schaal, S. & Ziegler, W. (eds.) (1992). *Messel, an Insight into the History of Life and of the Earth*, English edn. Oxford: Clarendon Press.

Schlager, W. & Philip, J. (1990). Cretaceous carbonate platforms. In *Cretaceous Resources, Events and Rhythms*, ed. R.N. Ginsburg & B. Beaudoin, pp. 173–95. Dordrecht: Kluwer Academic Publishers.

Schlanger, S.O. (1986). High frequency sea-level fluctuations in Cretaceous time; an emerging geophysical problem. In *Mesozoic and Cenozoic oceans*, ed. K.J. Hsu. American Geophysical Union, Geodynamic Series, **15**, 61–74.

Schlanger, S.O. & Jenkyns, H.C. (1976). Cretaceous anoxic events: causes and consequences. *Geologie en Mijnbouw*, **55**, 179–84.

Schlee, J.S. (1981). Seismic stratigraphy of the Baltimore Canyon Trough. *American Association of Petroleum Geologists Bulletin*, **65**, 26–53.

Schlichter, D. (1991). A perforated gastrovascular cavity in Leptoseris fragilis. A new improved strategy to improve heterotrophic nutrition in corals. *Naturwissenschaften*, **78**, 467–9.

Schmitz, B., Keller, G. & Stenvall, O. (1992). Stable isotope and foraminiferal changes across the Cretaceous-Tertiary boundary at Stevns Klint, Denmark: arguments for long-term oceanic instability before and after bolide-impact event. *Palaeogeography, Palaeoclimatology, Palaeoecology*, **96**, 233–60.

Scholle, P.A. & Arthur, M.A. (1980). Carbon isotope fluctuations in Cretaceous pelagic limestones: potential stratigraphic and petroleum exploration tool. *American Association of Petroleum Geologists Bulletin*, **64**, 67–87.

Schopf, T.J. M. (1970). Taxonomic diversity gradients of ectoprocts and bivalves and their geologic implications. *Geological Society of America Bulletin*, **81**, 3765–68.

Schreiber, B.C. & Ryan, W.B.F. (1995). Variations in the formative conditions of Messinian evaporites, water depths, and depositional cyclicity. *Abstracts, International Conference on the Biotic and Climatic Effects of the Mediterranean Event on the Circum Mediterranean, Benghazi, 1995*, pp. 55–56.

Schröder, T. (1992). A palynological zonation for the Paleocene of the North Sea Basin. *Journal of Micropalaeontology*, **11**, 113–26.

Schuhmacher, H. & Zibrowius, H. (1985). What is hermatypic? A redefinition of ecological groups in corals and other organisms. *Coral Reefs*, **4**, 1–9.

Schulter, C. (1992). Foraminiferal ontogeny and environmental variation across the K/T boundary at Nye Kløv, Denmark. Senior Thesis, Princeton University.

Schuster, F, Wielandt, U. & Luterbacher, H.P. (1993). Paleocene and Eocene corals and associated sediments in the Western Desert, Egypt [abstract]. In *First European Regional Meeting of the International Society for Reef Studies, Vienna, 16.-20. December 1993. Abstracts*, p. 90.

Schweitzer, P.N. & Lohmann, G.P. (1991). Ontogeny and habitat of modern menardiiform planktonic foraminifera. *Journal of Foraminiferal Research*, **21**, 332–46.

Schwert, D.P. (1992). Faunal transitions in response to an iceage: the late Wisconsinan record of Coleoptera in the north-central United States. *Coleopterists Bulletin*, **39**, 67–79.

Schwert, D.P. & Ashworth, A.C. (1988). Late Quaternary history of the northern beetle fauna of North America: a synthesis of fossil distributional evidence. *Memoirs of the Entomological Society of Canada*, **144**, 93–107.

Scott, A.C., Stephenson, J. & Collinson, M.E. (1994). The fossil record of plant galls. In *Plant Galls: Organisms Interactions, Populations*, ed. M.A.J. Williams. Systematics Association, Publication **49**, 441–63.

Scott, R.A. & Smiley, C.J. (1979). Some Cretaceous plant megafossils and microfossils from the Nanushuk Group, Northern Alaska, A preliminary report. *U.S. Geological Survey Circular*, **749**, 89–111.

Scott, R.W. (1990). Models and stratigraphy of mid-Cretaceous reef communities, Gulf of Mexico. *SEPM (Society for Sedimentary Geology) Concepts in Sedimentology and Paleontology*, **2**, 1–102.

Scott, R.W. (1995). Global environmental controls on Cretaceous reefal ecosystems. *Palaeogeography, Palaeoclimatology, Palaeoecology*, **119**, 187–99.

Seilacher, A. (1984). Constructional morphology of bivalves: evolutionary pathways in primary versus secondary soft-bottom dwellers. *Palaeontology*, **27**, 207–37.

Sellwood, B.W. & Price, G.D. (1993). Sedimentary facies as indicators of Mesozoic palaeoclimate. *Philosophical Transactions of the Royal Society of London*, B, **341**, 225–33.

Sepkoski, J.J. (1981). A factor analytic description of the Phanerozoic marine fossil record. *Paleobiology*, **7**, 35–53.

Sepkoski, J.J. (1986). Phanerozoic overview of mass extinction. In *Patterns and Processes in the History of Life*, ed. D.M. Raup & D. Jablonski, pp. 277–94. Berlin, Heidelberg: Springer-Verlag.

Sepkoski, J.J., Jr (1988). Alpha, beta, or gamma: where does all the diversity go? *Paleobiology*, **14**, 221–34.

Sepkoski, J.J., Jr. (1993). Ten years in the library: new data confirm paleontological patterns. *Paleobiology*, **19**, 43–51.

Sereno, P.C. (1990). Psittacosauridae. In *The Dinosauria*, ed. D.B. Weishampel, P. Dodson & H. Osmólska, pp. 579–92. Berkeley: University of California Press.

Sereno, P.C., Beck, A.L., Dutheil, D., Grado, B., Larsson, H.C.E., Lyon, G.H., Marcot, J.D., Rauhut, O.W.M., Sadleir, R.W., Sidor, C.A., Varricchio, D., Wilson, G.P. & Wilson, J.A. (1988). A long-snouted predatory dinosaur from Africa and the evolution of spinosaurids. *Science*, **282**, 1298–302.

Sereno, P.C, Dutheil, D.B., Iarochene, M., Larsson, H.C.E., Lyon, G.H., Magwene, P.M., Sidor, C.A., Varricchio, D.J. & Wilson, J.A. (1996). Predatory dinosaurs from the Sahara and Late Cretaceous faunal differentiation. *Science*, **272**, 986–91.

Shackleton, N.J. (1986). Paleogene stable isotope events. *Palaeogeography, Palaeoclimatology, Palaeoecology*, **57**, 91–102.

Shackleton, N.J., Berger, A. & Pettier, W.R. (1990). An alternative astronomical calibration of the lower Pleistocene timescale based on ODP Site 677. *Transactions of the Royal Society of Edinburgh, Earth Sciences*, **81**, 251–61.

Shackleton, N.J., Imbrie, J. & Pisias, N.G. (1988). The evolution of oceanic oxygen-isotope variability in the North Atlantic over the past three million years. *Philosophical Transactions of the Royal Society of London*, B, **318**, 679–88.

Shackleton, N.J. and Leg 81 Shipboard Scientific Party (1984). Oxygen isotopic calibration of the onset of ice-rafting and history of glaciation in the North Atlantic Region. *Nature*, **307**, 620–3.

Shackleton, N.J. & Opdyke, N.D. (1973). Oxygen isotope and palaeomagnetic stratigraphy of equatorial Pacific core V28–238: oxygen isotope temperatures and ice volumes on a 10^5 and 10^6 year scale. *Quaternary Research*, **3**, 39–55.

Shackleton, N.J. & Opdyke, N.D. (1977). Oxygen isotopic and paleomagnetic evidence for early Northern Hemisphere glaciation. *Nature*, **270**, 216–19.

Shafik, S. (1990). Late Cretaceous nannofossil biostratigraphy and biogeography of the Australian western margin. *BMR, Geology and Geophysics*, Report **295**, 164pp.

Sharpton, V.L. & Ward, P.D. (ed.) (1990). *Global Catastrophies in Earth History: an Interdisciplinary Conference on Impacts, Volcanism, and Mass Mortality.* Geological Society of America Special Paper, **247**, 1–631.

Shchepetov, S.V., Herman, A.B. & Belaya B.V. (1992). Mid-Cretaceous flora of the right bank of the Anadyr River: stratigraphic setting, systematic composition, atlas of fossil plants). SVKNII DVOAN SSSR, Magadan, 166pp. [in Russian].

Shearman, D.J. & Smith, A.J. (1985). Ikaite, the parent mineral of jarrowite-type pseudo-morphs. *Proceedings of the Geologists' Association*, **96**, 305–14.

Sheehan, P.M. (1985). Reefs are not so different – they follow the evolutionary pattern of level-bottom communities. *Geology*, **13**, 46–9.

Sher, A.V., Kaplina, T.N., Giterman, R.E., Lozhkin, A.V., Archangelov, A.A., Kiselyov, S.V., Kouznetsov, Y.V., Virina, E.I. & Zazhigin, V.S. (1979). Scientific excursion on problems 'late Caenozoic of the Kolyma Lowland'. *XIV Pacific Science Congress*, NAUKA, Moscow, pp. 1–116.

Shipman, P., Walker, A., Van Couvering, J.A., Hooker, P.J. & Miller, J.A. (1981). The Fort Ternan hominoid site, Kenya: Geology, age, taphonomy and palaeoecology. *Journal of Human Evolution*, **10**, 49–72.

Shirai, S. (1996). Phylogenetic interrelationships of neoselachians (Chondrichthyes, Euselachii). In *Interrelationships of Fishes*, ed. M. Stiassny, G.D. Johnson & L. Parenti, pp. 9–34. New York: Academic Press.

Shoemaker, E.M., Wolfe, R.F. & Shoemaker, C.S. (1990). Asteroid and comet flux in the neighbourhood of the Earth. In *Global Catastrophies in Earth History: an Interdisciplinary Conference on Impacts, Volcanism, and Mass Mortality*, ed. V.L. Sharpton & P.D. Ward, P.D. Geological Society of America Special Paper, **247**, 155–170.

Shotton, F.W., Blundel, P.J. & Williams, R.E.G. (1970). Birmingham University radiocarbon dates IV. *Radiocarbon*, **12**, 141–56.

Shreeve, J. (1996). Sunset on the savanna. *Discover* July 1996, 116–25.

Siesser, W.G. (1995). Paleoproductivity of the Indian Ocean during the Tertiary period. *Global and Planetary Change*, **11**, 71–88.

Signor, P.W. III (1985). Real and apparent trends in species richness through time. In *Phanerozoic Diversity Patterns*, ed. J.W. Valentine, pp. 129–50. Princeton: Princeton University Press.

Signor, P.W. III (1990). The geologic history of diversity. *Annual Review of Ecology and Systematics*, **21**, 509–39.

Skelton, P.W. & Benton, M.J. (1993). Mollusca: Rostroconchia, Scaphopoda and Bivalvia. In *The Fossil Record 2*, ed. M.J. Benton, pp. 237–63. London: Chapman & Hall.

Skelton, P.W., Crame, J.A., Morris, N.J. & Harper, E.M. (1990). Adaptive divergence and taxonomic radiation in post-Palaeozoic bivalves. In *Major Evolutionary Radiations*, ed. P.D. Taylor & G.P. Larwood. Systematics Association Special Volume No. **42**, 91–117.

Skelton, P.W. & Donovan, D.K. (1997). Middle Cretaceous reef collapse linked to ocean heat transport: comment and reply. *Geology*, **25**, 477.

Skelton, P.W., Gili, E., Rosen, B.R. & Valldeperas, F.X. (1997). Corals and rudists in the Late Cretaceous: a critique of the hypothesis of competitive displacement. In *VII International Symposium on Fossil Cnidaria and Porifera, Madrid, Spain, September 12–15, 1995. Boletin de la Real Sociedad Espaniola de Historia Naturel. Seccion Geologica*, **92**, 225–39.

Skelton, P.W. & Wright, V.P. (1987). A Carribean rudist bivalve in Oman: island hopping across the Pacific in the late Cretaceous. *Palaeontology*, **30**, 505–29.

Sliter, W.V. (1995). Cretaceous planktonic foraminfers from sites 865, 866 and 869: a synthesis of Cretaceous pelagic sedimentation in the Central Pacific Ocean Basin. *Proceedings of the Ocean Drilling Program, Scientific Results*, **143**, 15–30.

Sliter, W.V. & Brown, G.R. (1993). Shatsky Rise: seismic stratigraphy and sedimentary record of Pacific paleoceanography since the Early Cretaceous. *Proceedings of the Ocean Drilling Program, Scientific Results*, **132**, 3–13.

Sloan, L.C. & Barron, E.J. (1992). Paleogene climatic evolution: a climate model investigation of the influence of continental elevation and sea surface temperature upon continental climate. In *Eocene–Oligocene Climatic and Biotic Evolution*, ed. D.R. Prothero & W.A. Berggren, pp. 202–17. Princeton: Princeton University Press.

Smetacek, V., Bathmann, U., Nöthig, E-M. & Scharek, R. (1991). Coastal Eutrophication: Causes and Consequences. In *Ocean Margin Processes in Global Change*, ed. R.F.C. Mantoura, J-M. Martin & R. Wollast, pp. 251–79. Chichester, UK: John Wiley.

Smiley, C.J. (1966). Cretaceous floras from the Kuk River Area, Alaska, Stratigraphic and climatic interpretations. *Geological Society of America Bulletin*, **77**, 1–14.

Smiley, C.J. (1967). Paleoclimatic interpretations of some Mesozoic floral sequences. *American Association of Petroleum Geologists Bulletin*, **51**, 849–63.

Smiley, C.J. (1969a). Cretaceous floras of the Chandler-Colville Region, Alaska – Stratigraphy and preliminary floristics. *American Association of Petroleum Geologists Bulletin*, **53**, 482–502.

Smiley, C.J. (1969b). Floral zones and correlations of Cretaceous Kukpowruk and Corwin Formations, Northwestern Alaska. *American Association of Petroleum Geologists Bulletin*, **53**, 2079–93.

Smit, J. (1996). The K/T boundary Chicxulub impact event: a review. In *The Role of Impact Processes in the Geological and Biological Evolution of Planet Earth*, ed. K. Drobne, S. Gorican & B. Kotnik. Scientific Research Centre, SAZU, Ljubljana, pp. 83–84.

Smith, A.B. (1984). *Echinoid Palaeobiology*. London: Allen & Unwin.

Smith, A.B. (1994). *Systematics and the Fossil Record*. Oxford: Blackwell.

Smith, A.B. (1995). Late Campanian–Maastrichtian echinoids from the United Arab Emirates–Oman border region. *Bulletin of the Natural History Museum London (Geology)*, **51**, 121–240.

Smith, A.B. & Jeffery, C.H. (1998). Selectivity of extinction among sea urchins at the end of the Cretaceous Period. *Nature*, **392**, 69–71.

Smith, A.B., Morris, N.J., Gale, A.S. & Kennedy, W.J. (1995a). Late Cretaceous carbonate platform faunas of the United Arab Emirates–Oman border region. *Bulletin of the Natural History Museum London (Geology)*, **51**, 91–119.

Smith, A.B., Morris, N.J., Gale, A.S. & Rosen, B.R. (1995b). Late Cretaceous (Maastrichtian) echinoid–mollusc–coral assemblages and palaeoenvironments from a Tethyan carbonate platform succession, northern Oman Mountains. *Palaeogeography, Palaeoclimatology, Palaeoecology*, **119**, 155–68.

Smith, A.B. & Patterson, C. (1988). The influence of taxonomic method on the perception of patterns of evolution. *Evolutionary Biology*, **23**, 127–216.

Smith, A.B., Paul, C.R.C., Gale, A.S. & Donovan, S.K. (1988). Cenomanian and Lower Turonian echinoderms from Wilmington, south-east Devon, England. *Bulletin of the British Museum (Natural History), Geology Series*, **42**, 1–245.

Smith, A.B. & Wright, C.W. (1989–1996). British Cretaceous Echinoids. *Palaeontographical Society Monograph*.

Smith, A.G. (1970). The influence of Mesolithic and Neolithic Man on the British Vegetation. In *Studies in the Vegetational History of the British Isles*, ed. D. Walker & R.G. West, pp. 81–96. Cambridge: Cambridge University Press.

Smith, A.G. (1984). Newferry and the Boreal–Atlantic transition. *New Phytologist*, **98**, 35–55.

Smith, A.G. & Cloutman, E.W. (1988). Reconstruction of Holocene vegetation history in three dimensions at Waun-Fignen-Felen, an upland site in Wales. *Philosophical Transactions of the Royal Society of London*, B, **322**, 159–219.

Smith, A.G., Smith, D.G. & Funnell, B.M. (1994). *Atlas of Mesozoic and Cenozoic Coastlines*. Cambridge: Cambridge University Press.

Smith, G.R. (1992). Phylogeny and biogeography of the Catostomidae, freshwater fishes of North America and Asia. In *Systematics, Historical Ecology and North American Freshwater Fishes*, ed. R.L. Mayden, pp. 778–826. Stanford: Stanford University Press.

Smith, G.R. & Koehn, R.K. (1971). Phenetic and cladistic studies of biochemical and mor-phological characteristics of *Catostomus*. *Systematic Zoology*, **20**, 282–97.

Snyder, S.W. (1988). Synthesis of biostratigraphic and paleoenvironmental interpretations of Miocene sediments from the shallow subsurface of Onslow Bay, North Carolina continental shelf. In *Micropaleontology of Miocene Sediments in the Shallow Subsurface of Onslow Bay, North Carolina Continental Shelf*, ed. S.W. Snyder. Cushman Foundation for Foraminiferal Research, Special Publication, **25**, 179–89.

Sohl, N.F. (1964). Neogastropoda, Opisthobranchia and Basommatophora from the Ripley, Owl Creek, and Prairie Bluff Formations. *U.S. Geological Survey Professional Paper*, **331**, 153–344.

Sohl, N.F. (1987). Presidential Address. Cretaceous gastropods: contrasts between Tethys and the temperate provinces. *Journal of Paleontology*, **61**, 1085–111.

Solem, A. & Yochelson, E.L. (1979). North American land snails with a summary of other Paleozoic non-marine snails. *U.S. Geological Survey Professional Paper*, **1072**, 1–39.

Sorbini, L. & Tirapelle Rancan, R. (1979). Messinian fossil fish of the Mediterranean. *Palaeogeography, Palaeoclimatology, Palaeoecology*, **29**, 143–54.

Speijer, R.P. & Van der Zwaan, G.J. (1996). Extinction and survivorship of southern Tethyan benthic foraminifera across the Cretaceous/Palaeogene boundary. In *Benthic Recovery from Mass Extinction Events*, ed. M.B. Hart. Geological Society Special Publication No. **102**, 245–58.

Spencer Davies, P. (1992). Endosymbiosis in marine cnidarians. In *Plant–Animal Interactions in the Marine Benthos*, ed. D.M. John, S.J. Hawkins & J.H. Price. Systematics Association Special Volume No. **46**, 511–40.

Spicer, R.A. (1980). The importance of depositional sorting to the biostratigraphy of fossil plants. In *Biostratigraphy of Fossil Plants*, ed. D.L. Dilcher & T.N. Taylor, pp. 171–83. Stroudsburg, PA: Dowden, Hutchinson and Ross Inc.

Spicer, R.A. (1981). The sorting and deposition of allochthonous plant material in a modern environment at Silwood Lake, Silwood Park, Berkshire, England. *U.S. Geological Survey Professional Paper*, **1143**, 1–77.

Spicer, R.A. (1987). The significance of the Cretaceous Flora of Northern Alaska for the reconstruction of the Climate of the Cretaceous. *Geologisches Jahrbuch*, **96**, 265–92.

Spicer, R.A. (1990). Feuding over the nearest living relatives. *Journal of Biogeography*, **17**, 335–6.

Spicer R.A. & Chapman J.L. (1990). Climate change and the evolution of high latitude terrestrial vegetation and floras. *TREE*, **5**, 279–84.

Spicer, R.A. & Corfield, R.M. (1992). Review of terrestrial and marine climates in the Cretaceous with implications for modelling the 'Greenhouse Earth'. *Geological Magazine*, **129**, 169–80.

Spicer, R.A. & Herman, A.B. (1997). New Quantitative Palaeoclimate data for the Late Cretaceous Arctic: evidence for a warm polar ocean. *Palaeogeography, Palaeoclimatology, Palaeoecology*, **128**, 227–51.

Spicer, R.A., Herman, A.B. & Valdes, P.J. (1996). Mid and late Cretaceous Climate of Asia and Northern Alaska using CLAMP analysis. In *Memorial Conference Dedicated to Vsevolod Andreevich Vakhrameev (Abstracts and Proceedings) November 13–14th, 1996* M.: GEOS, 62–7.

Spicer, R.A. & Hill, C.R. (1979). Principal components and Correspondence Analyses of quantitative data from a Jurassic plant bed. *Review of Paleobotany and Palynology*, **28**, 273–99.

Spicer, R.A. & Parrish, J.T. (1986). Paleobotanical evidence for cool north polar climates in middle Cretaceous (Albian–Cenomanian) time. *Geology*, **14**, 703–6.

Spicer, R.A. & Parrish, J.T. (1990). Late Cretaceous–Early Tertiary palaeoclimates of northern high latitudes: a quantitative view. *Journal of the Geological Society, London*, **147**, 329–41.

Spicer, R.A., Parrish, J.T. & Grant, P.R. (1992). Evolution of vegetation and coal forming environments in the Late Cretaceous of the north slope of Alaska: a model for polar coal deposition at times of global warmth. In *Controls on the Deposition of Cretaceous Coal*, ed. P.J. McCabe & J.T. Parrish. Geological Society of America Special Paper, **267**, 177–92.

Spicer, R.A., Rees, P.McA. & Chapman, J.L. (1993). Cretaceous phytogeography and climate signals. *Philosophical Transactions of the Royal Society of London*, B, **341**, 277–86.

Stanley, G.D., Jr. (1992). Tropical reef ecosystems and their evolution. In *Encyclopedia of Earth System Science*, pp. 375–88. New York: Academic Press.

Stanley, G.D., Jr. & Cairns, S.D. (1988). Constructional azooxanthellate coral communities: an overview with implications for the fossil record. *Palaios*, **3**, 233–42.

Stanley, G.D., Jr. & Swart, P.K. (1995). Evolution of the coral–zooxanthellae symbiosis during the Triassic: a geochemical approach. *Paleobiology*, **21**, 179–99.

Stanley, S.M. (1977). Trends, rates and patterns of evolution in the Bivalvia. In *Patterns of Evolution as Illustrated by the Fossil Record*, ed. A. Hallam, pp. 209–50. Amsterdam: Elsevier.

Stanley, S.M. (1985). Rates of evolution. *Paleobiology*, **11**, 13–26.

Stanley, S.M. (1986). Anatomy of a regional mass extinction: Plio-Pleistocene decimation of the western Atlantic bivalve fauna. *Palaios*, **1**, 17–36.

Stanley, S.M., Wetmore, K.L. & Kennett, J.P. (1988). Macroevolutionary differences between two major clades of Neogene planktonic foraminifera. *Paleobiology*, **14**, 235–49.

Stauffer, B., Lochbronner, E., Oeschger, H. & Schwander, J. (1988). Methane concentration in the glacial atmosphere was only half that of pre-industrial Holocene. *Nature*, **332**, 812–14.

Stauffer, P.H., Nishimura, S. & Batchelor, B.C. (1980). Volcanic ash in Malaya from a catastrophic eruption of Toba, Sumatra, 30,000 years ago. In *Physical Geology of the Indonesian Island Arcs*, ed. S. Nishimura, pp. 156–64. Kyoto: Kyoto University Press.

Stearn, C.W. (1982). The shapes of Paleozoic and modern reef-builders: a critical review. *Paleobiology*, **8**, 228–41.

Stearns, S.C. (1992). *The Evolution of Life Histories*. Oxford: Oxford University Press.

Stebbins, G.L. (1974). *Flowering Plants; Evolution above the Species Level*. Cambridge MA: Harvard University Press.

Stebbins, G.L. (1981). Why are there so many species of flowering plants? *Bioscience*, **31**, 573–7.

Stehli, F.G., McAlester, A.L. & Helsley, C.E. (1967). Taxonomic diversity of Recent bivalves and some implications for geology. *Geological Society of America Bulletin*, **78**, 455–66.

Steininger, F.F., Bernor, R.L. & Fahlbusch, V. (1990). European Neogene marine/continental chronologic correlations. In *European Neogene Mammal Chronology*, ed. E.H. Lindsay, V. Fahlbusch & P. Mein, pp. 15–46. New York: Plenum Press.

Stewart, W.N. & Rothwell, G.W. (1993). *Paleobotany and Evolution of Plants*, 2nd edn. Cambridge: Cambridge University Press.

Stille, P. (1992). Nd-Sr isotopic evidence for dramatic changes of paleocurrents in the Atlantic Ocean during the past 80 m.y. *Geology*, **20**, 387–90.

Stockey, R.A., Hoffman, G.L. & Rothwell, G.R. (1997). The fossil monocotyledon *Limnobiophyllum scutatum*: resolving the phylogeny of the Lemnaceae. *American Journal of Botany*, **84**, 355–68.

Stoll, H.M. & Schrag, D.P. (1996). Evidence for glacial control of rapid sea-level changes in the Early Cretaceous. *Science*, **272**, 1771–74.

Storch, G. & Schaarschmidt, F. (1992). The Messel fauna and flora: a biogeographical puzzle. In *Messel, an Insight into the History of Life and of the Earth*, ed. S. Schaal & W. Ziegler, pp. 293–7. English edn. Oxford: Clarendon Press.

Storrs, G. & Gower, D. (1993). The earliest possible choristodere (Diapsida) and gaps in the fossil record of semi-aquatic reptiles. *Journal of the Geological Society, London*, **150**, 1103–7.

Stott, L.D. (1992). Higher temperatures and lower oceanic pCO_2: a climate enigma at the end of the Paleocene epoch. *Paleoceanography*, **7**, 395–404.

Stott, L.D. & Kennett, J.P. (1990). The paleoceanographic and climatic signature of the Cretaceous/Paleogene boundary in the Antarctic: stable isotopic results from ODP Leg 113. *Proceedings of the Ocean Drilling Program, Scientific Results*, **113**, 829–48.

Stranks, L. (1996). Physiognomic and Taphonomic studies in New Zealand and Australia; implications for the use of palaeobotany as a tool for palaeoclimate estimation. Unpublished DPhil thesis, University of Oxford.

Stranks, L. & England, P.C. (1997). The use of resemblance functions in the measurement of climatic parameters from the physiognomy of woody dicotyledons. *Palaeogeography, Palaeoclimatology, Palaeoecology*, **131**, 15–28.

Strathmann, R.R. (1978). The evolution and loss of feeding larval stages of marine invertebrates. *Evolution*, **32**, 894–906.

Street-Perrott, F.A. & Perrott, R.A. (1990). Abrupt climatic fluctuations in the tropics: the influence of Atlantic Ocean circulation. *Nature*, **343**, 607–12.

Stringer, C. (1986). The credibility of *Homo habilis*. In *Major Topics in Primate and Human Evolution*, ed. B. Wood, L. Martin & P. Andrews, pp. 266–94. Cambridge: Cambridge University Press.

Stuart, A.J. (1982). *Pleistocene Vertebrates in the British Isles*. London and New York: Longman.

Stucky, R.K. & McKenna, M.C. (1993). Mammalia. In *The Fossil Record 2*, ed. M. Benton, pp. 739–71. London: Chapman & Hall.

Sudre, J., Russell, Louis, P. *et al.* (1983). Les artiodactyles de l'Eocène inférieur d'Europe. *Bulletin du Muséum National d'Histoire Naturelle, Paris*, (4C) **5**, 281–333, 339–65.

Sullivan, R.M. (1987). A reassessment of reptilian diversity across the Cretaceous–Tertiary Boundary. *Contributions in Science, Los Angeles County Museum*, No **391**, 26 pp.

Surfer (1987). Golden Software, Golden, CO.

Sutcliffe, A.J. (1960). Joint Mitnor Cave, Buckfastleigh. *Transactions of the Torquay Natural History Society*, **13**, 1–26.

Sutcliffe, A.J. (1985). *On the Track of Ice Age Mammals*. London: British Museum (Natural History).

Sweet, A.R. & Braman, D.R. (1992). The K–T boundary and contiguous strata in western Canada: interactions between palaeoenvironments and palynological assemblages. *Cretaceous Research*, **13**, 31–79.

Swisher, C.C. III, Curtis, G.H., Jacob T., Getty, A.G., Suprijo, A. & Widiasmoro (1994). Age of the earliest known Hominids in Java, Indonesia. *Science*, **263**, 1118–21.

Swisher, C.C. III, Grajales-Nishimura, J.M., Montanari, A. *et al.* (1992). Coeval $^{40}Ar/^{39}Ar$ ages of 65.0 million years ago from Chicxulub crater melt rock and Cretaceous–Tertiary boundary tectites. *Science*, **257**, 954–8.

Swisher, C.C. III, Rink, W.J., Antón, S.C., Schwarcz, H.P., Curtis, G.H., Suprijo, A. & Widiasmoro (1996). Latest *Homo erectus* of Java: potential contemporaneity with *Homo sapiens* in Southeast Asia. *Science*, **274**, 1870–4.

Szalay, F. & Delson, E. (eds.) (1979). *Evolutionary History of the Primates*. New York: Academic Press.

Takahata, N. (1995). A genetic perspective on the origin and history of humans. *Annual Review of Ecology and Systematics*, **26**, 343–72.

Takhtajan, A. (1969). *Flowering Plants: Origin and Dispersal*. Edinburgh: Oliver and Boyd.

Talent, J.A. (1988). Organic reef-building: episodes of extinction and symbiosis? *Senckenbergiana Lethaea*, **69**, 315–68.

Tappan, H. & Loeblich, A.R., Jr. (1988). Foraminiferal evolution, diversification, and extinction. *Journal of Paleontology*, **62**, 695–714.

Tarduno, J.A., Sliter, W.V., Kroenke, L. *et al.* (1991). Rapid formation of Ontong Java Plateau by Aptian mantle plume volcanism. *Science*, **254**, 399–403.

Taylor, B. & Natland, J. (ed.) (1995). *Active Margins and Marginal Basins of the Western Pacific*. American Geophysical Union, Geophysical Monograph, **88**.

Taylor, J.D. (1973). The structural evolution of the bivalve shell. *Palaeontology*, **16**, 519–34.

Taylor, J.D. (1989). The diet of coral-reef Mitridae (Gastropoda) from Guam; with a review of other species in the family. *Journal of Natural History*, **23**, 261–78.

Taylor, J.D., Cleevely, R.J. & Morris, N.J. (1983). Predatory gastropods and their activities in the Blackdown Greensand (Albian) of England. *Palaeontology*, **26**, 521–53.

Taylor, J.D. & Jensen, K.R. (1962). Massive abundance of *Philine orientalis* A.Adams in the sublittoral benthos of Hong Kong, 1889. In *The Marine Flora and Fauna of Hong Kong and Southern China III*, ed. B. Morton, pp. 517–26. Hong Kong: Hong Kong University Press.

Taylor, J.D., Kantor, Y.I. & Sysoev, A.V. (1993). Foregut anatomy, feeding mechanisms, relationships and classification of the Conoidea (=Toxoglossa) (Gastropoda). *Bulletin of the Natural History Museum, London (Zoology)*, **59**, 125–70.

Taylor, J.D., Morris, N.J. & Taylor, C.N. (1980). Food specialization and the evolution of predatory gastropods. *Palaeontology*, **23**, 375–409.

Taylor, J.D. & Taylor, C.N. (1977). Latitudinal distribution of predatory gastropods on the eastern Atlantic shelf. *Journal of Biogeography*, **4**, 73–81.

Taylor, K.C., Lamorey, G.W., Doyle, G.A., Alley, R.B., Grootes, P.M., Mayewski, P.A., White, J.W.C. & Barlow, L.K. (1993). The 'flicking switch' of late Pleistocene climate change. *Nature*, **361**, 432–5.

Taylor, P.D. (1988). Major radiation of cheilostome bryozoans: triggered by the evolution of a new larval type? *Historical Biology*, **1**, 45–64.

Taylor, P.D. (1993). Bryozoa. In *The Fossil Record 2*, ed. M.J. Benton, pp. 465–89. London: Chapman & Hall.

Taylor, P.D. (1994a). An early cheilostome bryozoan from the Upper Jurassic of Yemen. *Neues Jahrbuch für Geologie und Paläontologie Abhandlungen*, **191**, 331–44.

Taylor, P.D. (1994b). Evolutionary palaeoecology of symbioses between bryozoans and hermit crabs. *Historical Biology*, **9**, 157–205.

Taylor, P.D. & Larwood, G.P. (1990). Major evolutionary radiations in the Bryozoa. In *Major Evolutionary Radiations*, ed. P.D. Taylor & G.P. Larwood, pp. 209–33. Oxford: Clarendon Press.

Tchechmedjieva, V.L. (1986). Paléoécologie des Madréporaires du Crétacé supérieur dans le Srednogorié de l'Ouest (Bulgarie occidentale). *Geologica Balcanica*, **16**, 55–81.

Tchernov, E., Polcyn, M.J. & Jacobs, L.L. (1996). Snakes with legs: the Cenomanian fauna of 'Ein Yabrud, Israel. *Journal of Vertebrate Paleontology*, **16** (3, suppl.), 68A.

Tedford, R.H., Flynn, L.J., Zhanxiang, Q., Opdyke, N.D. & Downs, W.R. (1991). Yushe Basin, China; palaeomagnetically calibrated mammalian biostratigraphic standards for the late Neogene of eastern Asia. *Journal of Vertebrate Paleontology*, **11**, 519–26.

Ter Braak, C.J.F. (1986). Canonical Correspondence Analysis: a new eigenvector technique for multivariate direct gradient analysis. *Ecology*, **67**, 1167–79.

Ter Braak, C.J.F. (1987–1992). *CANOCO – a FORTRAN program for Canonical Correspondence Ordination*. Ithaca, NY: Microcomputer Power.

Tewari, A., Hart, M.B. & Watkinson, M.P. (1996). Foraminiferal recovery after the mid-Cretaceous oceanic anoxic events (OAEs) in the Cauvery Basin, Southeast India. In *Biotic Recovery from Mass Extinction Events*, ed. M.B. Hart. Geological Society Special Publication No. **102**, 237–44.

Thewissen, J.G.M. (1990). Comment on 'Paleontological view of the ages of the Deccan Traps, the Cretaceous/Tertiary boundary, and the India–Asia collision'. *Geology*, **18**, 185.

Thierry, J. & Néraudeau, D. (1994). Variations in Jurassic echinoid biodiversity at ammonite zone levels: stratigraphic and palaeoecological significance. In *Echinoderms through Time*, ed. B. David, A. Guille, J-P. Feral & M. Roux, pp. 901–9. Rotterdam: A.A. Balkema.

Thierstein, H.R. (1976). Mesozoic calcareous nannoplankton biostratigraphy of marine sediments. *Marine Micropaleontology*, **1**, 325–62.

Thierstein, H.R., Geitzenauer, K.R., Molfino, B. & Shackleton, N.J. (1977). Global synchroneity of late Quaternary coccolith datum levels: validation by oxygen isotopes. *Geology*, **5**, 400–4.

Thomas, E. (1990*a*). Late Cretaceous–Early Eocene mass extinction in the deep sea. In *Global Catastrophes in Earth History*, ed. V.L. Sharpton & P.D. Ward. Geological Society of America, Special Paper, **247**, 481–95.

Thomas, E. (1990*b*). Late Cretaceous through Neogene deep-sea benthic foraminifers, Maud Rise, Weddell Sea, Antarctica, *Proceedings of the Ocean Drilling Program, Scientific Results*, **113**, 571–94.

Thomas, E. (1992). Cenozoic deep-sea circulation: evidence from deep-sea benthic foraminifera. In *The Antarctic Paleoenvironment: A Perspective on Global Change*, ed. J.P. Kennett & D.A. Warnke. American Geophysical Union, Geophysical Monograph, **56**, 141–65.

Thomas, E. & Shackleton, N.J. (1996). The Late Paleocene benthic foraminiferal extinction and stable isotope anomalies. In *Correlation of the Early Paleogene in Northwest Europe*, ed. R.W.O.B. Knox, R. Corfield & R.E. Dunay. Geological Society Special Publication No. **101**, 401–41.

Thomas, H. (1979). Le rôle de barrière écologique de la ceinture saharo-arabique au Miocène: arguments paléontologiques. *Bulletin du Muséum national d'Histoire naturelle, Paris*, **1**, 127–35.

Thomas, H. (1985). The early and middle Miocene land connection of the Afro-Arabian plate and Asia: a major event for hominid dispersal? In *Ancestors, The Hard Evidence*, ed. E. Delson, pp. 42–50. New York: Alan R. Liss.

Thomas, H., Sen, S., Khan, M., Battail, B. & Ligabue, G. (1982). The Lower Miocene fauna of Al-Sarrar (Eastern Province, Saudi Arabia). *ATLAL, The Journal of Saudi Arabian Archaeology*, **5**, 109–36.

Thompson, R.D. (1992). The changing atmosphere and its impact on planet Earth. In *Environmental Issues in the 1990s*, ed. A.M. Mannion & S.R. Bowlby, pp. 61–96. Chichester, UK: John Wiley.

Thurman, H.V. (1985). *Introductory Oceanography*. Columbus, OH: Charles E. Merrill Publishing Co.

Tiedemann, R., Sarnthein, M. & Shackleton, N.J. (1994). Astronomic timescale for the Pliocene Atlantic d^{18}O and dust flux records of Ocean Drilling Program Site 659. *Paleoceanography*, **9**, 619–38.

Tiffney, B.H. (1985). Perspectives on the origin of the floristic similarity between eastern Asia and Eastern North America. *Journal of the Arnold Arboretum*, **66**, 73–94.

Tiffney, B.H. (1986). Fruit and seed dispersal and the evolution of the Hamamelidae. *Annals of the Missouri Botanical Garden*, **73**, 394–416.

Tillier, S., Masselot, M. & Tillier, A. (1996). Phylogenetic relationships of the pulmonate gastropods from rRNA sequences, and tempo and age of the stylommatophoran radiation. In *Origin and Evolutionary Radiation of the Mollusca*, ed. J.D. Taylor, pp. 267–84. Oxford: Oxford University Press.

Toumarkine, M. & Luterbacher, H. (1985). Paleocene and Eocene planktic foraminifera. In *Plankton Stratigraphy*, ed. H.M. Bolli, J.B. Saunders & K. Perch-Nielsen, pp. 87–154. Cambridge: Cambridge University Press.

Traverse, A. (1988). *Palaeopalynology*. London: Unwin Hyman.

Trinkaus, E., Stringer, C., Ruff, C. *et al.* (1996). The Boxgrove tibia. *American Journal of Physical Anthropology*, Meetings Supplement **22**, 230–1.

Tumanova, T. (1983). [The first ankylosaur from the Lower Cretaceous of Mongolia.] *Trudy Sovmestnaya Sovetsko-Mongol'skaya Paleontologicheskaya Ekspediya*, **24,** 110–20 [In Russian].

Turner, A. (1992). Large carnivores and earliest European hominids: changing determinants of resource availability during the Lower and Middle Pleistocene. *Journal of Human Evolution*, **22**, 109–26.

Turner, S. & Young, G.C. (1987). Shark teeth from the Early-Middle Devonian Cravens Peak Beds, Georgina Basin, Queensland. *Alcheringa*, **11**, 233–44.

Uhen, M.D. & Gingerich, P.D. (1995). Evolution of *Coryphodon* (Mammalia, Pantodonta) in the late Paleocene and early Eocene of northwestern Wyoming. *Contributions from the Museum of Paleontology, University of Michigan*, **29**, 259–89.

Underwood, M.B., Ballance, P.F., Clift, P., Hiscott, R.N., Marsaglia, K.M., Pickering, K.T. & Reid, R.P. (1995). Sedimentation in Forearc basins, trenches, and collision zones of the Western Pacific: a summary of results from the Ocean Drilling Program, WPAC. In *Active Margins and Marginal Basins of the Western Pacific*, ed. B. Taylor & J. Natland. American Geophysical Union, Geophysical Monograph, **88**, 315–53.

Ungar, P. (1996). Dental microwear of European Miocene catarrhines: evidence for diets and tooth use. *Journal of Human Evolution*, **31**, 335–66.

Upchurch, G.R., Jr. (1984). Cuticle evolution in Early Cretaceous angiosperms from the Potomac Group of Virginia and Maryland. *Annals of the Missouri Botanical Garden*, **71**, 522–50.

Upchurch, G.R. & Wolfe, J.A. (1987). Mid-Cretaceous to early Tertiary vegetation and climate: evidence from fossil leaves and wood. In *The Origins of Angiosperms and their Biological Consequences*, ed. E.M. Friis, W.G. Chaloner & P.R. Crane, pp. 75–105. Cambridge: Cambridge University Press.

Vakhrameev, V.A. (1991). *Jurassic and Cretaceous Floras and Climates of the Earth*. Cambridge: Cambridge University Press.

Valentine, J.W. (1973). *Evolutionary Paleoecology of the Marine Biosphere*. Englewood Cliffs, NJ: Prentice-Hall.

Valentine, J.W. (1989). How good was the fossil record? Clues from the Californian Pleistocene. *Paleobiology*, **15**, 83–94.

Valentine, J.W. Foin, T.C. & Peart, D. (1978). A provincial model of Phanerozoic marine diversity. *Paleobiology*, **4**, 55–66.

Van Bergen, P.F., Collinson, M.E., Briggs, D.E.G. *et al.* (1995). Resistant biomacromolecules in the fossil record. *Acta Botanica Neerlandica*, **44**, 319–42.

Van Couvering, J.A.H. (1980). Community evolution in East Africa during the late Cenozoic. In *Fossils in the Making, Vertebrate Taphonomy, and Paleoecology*, ed. A.K. Behrensmeyer & A. Hill, pp. 272–98. Chicago: University of Chicago Press.

Van Valen, L.A. (1973). A new evolutionary law. *Evolutionary Theory*, **1**, 1–3.

Varney, M. (1996). The marine carbonate system. In *Oceanography: An Illustrated Guide*, ed. C.P. Summerhayes & S.A. Thorpe, pp. 182–94. London: Manson Publishing.

Vaughan, A.P.M. (1995). Circum-Pacific mid-Cretaceous deformation and uplift: a super-plume-related event? *Geology*, **23**, 491–4.

Vaughan, T.W. & Wells, J.W. (1943). Revision of the suborders families, and genera of the Scleractinia. *Geological Society of America Special Paper*, **44**, 1–363.

Vávra, N. (1980). Tropische Faunenelemente in den Bryozoenfaunen des Badenien (Mittelmiozän) der Zentralen Paratethys. *Sitzungsberichten der Österreichischen Akademie der Wissenschaften. Mathematisch-Naturwissenschaftliche Klasse*, **189**, 49–63.

Vermeij, G.J. (1977). The Mesozoic marine revolution: evidence from snails, predators and grazers. *Paleobiology*, **3**, 245–58.

Vermeij, G.J. (1978). *Biogeography and Adaptation*. Cambridge: Harvard University Press.

Vermeij, G.J. (1987). *Evolution and Escalation. An Ecological History of Life*. Princeton: Princeton University Press.

Vermeij, G.J. (1991). Anatomy of an invasion: the trans-Arctic interchange. *Paleobiology*, **17**, 335–56.

Vermeij, G.J. (1993*a*). *A Natural History of Shells*. Princeton: Princeton University Press.

Vermeij, G.J. (1993*b*). *Spinucella*, a new genus of Miocene to Pleistocene muricid gastropods from the eastern Atlantic. *Contributions to Tertiary and Quaternary Geology*, **30**, 19–27.

Vermeij, G.J. & Rosenberg, G. (1993). Giving and receiving: the tropical Atlantic as donor and recipient region for invading species. *American Malacological Bulletin*, **10**, 181–94.

Vermeij, G.J., Schindel, D.E. & Zipser, E. (1981). Predation through geological time: evidence from gastropod shell repair. *Science*, **214**, 1024–26.

Veron, J.E.N. (1995). *Corals in Space in Time. The Biogeography and Evolution of the Scleractinia*. Sydney: University of New South Wales Press.

Veron, J.E.N. & Kelley, R. (1988). Species stability in reef corals of Papua New Guinea and the Indo Pacific. *Memoirs of the Association of Australasian Palaeontologists*, **6**, 1–69.

Veron, J.E.N., Odorico, D.M., Chen, C.A. & Miller, D.J. (1996). Reassessing evolutionary relationships of scleractinian corals. *Coral Reefs*, **15**, 1–9.

Veum, T., Jansen, E., Arnold, M., Beyer, I. & Duplessy, J-C. (1992). Water mass exchange between the North Atlantic and the Norwegian Sea during the past 28,000 years. *Nature*, **356**, 783–5.

Viles, H.A. & Goudie, A.S. (1990). Tufas, travertines and allied carbonate deposits. *Progress in Physical Geography*, **14**, 19–41.

Viskova, L.A. (1994). The dynamics of diversity of Gymnolaemata around the Cretaceous–Paleogene crisis. In *Fossil and Living Bryozoa of the Globe*, p. 61. Perm: All-Russian Paleontological Society.

Vogel, K. (1975). Endosymbiotic algae in rudists? *Palaeogeography, Palaeoclimatology and Palaeoecology*, **17**, 327–32.

Voigt, E. (1981). Upper Cretaceous bryozoan-seagrass association in the Maastrichtian of the Netherlands. In *Recent and Fossil Bryozoa*, ed. G.P. Larwood & C. Nielsen, pp. 281–98. Fredensborg: Olsen & Olsen.

Volk, T. (1989). Sensitivity of climate and atmospheric CO_2 to deep-ocean and shallow-ocean carbonate burial. *Nature*, **337**, 637–40.

Vrba, E.S. (1995*a*). On the connections between paleoclimate and evolution. In *Paleoclimate and Evolution, with Emphasis on Human Origins*, ed. E.S. Vrba, G.H. Denton, T.C. Partridge & L.H. Burckle, pp. 24–45. New Haven: Yale University Press.

Vrba, E.S. (1995*b*). The fossil record of African antelopes (Mammalia, Bovidae) in relation to human evolution and paleoclimate. In *Paleoclimate and Evolution, with Emphasis on Human Origins*, ed. E.S. Vrba, G.H. Denton, T.C. Partridge & L.H. Burckle, pp. 385–424. New Haven: Yale University Press.

Vrba, E., Denton, G.H., Partridge, T. & Burckle, L.H. (ed.) (1995). *Paleoclimate and Evolution, with Emphasis on Human origins*. New Haven: Yale University Press.

Wake, D.B. (1966). Comparative osteology and evolution of the lungless salamanders, family Plethodontidae. *Memoirs of the Southern California Academy of Sciences*, **4**, 1–111.

Wake, D.B. & Lynch, J.F. (1976). The distribution, ecology and evolutionary history of plethodontid salamanders in tropical America. *Science Bulletin of the Natural History Museum of Los Angeles County*, **25**, 1–65.

Walker, I.R. (1987). Chironomidae (Diptera) in Palaeoecology. *Quaternary Science Reviews*, **6**, 29–40.

Walker, I.R. (1995). Chironomids as indicators of past environmental change. In *The Chironomidae: Biology and Ecology of Non-biting Midges*, ed. P.S. Armitage, P.S. Cranston & L.C.V. Pinder. London: Chapman & Hall.

Walker, I.R. & Mathewes, R.W. (1987). Chironomids, lake trophic status and climate. *Quaternary Research*, **28**, 431–7.

Walker, I.R. & Mathewes, R.W. (1989). Much ado about dead Diptera. *Journal of Paleolimnology*, **2**, 19–22.

Walker, I.R., Smol, J.P., Engstrom, D.R. & Birks, H.J.B. (1991). An assessment of Chironomidae as quantitative indicators of past climatic change. *Canadian Journal of Fisheries and Aquatic Science*, **48**, 975–87.

Walker, I.R., Smol, J.P., Engstrom, D.R. & Birks, H.J.B. (1992). Aquatic invertebrates, climate, scale, and statistical hypothesis testing: a response to Hann, Warner, and Warwick. *Canadian Journal of Fisheries and Aquatic Science*, **49**, 1276–80.

Walker, I.R., Wilson, S.E. & Smol, J.P. (1995). Chironomidae (Diptera): quantitative paleo-salinity indicators for lakes in western Canada. *Canadian Journal of Fisheries and Aquatic Science*, **52**, 950–60.

Walker, J.W. (1976). Comparative pollen morphology and phylogeny of the ranalean complex. In *Origin and Early Evolution of Angiosperms*, ed. C.B. Beck, pp. 241–99. New York: Columbia University Press.

Walker, J.W. & Doyle, J.A. (1975). The bases of angiosperm phylogeny: palynology. *Annals of the Missouri Botanical Garden*, **62**, 664–723.

Walker, J.W. & Walker, A.G. (1984). Ultrastructure of Lower Cretaceous angiosperm pollen and the origin and early evolution of flowering plants. *Annals of the Missouri Botanical Garden*, **71**, 464–521.

Walker, M.J.C. (1995). Climatic changes in Europe during the last glacial/interglacial transition. *Quaternary International*, **28**, 63–70.

Wallace, A.R. (1876). *The Geographical Distribution of Animals with a Study of the Relations of Living and Extinct Faunas as Elucidating the Past Changes of the Earth's Surface*. London: Macmillan and Co.

Walter, H. (1979). *Vegetation of the Earth and Ecological Systems of the Geo-biosphere*, 2nd edn. New York: Springer-Verlag.

Ward, L.W. (1984). Stratigraphy and outcropping Tertiary beds along the Pamunkey River - Central Virginia Coastal Plain. In *Stratigraphy and Paleontology of the Outcropping Tertiary Beds in the Pamunkey River Region, Central Virginia Coastal Plain*, ed. L.W. Ward & K. Krafft, Guidebook for Atlantic Coastal Plain Geological Association 1984 Field Trip, Atlantic Coastal Plain Geological Association, pp. 11–77.

Ward, L.W. & Blackwelder, B.W. (1980). Stratigraphic revision of upper Miocene and lower Pliocene beds of the Chesapeake Group, middle Atlantic Coastal Plain. *U.S. Geological Survey Bulletin* **1482-D**, 1–61.

Ward, L.W. & Powars, D.S. (1991). Tertiary lithology and paleontology, Chesapeake Bay Region. In *Geologic Evolution of the Eastern United States*, ed. A. Schultz & E. Compton-Gooding, pp. 161–93. Virginia Museum of Natural History, Guidebook No. **2**.

Ward, W.C., Keller, G., Stinnesbeck, W. & Adatte, T. (1995). Yucatan subsurface stratigraphy: implications and constraints for the Chicxulub impact. *Geology*, **23**, 873–6.

Warén, A. & Bouchet, P. (1993). New records, species, genera and a new family of gastropods from hydrothermal vents and hydrocarbon seeps. *Zoologica Scripta*, **22**, 1–90.

Warheit, K.I. (1992). A review of the fossil seabirds from the Tertiary of the North Pacific: plate tectonics, paleooceanography, and faunal change. *Paleobiology*, **18**, 401–24.

Warner, B.G. & Hann, B.J. (1987). Aquatic invertebrates as paleoclimatic indicators. *Quaternary Research*, **28**, 427–30.

Warwick, W.F. (1980). Palaeolimnology of the Bay of Quinte, Lake Ontario: 2800 years of cultural influence. *Canadian Bulletin of Fisheries and Aquatic Sciences*, **206**, 1–118.

Warwick, W.F. (1985). Morphological abnormalities in Chironomidae (Diptera) larvae as measures of toxic stress in freshwater ecosystems: indexing antennal deformities in *Chironomus* Meigen. *Canadian Journal of Fisheries and Aquatic Sciences*, **42**, 1881–914.

Warwick, W.F. (1989). Chironomids, lake development and climate: a commentary. *Journal of Paleolimnology*, **2**, 15–17.

Warwick, W.F. (1991). Indexing deformities in ligulae and antennae of Procladius larvae (Diptera: Chironomidae): application to contaminant-stressed environments. *Canadian Journal of Fisheries and Aquatic Sciences*, **48**, 1151–66.

Watabe, N. & Wilbur, K.M. (1966). Effects of temperature on growth, calcification, and coccolith form in *Coccolithus huxleyi* (Coccolithineae). *Limnology and Oceanography*, **11**, 567–75.

Watkins, D.K. (1989). Nannoplankton productivity fluctuations and rhythmically-bedded pelagic carbonates of the Greenhorn Limestone (Upper Cretaceous). *Palaeogeography, Palaeoclimatology, Palaeoecology*, **74**, 75–86.

Watkins, D.K., Wise, S.W., Pospichal, J.J. & Crux, J. (1996). Upper Cretaceous calcareous nannofossil biostratigraphy and paleoceanography of the Southern Ocean. In *Microfossils and Oceanic Environments*, ed. A. Moguilevsky & R. Whatley, pp. 355–381. Aberystwyth: University of Wales Press.

Waton, P.V. (1982). Man's impact on the chalklands, some new pollen evidence. In *Archaeological Aspects of Woodland Ecology*, ed. S. Limbrey & M. Bell. *British Archaeological Reports, International Series*, **146**, 75–91.

Watson, J. (1988). The Cheirolepidiaceae. In *Origin and Evolution of Gymnosperms*, ed. C.B. Beck, pp. 382–447. New York: Columbia University Press.

Weaver, P.P.E. & Clement, B.M. (1986). Synchroneity of Pliocene planktonic foraminiferal datums in the North Atlantic. *Marine Micropaleontology*, **10**, 295–307.

Webb, G. (1996). Was Phanerozoic reef history controlled by the distribution of non-enzymatically secreted reef carbonates (microbial carbonate and biologically induced cement)? *Sedimentology*, **43**, 947–71.

Webb, L.J. (1959). Physiognomic classification of Australian rain forests. *Journal of Ecology*, **47**, 551–70.

Webb, P.N., Harwood, D.M., McKelvey, B.C., Merler, J.H. & Stott, L.D. (1984). Cenozoic marine sedimentation and ice-volume variation on the East Antarctic craton. *Geology*, **12**, 287–91.

Wei, K.Y. & Kennett, J.P. (1983). Non-constant extinction rates of Neogene planktonic foraminifera. *Nature*, **305**, 218–20.

Wei, K.Y. & Kennett, J.P. (1986). Taxonomic evolution of Neogene planktonic foraminifera and paleoceanographic relations. *Paleoceanography*, **1**, 67–84.

Wei, W. & Wise, S.W. (1990). Biogeographic gradients of middle Eocene–Oligocene calcareous nannoplankton in the South Atlantic Ocean. *Palaeogeography, Palaeoclimatology, Palaeoecology*, **79**, 29–61.

Weissert, H. & Lini., A. (1991). Ice age interludes during the Cretaceous greenhouse climate? In *Controversies in Modern Geology*, ed. D.W. Müller, J.A. McKenzie & H. Weissert, pp. 173–91. London: Academic Press.

Wellnhofer, P. (1991). *The Illustrated Encyclopaedia of Pterosaurs*. London: Salamander Books.

Wells, J.W. (1956). Scleractinia. In *Treatise on Invertebrate Paleontology. Part F. Coelenterata*, ed. R.C. Moore, pp. F328-F444. Boulder, CO and Lawrence, KA: Geological Society of America and University of Kansas Press.

Werner, C. & Rage, J-C. (1995). Mid-Cretaceous snakes from Sudan. A preliminary report on an unexpectedly diverse snake fauna. *Comptes Rendus de Académie des Sciences, Paris*, **319**, 247–52.

Westbroek, P., Brown, C., van Bleijswijk, J. *et al.* (1993). A model system approach to biological climate forcing. The example of *Emiliania huxleyi*. *Global and Planetary Change*, **8**, 27–46.

Westbroek, P., Buddemeir, B., Coleman, M. *et al.* (1994). Strategies for the study of climate forcing by calcification. In *Past and Present Biomineralization Processes. Considerations about the Carbonate Cycle*, ed. F. Doumenge. *Bulletin de l'Institut océanographique, Monaco, numéro spécial* **13**, 37–60.

Westermann, G.E.G. (1996). Ammonoid life and habitat. In *Ammonoid Paleobiology*, ed. N.H. Landman, K. Tanabe & R.A. Davis, pp. 608–707. New York and London: Plenum Press.

Westphal, F. (1967). Erster Nachweis des Riesensalamanders (*Andrias*, Urodela, Amphibia) im europäischen Jung pliozän. *Neues Jahrbuch für Geologie und Paläontologie, Monatshefte*, **1967**, 67–73.

Whalley, P. (1988). Insect evolution during the extinction of the Dinosauria. *Entomologia Generalis*, **13**, 119–24.

Wheeler, E.A. & Baas, P. (1991). A survey of the fossil record for dicotyledonous wood and its significance for evolutionary and ecological wood anatomy. *International Association of Wood Anatomists Bulletin*, **12**, 275–332.

Wheeler, E.A. & Baas, P. (1993). The potentials and limitations of dicotyledonous wood anatomy for climatic reconstructions. *Paleobiology*, **19**, 487–98.

Wheeler, E.A., Lehman, T.M. & Gasson, P.E. (1994). Javelinoxylon, an Upper Cretaceous dicotyledonous tree from Big Bend National Park, Texas, with presumed malvalean affinities. *American Journal of Botany*, **81**, 703–10.

Wheeler, P.E. (1993). The influence of stature and body form on hominid energy and water budgets; a comparison of *Australopithecus* and early *Homo* physiques. *Journal of Human Evolution*, **24**, 13–28.

White, B. (1987). Anoxic events and allopatric speciation in the deep sea. *Biological Oceanography*, **5**, 243–59.

White, T. (1995). African omnivores: global climatic change and Plio-Pleistocene hominids and suids. In *Paleoclimate and Evolution, with Emphasis on Human Origins*, ed. E.Vrba, G. Denton, T. Partridge & L. Burckle, pp. 369–84. New Haven: Yale University Press.

White, T.D., Suwa, G. & Asfaw, B. (1994, 1995). *Australopithecus ramidus*, a new species of early hominid from Aramis, Ethiopia. *Nature*, **371**, 306–12; **375**, 88.

Whybrow, P.J. (1984). Geological and faunal evidence from Arabia for mammal 'migrations' between Asia and Africa during the Miocene. *Courier Forschungsinstitüt Senckenberg*, **69**, 189–98.

Whybrow, P.J. (1987). Miocene geology and palaeontology of Ad Dabtiyah, Saudi Arabia. *Bulletin of the British Museum (Natural History), Geology*, **41**, 367–457.

Whybrow, P.J. (1992). Land movements and species dispersal. In *The Cambridge Encyclopedia of Human Evolution*, ed. S. Jones, R. Martin & D. Pilbeam, pp. 169–73. Cambridge: Cambridge University Press.

Whybrow, P.J., Collinson, M.E., Daams, R., Gentry, A.W. & McClure, H.A. (1982). Geology, fauna (Bovidae, Rodentia) and flora from the Early Miocene of eastern Saudi Arabia. *Tertiary Research*, **4**, 105–20.

Whybrow, P.J. & Hill, A. (eds.) (1999). *Fossil Vertebrates of Arabia*. New Haven: Yale University Press.

Whybrow, P.J., Hill, A., Yasin al-Tikriti, W. & Hailwood, E.A. (1990). Late Miocene primate fauna, flora and initial palaeomagnetic data from the Emirate of Abu Dhabi, United Arab Emirates. *Journal of Human Evolution*, **19**, 583–8.

Widmark, J. & Malmgren, B. (1992). Benthic foraminiferal changes across the Cretaceous–Tertiary boundary in the deep sea; DSDP sites 525, 527, and 465. *Journal of Foraminiferal Research*, **22**, 81–113.

Wiedmann, J. (1988). Plate Tectonics, Sea Level Changes, Climate -and the Relationship to Ammonite Evolution, Provincialism, and Mode of Life. In *Cephalopods Present and Past*, ed. J. Wiedmann & J. Kullmann, pp. 737–65. Stuttgart: E. Schweizerbart'sche Verlagsbuchhandlung.

Wigley, T.M.L. (1976). Spectral analysis: astronomical theory of climate change. *Nature*, **264**, 629–31.

Wilde, V. (1989). Untersuchungen zur Systematik der blattreste ause dem Mitteleozän der Grube Messel bei Darmstadt (Hessen, Bundesrepublik Deutschland). *Courier Forschungsinstitüt Senckenberg*, **115**, 1–213.

Wiley, E.O. (1976). Phylogeny and biogeography of fossil and recent gars (Actinopterygii: Lepisosteidae). *University of Kansas, Museum of Natural History, Miscellaneous publications*, **64**, 1–111.

Wiley, E.O, & Mayden, R.L. (1985). Species and speciation in phylogenetic systematics, with examples from the North American fish fauna. *Annals of the Missouri Botanical Garden*, **72**, 596–635.

Wilkinson, B.J. (1984). Interpretation of past environments from sub-fossil caddis larvae. In *Proceedings of the 4th International Symposium on Trichoptera*, ed. J.C.Morse, pp.447–52. The Hague: Dr W. Junk (*Series Entomologicae* 30).

Wilkinson, K.N., Berzins, V., Branch, N.P., Fairburn, A.S. & Lowe, J.J. (1994). Great Rissington/Bourton Link Sewerage scheme: a study of the Late Devensian and early to middle Flandrian environments from material recorded during the watching brief, Report for Thames Water Ltd/Cotswold Archaeological Trust, Institute of Archaeology, University of London, 38pp.

Williams, N.E. (1988). The use of caddisflies (Trichoptera) in palaeoecology. *Palaeogeography, Palaeoclimatology, Palaeoecology*, **62**, 493–500.

Wilson, J.J. (1963). Cretaceous stratigraphy of the central Andes of Peru. *Bulletin of the American Association of Petroleum Geologists*, **47**, 1–34.

Wilson, M.E.J. & Rosen, B.R. (1998). Implications of paucity of corals in the Paleogene of SE Asia: plate tectonics or Centre of Origin? In *Biogeography and Geological Evolution of SE Asia*, ed. R. Hall & J.L. Holloway, pp. 165–95. Leiden: Backhuys Publishers.

Wilson, M.V.H. & Williams, R.R.G. (1992). Phylogenetic, biogeographic, and ecological significance of early fossil records of North American Freshwater fishes. In *Systematics, Historical Ecology and North American Freshwater Fishes*, ed. R.L. Mayden, pp. 224–44. Stanford: Stanford University Press.

Wind, F.H. (1979). Maestrichtian–Campanian nannofloral provinces of the southern Atlantic and Indian Oceans. In *Deep Drilling Results in the Atlantic Ocean: Continental Margins and Paleoenvironment*, ed. M. Talwani, W.W. Hay & W.B.F. Ryan. Maurice Ewing Series, **3**, 123–37.

Windley, D.E. (1995). Calcareous nannofossil applications in the study of cyclic sediments of the Cenomanian. Unpublished PhD thesis, University of London.

Wing, S.L. (1998). Tertiary vegetation of North America as a context for mammalian evolution. In *Evolution of Tertiary Mammals of North America*, Volume 1, ed. C.M. Janis, K.M. Scott & L.L. Jacobs, pp. 37–64. Cambridge: Cambridge University Press.

Wing, S.L., Alroy, J. & Hickey, L.J. (1995). Plant and mammal diversity in the Paleocene to Early Eocene of the Bighorn Basin. *Palaeogeography, Palaeoclimatology, Palaeoecology*, **115**, 117–55.

Wing, S.L. & Boucher, L.D. (1998). Ecological aspects of the Cretaceous flowering plant radiation. *Annual Review of Earth and Planetary Sciences*, **26**, 379–421.

Wing, S.L. & Greenwood, D.R. (1993). Fossils and fossil climate: the case for equable continental interiors in the Eocene. *Philosophical Transactions of the Royal Society of London*, B, **341**, 243–52.

Wing, S.L., Hickey, L.J. & Swisher, C.C. (1993). Implications of an exceptional fossil flora for Late Cretaceous vegetation. *Nature*, **363**, 342–4.

Wing, S.L., Sues, H-D. *et al.* (1992). Mesozoic and Early Cenozoic terrestrial ecosystems. In *Terrestrial Ecosystems through Time*, ed. A.K. Behrensmeyer, J.D. Damuth, W.A. DiMichele, R. Potts, H-D. Sues & S.L. Wing, pp. 327–416. Chicago: University of Chicago Press.

Wing, S.L. & Tiffney, B.H. (1987). The reciprocol interaction of angiosperm evolution and tetrapod herbivory. *Review of Palaeobotany and Palynology*, **50**, 179–210.

Winter, A., Jordan, R.W. & Roth, P.H. (1994). Biogeography of living coccolithophores in ocean waters. In *Coccolithophores*, ed. A. Winter & W.G. Siesser, pp. 161–78. Cambridge: Cambridge University Press.

Winterer, E.L. (1991). The Tethyan Pacific during Late Jurassic and Cretaceous times. *Palaeogeography, Palaeoclimatology, Palaeoecology*, **87**, 253–65.

Wise, S.W. (1988). Mesozoic–Cenozoic history of calcareous nannofossils in the region of the Southern Ocean. *Palaeogeography, Palaeoclimatology, Palaeoecology*, **67**, 157–79.

Wise, S.W. & Wind, F.H. (1977). Mesozoic and Cenozoic calcareous nannofossils recovered by DSDP Leg 36 on the Falkland Plateau, southwest Atlantic sector of the Southern Ocean. *Initial Reports of the Deep Sea Drilling Project*, **36**, 269–492.

Wolfe, J.A. (1978). A paleobotanical interpretation of Tertiary climates in the Northern Hemisphere. *American Scientist*, **66**, 694–703.

Wolfe, J.A. (1979). Temperature parameters of humid to mesic forests of eastern Asia and relation to forests of other regions of the northern hemisphere and Australasia. *U.S. Geological Survey Professional Paper*, **1106**, 1–37.

Wolfe, J.A. (1985). Distribution of major vegetational types during the Tertiary. In *The Carbon Cycle and atmospheric CO₂: Natural Variations Archean to Present*. American Geophysical Union, Geophysical Monograph, **32**, 357–75.

Wolfe, J.A. (1987). Late Cretaceous–Cenozoic history of deciduousness and the terminal Cretaceous event. *Paleobiology*, **13**, 215–26.

Wolfe, J.A. (1990). Palaeobotanical evidence for a marked temperature increase following the Cretaceous/Tertiary boundary. *Nature*, **343**, 153–6.

Wolfe, J.A. (1991). Palaeobotanical evidence for a June 'impact winter' at the Cretaceous/Tertiary boundary. *Nature*, **352**, 420–3.

Wolfe, J.A. (1993). A method of obtaining climatic parameters from leaf assemblages. *U.S. Geological Survey Bulletin*, **2040**, 73 pp.

Wolfe, J.A. (1995). Paleoclimatic estimates from Tertiary leaf assemblages. *Annual Reviews of Earth and Planetary Science*, **23**, 119–42.

Wolfe, J.A., Doyle, J.A. & Page, V.M. (1975). The bases of angiosperm phylogeny: paleobotany. *Annals of the Missouri Botanical Garden*, **62**, 801–24.

Wolfe, J.A. & Upchurch, G.R., Jr. (1987). North American nonmarine climates and vegetation during the late Cretaceous. *Palaeogeography, Palaeoclimatology, Palaeoecology*, **61**, 33–77.

Wollast, R. (1994). The relative importance of biomineralization and dissolution of CaCO$_3$ in the global carbon cycle. In *Past and Present Biomineralization Processes. Considerations about the Carbonate Cycle*, ed. F. Doumenge. *Bulletin de l'Institut océanographique, Monaco, numéro spécial*, **13**, 37–60.

Woo, K-S., Anderson, T.F., Railsback, L.B. & Sandberg, P.A. (1992). Oxygen isotope evidence for high salinity surface seawater in the mid-Cretaceous Gulf of Mexico: implications for warm saline deepwater formation. *Paleoceanography*, **9**, 353–87.

Woo, K-S., Railsback, L.B. & Sandberg, P.A. (1992). Oxygen isotope evidence for high-salinity surface seawater in the mid-Cretaceous Gulf of Mexico: implications for warm, saline deepwater formation. *Paleoceanography*, **7** (5), 673–85.

Wood, B. (1992). Origin and evolution of the genus *Homo*. *Nature*, **355**, 783–90.

Wood, R. (1993). Nutrients, predation and the history of reef-building. *Palaios*, **8**, 526–43.

Wood, R. (1995). The changing biology of reef-building. *Palaios*, **10**, 517–29.

Woodburne, M.O. (ed.) (1987). *Cenozoic Mammals of North America, Geochronology and Biostratigraphy*. Berkeley: University of California Press.

Woodburne, M.O. (1996). Precision and resolution in mammalian chronstratigraphy: principles, practices, examples. *Journal of Vertebrate Paleontology*, **16**, 531–55.

Woodburne, M.O. & Case, J.A. (1996). Dispersal, vicariance, and the Late Cretaceous to Early Tertiary land mammal biogeography from South America to Australia. *Journal of Mammalian Evolution*, **3**, 121–61.

Woodward, F.I. (1987). Stomatal numbers are sensitive to increases in CO$_2$ concentration from pre-industrial levels. *Nature*, **327**, 617–8.

Woolbach, W.S., Lewis, R.S. & Anders, E. (1985). Cretaceous extinctions: evidence for wildfires and search for meteoric material. *Science*, **230**, 167–70.

Wray, G.A. (1996). Parallel evolution of nonfeeding larvae in echinoids. *Systematic Biology*, **45**, 308–22.

Wright, C.W. (1996). *Treatise on Invertebrate Paleontology Part L Mollusca 4 (Revised). Volume 4 Cretaceous Ammonoidea*, (With contributions by J.H. Callomon & M.K. Howarth), Boulder, CO and Lawrence, KA: Geological Society of America and The University of Kansas Press.

Xiuming Lui, Shaw, J., Tungheng Lui, Heller, F. & Baoyin Yuan, (1992). Magnetic mineralogy of Chinese loess and its significance. *Geophysical Journal International*, **108**, 301–8.

Yonge, C.M. (1940). The biology of reef-building corals. *Great Barrier Reef Expedition 1928–29 Scientific Reports*, **1**, 353–91.

Young, G.C. (1986). Cladistic methods in Paleozoic continental reconstruction. *Journal of Geology*, **94**, 523–37.

Young, G.C. (1990). Devonian vertebrate distribution patterns and cladistic analysis of palaeogeographic hypotheses. In *Palaeogeography and Biogeography*, ed. W.S. McKerrow & C.R. Scotese. Geological Society, London, Memoir No. **12**, 243–55.

Young, J.R. (1994). Functions of coccoliths. In *Coccolithophores*, ed. A. Winter & W.G. Siesser, pp. 63–82. Cambridge: Cambridge University Press.

Zabala, M. & Malaquer, P. (1988). Illustrated keys for the classification of Mediterranean Bryozoa. *Treballs del Museu de Zoologia, Barcelona*, **4**, 1–294.

Zachariasse, W.J., Zijderveld, J.D.A., Langereis, C.G., Hilgen, F.G. & Verhallen, P.J.J.M. (1989). Early Late Pliocene biochronology and surface water temperature variations in the Mediterranean. *Marine Micropaleontology*, **14**, 339–55.

Zachos, J.C., Berggren, W.A., Aubry, M-P. & Mackensen, A. (1992). Eocene-Oligocene climatic and abyssal circulation history of the southern Indian Ocean. *Proceedings of the Ocean Drilling Program, Scientific Results*, **120**, 839–54.

Zachos, J.C., Lohmann, K.C., Walker, J.C.G. *et al.* (1993). Abrupt climate change and transient climates during the Paleogene: a marine perspective. *Journal of Geology*, **101**, 191–213.

Zachos, J.C., Rea, D., Seto, K. *et al.* (1992). Paleogene and early Neogene deep water paleoceanography of the Indian Ocean as determined from benthic foraminifer stable carbon and oxygen isotope records. American Geophysical Union, Geophysical Monograph, **70**, 351–86.

Zachos, J.C., Stott, L.D. & Lohmann, K.C. (1994). Evolution of early Cenozoic temperatures. *Paleoceanography*, **9**, 353–87.

Zahnie, K. & Grinspoon, D. (1990). Comet dust as a source of amino acids at the Cretaceous/Tertiary boundary. *Nature*, **348**, 157–60.

Zeuner, F.E. (1945). *The Pleistocene Period*. London: Ray Society.

Zherikhin, V.V. (1993). Possible evolutionary effects of ecological crisis: paleontological and contemporary data. In *Aerial Pollution in Kola Peninsula*, ed. M.V. Kozlov, E. Haukioja & V.T. Yarmishko, Proceedings of the International Workshop, 1992, 53–60.

Zimmerman, H.B., Boersma, A. & McCoy, F.W. (1987). Carbonaceous sediments and palaeoenvironment of the Cretaceous South Atlantic Ocean. In *Marine Petroleum Source Rocks*, ed. J. Brooks & A.J. Fleet. Geological Society Special Publication No. **26**, 271–86.

Index

Page numbers in *italics* refer to figures, page numbers in **bold** refer to tables.

sepiolite 15
Seribiscutum primitivum 42, *46*
  movement into intermediate latitudes 47, 49
shelf deposits, Cenozoic 126–9
shelf seas, Mesozoic, spread of ammonites 101
Siderastrea 174, *175*, 178
Sidlings Copse 279, 285
  changes in woodland 275
  expansion of pioneer arboreal vegetation 275, *276–7*
  pre-clearance climax woodland *276–7*, 278
  woodland clearance 284
Sirenia 335–6, 355
Siwalik Formation
  Mid Miocene faunal trends 364
  record of Asian Neogene life 357
social wasps 290
soil carbonates, recording Late Paleocene carbon excursion 340–1
South Africa, early *Homo sapiens* 386
South America 334
  Late Eocene, probable arrival of mammals from Africa 342
South America–Africa connection, Paleogene, a biogeographical necessity 326–7
South America–Antarctica–Australia block, breakup of 322
South Atlantic Ocean 6
South Atlantic Seaway, aided ammonite migration 101
South China, intracontinental orogeny 6
Spartum Fen *270–1*
  changes in woodland 275
  dominant *Alnus* woodland 278, 279
  expansion of pioneer arboreal vegetation 275
  hazel maxima 277
  high frequency of cereal pollen *270–1*, 284
  pre-clearance climax woodland *270–1*, 278
  response to climate amelioration 269
  rise in water table 285
  tundra vegetation pollen 268
  woodland clearance 284, 286–7
spatial analysis
  of African Mid Miocene faunas *358*, *359*, 361–2
  based on Harrison concept 353

species abundance records, environmental change impact and glacial–interglacial climatic oscillation 79–80
species diversity/richness 347, *348*
  and abundance
    angiosperms, Cretaceous, mid-palaeolatitudes 211–13
    non-angiosperms, Cretaceous, mid-palaeolatitudes 213–15
  angiosperms, trends in *210*
  Britain
    Last Cold Stage mammals 373, 375–7
    Last Interglacial mammals 372–3, *374*
  Cenozoic, patterns of and floral diversity seem to reflect climate change 242
  and environmental change 129–30
  and floral biology 230–1
  gastropods 155–7
  and geography, mutual dependency 88
  latitudinal diversity gradient, bivalves 140–5
  planktonic foraminifera, declined 59
Sphagnium 268
Sphenolithus 43
spores 265
  see also plants, free-sporing
stable carbon isotope technique, use of 366
Stegaster, extinction of 193–4
Steller's Sea Cow 335–6
Stephanocoenia 174, 178
Stephanorhinus hemitoechus 370
^{87}Sr/^{86}Sr ratio 17, 57
  overall increase in during the Cenozoic 27
Stratiotes 229
Subbotina linaperta 69, 72–3
  morphological response accompanied by habitat shift 72, 74
  shape variations 72, *72*, *73*
  significant decrease in test size 69, 72, *72*
subduction
  of the East Pacific Rise, effects of 21
  and the rise of the Andes 6
sulphate aerosols, formation of 38
sulphur, emitted by volcanic activity 23
superplume, mid-Cretaceous 6, 7, 9, 17
Supertethys hypothesis 16
Sus scrofa 370

Talpa europaea 370
taphonomy, imposing limitations on CLAMP 259–60